Brickwork Level 2

Brickwork Level 2

For CAA Construction Diploma and NVQs

Malcolm Thorpe

ELSEVIER

AMSTERDAM • BOSTON • HEIDELBERG • LONDON • NEW YORK • OXFORD
PARIS • SAN DIEGO • SAN FRANCISCO • SINGAPORE • SYDNEY • TOKYO

Butterworth-Heinemann is an imprint of Elsevier

B H

Butterworth-Heinemann is an imprint of Elsevier
The Boulevard, Langford Lane, Kidlington, Oxford OX5 1GB, UK
30 Corporate Drive, Suite 400, Burlington, MA 01803, USA

First edition 2010

British Library Cataloguing in Publication Data
A catalogue record for this book is available from the British Library

Library of Congress Cataloging-in-Publication Data
A catalog record for this book is available from the Library of Congress

ISBN: 978-1-85617-765-8

For information on all Butterworth-Heinemann publications visit our web site at books.elsevier.com

Printed and bound in China

10 11 12 13 14 15 10 9 8 7 6 5 4 3 2 1

Contents

Preface

The changes in construction training have led to the need to produce this series of books which incorporate both National Vocational Qualifications (NVQs) and Diplomas.

The content of each book follows both routes and provides the necessary information for the various job knowledge tests.

After the initial chapter, which gives the construction student an insight into the industry they are entering, each chapter follows very closely the NVQ and Diploma units.

The aim of each book is to provide an information resource and student workbook for all building craft students. It can be used to provide teaching and assessment material, or used simply to reinforce college lectures.

Each chapter has a set of multiple-choice questions designed to test your level of knowledge before moving on to the next chapter.

Malcolm Thorpe

CHAPTER 1

The Construction Industry

This chapter will cover the following NVQ and Diploma units:

- NVQ All
- CC All

This chapter is about:

- The construction industry
- Types of communal building
- Source of construction work
- Range of activities
- The building team
- Jobs and careers

The following NVQ performance criteria will be covered:

This chapter has no comparable Level 2 NVQ units but it gives the student an early introduction to the construction industry.

The following Diploma outcomes will be covered:

This chapter has no comparable Level 2 Diploma units but it gives the student an early introduction to the construction industry.

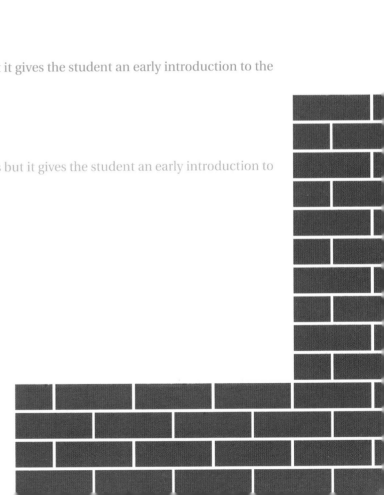

Introduction

When students are thinking of entering the building trade they may ask many questions. The three main questions are:

- What is the construction industry?
- What can the construction industry offer me?
- What type of education will I need?

The construction industry

The construction industry is one of the largest employers of labour in the country, with a labour force of just over one million, a figure that has dropped steadily over the past years.

Construction means creating, not only the houses we need to live in but many other buildings such as schools, hospitals and shopping centres.

The majority of buildings and structures are designed and constructed for a specific purpose. The use of the building will determine the size, shape, style and ultimately the cost.

Every person employed in the construction industry makes a direct contribution to the community in general but also to the nation.

The industry is made up of a large number of firms which can be classified as:

- builders
- contractors
- subcontractors, etc.

The firms range in size from one-person firms to multinational companies.

- A small company is defined as having between one and 49 employees.
- A medium company is defined as having between 50 and 249 employees.
- A large company is defined as having more than 250 employees.

There are also several different types of construction work to consider when thinking of joining the construction industry.

The whole industry can be further divided into four:

- *New work* refers to all types of building work and services which are about to start.
- *Maintenance work* refers to any work on an existing building which requires damaged or out-of-date items to be brought up to an acceptable standard.

Examples of maintenance work include new kitchen and bathroom units, or external brickwork requiring repointing, etc.

- *Refurbishment* is when an old building has been allowed to fall into a state of disrepair and it needs to be brought back to standard. Changing existing buildings for another purpose is also classed as refurbishment.

An example is when an old warehouse has been changed into a block of flats.

- *Restoration work* is when an old building is brought back to its original state.

Examples are old historic buildings bought by the National Trust and then painstakingly restored to their former glory.

The construction process

The construction process is said to be the most complex of all industries.

People employed in manufacturing industries travel each day to the same place and do the same type of work.

In the construction industry the employees move to a different place of work as soon as the particular job has been completed. The distance depends on the nature and size of the contractor and the availability of work.

No two construction sites are ever the same, and it is seldom that more than a few dwellings are the same.

The construction industry does not lend itself to production-line methods, so it is very labour intensive.

The construction team is therefore comprised of people possessing a vast range of skills, from the tradespeople to the professionals.

The construction industry differs from other industries in the following ways:

- Work is carried out in the open and is subject to stoppages from the weather.
- Every day the plan of work is different.
- The labour force is not static and can change daily.
- Great distances have to be travelled by employees, so they are often many miles from the head office.
- Every job is different, so there is no repetition through which employees can produce more after gaining experience.
- Many of the contracts are completed by one person after being designed by another person.
- Safety in the industry has a very poor record.
- The industry is very labour intensive.

All the above statements can cause many problems and it is very difficult for any one person to rectify them; therefore the construction team becomes very important indeed.

Types of building

Many different types of construction are required to fulfil the needs of today's ever-demanding society.

These consist of the following and are shown in Figure 1.1:

- dwelling units – for people to live in
- communal buildings – for all people to share
- industrial units – for people to work in
- recreational units – for people to relax in
- communications – roads, rail, sea and air networks – to allow people to move from one place to another.

Community buildings

For the purpose of Level 2, only buildings for communal purposes will be dealt with.

These buildings cover those found in recreational, educational, spiritual and leisure sectors. Buildings in this category are used for the benefit of the community. They are used to accommodate the facilities that society provides for:

- health care
- welfare
- education
- entertainment.

Examples of buildings that provide for the spiritual needs of the community are:

- churches
- chapels
- cathedrals.

Examples of community buildings include:

- shopping centres
- hospitals
- police and fire stations.

Other buildings in this category provide buildings for leisure, such as:

- public houses
- dance halls

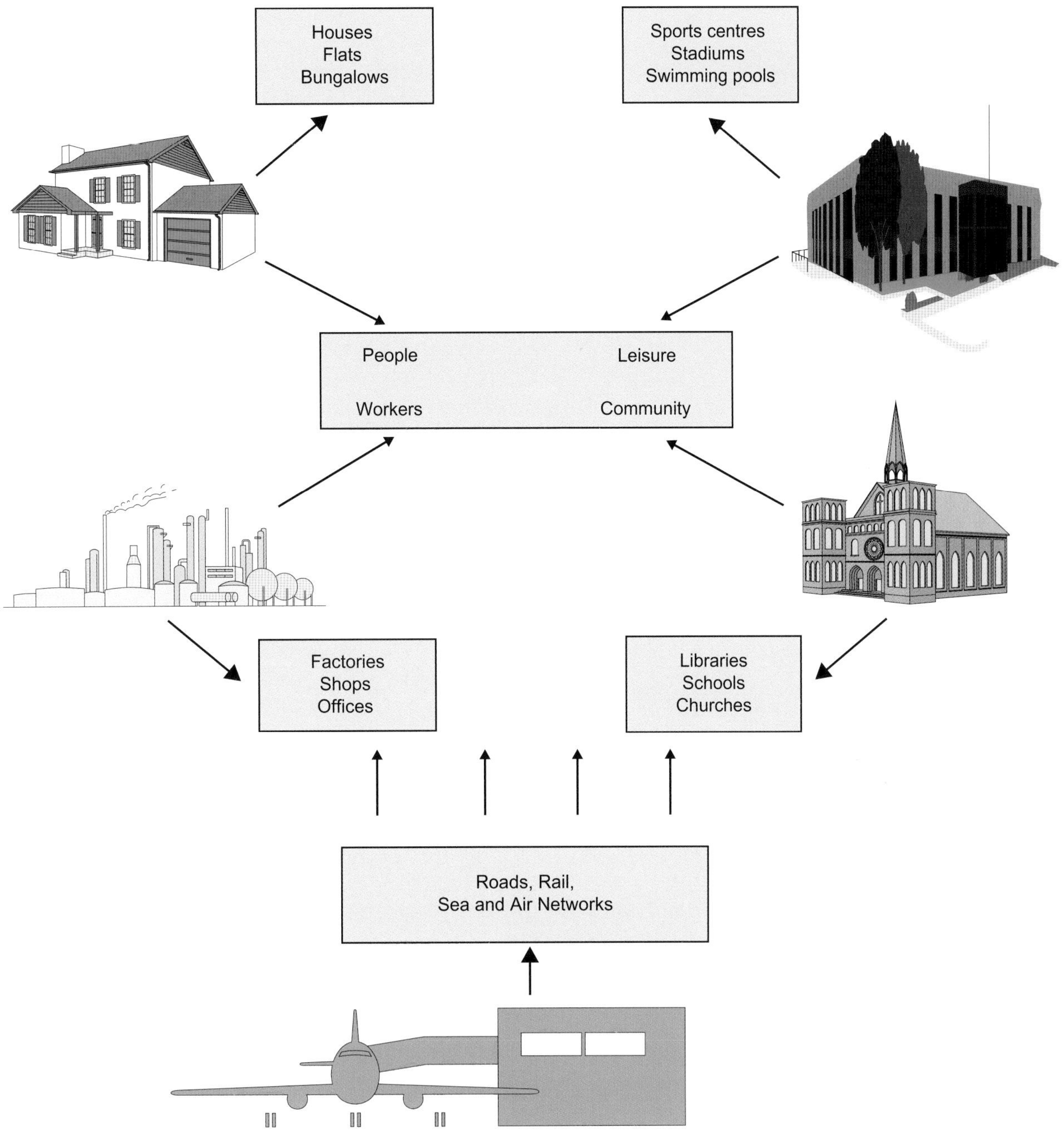
Houses
Flats
Bungalows
Sports centres
Stadiums
Swimming pools
People
Leisure
Workers
Community
Factories
Shops
Offices
Libraries
Schools
Churches
Roads, Rail,
Sea and Air Networks

FIGURE 1.1
Types of construction

- theatres
- cinemas
- concert halls.

Buildings designed and constructed for education include:

- schools
- colleges
- libraries.

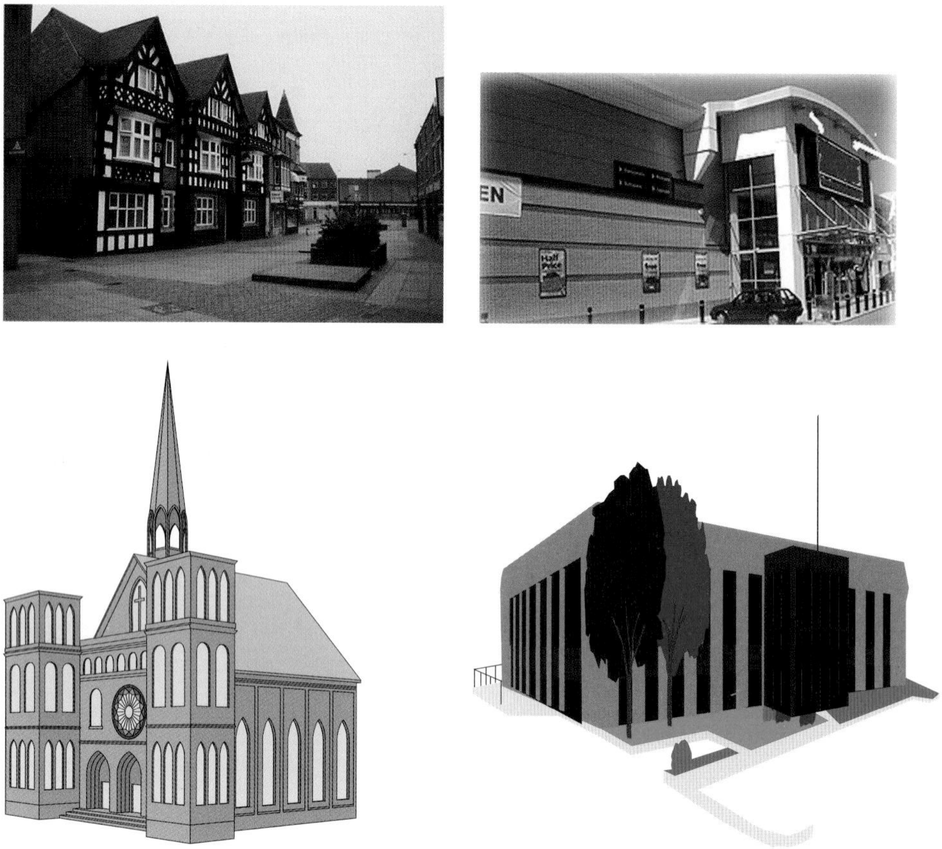

FIGURE 1.2
Communal buildings

Buildings in this category are designed and built for the use of all the community. The list is by no means exhaustive.

CLASSIFICATION OF COMMUNAL BUILDINGS

All buildings are designed to fulfil a role, none more so than community buildings. This category covers a very wide range of buildings, some of which are shown in Figure 1.2.

The source of construction work

The client

The client is the employer or building owner who employs an architect to design a building or a contractor to construct a building, and has overall responsibility for the financing of the project.

A client can be an individual or an organization. For example:

- private individual – Mr Smith
- partnership – Mr and Mrs Smith
- limited company – Smith Ltd

- local authority – Oxford District Council
- government department – Department of the Environment
- statutory authority – police
- public undertaking – private hospital.

The above clients can also be divided into two main categories:

- the private sector
- the public sector.

THE PRIVATE SECTOR

The private sector consists of work financed by an individual client such as:

- private individual
- partnership
- limited company.

Examples of the above are:

- Private individual – a small garage owner requires a boundary wall to be constructed around his premises.
- Partnership – two taxi cab proprietors need a new garage to be designed for their fleet of cars.
- Limited company – a large organization may contact a contractor to provide a large office accommodation block.

THE PUBLIC SECTOR

The public sector consists of work engaged by publicly financed sources such as:

- local authority
- government department
- statutory authority
- public undertaking.

The following are examples of clients within each group:

- Local authority – Mansfield District Council may require a new library to be constructed.
- Government department – the Department of Health may require a new office accommodation block for their new department.
- Statutory undertaking – a police authority may require a police station to be built in a new area.
- Public undertaking – an open-cast coal mine may require a landscaping project.

Range of activities

The construction industry undertakes an enormous amount of work including a variety of activities. It would be impossible to find a firm in the industry that could carry out all of the activities required.

The following activities are undertaken in the construction industry (Figure 1.3):

- erecting buildings
- repairing and maintaining buildings
- constructing roads and bridges
- erecting steel and concrete structures
- civil engineering work, including sewers, gas and water mains, electricity cables, etc.
- open-cast developments
- demolition work
- plant hire
- work carried out by specialist firms.

The range is very wide, which reinforces the fact that the industry is very complex.

The building team

When you start work in the construction industry you will meet a lot of other people.

Constructing buildings is not a task that can be completed by one person alone; it takes a whole group of people, who are known as the building team. The members of the building team will change according to the type of building.

With community buildings the main members are as follows:

Client

This person is the most important member of the building team. Without the client no buildings would be required. The client provides the money in exchange for a completed building.

Architect

The architect is the client's agent and is considered the leader of the design team. The main role of the architect is to put into reality the client's ideas for a building.

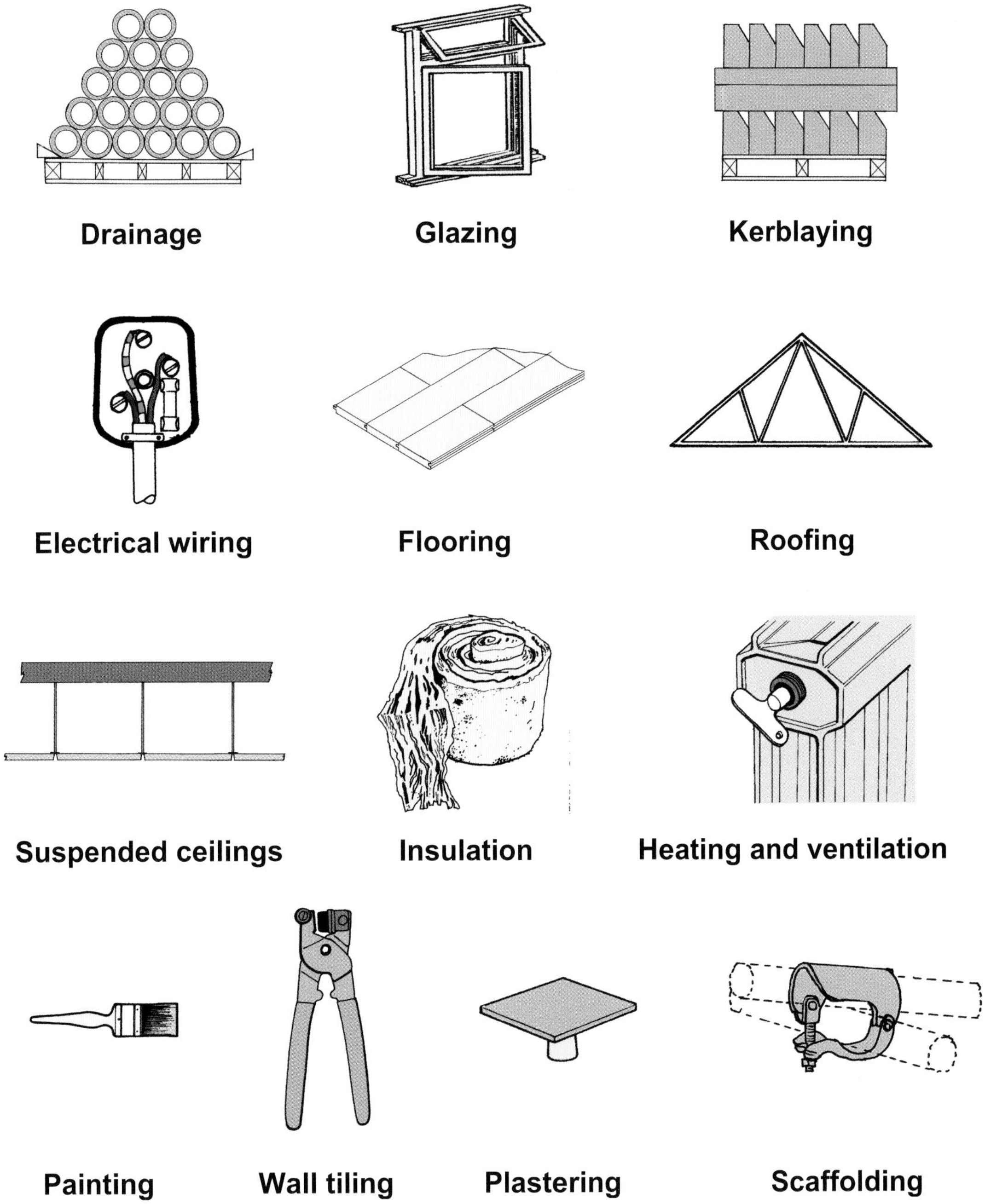
Drainage
Glazing
Kerblaying
Electrical wiring
Flooring
Roofing
Suspended ceilings
Insulation
Heating and ventilation
Painting
Wall tiling
Plastering
Scaffolding

FIGURE 1.3
Range of activities

Private quantity surveyor

The private quantity surveyor is employed by the client, but generally on the recommendations of the architect.

The quantity surveyor is employed as early on in the contract as possible – normally at design stage – so that advice can be given to the client, with the approval of the architect, on the approximate cost of the contract.

During the work the private quantity surveyor will meet on site with the contractors' surveyor to agree interim payments for the work completed, and at the end of the contract agree a final account for presentation to the client.

Estimator

The role of the estimator is to arrive at an overall cost for the complete contract.

He or she will normally break down each item in the bill of quantities into three main parts (labour, materials and plant) and apply the firm's rate to each item to produce the amount it will cost the contractor to complete each item. The estimator will also add to these prices the cost of overheads and profit.

Clerk of works

The clerk of works is nominated or approved by the architect, but is employed by the client.

The clerk of works acts as the client's representative on site under the direction of the architect. He or she may be resident on larger sites or be a regular visitor on smaller ones.

Their primary duty is to ensure that the constructed works conform to the specifications laid down in the contract documents.

Main contractor

In the construction industry a firm operates either as the main contractor, directing all the work on site, or as a nominated or domestic subcontractor.

The main contractor agrees to complete the building for an agreed sum of money.

The organization of a firm can grow as the workload is increased, i.e. extra posts can be created or departments can be introduced, such as joinery works.

The structure can therefore differ considerably between firms. The roles and responsibilities will therefore vary to a large degree from firm to firm and even from site to site.

The structure of a firm could be divided into office staff and site staff.

Office staff	**Site staff**
Office manager	Contracts manager
Estimator	Site agent
Quantity surveyor	General supervisor
Plant manager	Trades supervisor
Personnel officer	Tradespeople
Buyer	Safety officer

Site agent

Sometimes called a site manager, the site agent is the contractor's senior representative who is resident on site and has complete responsibility for the contract from start to finish.

The site agent carries out similar duties to those of the contracts manager, with the addition of:

- providing the head office with daily and weekly reports recording progress on site
- providing a safe site
- liaising with all visitors
- recording architects' variations and additional work.

Contracts manager

The contracts manager is classified as the leader of the site management team on a number of projects.

He or she normally travels around the various sites and is not normally resident on a particular site.

The contracts manager is responsible for overall planning, management and building operations, and should keep in constant contact with head office and site staff.

General supervisor

On large contracts a general supervisor may be employed who works under the contracts manager and is responsible for organizing and controlling the work of all the trades supervisors on site. This role is more supervisory when working under the contracts manager.

On small sites the role of the general supervisor becomes more important. He or she has overall control of the site and carries out both the duties of the contracts manager and his or her own duties.

Trades supervisors

It is normal for each one of the trades on site to have its own trades supervisor to assist in the day-to-day control of the various trade teams.

Each particular trade may have a supervisor to control them.

Ganger

On large construction sites with many general operatives it is advisable to employ a person to organize and control them.

The ganger has responsibility over all the semi-skilled labourers, plant operators, drain layers, concreters, etc., and performs a very similar job roles to that of the trades supervisor.

Tradespeople

The operatives on site are those who actually carry out the physical work – bricklayers, joiners, plumbers, painters, etc.

It would be impossible to carry out any contract without some contribution from these people.

General operatives

These are employees who perform tasks that do not require the levels of skill that a trades person should possess.

Some of them are semi-skilled and perform such tasks as concreting and drain laying.

Safety officer

It is the responsibility of each employer who employs more than 20 personnel to appoint, in writing, a safety officer.

The safety officer ensures that the firm complies with various pieces of legislation.

This person must be experienced and knowledgeable on safety, and if possible should be allowed to devote all of his or her time to this area of work.

They should be able to advise on all matters of safety to the contracts manager, carry out safety inspections, keep records, investigate accidents and arrange for staff to have adequate safety training.

It is uneconomical for firms to employ full-time safety officers. An alternative is to join a safety group where the safety officer is shared.

Subcontractor

The construction industry undertakes a wide variety of tasks.

The vast majority of main contractors only employ tradespeople in the main trades, such as bricklayers, labourers, painters and joiners.

If a contract has been won by a firm that requires other trades, then the main contractor will contact one of the many subcontractors who specialize in various trades. The subcontract work may account for a substantial amount of the contract. Subcontractors are a very important part of the industry and cover a wide range of activities not covered by the main contracting firms.

This system of subcontracting work out generally operates very well, because the firms become experts in their own field, resulting in a high standard of work at very competitive prices.

The main difficulty in subcontracting work is the problem of controlling and co-ordinating the numerous contractors into the main contractor's programme.

Note

This set of notes on personnel in the construction industry, either in the head office or on site, is by no means conclusive.

Please add to these notes any other personnel about which you have found information.

Jobs and careers

Because the production of the built environment is such a large-scale and complex operation, the construction industry demands a wide range of knowledge and skills.

Traditionally, employers recruit people for work in the industry by carefully defining the tasks to be performed and then matching them to the apparent abilities of candidates who present themselves for the appointment.

These abilities are reflected in the candidate's:

- qualifications – obtained through specific training programmes or educational courses
- experience – with evidence in the form of references provided by previous employers, describing their impressions of the candidate's performance at work.

According to the kind of qualification achieved, employees have traditionally played fixed and distant roles in the construction industry:

- as an operative, working on site constructing the built environment
- as a technician, communicating information about the built environment
- as a professional, making decisions about the built environment.

Through increased experience within each category it has traditionally been possible for an employee to improve the quality of their output so that they become more valuable to a firm. In return, they may be given an increase in wages and receive more responsibility.

Such development is described as a career in the construction industry.

Qualifications

In this day and age it is becoming very important to achieve qualifications and recognition for work done.

When you have successfully passed certain examinations you are entitled to use various letters after your name. For example, after completing the Higher National Certificate in Building the letters HNC can be used.

Multiple-choice questions

Self-assessment

This section of the book is designed to allow you to check your level of knowledge. The section consists of revision questions for this chapter. The questions are all multiple choice and have four possible answers. The answers are to be found at the end of the book.

The main type of multiple-choice question will be the four-option multiple-choice question. This will consist of a question or statement, known as the stem, followed by a choice of four different answers, called the responses. Only one of these responses is the correct answer; the others are incorrect and are known as distracters.

You should attempt to answer the questions by choosing either (a), (b), (c) or (d).

Example

The person employed by the local authority to ensure that the Building Regulations are observed is called the:

(a) clerk of works
(b) building control officer
(c) council inspector
(d) safety officer

The correct answer is the building control officer, and therefore (b) would be the correct response.

The construction industry

Question 1 Construction firms with fewer than 50 employees are known as:

(a) small firms
(b) large firms
(c) medium firms
(d) intermediate firms

Question 2 State the type of community unit shown.

(a) library
(b) school
(c) church
(d) theatre

Question 3 Groups of shops all under one roof are known as:

(a) high-street shops
(b) shopping centres
(c) community centres
(d) arenas

Question 4 The main function of the client is to:

(a) design the building
(b) build it
(c) cost the building
(d) finance the building

Question 5 A council that requires a new library to be constructed is known as:

(a) a partnership
(b) a private individual
(c) a limited company
(d) a local authority

Question 6 Which of the following communal buildings has been designed for educational needs?

(a) church
(b) hospital
(c) school
(d) football stadium

Question 7 A company that supplies various trades to a construction project is known as the:

(a) subcontractor
(b) main contractor
(c) specialist contractor
(d) jobbing builder

Question 8 A person who designs a communal building is known as the:

(a) architect
(b) surveyor
(c) estimator
(d) client

CHAPTER 2

Health and Safety in the Construction Industry

This chapter will cover the following NVQ and Diploma units:

- NVQ VR01
- CC 2001K

This chapter is about:

- Awareness or relevant current statutory requirements and official guidance
- Personal responsibilities relating to workplace safety, wearing appropriate personal protective equipment and compliance with warning/safety signs
- Personal behaviour in the workplace
- Security in the workplace
- Relationships

The following NVQ performance criteria will be covered:

- Performance criterion 1: Identification of hazards
- Performance criterion 2: Workplace safety
- Performance criterion 3: Security arrangements
- Performance criterion 4: Emergency procedures

The following Diploma outcomes will be covered:

- Know the health and safety regulations, roles and responsibilities
- Accident, first aid and emergency procedures
- Identify hazards
- Health and hygiene
- Safe handling of materials
- Working platforms
- Electricity
- Personal protective equipment
- Emergency procedures
- Signs and notices

Safety legislation

Safety in the workplace will be covered only briefly in this chapter as it has already been covered in depth in Chapter 2 of Level 1. If more detailed information is required please refer to Level 1.

The construction industry is often involved in very difficult and often hazardous sites. It is therefore very important that the new recruit is aware of these dangers and that there are various regulations in place to control and reduce these possible hazards.

Prevention of hazards in the workplace

Hazards within a workplace can occur because of several circumstances. There may be faults in equipment, tools, stored substances, dangerously stacked materials, materials obstructing safe access, or simply a lack of site safety.

The health and safety of employees at their workplace and any other persons at risk through work activities are covered through various Acts of legislation and regulations.

These include the following:

- The Health and Safety at Work Act 1974
- The Control of Substances Hazardous to Health Regulations 2002 (COSHH)
- The Noise at Work Regulations 2005
- Work at Height Regulations 2005
- Reporting of Injuries, Diseases and Dangerous Occurrences Regulations 1995 (RIDDOR)
- The Personal Protective Equipment at Work Regulations 1992
- The Fire Precautions (Workplace) Regulations 1997
- Provision of the Use of Work Equipment Regulations 1998 (PUWER)
- The Electricity at Work Regulations 1989.

The main health and safety legislation applicable to building sites and workshops is covered by the Health and Safety at Work Act 1974.

HEALTH AND SAFETY AT WORK ACT 1974

The four main objectives of the HASAWA are:

- To secure the health, safety and welfare of all persons at work.
- To protect the general public from risks to health and safety arising from out of work activities.

- To control the use, handling, storage and transportation of explosives and highly flammable substances.
- To control the release of noxious or offensive substances into the atmosphere.

This Act requires employers to ensure so far as is reasonably practicable the health and safety of their employees, other people at work and members of the public who may be affected by their work.

Employees have to co-operate with their employer on health and safety matters and not do anything that puts them or others at risk.

Employees should be trained and clearly instructed in their duties.

The main purpose of this Act is to cover all aspects of safety.

The framework promotes, stimulates and encourages high standards of health and safety in the workplace.

The Act involves everyone, management, employees, the self-employed, the employees' representatives, the controllers of premises, and the manufacturers of plant, equipment and materials, in matters of health and safety.

The Act also deals with the protection of the public, where they may be affected by the activities of people at work.

Outline of the Act

The Act itself is very complex and is an extensive document with numerous parts and sections. There are four main parts.

Part 1 of the Act describes:

Employers' and management duties:

1. Provide and maintain a safe working environment.
2. Ensure safe access to and from the workplace.
3. Provide and maintain safe machines, equipment and methods of work.
4. Ensure the safe handling, transport and storage of all machinery, equipment and materials.
5. Provide their employees with the necessary information, instruction, training and supervision to ensure safe working.
6. Prepare, issue to employees and update as required a written statement of the firm's safety policy.
7. Involve trade union safety representatives (where appointed) with all matters concerning the development, promotion and maintenance of health and safety requirements.

Employees' duties:

1. Take care at all times and ensure that they do not put themselves, their workmates or any other person at risk by their actions.

2. Co-operate with their employers to enable them to fulfil the employer's health and safety duties.
3. Use the equipment and safeguards provided by the employers.
4. Never misuse or interfere with anything provided for health and safety.

Safety procedures and documentation

In order to comply with the various safety legislation, an employer is required:

- To display notices and certificates – e.g. a copy of a valid fire certificate.
- To notify relevant records – e.g. commencement of any building works likely to last in excess of six weeks has to be notified to the relevant authority.
- To keep relevant records –e.g. the accident book in which details of *all* accidents are recorded.

THE CONTROL OF SUBSTANCES HAZARDOUS TO HEALTH (COSHH)

The Control of Substances Hazardous to Health 2002 (COSHH) must be consulted when dealing with the handling, moving, storage and finally disposal of potentially hazardous materials or products.

Alongside these regulations it is important to use any detailed codes of practice and manufacturer's advice which is often found on the packaging of the material.

A material or product that is hazardous to health could be anything that could affect your health. Dangers can arise from ingestion, absorption, exposure and inhalation of poisonous materials.

THE NOISE AT WORK REGULATIONS 2005

Building sites can be noisy places to work in. They should display signs as shown in Figure 2.1.

Excessive noise can be harmful, either by causing hearing damage or by creating nuisance, which may lead to stress.

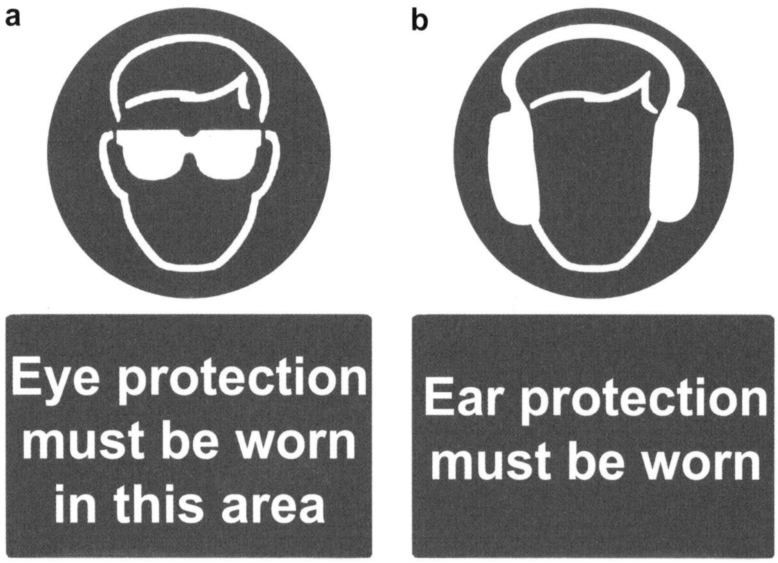
a
b
Eye protection must be worn in this area
Ear protection must be worn

FIGURE 2.1
Eye and Ear protection signs

Loud noise can cause a temporary partial loss of hearing, with recovery time varying from around 15 minutes after the noise stops to a few days, depending on the level of the noise.

Employers should do as much as possible to reduce noise. However, it may not be possible to quieten all machines enough to ensure that there is no hazard, and therefore proper ear protection should also be available.

WORK AT HEIGHT REGULATIONS 2005

This regulation has been put in place to protect the workforce from injury or death from working at heights.

Your employer must provide the necessary equipment for working at heights, such as ladders, working platforms and scaffolding.

As an employee you must follow any instructions and training received while using any equipment provided and report any possible hazards to your supervisor.

More detailed information will be given in the section on working platforms.

REPORTING OF INJURIES, DISEASES AND DANGEROUS OCCURRENCES REGULATIONS 1995 (RIDDOR)

All employers have a duty under RIDDOR to report accidents, diseases and dangerous occurrences. Reporting accidents and ill-health is a legal requirement.

This information helps the Health and Safety Executive (HSE) and local authorities to identify where and how risks arise, and to investigate serious accidents. They can then help with providing advice on how to reduce injury and ill-health in the workplace.

PERSONAL PROTECTIVE EQUIPMENT AT WORK REGULATIONS 1992

Personal protective equipment (PPE) is defined as 'all equipment which is intended to be worn or held by the person at work and which protects him/her against one or more risks to health or safety'.

The main requirement of the regulation is that PPE is to be supplied and used at work whenever there are risks to health and safety that cannot be adequately controlled in other ways.

To allow the correct type of PPE to be chosen the employer must carefully consider the different hazards in the workplace. This will enable the employer to assess which types of PPE are suitable to protect the workers from the hazard and allow the work to be completed in a safe manner.

Types of PPE include gloves, masks and goggles.

THE FIRE PRECAUTIONS (WORKPLACE) REGULATIONS 1997

The Fire Precautions (Workplace) Regulations 1997, as amended, cover places of work where one or more person is employed, e.g. commercial premises, universities, hospitals, shops, hotels, offices and building sites.

The regulations state that premises with five or more workers must have a written fire risk assessment detailing the appropriate fire safety work required, although some premises can be exempt.

Employers must train employees in fire safety following the written risk assessment.

An emergency plan may have to be prepared and sufficient workers trained and equipped to carry out their functions within any such plan.

All equipment and facilities such as fire extinguishers, alarm systems and emergency doors should be regularly maintained and faults rectified as soon as possible. Defects and repairs must be recorded.

Employers must plan, organize, control, monitor and review the measures taken to protect employees from fire while at work. If there are five or more employees, then a record must be maintained.

PROVISION AND USE OF WORK EQUIPMENT REGULATIONS 1998

These regulations require risks to people's health and safety, from equipment that they use at work, to be prevented or controlled.

The regulations require that the equipment provided for use at work is:

- suitable for the job it has to do
- regularly maintained to ensure that it is safe to use.

They also require that:

- training is provided for those who use it.

In general, any equipment that is used by an employee at work is covered.

DO NOT CHANGE GRINDING
WHEEL UNLESS AUTHORIZED
TO DO SO

FIGURE 2.2
Abrasive wheels sign

Mechanical tools and equipment

These have exposed moving parts such as grinding wheels, sanders, drills, chipping hammers, portable saws, rotary wire brushes and air compressors. A typical tool and sign are shown in Figure 2.2.

There are special regulations concerning grinding wheels and portable saws, and people must be authorized to use them.

- Never use any mechanical equipment that is unfamiliar.
- First read the manufacturer's instructions or seek advice.
- Never lay a tool down while it is still rotating.
- Never wear loose-fitting clothes when using tools with fast moving parts.

Most accidents are caused by lack of knowledge, misuse, makeshift repairs or using faulty tools and equipment.

All brickwork apprentices have to achieve the Abrasive Wheels Certificate as part of their programme.

FIGURE 2.3
Caution sign for electricity

The Electricity at Work Regulations 1989

Any equipment that uses electricity is covered by the Electricity at Work Regulations. Appropriate signs should be positioned as shown in Figure 2.3.

Your employer has a duty to make sure that the whole construction site is a safe area and that there is no chance of coming into contact with a live electrical current.

ELECTRICITY

Electricity is something that you cannot see or hear. Strict rules are laid down for its use and *must* be rigidly obeyed by everyone.

All tools should comply with British Standard 2769 and, except for 'double-insulated' tools, must be effectively earthed. Double-insulated tools, which have their own built-in safety system and bear the 'kite mark' and 'squares symbol', do not require an earth lead (Figure 2.4).

Any electrical tool or equipment must be operated at a reduced voltage of 110 V (volts), or lower if possible. Below 60 V the risk of death is greatly reduced.

Workplace safety

Accidents

When joining the construction industry it is important to remember that you will be joining an industry with one of the highest injury and accident ratings. It therefore cannot be stressed enough that you could be at constant risk unless you start as you intend to continue, with a good safety attitude.

Any type of work carried out by the construction industry is often difficult and hazardous. Every site will be different, and therefore every site will bring possible new dangers.

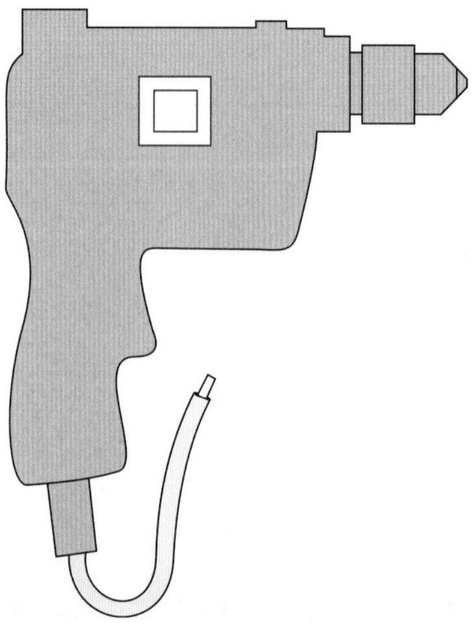

FIGURE 2.4
Double-insulated drill

It is of utmost importance that *all* trainees are capable of using hand tools and equipment efficiently and safely at an early stage in their development.

Furthermore, they should be aware of the causes of accidents and be able to take action and deal with any accident that may occur.

An accident is an unexpected or unplanned happening which results in personal injury or damage, sometimes death.

Reported accidents are those which result in death, major injury and more than three days' absence from work or are caused by dangerous occurrences reported to the HSE.

Every day a large number of the industrial accidents that are reported involve construction workers.

ACCIDENTS DO NOT JUST HAPPEN, THEY ARE CAUSED.

Learning to spot a dangerous situation is not as difficult as it sounds, because accidents follow a regular pattern. The same kind of accident happens over and over again. Every day of the year, all over the country, the same set of dangerous conditions builds up and the same unsafe acts take place.

Do any of the things you normally see and do at work add up to a source of danger? Next time you are tempted to take a risk – STOP and THINK again.

Remember

It is your responsibility to act in a safe manner.

Types of hazard

Everyone involved in the construction industry should be aware of the possible dangers and hazards on the construction site.

Site safety will be improved if everyone is safety conscious.

Types of hazard include the following:

- falling objects
- falls of operatives
- transportation of plant and materials
- electricity
- machinery and equipment
- fire and explosions.

Personal protective equipment

Depending on the type of workshop or site situation, the wearing of correct safety clothing and safe working practices are the best methods of avoiding accidents or injury. On some sites certain PPE is compulsory.

All construction operatives have a responsibility to safeguard themselves and others. Making provision to protect oneself often means wearing the correct protective clothing and safety equipment.

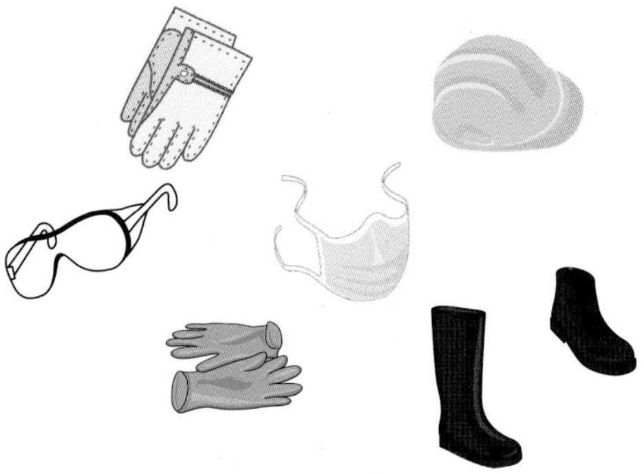

FIGURE 2.5
A selection of personal protective equipment

Your employer is obliged by law to provide PPE, a selection of which is shown in Figure 2.5.

Safety signs

As you go about your work on the building or construction site you will see various signs and notices. Your employer will give you instruction on what they mean and what you should do when you see one.

Safety signs fall into four separate categories, which can be recognized by their shape and colour. Sometimes they may be just a symbol; others may include letters or figures and provide extra information such as the clearance height of an obstacle or the safe working load of a crane.

The four categories are:

PROHIBITION SIGNS

- Shape – circular
- colour – red border and cross bar; black symbol on white background
- meaning – shows what must not be done
- example – no smoking.

MANDATORY SIGNS

- Shape – circular
- colour – white symbol on blue background

- meaning – shows what must be done
- example – wear hand protection.

WARNING SIGNS

- Shape – triangular
- colour – yellow background with black border and symbol
- meaning – warns of hazard or danger
- example – caution, fork-lift truck working.

INFORMATION SIGNS

- Shape – square or oblong
- colour – white symbols on green background
- meaning – indicates or gives information of safety provision
- example – first aid point.

Security arrangements

It is the responsibility of everyone on the work site to ensure that the security of that site is maintained. Security can take many forms and they are all equally important.

Visual security

- Alarms – positioned in an accessible place within view of the general public
- bars, mesh and locks – fitted to glass-panelled doors and windows
- padlocks, padlock and chains – fitted to compound gates, pieces of plant and machinery
- lighting – floodlights and movement-activated lights
- security firms.

Individual security

It is the responsibility of all employees to contribute to the overall security of the firm, for example:

- Tidiness – Do not invite crime by leaving tools and equipment where they may be easily seen. If there is a secure store, lock them away.
- Plant – If possible, return all plant to a secure compound, or if necessary immobilize.
- Unauthorized access – It is the responsibility of all employees on site to challenge anyone who they feel has no authorization to be within a particular area. (Politeness is the best approach.)

If, despite the security measures taken, your site is breached, there are certain procedures you should follow. These should be given to you by your line manager, and may include:

- reporting the incident to the site supervisor
- reporting the incident to the police
- checking the inventory to find out what has been taken
- recording damage done to the premises and/or equipment.

Emergency procedures

Responding to emergencies

It is important that all new trainees are aware of the emergency procedures, which could change according to the type of site or workplace they are in.

From day one you should be aware of what to do in the event of a fire or an accident. Should an emergency happen you should be able to:

- Know what to do – acting quickly and calmly, carry out the correct procedure
- Follow the fire procedure – take the correct action in the event of discovering a fire:
    1. Select and use the correct type of fire extinguisher. (Only if the fire is small enough for you to put it out.)
    2. Call for help, sound the alarm.

3. Telephone the Fire Service: 999. Give the correct address of the building.
4. Leave the building by the nearest exit.
5. Go directly to the assembly area. Await the roll call.

Accidents

An accident involving injury to a person can happen at any time. It may be a workmate who has fallen off a ladder or someone with a burn or a cut, or who has fainted. To help them when they most need it you should know what to do!

IMMEDIATE ACTION

1. Unless you are a fully trained first aider – *do not attempt to treat the injured person.* (Only move an injured person if their life is in danger, e.g. danger from fire.)
2. GET HELP. Report the accident to a person in charge or:
3. Telephone the Emergency Services: 999.
   When your call is answered you should have the following information at hand:
   - type or types of services required – fire, ambulance, police
   - type of accident
   - location/address at which it has happened
   - telephone number you are calling from
   - your name.

Remember

An accident is an event causing injury or damage that could have been avoided by following correct methods and procedures.

Risk assessments

The employer has a duty to protect the workforce as far as is reasonably practicable. Risk assessment is a very important part of protecting all site operatives.

Risk assessment is simply a careful examination of what could cause harm to the workforce. The workforce has a right to be protected from harm caused by a failure to take reasonable control measures.

In the construction industry risk assessments are carried out by experienced people who have been taught to identify what risks are possible when carrying out tasks.

There are five main steps to risk assessments:

- Step 1 – Identify the hazard.
- Step 2 – Decide who might be at risk and how.
- Step 3 – Evaluate the risks and decide on the best precautions.

- Step 4 – Record your findings and implement them.
- Step 5 – Review your assessment and update if necessary.

Health and hygiene

Certain precautions must be taken to ensure that the health of employees in construction firms is protected against hazards, as mentioned in the previous section. As far as is practicable, their health must also be protected.

Vulnerable parts of the body

The health of the site operatives can be divided into the following areas of the human body:

- Skin – one of the most common problems with the skin is dermatitis. This is caused by contact between the skin and the many cements and plasters on site. To reduce the problem barrier creams could be used or appropriate gloves worn.
- Eyes – protection of the eyes has been mentioned previously, but as they are the only ones you have it is important to take extra care and use the appropriate glasses or goggles for the job.
- Ears – again, most sites provide a selection of ear protectors or plugs to be used when working with or close to noise.
- Lungs – many construction operations involve dust. It is therefore very important to protect yourself against inhaling any harmful dust. Protective breathing apparatus or simple disposable masks should be available.

PERSONAL HYGIENE

Always keep yourself clean and tidy; just because you are in one of the dirtiest occupations, there is no need to look untidy.

If you are working in a client's home you need to present yourself correctly.

Always wash regularly and have your work clothes regularly washed.

Wash your hands after going to the toilet and before eating and drinking.

Handling materials and components

The Manual Handling Operations Regulations 1992 outline how to deal with risks to the safety and health of construction site operatives.

The site operative should be able to select and use appropriate safety equipment and protective clothing when handling different materials.

Remember

Do not overcomplicate the process.

If you run a small business and you are confident you understand what is involved, you could do the assessment yourself. You do not have to be a health and safety expert.

Most of the risks come from tripping, slipping and moving heavy loads.

If you are an employer of a large company you should ask a health and safety advisor to help out.

Remember

A hazard is anything that can cause harm, such as chemicals, electricity or working from ladders.

The risk is the chance, high or low, that an employee could be harmed by these and other hazards.

If the item to be moved is an awkward size or shape the site operative should be able to select and use appropriate equipment or aids to carry materials.

They should also be able to demonstrate safe manual handling techniques.

Handling, moving and storing materials are dealt with in more detail in Chapter 5.

Working platforms

The Work at Height Regulations 2005 applies to all work at height where there is a risk of a fall liable to cause personal injury.

There are on average 65 fatal accidents and over 4000 major injuries in the construction industry each year. They remain the single largest cause of workplace deaths and one of the main causes of injury.

This chapter will only deal with working platforms up to 2 m high.

These regulations place a duty on all employers to provide a risk-free environment in which to work.

- Duty holders must avoid working at height where possible.
- Where working at height cannot be avoided, equipment or measures to prevent falls must be in place.
- If there is a risk of a fall, equipment to minimize the distance of the fall must be used.

Remember

Before any work is carried out at height a risk assessment must be produced.

A scaffold is a temporary staging to assist bricklayers and other tradespeople to construct a building.

The scaffold must be spacious and strong enough to support people and materials during construction.

As explained before, many accidents are due to simple faults such as misuse of tools, untied ladders, a missing toeboard, etc.

The three basic requirements for scaffolds are:

- They should be suitable for the purpose.
- They should be safe.
- They should comply with the regulations.

Before work starts

Scaffolding should not be erected, substantially added to, altered or dismantled except under the immediate supervision of a *competent person* and, as far as possible, by *competent workers* possessing adequate experience of such work.

The competent person should also be given sufficient and sound materials for the job. It is false economy and highly dangerous to skimp on materials, and if faulty materials are provided the dangers may be hidden from those who use them.

Before they are used, tubes, couplers and boards should be inspected by someone who knows what defects to look for. Tubes that are bent, or weakened by rust, and damaged couplers and boards with bad splits or knots should be discarded.

Trestle platforms

There are several types of trestle scaffolds which are widely used by all trades to provide working platforms up to 2 m in height in confined spaces. All trestle platforms should stand on a firm level base.

Guardrails and toeboards are not normally required unless the platform exceeds 2 m in height.

They have the advantage over other forms of scaffold in that they are quickly and easily erected and dismantled.

ADJUSTABLE STEEL TRESTLE

There are numerous designs of steel trestles; one is shown in Figure 2.6.

Trestles are positioned to suit batten or staging thickness.

Boards normally used on sites are 225 × 38 × 3900 mm, which should be supported every 1.5 m for British Standard (BS) boards, or every 1.2 m for others.

Some patented staging can span up to 3 m.

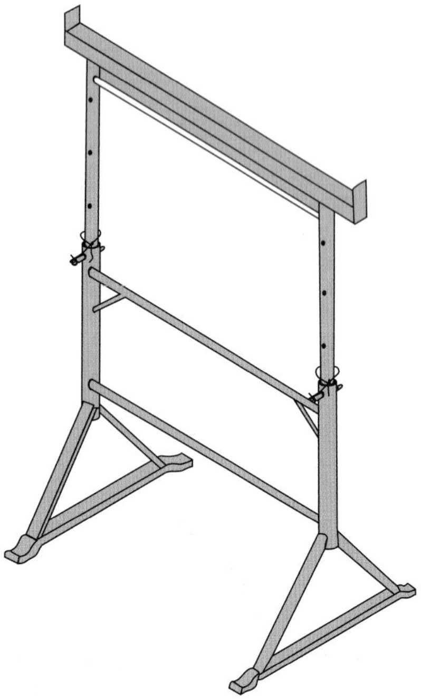

FIGURE 2.6
Typical steel trestle

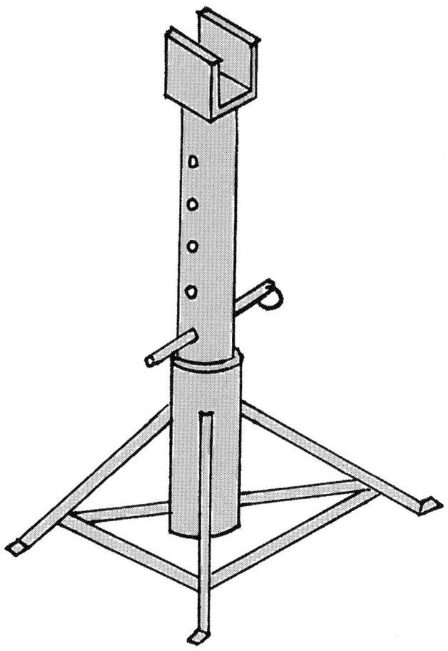

FIGURE 2.7
Splithead

Splitheads

Metal splitheads support scaffolding and provide a continuous platform for working.

The height of the platform ranges from 700 mm to 2 m.

Splitheads are supported on a tripod base. A pin-and-hole method is used for main adjustment; fine adjustment can be achieved using a screwjack.

An example is shown in Figure 2.7.

Scaffold boards

All scaffold boards should be made to BS specifications or they should be 'specials'.

To prevent boards from splitting, the ends should be bound with a galvanized metal band (Figure 2.8). Sometimes the board ends are cut at an angle to reduce the risk of damage.

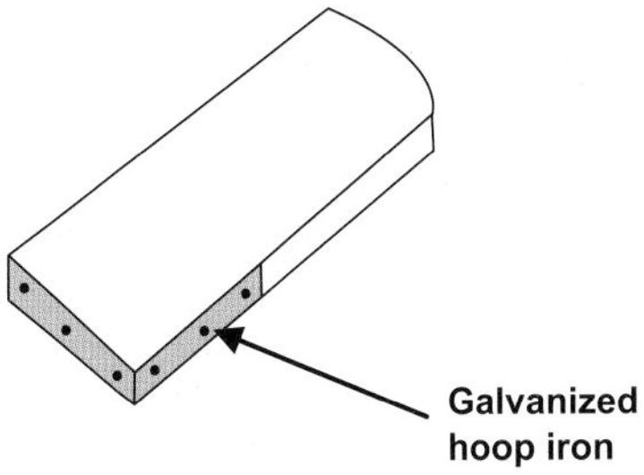
Galvanized
hoop iron

FIGURE 2.8
Boards prevented from splitting

Scaffold boards must be:

- made from straight-grained timber
- free from knots and shakes
- free from decay
- clean and free from grease and thick paint, etc.

Scaffold boards must not be twisted or warped or have split ends.

The distance between the supports governs the thickness of the board used:

- 1.2 m for graded boards
- 1.5 m for BS boards.

OVERHANG

No board should overhang its supports by more than four times the board thickness, or less than 50 mm.

Note

Two scaffold boards must not be used one on top of the other.

Stepladders

Stepladders are one of the most commonly used items of equipment in the construction industry.

Recently, the Work at Height Regulations has brought in stringent guidelines for their safe use. Ladders and stepladders have not been banned by these regulations, but require consideration to be given for their use.

They are only recommended for short duration work, i.e. a maximum of 30 minutes, and only for light work.

- Never work from the top two steps unless there is a safe handhold.
- Do not overreach.

The use of stepladders is also one of the most common subjects in a toolbox talk on site.

Every year an average of 14 people die and a further 1200 are injured by falling from a ladder or stepladder.

Stepladders are sold by the number of treads and the main sizes available have from five to 14 treads.

WOODEN STEPLADDERS

These can be made from various redwoods (Figure 2.9).

You should always inspect a stepladder before using it.

FIGURE 2.9
Wooden stepladder

Parts of wooden stepladders

- Stiles – these usually taper towards the top and are wide enough to take one 250 mm wide scaffold board.
- Treads – treads should be at least 90 mm deep and spaced at 250 mm intervals.
- Locking device – fitted to limit the degree of opening and prevent collapse.

Warning

Do not paint timber steps – paint can hide defects.

ALUMINIUM STEPS

Aluminium steps are made of aluminium alloy. They are lighter than timber steps, very strong and rot proof, and will not twist, warp, burn or rust. A typical aluminium stepladder is shown in Figure 2.10.

As with wooden stepladders, always check before using.

Parts of aluminium stepladders

- Treads – treads should be at least 90 mm deep and spaced at 250 mm intervals, and have a non-slip surface.
- Locking bar – this is fitted to limit the degree of opening and prevent collapse.

Standing ladders

Single section ladders are available up to 9 m long and made from timber or aluminium. They are the most common means of access to scaffolding.

FIGURE 2.10
Aluminium stepladder

DOUBLE EXTENSION LADDERS

These have two sections similar to standing ladders with a position for coupling them together:

- without ropes – up to 4.9 m long when closed, extending to approximately 9 m
- with ropes – up to 7.3 m long when closed, opening to approximately 12 m.

TRIPLE EXTENSION LADDERS

These are similar to double extension ladders, but with three sections:

- without ropes – up to 7.3 m long when closed, extending to approximately 19 m
- with ropes – up to 3 m long when closed, extending to approximately 7.5 m.

A selection of ladders is shown in Figure 2.11.

POLE LADDERS

These are single-section ladders with the stiles made from one straight tree trunk cut down the middle. This ensures even strength and flexibility. They are available up to 12 m long and are used mainly for access to tubular scaffolding.

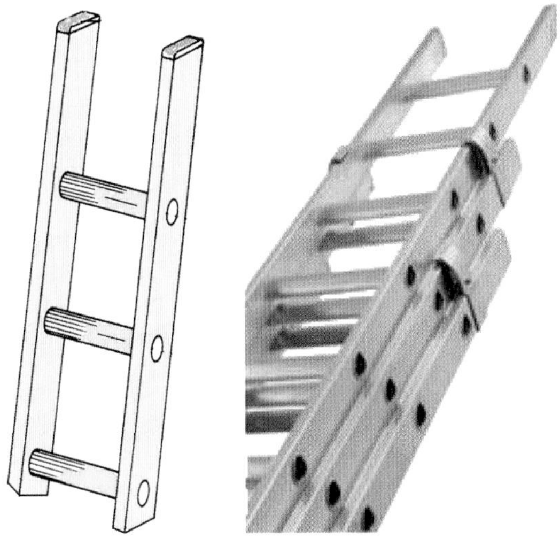

FIGURE 2.11
Types of ladders

Safe use of ladders

RAISING AND LOWERING LADDERS

Ladders should be raised with the section closed.

Extension ladders with long sections are raised one section at a time and slotted into position before use.

Two site operatives are required to raise and lower heavier types of ladders.

When erected, the correct safety angle is 75° or a ratio of four up to one out (Figure 2.12).

Lighter ladders may be raised by one person, but the bottom must be placed against a firm stop before lifting begins.

ACCESS

It is usual to gain access to a scaffold from a ladder which should be firmly secured inside the scaffolding (Figure 2.13).

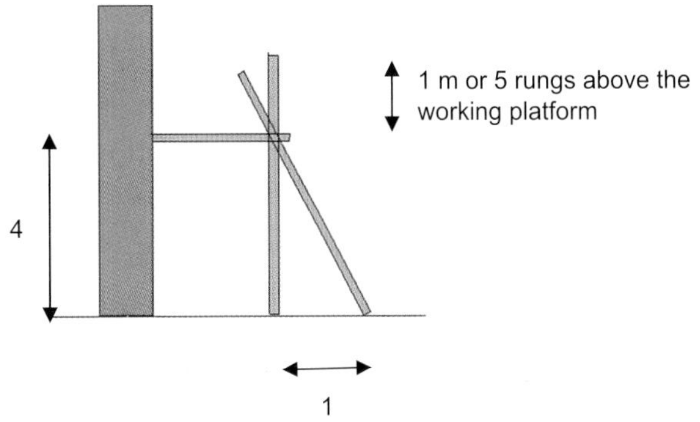
1 m or 5 rungs above the working platform
4
1

FIGURE 2.12
Correct angle of ladder

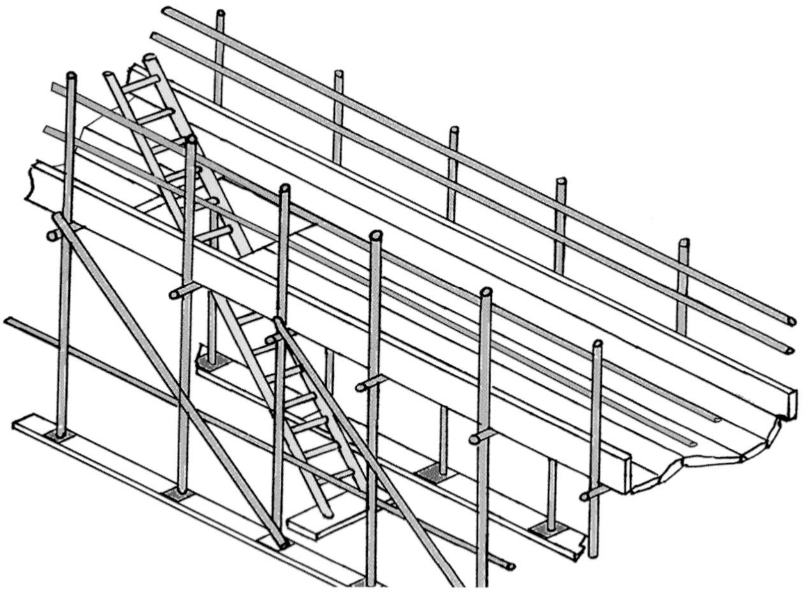

FIGURE 2.13
Access to working platform

The final rung of the ladder from which the operative steps onto the working platform should be just above the surface of the platform. There must be at least five rungs above the working platform.

Access platforms more than 2 m high must have guardrails and toeboards.

The risk of falling materials causing injury should be kept to a minimum by keeping working platforms free from loose materials and debris.

LOADS ON SCAFFOLDS

The bricklayer's scaffold should be wide enough to allow the stacking of materials, working space for the bricklayer and enough space for the passage of other workers and materials (Figure 2.14).

TYING LADDERS

The appropriate regulations state that ladders must have a firm and level base on which to stand, and if over 3 m long, they must be fixed at the top, or if this is not possible at the bottom. If neither way is possible, a person must 'foot' the ladder, that is, stand with one foot on the bottom rung and the other firmly on the ground.

The assistant must hold both stiles and pay attention all the time. This prevents the base from slipping outwards and the ladder from falling sideways.

Figure 2.15 shows a method of fixing ladders to a scaffold.

LIFTING AND CARRYING LADDERS

To lift and carry ladders over short distances rest the ladders on the shoulder, and lift them on the shoulder.

Note

NEVER tie ladders to pipes or gutters.

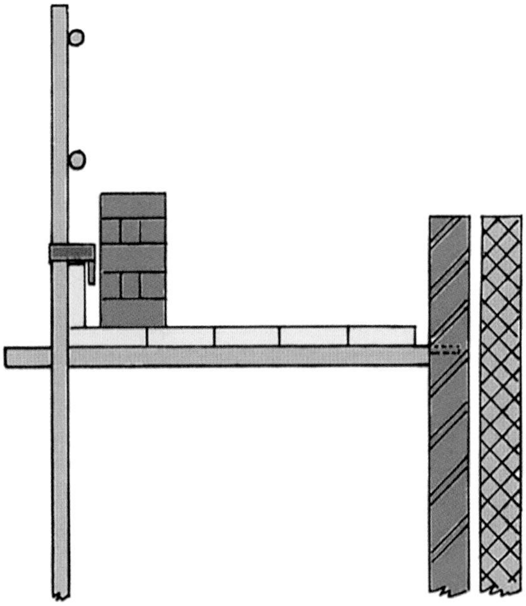

FIGURE 2.14
Loads on working platforms

Lift vertically by grasping the rung just below normal reach.

The correct balance and angle must be found before moving.

Remember

Take care when raising to avoid overhead obstructions, such as electric cables.

Tower scaffolds

These may be either mobile or static and are suitable for both internal and external work up to approximately 6 m.

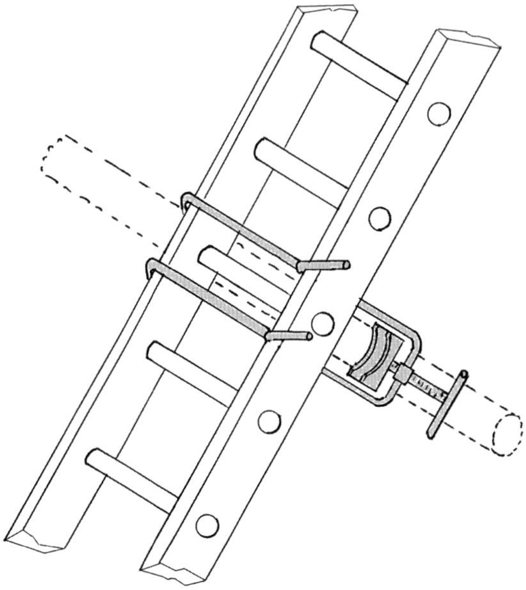

FIGURE 2.15
Ladder clip

Towers may be constructed from individual scaffold components or proprietary units. This unit will only deal with proprietary mobile towers. A typical mobile tower is shown in Figure 2.16.

They usually support a single working platform, not projecting beyond the base area, and are provided with toeboards and handrails.

Access to the working platform is by a ladder, which may be fastened inside or outside the structure.

Light-duty access towers are very common in the building industry for lightweight work such as maintenance of gutters, painting, etc.

They will not support a load greater than 1.5 kN/m^2. This is approximately equivalent to two people standing per square metre.

The safe working load should be clearly displayed on the working platform.

Tower scaffolds should always be vertical, even if erected on sloping ground. Mobile towers should only be moved on firm, level ground.

Towers should never be moved with people or materials on them.

They should always be pushed at the lowest practical point.

If extra working height is required, then the base measurement can be increased by the use of outriggers. These are tubes or special units that connect to the bottom of the tower, at the corners, giving greater overall base measurements. Outriggers also help to stabilize a scaffold tower and are usually used for this purpose as well as giving extra working height.

Remember

Access must be provided through the working platform.

NEVER lean a ladder against the tower scaffold.

FIGURE 2.16
Mobile tower scaffold

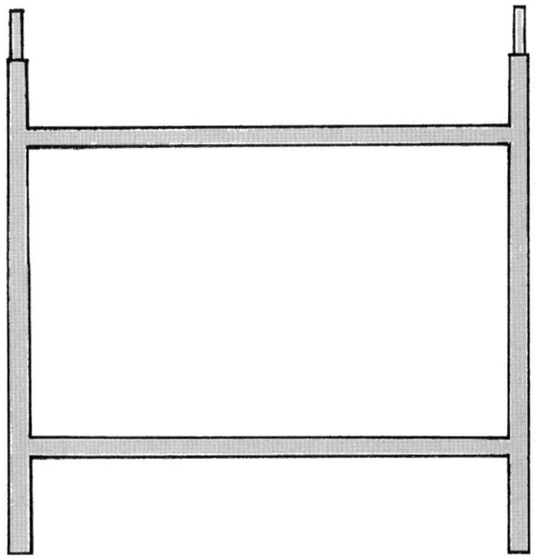

FIGURE 2.17
System tower scaffold

System scaffold

These systems are proprietary systems consisting of special units that fit into each other (Figure 2.17). They can be used to produce an access tower or a hop-up in a very short time.

This type of scaffold is increasingly being used for quick, straightforward platforms.

There are many patent types of frame available, but basically each consists of two short tubes which act as uprights, and these are joined near the top and bottom by tubes with a welded joint at each end.

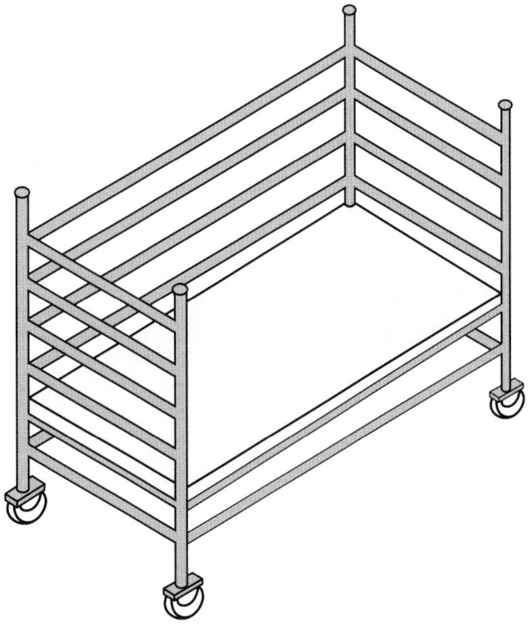

FIGURE 2.18
Portable low working platform

Low towers

Many manufacturers now produce various designs of working platforms to meet the current Work at Height Regulations.

They are restricted to a maximum of 2 m in height and usually have a safe working load of 150 kg. They arrive ready assembled, are very lightweight and are easy to move.

A low working platform is shown in Figure 2.18.

Multiple-choice questions

Self-assessment

This section of the book is designed to allow you to check your level of knowledge. The section consists of revision questions for this chapter. The questions are all multiple choice and have four possible answers. The answers are to be found at the end of the book.

The main type of multiple-choice question will be the four-option multiple-choice question. This will consist of a question or statement, known as the stem, followed by a choice of four different answers, called the responses. Only one of these responses is the correct answer; the others are incorrect and are known as distracters.

You should attempt to answer the questions by choosing either (a), (b), (c) or (d).

Example

The person employed by the local authority to ensure that the Building Regulations are observed is called the:

(a) clerk of works
(b) building control officer
(c) council inspector
(d) safety officer

Health and safety in the construction industry

Question 1 The official body that enforces health and safety is:

(a) the local authority
(b) the public health authority
(c) the Health and Safety Executive
(d) the employer

Question 2 When should you wear safety goggles on site?

(a) at all times
(b) in strong sunlight
(c) when cutting bricks
(d) when told to

Question 3 Why shouldn't you paint ladders?

(a) the paint might hide a defect
(b) the paint will make the ladder slippery
(c) the paint will not last
(d) regular repainting will be required

Question 4 What should you do when you hear the fire alarm ringing?

(a) leave the site

(b) go to the fire assembly point

(c) go to the canteen

(d) go to see the fire

Question 5 What shape should an information sign be?

(a) circular

(b) oval

(c) square or oblong

(d) triangular

Question 6 Which of the following current regulations covers working in dust?

(a) Health and Safety at Work Act

(b) Control of Substances Hazardous to Health

(c) Work at Height Regulations

(d) Personal Protective Equipment Regulations

Question 7 How many rungs should a ladder be above a working platform?

(a) 3

(b) 4

(c) 5

(d) 6

Question 8 A scaffold tower should only be erected by:

(a) a competent person

(b) the trainee bricklayer

(c) the site supervisor

(d) the hire company

CHAPTER 3

Communication

This chapter will cover the following NVQ and Diploma units:

- NVQ VR02
- CC 2002K

This chapter is about:

- Interpreting building information
- Determining quantities
- Relaying information

The following NVQ performance criteria will be covered:

- Performance criterion 1: Communicate with others
- Performance criterion 2: Work relationships

The following Diploma outcomes will be covered:

- Know how to interpret building information
- Know how to determine quantities of resources
- Know how to communicate workplace requirements efficiently

Types of information

This chapter deals with extracting and interpreting information and then correctly relaying it to other people.

Throughout your working life you will have to consult various sources of information and during this chapter you will be required to make decisions and solve problems from the information given or extracted.

Information sources

There are numerous sources of information available and they include the following: drawings, programmes of work, schedules, specifications, policies, mission statements, manufacturers' technical information, organizational documentation, and training and development records and documents.

DRAWINGS

The design team will be required to produce working drawings for the builder to use on the site.

Drawings, schedules and specifications will have to be prepared, explaining how the design team requires the building to be constructed. To be able to read these drawings it is essential that the trainee is able to understand them.

Drawings should be produced according to:

> *BRITISH STANDARD RECOMMENDATIONS FOR DRAWING OFFICE PRACTICE BS 1192*

These recommendations apply to the sizes of drawings, the thickness of lines, dimension of lettering, scales, various projections, graphical symbols, etc.

The person carrying out a task should be able to read drawings and extract the required information.

Information concerning a project is normally given on drawings and written on printed sheets.

Drawings should only contain information that is appropriate to the reader; other information should be produced on schedules, specifications or information sheets.

Programmes of work

There are several methods, but the main one used by most construction firms is the bar chart

The bar chart is simple in concept and equally simple to understand. It is hardly surprising therefore that most site supervisors prefer this method.

The activities are listed down the left-hand side in the sequence in which they will take place on site.

The time scale is drawn horizontally and the bars represent the time when the work will proceed on the activities. Bars may be suitably shaded or coloured to distinguish individual trades.

It is usual for only half of the bar to be used for programming; the remainder is completed as progress is achieved. Alternatively, progress can be recorded immediately above or below the bar or in a separate chart specially provided for that purpose.

The person drawing the bar chart could decide the layout, or it could be company policy to adopt a standard format.

The minimum information on the bar chart is the activities and their duration, but plant, labour and material deliveries could also be included.

An example is shown in Figure 3.1. This example shows the activities in their correct sequence. Against each activity is a bar showing the anticipated duration. Each bar should produce a flow through the various activities until completion of the project.

SCHEDULE

This is a contract document that can be used to record repetitive design information about a range of similar components, such as:

- doors
- windows
- ironmongery
- decorative finishes
- inspection chambers (manholes).

A typical schedule for doors is shown in Table 3.1.

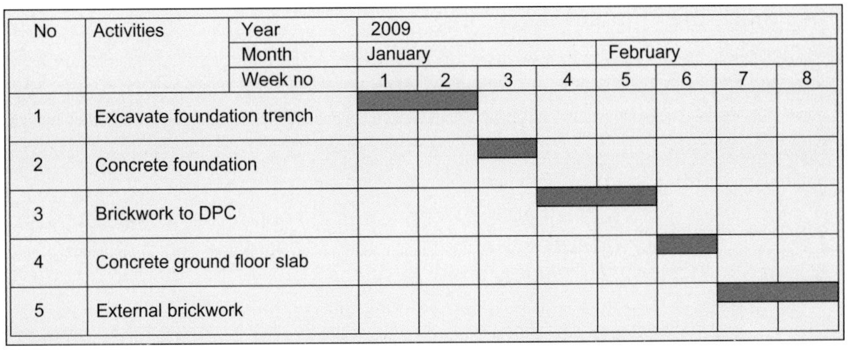
No
Activities
Year
Month
Week no
2009
January
February
1
2
3
4
5
6
7
8
1
Excavate foundation trench
2
Concrete foundation
3
Brickwork to DPC
4
Concrete ground floor slab
5
External brickwork

FIGURE 3.1
Typical bar chart

Table 3.1 Door schedule

Description	D1	D2	D3	D4	D5
External panelled	•				
Internal flush		• 2			
Internal panelled			•		
Internal half-glazed					
Internal glazed					•

SPECIFICATION

This is a document giving a written description of materials to be used, and construction methods to be employed in the construction of the building (Table 3.2).

POLICY

A construction company, whether it is a one-person firm or a large company, must have an aim to achieve. To achieve this aim objectives are produced for each person and written down. This group of objectives becomes the policy of the firm.

There will be a different set of policies for the site manager, who has to implement the company's general policy. Typical site policies could be for safety on site, security, quality control, etc.

MISSION STATEMENT

A mission statement is a formal description of the mission of the business. A typical mission statement is shown in Figure 3.2.

The mission statement might be published on promotional material and on display in the reception area of the company.

It should:

- be brief and easy to understand and remember
- be able to accommodate change
- make the company stand out.

Table 3.2 Specification

Item no.	Description
1	Block partition walls, 102.5 in stretcher bond in cement mortar (1:4)
2	Brick facework, 102.5, in stretcher bond using multicoloured facings in cement mortar (1:4) Pointing with half round joint as the work proceeds
3	Blockwork to inner skin of cavity wall, 102.5, in stretcher bond in cement mortar (1.5)
4	Form cavity in hollow wall Insert 50 mm polystyrene batts
5	Damp-proof course, 102.5, of single-layer hessian base bitumen felt and bedded in cement mortar

> *Our mission is to be the customer's first choice for affordable housing, delivering outstanding quality and great customer service at a competitive price.*

FIGURE 3.2
Mission statement

MANUFACTURER'S TECHNICAL INFORMATION

Any efficient office should have up-to-date information regarding new and existing products from the various manufacturers who produce materials and equipment of interest to them.

Many manufacturers produce technical information which is free for the asking.

The manufacturer's technical information consists of drawings and text which are provided to give the user details of the product. A typical manufacturer's leaflet shown in Figure 3.3 shows technical details for a small mixer.

Organizational documentation

Each type of company will create various organizational documents to be used, but they will usually fall into the following:

- time sheets
- job sheet/daywork sheets

SMALL MIXER

Models 100 litre & 150 litre

Small mixer

Easily transported in the back of a van or trailer.

Both models are easily maintained.

Available in solid and Pneumatic tyres.

DIMENSIONS		
TYPE	100 litre	150 litre
Length	1500 mm	1500 mm
Width	1200 mm	1200 mm
Height	1700 mm	1800 mm

Model	Engine	Wheels	Weight
100 litre	1.5 h.p diesel engine or electric motor	Two tubeless rubber tyres or solid tyres	600 kg
150 litre	1.5 h.p diesel engine or electric motor	Two tubeless rubber tyres or solid tyres	750 kg

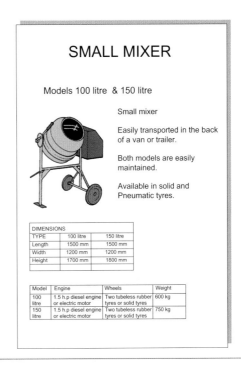

FIGURE 3.3
Manufacturer's leaflet

- material sheets
- site diary, etc.

TIME SHEETS

Time sheets should be completed weekly by each employee. Failure to complete and hand in a time sheet could result in a loss of wages. A typical time sheet is shown in Figure 3.4.

DAYWORK SHEETS

A job sheet or daywork sheet could also be handed out to each employee to record any work completed that was not originally planned.

The RIBA form of contract states:

> *'where work cannot be properly measured and valued, the Contractor shall be allowed daywork rates on the prices prevailing when such work is carried out'.*

Records must be kept on a weekly basis.

The site manager must record the number of hours worked, the materials used, and the use of any plant or plant hire, for the duration of the job. A typical daywork sheet is shown in Figure 3.5.

RECORDING INCOMING MATERIALS

There will be times when you will have to take responsibility to off-load materials, checking that they are exactly what was ordered and of appropriate quality, and then store them properly.

Always remember to check the delivery ticket before you start to unload the materials. A lorry driver could deliver to the wrong site.

WEST BUILDERS

Timesheet

Employee .... Site ....

Date	Start Time	Finish Time	Total Time	Travel Time	Other expenses
Mon					
Tue					
Wed					
Thu					
Fri					
Sat					
Sun					
Total					

FIGURE 3.4
Time sheet

Daywork Sheet **Contract** ....

No .... **Week ending**.....................................

Description of work

....

....

Labour

Name	Trade	Hours						Total hours	Rate	£	P	£	p
		M	T	W	T	F	S						

Materials

Description	Rate	£	p

FIGURE 3.5
Daywork sheet

Ensure they are exactly what were ordered by you or the company. A typical delivery sheet is shown in Figure 3.6.

After the driver has off-loaded the materials you will be asked to sign an acceptance slip stating that the order is complete and undamaged.

With certain materials, breakages can be reported to the supplier within a time limit (often 7–14 days) and they will be replaced, but shortages will not. It is therefore very important that shortages are checked before signing.

If any of the materials or components are missing or defective, they should immediately be identified and noted on both copies of the delivery ticket.

WEST BUILDERS

Contract : ....

Contract No : ....

Prepared by: ....

Date	Delivery Note No	Supplier	Materials	Rate	Total value

FIGURE 3.6
Material requisition sheet

Both tickets should be signed stating the items short or damaged.

If you have found any problems with any delivery you should contact the site supervisor immediately and he will (or will ask you to) contact the supplier to explain the problem.

SITE DIARY

The site supervisor should complete a site diary with various information important to the site. An example is shown in Figure 3.7.

This is one of the most important reports that a site supervisor has to produce during a contract.

It should include the following:

- the weather, morning and afternoon
- any visitors on site
- delays by subcontractors
- instructions from the clerk of works

SITE DIARY AND DAILY REPORT

Contract	Contract No:	Week ending: ....
Weather	Temperature AM	Temperature PM ....

Labour on site		Plant on site	
Own staff	Subcontractors	Own plant	Hired plant

Job stoppages	Man hours lost	Reason
Drawings received	Information received	Telephone calls
Visitors to site	Accidents	Reported accidents

Brief report of work and progress on site
Signed : Site manager .... Contracts manager ....

FIGURE 3.7
Site diary

- materials delivered
- number of staff on the site.

TRAINING AND DEVELOPMENT RECORDS

Any company should consider each employee's career development.

Whatever training has been given it should be documented according to company policy. Probably the first document a new trainee would come into contact with would be the contract of employment, which sets out their terms of employment. A typical contract is shown in Figure 3.8.

Storing information

When information has been found the last place to store it is in your head.

The relative information should be stored in a filing system which everyone can understand. The filing system could consist of something as simple as document files or a more elaborate computer program.

Any information found may be required over and over again for a particular contract, but can also be useful information for future programmes.

The storage of drawings on site can cause problems, especially with dampness. They could be laid flat in a drawing chest or rolled up and placed in racks.

Whichever method chosen they should be recorded when received on site. A typical drawings register is shown in Figure 3.9.

WEST BUILDERS
CONTRACT OF EMPLOYMENT

WB

Statement of main terms of employment

Name of Employer ....
Name of Employee ....
Title of Job ....
Statement issue date ....
Employment commencement date ....

Your hours of work, rates of pay, overtime, holiday entitlement and payment, pension scheme, disciplinary procedures, notice and termination of employment and disputes procedure are in accordance with the following documents:

1. **The National Working Rules**
2. **The Company Wages Register**
3. **The Company Handbook**

FIGURE 3.8
Contract of employment

WEST BUILDERS
Drawings Register

Contract : .... **Drawings from :** ....
Contract No : .... **Date commenced :** ....

Drawing No	Description	Scale	Number received	Date of issue	Amendments

FIGURE 3.9
Drawings register

INTERPRETING INFORMATION

Selecting information from simple drawings, specifications and schedules

It is important that you can understand drawings and extract the correct information from them. The following section gives you a chance to practise reading and understanding drawings, specifications and schedules.

DRAWINGS

You have received the site plan and location plan for a new office block, shown in Figure 3.10. Explain the items marked 1–7 on the plan by completing Table 3.3.

SPECIFICATIONS

These documents are prepared to specify the exact quality of materials and skill required throughout the contract.

It is essential to read and understand exactly what is required on a contract as this will affect the price of the work.

Specifications are used only on large contracts, where the drawings cannot contain all the information required by the contractor.

Drawings for small building works usually have the specification written on the drawings.

A typical specification is shown in Table 3.4. Read it carefully and complete the following questions.

Specification details

To show that you can understand specifications in Table 3.4, complete Table 3.5.

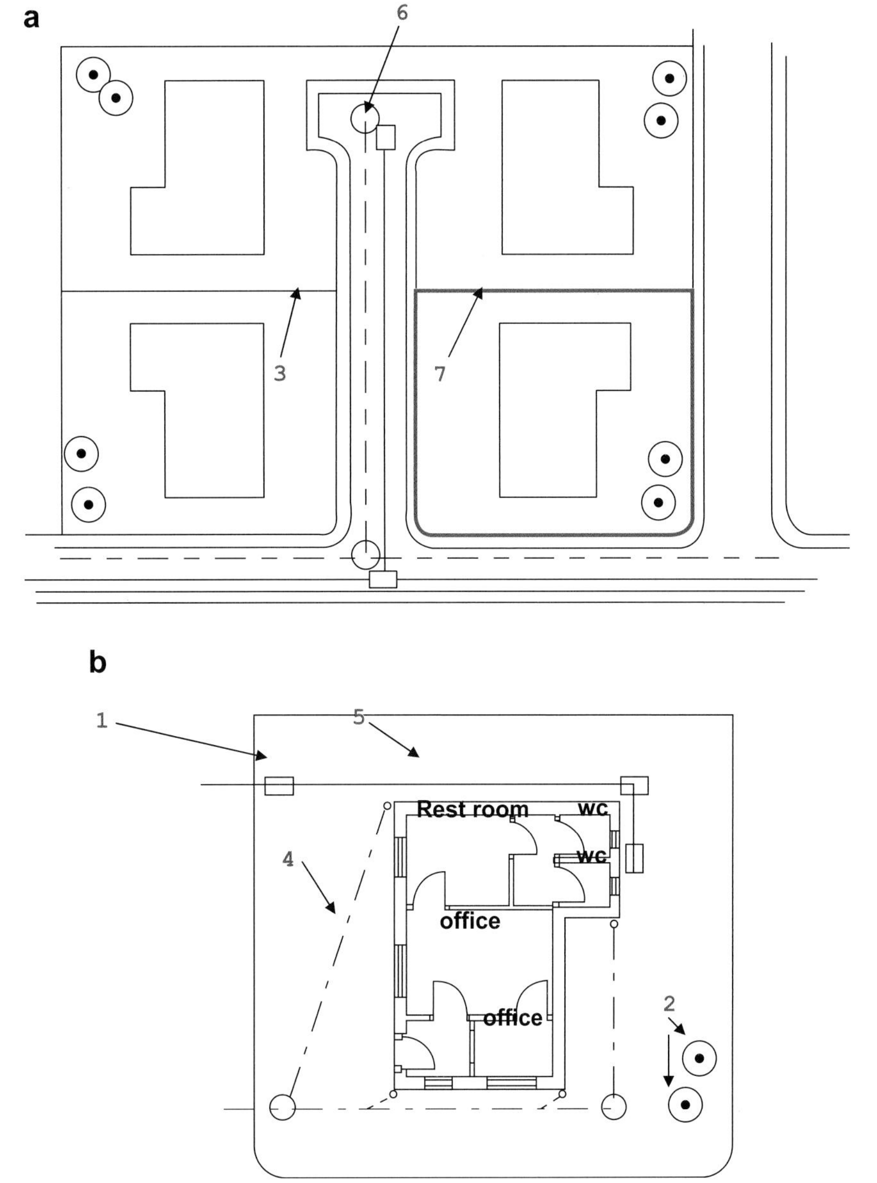
a
6
3
7
b
1
5
Rest room
wc
wc
4
office
office
2

FIGURE 3.10
(a) Site plan and (b) location plan for small office

Table 3.3 Definitions of items in the plan

No.	Definition
1	
2	
3	
4	
5	
6	
7	

Table 3.4 Specification

Item no.	Description
1	Block partition walls, 102.5, in stretcher bond in cement mortar (1:4)
2	Brick facework, 102.5, in stretcher bond using multicoloured facings in cement mortar (1:4) Pointing with half round joint as the work proceeds
3	Blockwork to inner skin of cavity wall, 102.5, in stretcher bond in cement mortar (1:5)
4	Form cavity in hollow wall and insert 50 mm polystyrene batts
5	Damp-proof course, 102.5, of single layer hessian base bitumen felt and bedded in cement mortar
6	Concrete floor slab 150 mm thick, 1:2:4 mix on 2000 gauge DPM on 150 mm consolidated hardcore
7	Screed to concrete floor, 50 mm thick with wooden float finish to receive thermoplastic tiles
8	Internal walls finished with three coat plasterwork
9	Timber flat roof using 200 × 50 joists on 100 × 50 wall plate
10	Window frames fitted according to window schedule as the work proceeds

Table 3.5 Specification details

Question no.	Question	Answer
1	What type of cavity insulation will you order?	
2	What type of damp-proof course is specified?	
3	How many m^2 of blockwork is required?	
4	If three wall ties are required per m^2 how many will you order for the cavity?	
5	Explain the type of mix required for the concrete floor	1: 2: 4:
6	What does DPM mean?	
7	What materials will you order for the screed to the floors?	
8	Will the screed finish be smooth or rough?	
9	Will jigs be required when fixing the windows?	
10	What will be required by the bricklayer when the wall plate is fixed?	

Producing simple working drawings

SCALE

Scales are used in drawings to enable large objects such as buildings to be drawn to a convenient size, which will fit onto the drawing paper, while still maintaining accurate proportions which can be drawn or measured as required.

Scales use ratios to relate measurements on a drawing to the real dimensions of the actual item being drawn.

Once these items have been drawn to a smaller size, which is known as a ratio to the real item, it is known as a scale drawing.

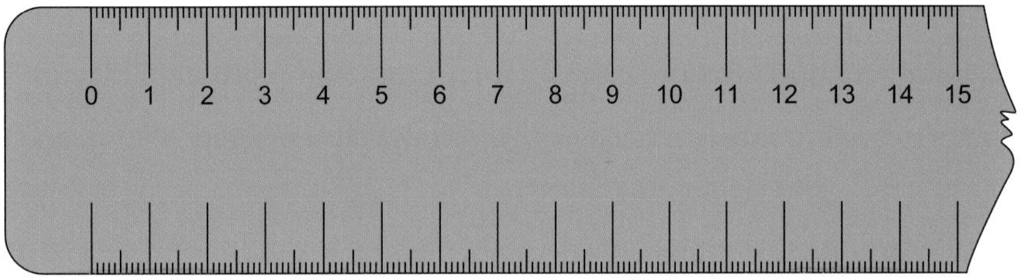
0
1
2
3
4
5
6
7
8
9
10
11
12
13
14
15

FIGURE 3.11
Ruler

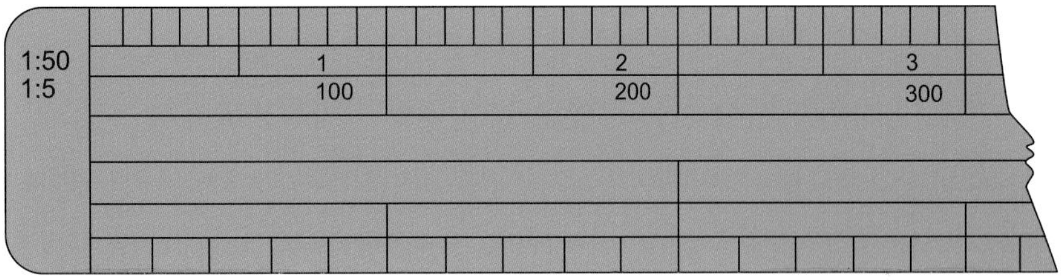
1:50
1:5
1
100
2
200
3
300

FIGURE 3.12
Scale rule

Scale rules

An ordinary ruler, 300 mm long (Figure 3.11), is useful for simple drawings, but scale drawings require a scale rule.

A scale rule (Figure 3.12) has a series of marks for measuring purposes. The scale rule is graduated in millimetres. Each scale represents a ratio of given units to one unit.

Table 3.6 lists the preferred scales for building drawings.

Use of scales

The choice of scale will depend on two things:

- the size of the object to be drawn
- the amount of detail that needs to be shown.

A scale is used to measure distances on drawings and for taking measurements of a drawing. If scaled dimensions and written dimensions disagree, then the written dimension should always be used.

Table 3.6 Preferred scales for building drawings

Type of drawing	Scales
Block plans	1:2500, 1:1250
Site plans	1:500, 1:200
Location drawings	1:200, 1:100, 1:50
Range drawings	1:100, 1:50, 1:20
Detail drawings	1:10, 1:5, 1:1
Assembly drawings	1:20, 1:10, 1:5

The ratio shows how many times bigger one quantity is than the other.

If the drawing is made to a scale of 1:100, 1 mm will represent 100 mm.

If the drawing is made to a scale of 1:20, 1 mm will represent 20 mm, and so on.

If the following drawing was drawn 150 mm long it would represent something much bigger in size:

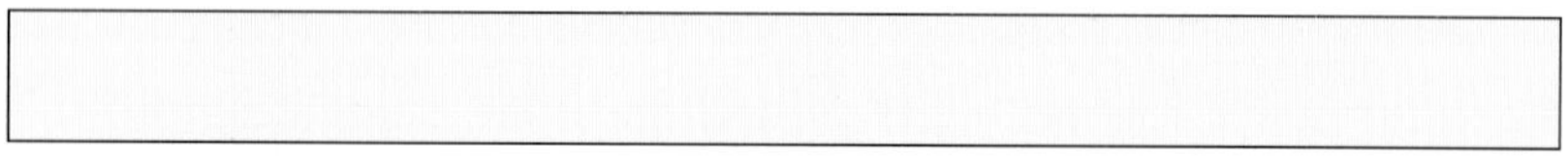

u0175 To a scale of 1:10 it would represent 1500 mm.

If it is drawn to a scale of 1:20 it would represent 3000 mm.

All you have to do is multiply the scale measurement by the scale ratio.

Scale rules can be used for both reading and preparing a drawing.

Producing simple drawings

There are many methods of drawing a particular object, each of which can portray a different feature of the object.

Remember

It is recommended that written dimensions on a drawing should always be used, rather than taking dimensions off drawings with a scale rule.

Before a drawing is started you must decide what you actually need to define.

You need to decide whether a front, side or back view is required, or even a plan of the object.

Differing perspectives are also possible, along with sections and exploded views.

A decision must also be made on whether the finished drawing will be a traditional line drawing or a freehand drawing.

BASIC DRAWING METHODS

Methods of projection

There are several methods of projection, the most common of which are:

- orthographic projection
- isometric projection
- oblique projection.

Orthographic projection is the method of showing solid objects which are actually three-dimensional in two-dimensional drawings by means of related views and plans, elevations and sections.

Most constructional drawings are of this nature. A typical example is shown in Figure 3.13.

The projected faces that represent the sides of the object are known as elevations. Those viewed looking down from above or up from below are known as plans.

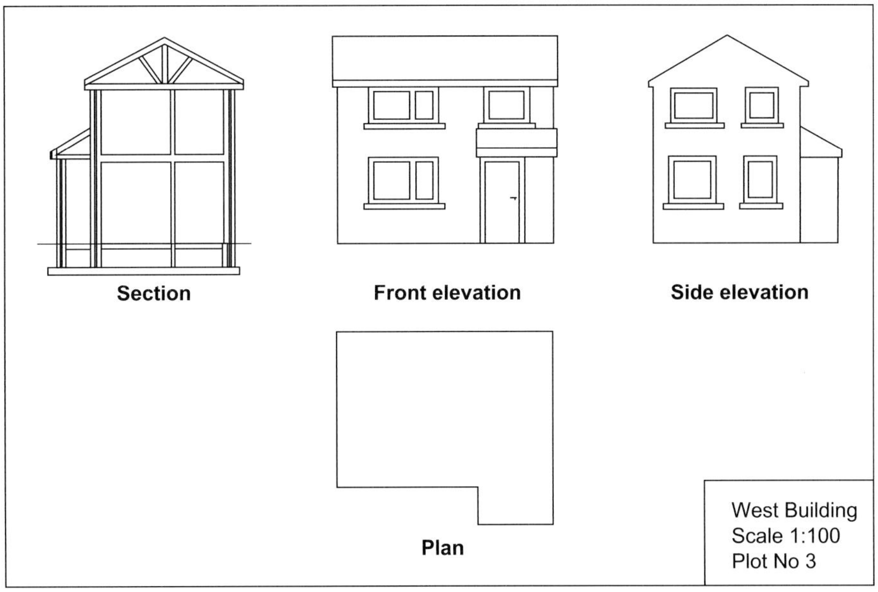
Section
Front elevation
Side elevation
Plan
West Building
Scale 1:100
Plot No 3

FIGURE 3.13
Orthographic projection

The outline of the building shown can be viewed in the direction of the arrows to show all the relevant details.

The views should be set out on drawing paper in a special format.

The front elevation is drawn first and the other views are related to it.

The plan of the building is always drawn directly underneath the front elevation.

The view from the left is placed to the right of the front elevation and the view from the right is placed to the left of the front elevation.

A view from the rear may be placed to the left or right depending on space and personal choice.

A section is placed on the opposite side to the intended view.

Isometric projection

This is a method of presenting the detail in pictorial format (Figure 3.14).

The definition of isometric projection is that all vertical lines remain the same but the horizontal lines are all drawn at 30 degrees to the horizontal.

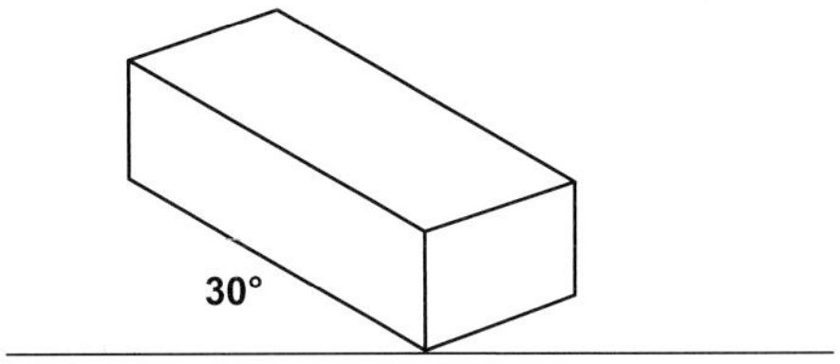
30°

FIGURE 3.14
Isometric projection

Oblique projection

This is another method of presenting the detail in pictorial format.

There are two main types of oblique projections, cavalier and cabinet (Figure 3.15). Both are drawn with the front face drawn to true shape and size, similar to a front elevation.

The faces that recede are drawn at an angle of 45 degrees, and the difference between the two methods is the scale of these receding lines. The cavalier method uses the full scale, whereas the cabinet uses half full size scale, which creates a more normal appearance.

SCALED DRAWINGS

The preceding pages have dealt with the various techniques involved in construction drawings. This section examines in more detail the types of construction drawings that you will have to prepare in your assignments.

All the points raised before will also apply to the drawings.

Restrictions have been placed on the figures by the page size; the actual sizes of production drawings are likely to be much larger.

The following drawings would be found within most construction drawings:

- elevations
- plans
- sections.

Most construction drawings are scale drawings, that is, the land, buildings or objects shown on them are not represented at their true size, but are larger or more often smaller in some proportion.

Drawings for a small garage are shown in Figure 3.16.

The position of drawings on the paper depends on the size of the contract, but in general the plans should always be underneath the elevations, with the sections at the side of the elevations.

On small contracts all the drawings could be positioned on one sheet of paper, but on larger contracts, where several hundreds of details are required, numerous drawing sheets are required. They should all be correctly numbered as received.

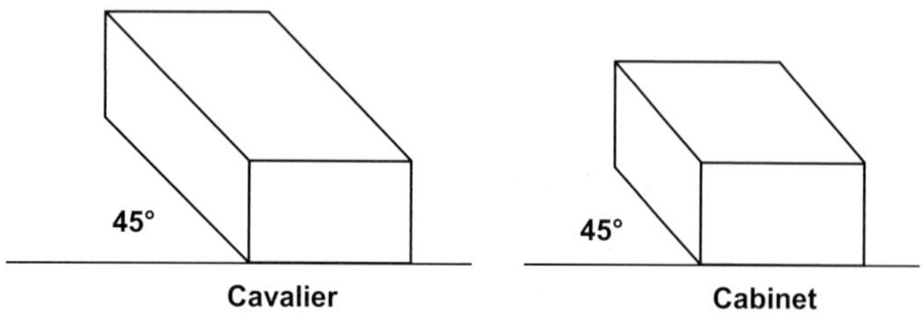
45°
Cavalier
45°
Cabinet

FIGURE 3.15
Oblique projections: (a) cavalier; (b) cabinet

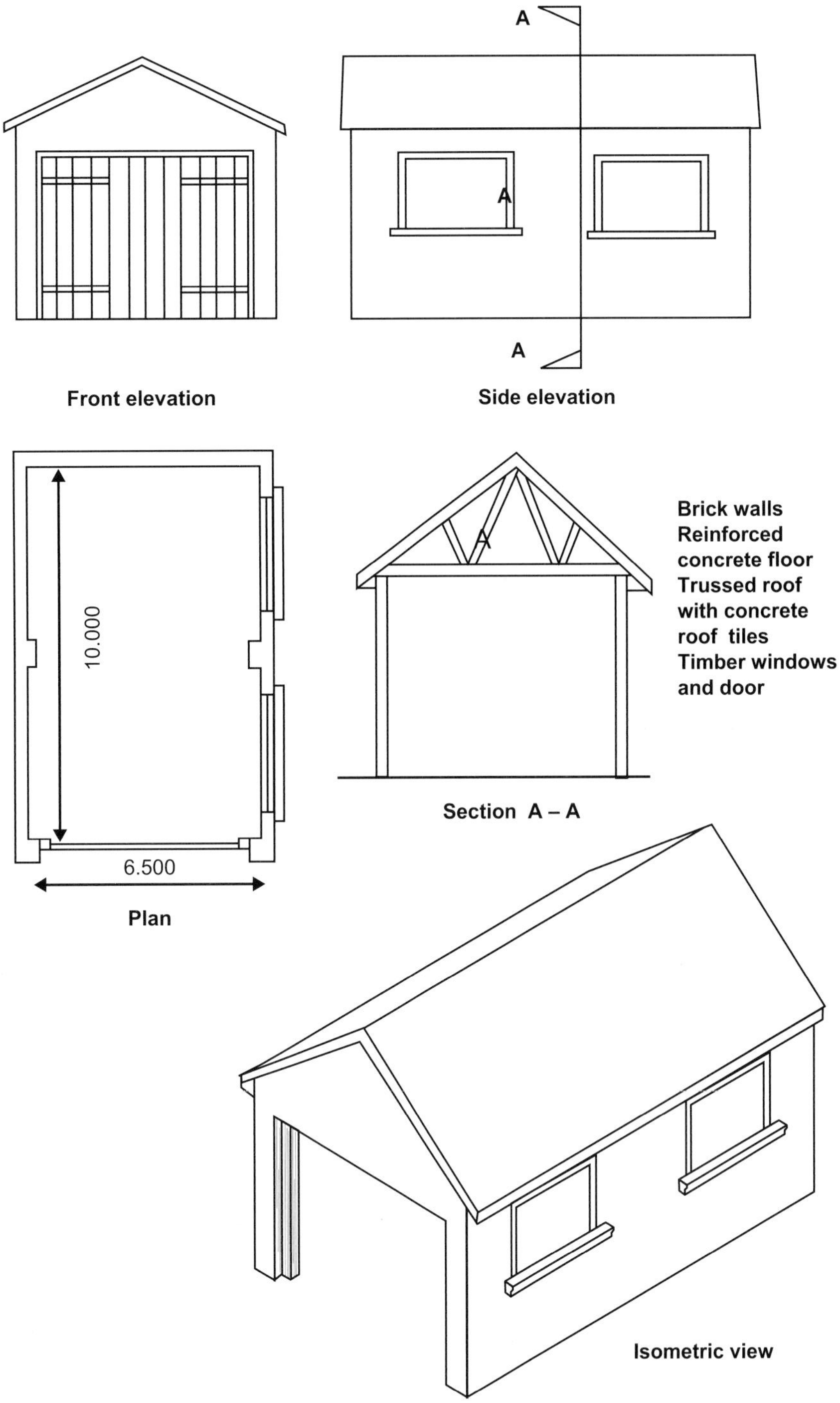
A
A
A
A
Front elevation
Side elevation
10.000
6.500
Plan
A
Section A – A
Brick walls
Reinforced concrete floor
Trussed roof with concrete roof tiles
Timber windows and door
Isometric view

FIGURE 3.16
Typical drawing layout

Elevations

These consist of front elevation, side elevation and occasionally rear elevation. Elevations of the garage are shown in Figure 3.17.

They are drawn to a scale of 1:200, 1:100 or 1:50, depending on the detail required.

The elevations give a visual effect of the finished building from all directions.

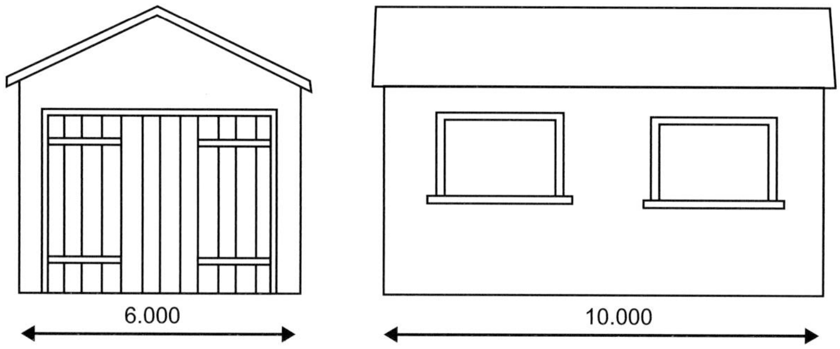
6.000
10.000

FIGURE 3.17
Elevations of the garage

They have limited dimensions on them as most information is gained from other drawings.

Plans

These are drawn to a scale of 1:200, 1:100 or 1:50, usually to match the elevations. The plan of the garage is shown in Figure 3.18.

As stated before, the ground floor plan should be drawn below the front elevation. If there is room, the first floor plan should be drawn below the ground floor plan.

One of the main purposes of the floor plans is to show the disposition of the various rooms, the position of entrances, circulation routes and the overall character of the building.

The size is still too small to add too much in the way of dimensions.

Notes, if included on such drawings, tend to be general descriptions.

Sections

Sections of the building are chosen to display a detail that otherwise would be difficult to see on the plans and elevations.

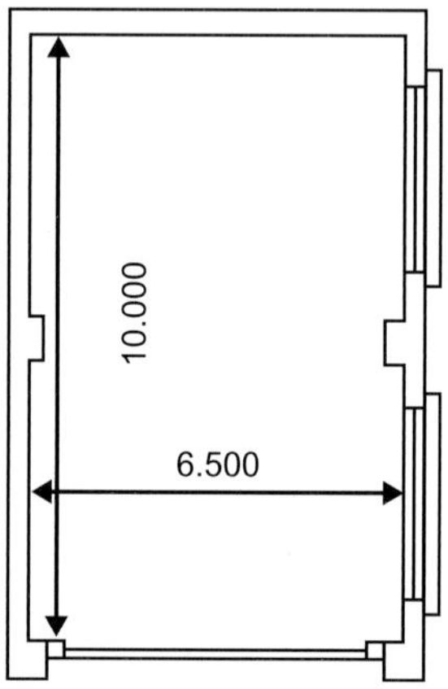
10.000
6.500

FIGURE 3.18
Plan of the garage

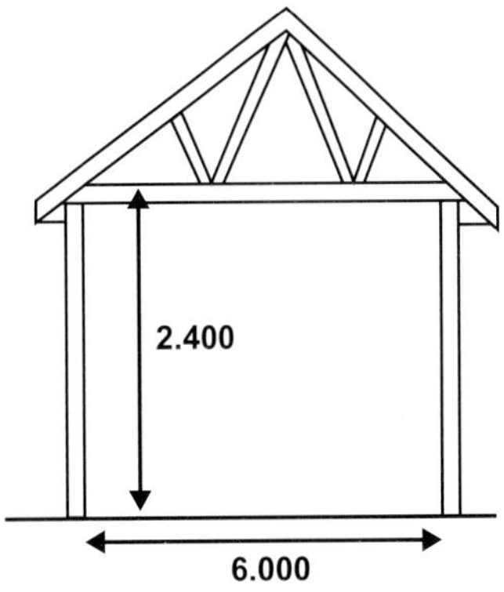
2.400
6.000

FIGURE 3.19
Section

These drawings can either be cross-sectional or longitudinal sections, and either full or part sections. A section of the garage is shown in Figure 3.19.

A full cross-section would show the full height of the building and its respective floor and ceiling levels.

Depending on the scale chosen, information would be restricted to that which could not be found elsewhere.

Quantities of materials

You will at some stage in your career have to determine quantities of materials required to carry out an activity. This will entail having to calculate areas and volumes.

The buyer will have the responsibility of taking off and scheduling materials from the bills of quantities, or on smaller jobs, the working drawings.

When the site supervisor is given this role he or she must be careful to ensure that correct quantities and details of quality of materials are extracted from the drawings and specifications.

Allowances should be made for materials wastage which, from experience, is usually 5–10 per cent.

A bricklayer should know how to work out the number of bricks needed to build a wall. This is a very simple matter using a pocket calculator. Do not panic at the very mention of arithmetic. Let the calculator take the strain and deal with the decimal point!

Method A

There is nothing wrong with counting up the number of bricks needed for one course, then multiplying by the total number of courses in the wall to get the answer.

This method works well for stretcher bond walling, but becomes difficult when applied to more complicated shapes that have doors and windows, or to cavity walling.

Method B

This method is based on finding the surface area of the wall. It makes deductions for doors and windows easy. Use the calculator to add, subtract and multiply. Do not strain the brain!

Types of calculation

Calculations in construction can be:

- numbers – one-off items such as chimney pots
- linear – lengths of materials, etc.
- superficial – areas of walls and floors, etc.
- cubic – volume of excavations, etc.
- time – used for calculating bills
- cost – used in pricing materials and labour.

NUMBER

Many building materials are measured by number.

Chimney pots, air bricks, doors and windows are all examples of materials that are measured by number.

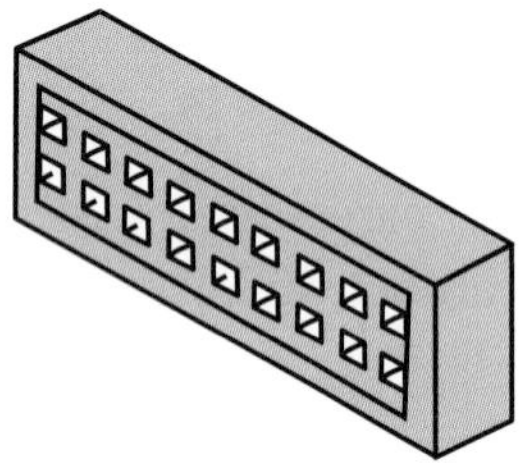

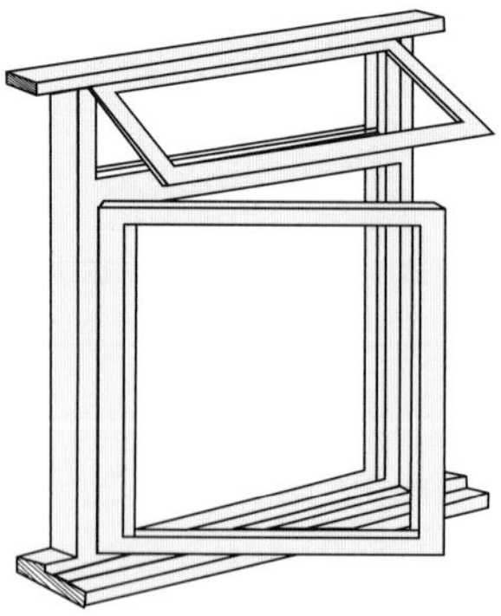

LINEAR

Many building materials are measured by length, but stating the width and thickness.

Timber is measured by length, for example 3 m length of 50 × 50 mm.

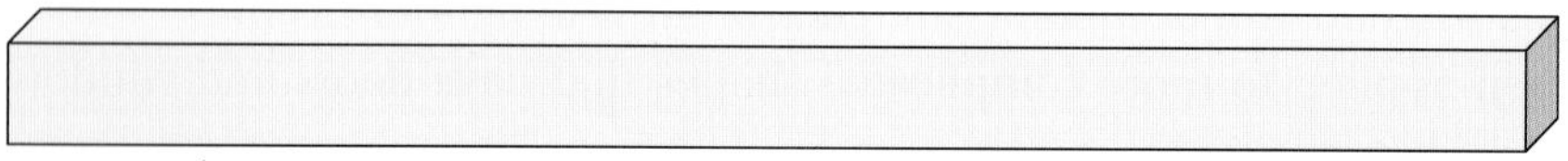

SUPERFICIAL

This is also known as square measurement, as the length and depth are multiplied to achieve the superficial area of the material required.

Examples are brick and blockwork.

The length of the wall is multiplied by the height to give the square area of brickwork required.

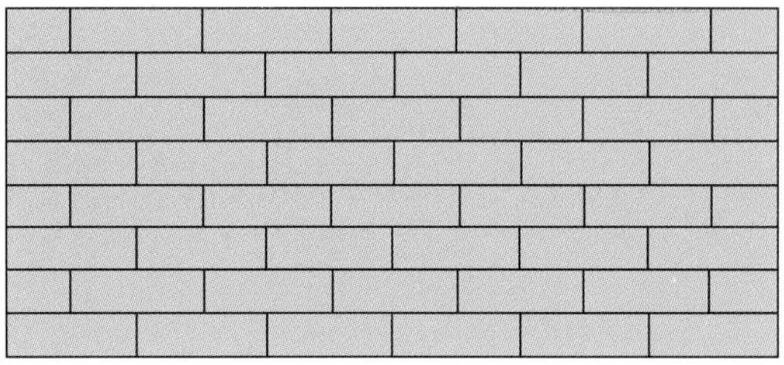

CUBIC

Cubic measurements are taken when there are three dimensions: length, width and depth.

Examples are found in excavating trenches and concrete for foundations.

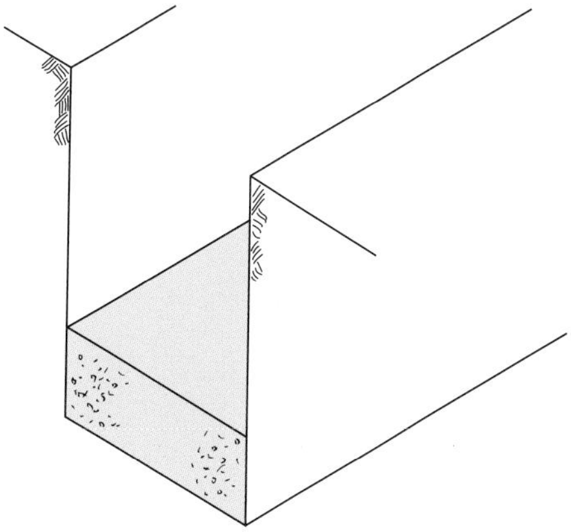

TIME

Time is used by all construction firms when calculating bills using information from time sheets completed by each employee.

COST

Estimating the cost of the work is impossible without knowledge of money. All employees in the construction industry are employed on a certain rate for the job. Calculating the cost for any work will involve both time and cost.

CALCULATORS

These are a very useful tool, *but* do not rely on the calculator at first.

A basic calculator with $+$, $-$, $\times$, $\div$, e is all that is needed.

Memory in a calculator can also be useful, but all steps need to be shown on paper.

Remember

All calculations should be shown and not done on bits of scrap paper or in your head. This will enable any mistakes to be found.

When end tests are being carried out marks are given for the correct method of setting out and not just for the answer.

It is particularly important to check answers when using calculators because it is easy to press the wrong button on small calculators.

DECIMALS

It is important that the decimal points are correctly placed. Misplacing a decimal point can make a result 10 times too large or small.

THE METRIC SYSTEM

The metric system has been devised so that standard units are used throughout the world. The metric system is much easier to use in calculations and measurements than the imperial system.

Unit of length

metre (m)

millimetre (mm)

kilometre (km)

1 km = 1000 m

1 m = 1000 mm

Therefore 1 km = 1 000 000 mm.

Unit of mass

kilogram (kg)

gram (g)

tonne (t)

1 t = 1000 kg

1 kg = 1000 g

Unit of force

Newton (N)

kilonewton (kN)

meganewton (MN)

1 MN = 1000 kN

1 kN = 1000 N

To convert mass (weight) to force = multiply by 9.81 m/s.

Examples

BRICKS

The drawing below shows the nominal sizes of a brick as:

215 mm × 102.5 mm × 65 mm

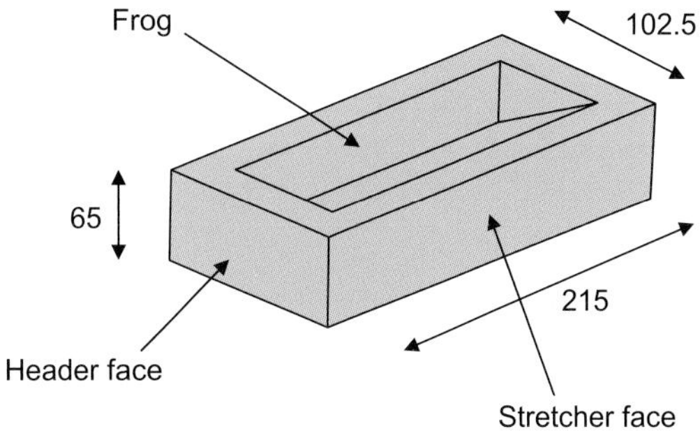
Frog
102.5
65
215
Header face
Stretcher face

These are the net sizes of the average brick and it is necessary to add 10 mm to the length, width and depth to produce a size when installed in the wall.

$$\text{Therefore: length} = 225\,\text{mm, depth} = 75\,\text{mm and width} = 112.5\,\text{mm.}$$

Bricks are laid to a gauge (number of courses to a set measurement) of four courses to 300 mm.

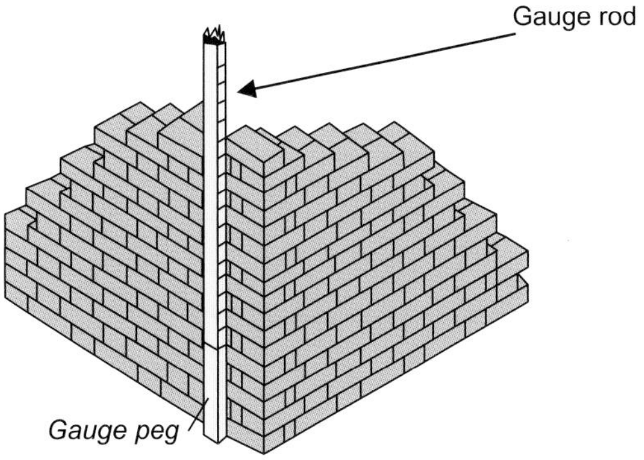
Gauge rod
Gauge peg

Brick calculations

The size of a brick including the joints:

$$= 225 \times 112.5 \times 75\,\text{mm}$$

If we find the face area of one brick and divide that into 1 square metre it will result in the brick requirement.

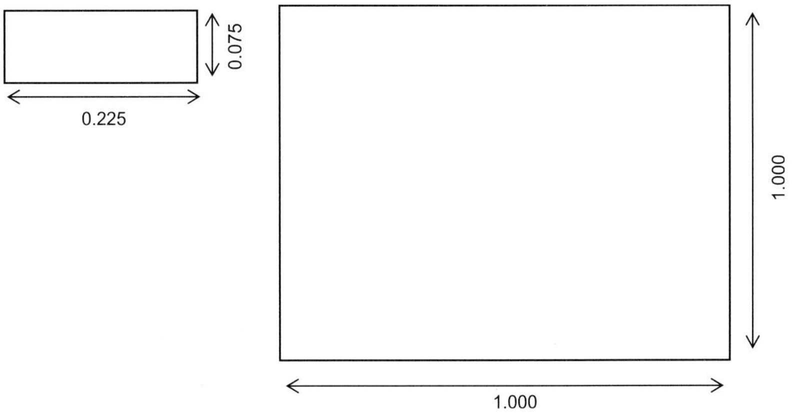
0.075
0.225
1.000
1.000

Area of the stretcher face of a brick:

$$= 0.225 \times 0.075 = 0.016875 \text{ m}^2$$

Area of a half-brick wall $= 1 \text{ m} \times 1 \text{ m} = 1 \text{ m}^2$

Therefore,

$$0.01678 \text{ m}^2 = 59.27 \text{ bricks}$$

Allowing for waste = 60 bricks per square metre of half-brick thick wall.

The other brick requirements are:

- 1 brick wall: 120 bricks
- 1½ brick wall: 180 bricks.

Mortar calculations

Nominal size for a brick is $215 \times 102.5 \times 65$ mm.

Nominal size for a joint is 10 mm.

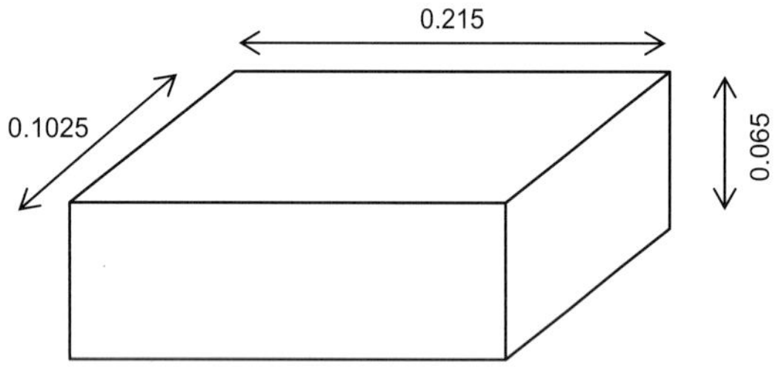
0.215
0.1025
0.065

If the brick size $= 215 \times 102.5 \times 65$ mm, then

Bed joint = Length × Width × Thickness of bed joint

$= 0.125 \times 0.1025 \times 0.01$

$= 0.00013 \text{ m}^3$

Cross-joint = Width × Depth × Thickness of cross-joint

$= 0.1025 \times 0.075 \times 0.01$

$= 0.00008 \text{ m}^3$

Therefore, the total for one bed joint and one cross-joint

$= 0.00013 + 0.00008$

$= 0.0002 \text{ m}^3$

If the calculation was for 1000 bricks,

Mortar required $= 0.0002 \times 1000$

$= 0.2 \text{ m}^3$

> **Note**
>
> Normal allowance is around 0.6 m^3 per 1000 bricks, which allows for the frogs and waste.
>
> Another method to simplify the calculation is to allow 1 kg of mortar for each brick.
>
> It follows that 1000 bricks would require 1000 kg of mortar, which is 1 tonne.

Therefore,

½ brick wall = 60 bricks + 60 kg mortar

1 brick wall = 120 bricks + 120 kg mortar

1½ brick wall = 180 bricks + 180 kg mortar

Example 1

Calculate the number of bricks required for the wall area.

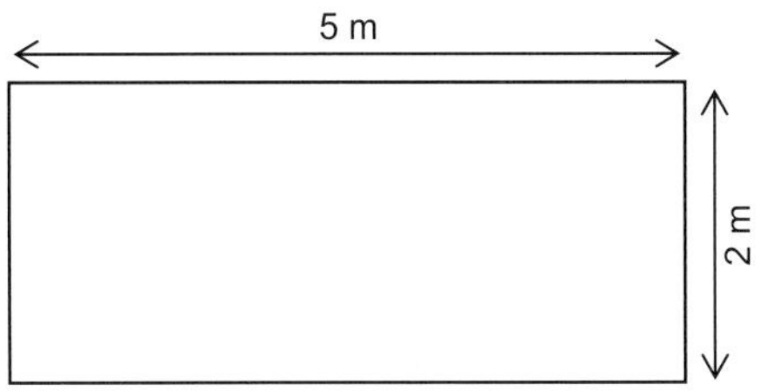

5 m
2 m

Answer 1

$$\text{Surface area} = 5 \times 2 = 10\ \text{m}^2\ \text{(square metres)}$$

You require 60 bricks (from previous calculations) to build 1 m^2 of ½ brick thick (102 mm) stretcher bond walling.

The surface area of the wall is 10 m^2, so you need $10 \times 60 = 600$ bricks.

Example 2

Calculate the number of bricks required for the wall area.

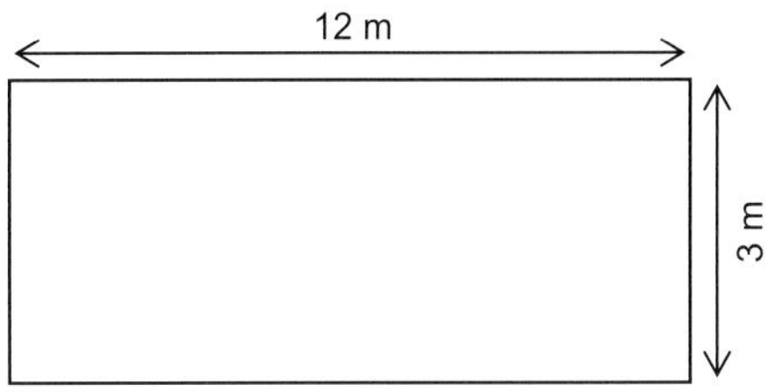

12 m
3 m

Answer 2

$$\text{Surface area} = 12 \times 3 = 36\ \text{m}^2\ \text{(square metres)}$$

You require 60 bricks to build 1 m^2 of ½ brick thick (102 mm) stretcher bond walling.

The surface area of the wall is 36 m^2, so you need $36 \times 60 = 2160$ bricks.

Walls one brick (1B) thick (215 mm) need 120 bricks to build 1 m^2 (1 m^2), and the same working method can be used as shown in Example 1.

> **Note**
>
> When doing any calculation always write down your working method as shown by the two lines in Example 2. This is worth it, because you can then check what you did. It will also ensure that you do not get in a muddle with the decimal point.

Example 3

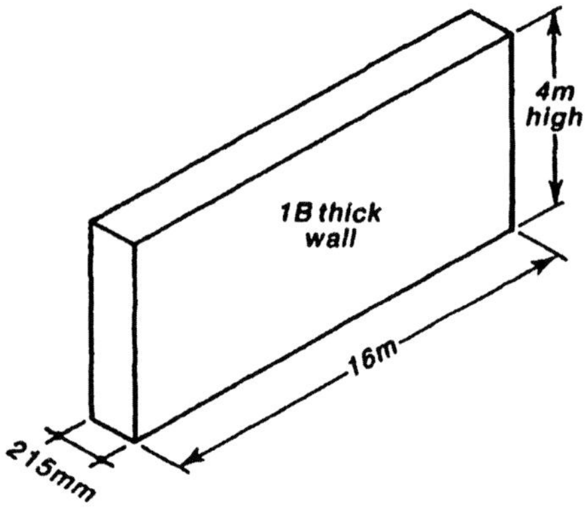
4m
high
1B thick
wall
16m
215mm

How many bricks are needed for this wall?

Answer 3

120 bricks per m^2

Two lines of working method:

Surface area $= 16 \times 4 = 64\,m^2$

Number of bricks needed $= 64 \times 120 = 7680$ bricks

Walls are not always straight elevations, and buildings have corners, but the same method of calculation can be used.

Example 4

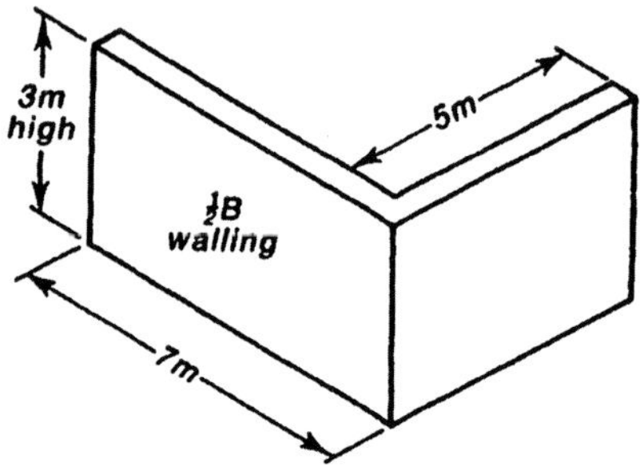
3m
high
½B
walling
5m
7m

How many bricks are required to build the wall shown?

Answer 4

60 bricks per m^2

Three lines of working method:

Total length of wall $= 7 + 5 = 12$ m

Surface area $= 12 \times 3 = 36\,m^2$

Number of bricks required $= 36 \times 60 = 2160$ bricks

Note

If calculations are not a happy thought for you, it is always a good idea to retrace your steps. Go back to Example 1. Just change the dimensions of that wall, invent your own figures, and then work out how many bricks will be needed this time.

Repeat what you have just done, but with Examples 2, 3 and 4, to calculate the number of bricks required for each example, but with different wall dimensions.

Example 5

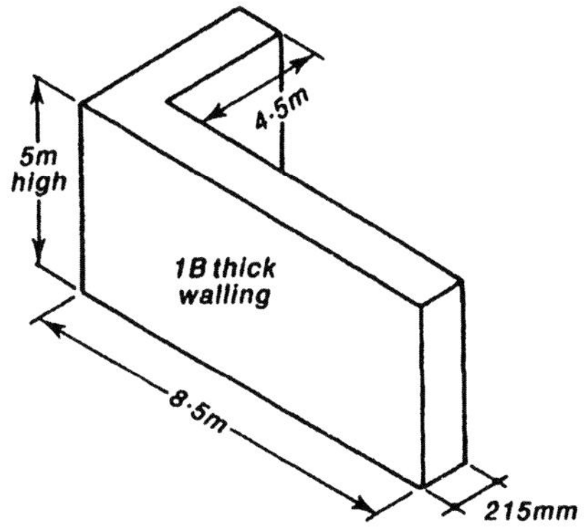
4·5m
5m
high
1B thick
walling
8·5m
215mm

How many bricks are required to build this wall?

Answer 5

120 bricks per m^2

Three lines of working method:

Total length $= 8.5 + 4.5 = 13$ m

Surface area $= 13 \times 5 = 65\,m^2$

Number of bricks required $= 65 \times 120 = 7800$ bricks

Example 6

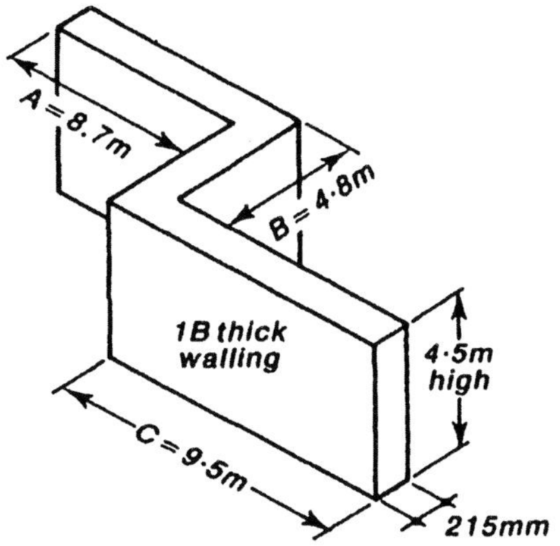
A=8·7m
B=4·8m
1B thick
walling
4·5m
high
C=9·5m
215mm

How many bricks are required to build this wall?

Answer 6

120 bricks per m^2

Three lines of working method

Total length $= A + B + C = 8.7 + 4.8 + 9.5 = 23$ m

Surface area = 23 × 4.5 = 103.5 m^2

Number of bricks required = 103.5 × 120 = 12420 bricks

BLOCKS

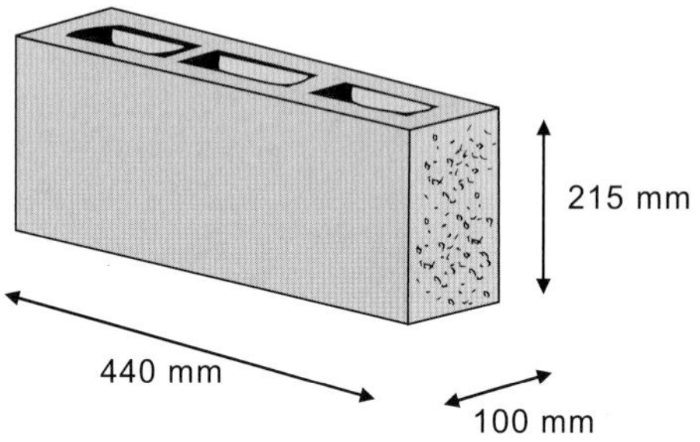
215 mm
440 mm
100 mm

Note

For additional practice, repeat Examples 5 and 6 after changing the dimensions, invent your own, and assume that the walls are only 102 mm thick.

The dimensions for blocks are shown as:

440 mm long × 215 mm high × 100 mm wide

These are net dimensions and the thickness of the joint is added to obtain the following dimensions:

450 mm long × 225 mm high × 100 mm wide

Blocks when laid are equal to three courses of bricks.

The calculations for blocks are very similar to brick calculations.

First calculate the area of one block then divide it into 1 square metre.

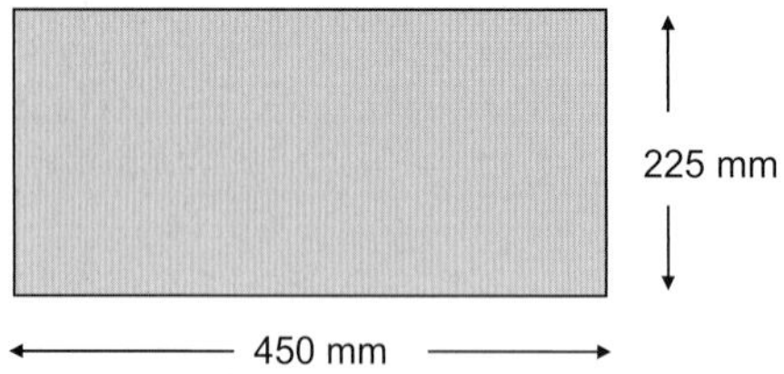
225 mm
450 mm

450 mm × 225 mm = 0.10125 mm^2

Number of blocks per square metre = 1.000/0.10125 = 9.87 blocks

It is usual to include waste and allow 10 blocks per square metre.

Blockwork
Example 7

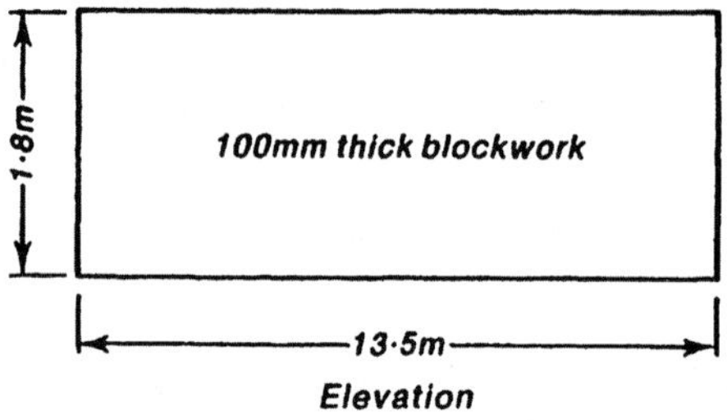
1·8m
100mm thick blockwork
13·5m
Elevation

Calculate how many standard size blocks are required to build this wall.

Answer 7

10 blocks per m^2 (from previous calculations)

Two lines of working method:

Surface area = $13.5 \times 1.8 = 24.3\,m^2$

Number of blocks = $24.3 \times 10 = 243$ blocks

Example 8

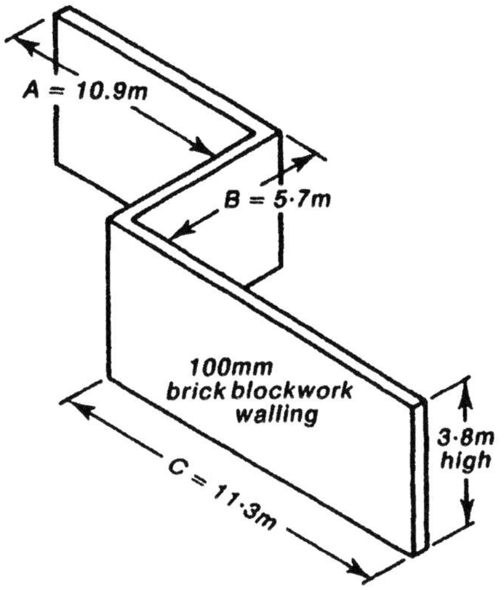
A = 10.9m
B = 5·7m
100mm
brick blockwork
walling
3·8m
high
C = 11·3m

How many standard size blocks are needed to build this wall?

Answer 8

10 blocks per m^2

Three lines of working method:

Total length of wall = A + B + C

= 10.9 + 5.7 + 11.3 = 27.9 m

Surface area = $27.9 \times 3.8 = 106.02\,m^2$

Number of blocks = $106.02 \times 10 = 1060.2$

Rounded off to 1061 blocks

Note

When ordering blocks for the inner leaf of cavity walls or for partition walls, take care to make clear the thickness you want, 100, 140, 150, 190 or 215 mm, as well as type, compressive strength, density and surface finish.

MORTAR

Allow 0.02 m^3 (cubic metre) (from previous calculations) of ready-mixed mortar to build 1 m^2 (square metre) of stretcher bonded brickwork. This is an average figure, bearing in mind that some bricks have frogs and others are perforated.

Table 3.7 shows approximately how much ready-mixed mortar is required for different wall thicknesses in brick and block.

Table 3.7 Materials per square metre of walling

	Stretcher bond			English bond			Flemish bond			English garden wall bond			Flemish garden wall bond			Header bond		
	F	C	M (m^3)	F	C	M (m^3)	F	C	M (m^3)	F	C	M (m^3)	F	C	M (m^3)	F	C	M (m^3)
½B 102 mm thick	60		0.02															
1B 215 mm thick	60[a]	60[a]	0.05	90	30	0.05	80	40	0.05	73[b]	47	0.05	67[b]	53	0.05	120		0.047
1½B 327 mm thick				90	90	0.08	80	100	0.08	73	107	0.08	67	113	0.08			
2B 440 mm thick				90	150	0.11	80	160	0.11	73	167	0.11	67	173	0.11			
100 mm standard size blocks	10		0.01															
140 mm standard size blocks	10		0.014															
150 mm standard size blocks	10		0.015															
190 mm standard size blocks	10		0.019															
215 mm standard size blocks	10		0.022															

F: facings; C: commons; M: mortar.

[a]215 mm thick walls can be built to show stretcher bond and give a face finish if required both sides if steel mesh bed joint reinforcement or butterfly wall ties are incorporated every fourth course.

[b]Garden wall bonds are intended for 215 mm thick free-standing boundary walls using 100% facing bricks.

Example 9

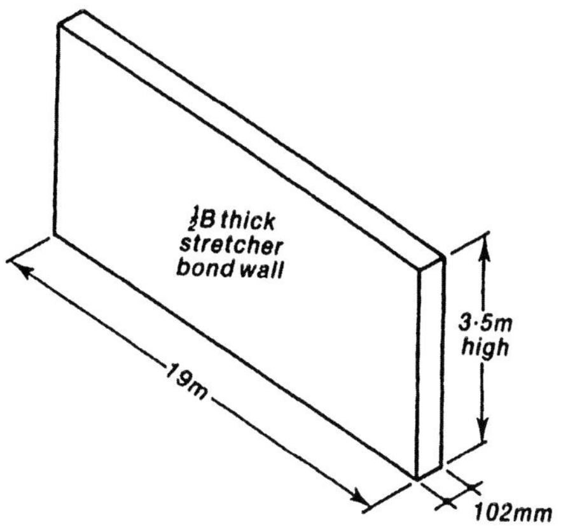
½B thick stretcher bond wall
19m
3·5m high
102mm

Calculate how much mortar will be required to build the brick wall.

Answer 9

0.02 m^3 of mortar per m^2 of walling (from previous calculations)

Two lines of working method:

Surface area of wall $= 19 \times 3.5 = 66.5\,m^2$

Volume of mortar required $= 66.5 \times 0.02 = 1.33\ m^3$

Note

For additional practice, go back to calculate how much mortar would be required to build the wall in Example 4.

Example 10

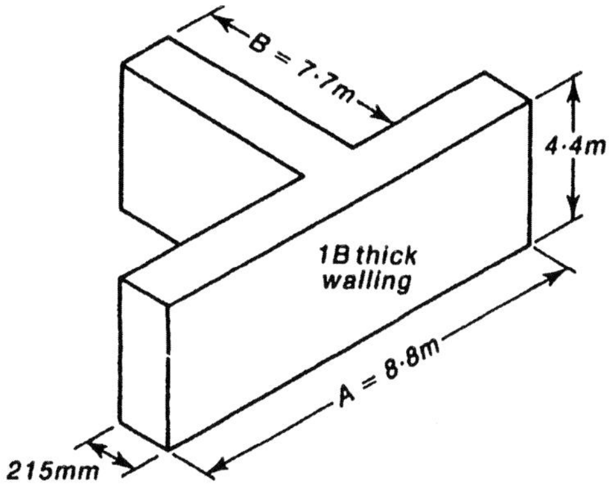
B = 7·7m
4·4m
1B thick walling
A = 8·8m
215mm

Calculate how many bricks and how much mortar will be required to build the following wall:

120 bricks and 0.05 m^3 of mortar per m^2

Answer 10

Four lines of working method:

Total length $= A + B = 8.8 + 7.7 = 16.5$ m

Surface area $= 16.5 \times 4.4 = 72.6\,m^2$

Number of bricks $= 72.6 \times 120 = 8712$ bricks

Volume of mortar $= 72.6 \times 0.05 = 3.63\ m^3$

Example 11

Calculate how many standard sized blocks and how much mortar will be required to build the following block wall:

10 blocks and 0.01 m^3 of mortar per m^2

Answer 11

Four lines of working method:

Total length = A + B + C + D

= 11.4 + 8.2 + 9.6 + 7.3 = 36.5 m

Surface area = 36.5 × 2.7 = 98.55 m^2

Number of blocks = 98.55 × 10 = 985.5 blocks

Rounded off to 986 blocks

Volume of mortar = 98.55 × 0.01 = 0.9855 m^3

Rounded off to 1 m^3

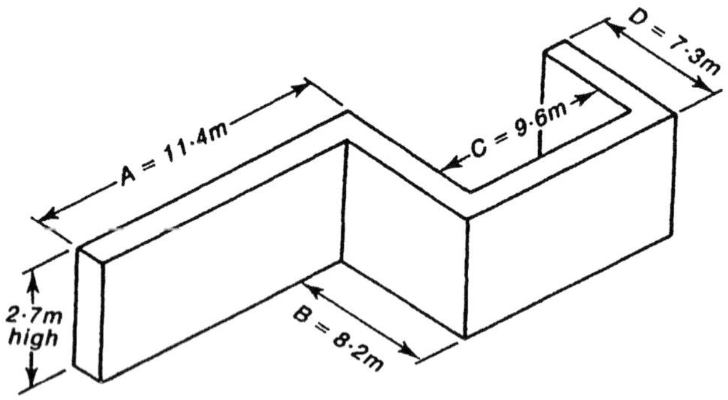
A = 11·4m
B = 8·2m
C = 9·6m
D = 7·3m
2·7m
high

ROUNDING OFF

With building calculations, it is customary to 'round off' decimal points, so as to tidy up quantities of materials into whole numbers where possible.

For example, it is not practical to order half a bag of cement, nor worry about 0.2 of a cubic metre of a bulk item like trench excavation.

It is sensible to round up with building materials, as site handling can result in damage or loss even in excess of the percentage allowances for cutting and waste.

Taking Example 11, 1 m^3 (1 cubic metre) is a less cumbersome figure than 0.9855 m^3 and involves fewer numbers with which to make mistakes.

OPENINGS

Most walls and buildings have door and window openings. This means that you must reduce your order for bricks and blocks, depending on the size and the number of openings.

The best way is to deduct the surface area of an opening from the overall wall area, before the final step of calculating the number of bricks or blocks needed (see Example 12).

Example 12

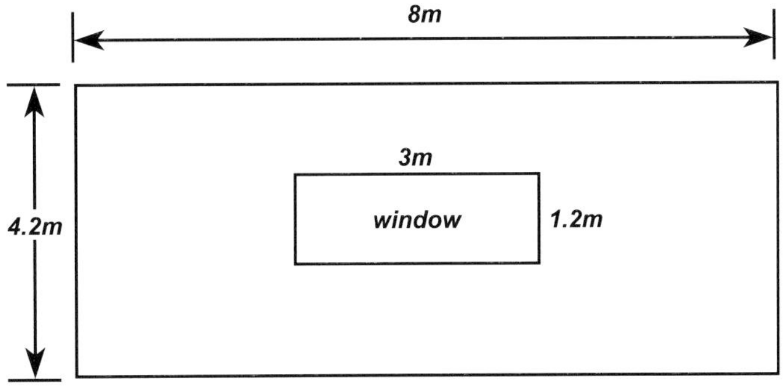
8m
3m
window
1.2m
4.2m

Elevation of $\frac{1}{2}$B wall

Calculate how many bricks will be required to build this wall, making the necessary deduction for the opening.

Answer 12

Using 60 bricks per m^2

Four lines of working method:

Overall surface area of the wall $= 8 \times 4.2 = 33.6\,m^2$

Area of window opening $= 3 \times 1.2 = 3.6\,m^2$

Net area of walling $= 33.6 - 3.6 = 30\,m^2$

Number of bricks required $= 30 \times 60 = 1800$ bricks

Example 13

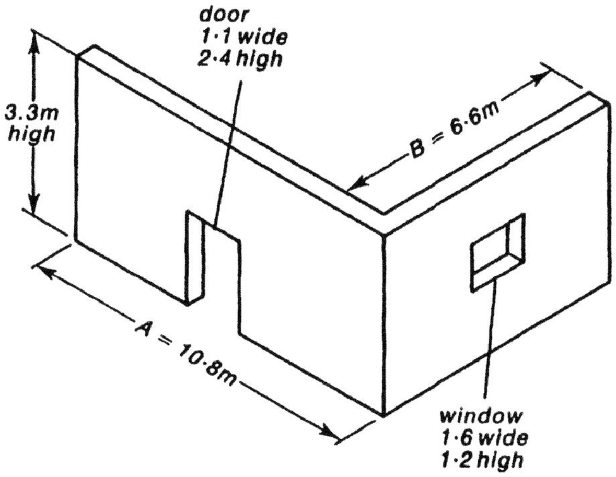
door
1·1 wide
2·4 high
3.3m
high
B = 6·6m
A = 10·8m
window
1·6 wide
1·2 high

How many bricks are required to build the wall shown, making allowances for openings?

Answer 13

Using 60 bricks per m^2

Five lines of working method:

Total length of wall $= A + B$

$= 10.8 + 6.6. = 17.4$ m

Surface area = $17.4 \times 3.3 = 57.42\,m^2$

Area of door opening = $1.1 \times 2.4 = 2.64\,m^2$

Area of window opening = $1.6 \times 1.2 = 1.92\,m^2$

Net area of walling = $57.42 - 4.56 = 52.86\,m^2$

Number of bricks required = $52.86 \times 60 = 3171.6$ bricks

Rounded off to 3172 bricks

CAVITY WALLING

When working out quantities of materials for cavity work, each leaf of the walling must be dealt with on its own, because you need separate totals for ordering bricks and blocks (see Example 14).

> **Note**
>
> For additional practice, calculate how much ready-mixed mortar will be required to build each wall shown in Examples 12 and 13. (Follow lines of working method given in Example 9.)

Example 14

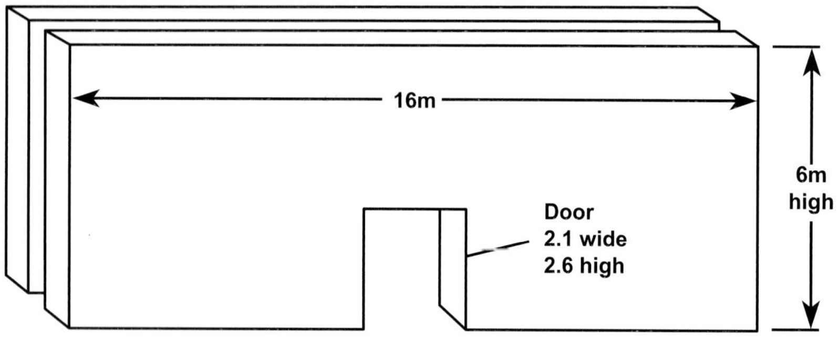
16m
6m
high
Door
2.1 wide
2.6 high

Elevation of 265mm thick cavity wall

Calculate how many facing bricks and blocks will be required to build the cavity shown, making due allowances for the opening.

Answer 14

Using 60 bricks and 10 blocks per m^2

Five lines of working method:

Overall surface area = $16 \times 6 = 96\,m^2$

Area of opening = $2.1 \times 2.6 = 5.46\,m^2$

Net area of walling = $96 - 5.46 = 90.54\,m^2$

Number of facing bricks = $90.54 \times 60 = 5432.4$

Rounded off to 5433 bricks

Number of blocks = $90.54 \times 10 = 905.4$

Rounded off to 906 blocks

PERIMETERS

$$\text{Perimeter of a rectangle} = (2 \times \text{Length}) + (2 \times \text{Width})$$

Example 15

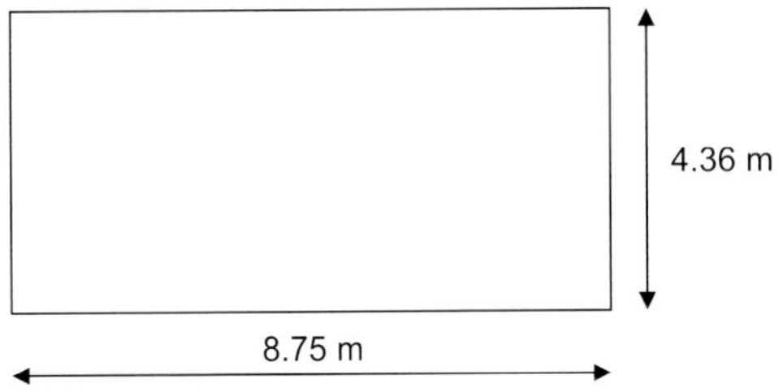
4.36 m
8.75 m

Find the perimeter of the area shown.

Answer 15

Perimeter $= 2 \times \text{L} + 2 \times \text{W}$

$= (2 \times 8.75) + (2 \times 4.36)$

$= 17.50 + 8.72$

$= 26.22$ m

Example 16

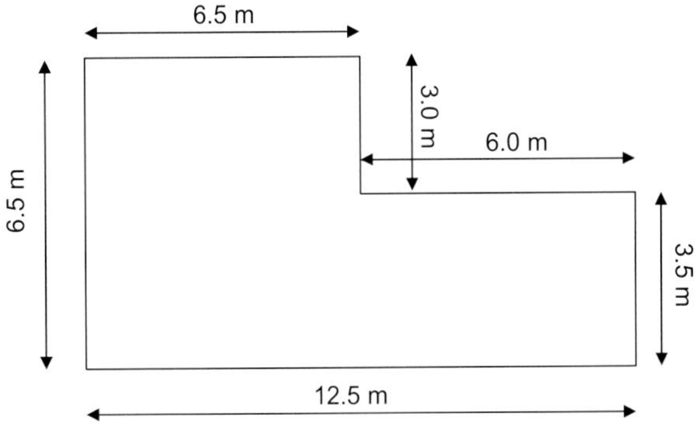
6.5 m
3.0 m
6.0 m
6.5 m
3.5 m
12.5 m

Find the perimeter of the area shown.

Answer 16

Perimeter $= 2\text{L} \times 2\text{W}$

$= (2 \times 12.5) + (2 \times 6.5)$

$= 25 + 13$

$= 38$ m

There is no need to add all the individual sides when the overall length and width is known. If the dimension was missing from the long side then the two smaller sides would have to be added together to give the overall length

$$= 6.5 + 6 = 12.5\,\text{m}$$

Note

The reason for always deducting or adding 4× the width of the walls is as follows:

Irrespective of the design of a building there will always be four external right angles.

Example 17

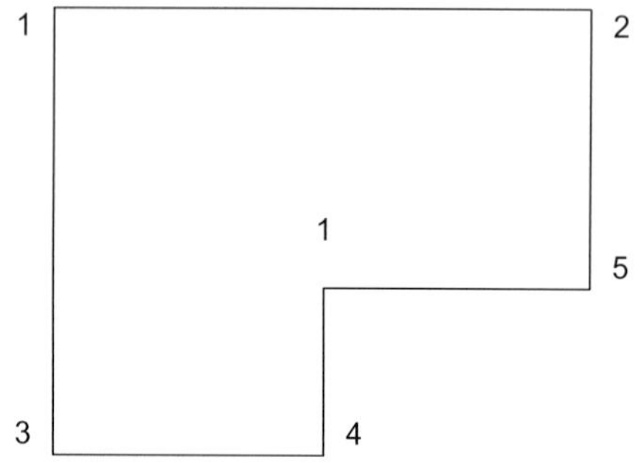
1
2
1
5
3
4

The first design shows five external corners and one internal corner. If you deduct the internal from external corners you are left with four external corners.

This rule applies to all rectangular designs.

Example 18

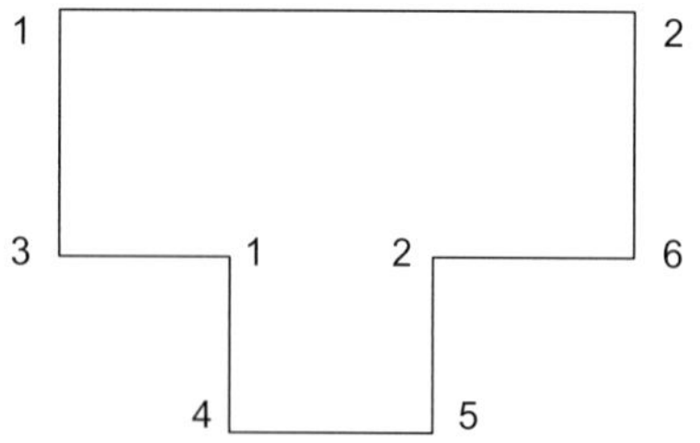
1
2
3
1
2
6
4
5

The second design shows six external and two internal corners. If you deduct the internal from external corners you are still left with four external corners.

This rule applies to all rectangular designs with corner insets.

Remember also that the perimeters of designs with the same overall dimensions are the same irrespective of the number of corner insets.

Example 19

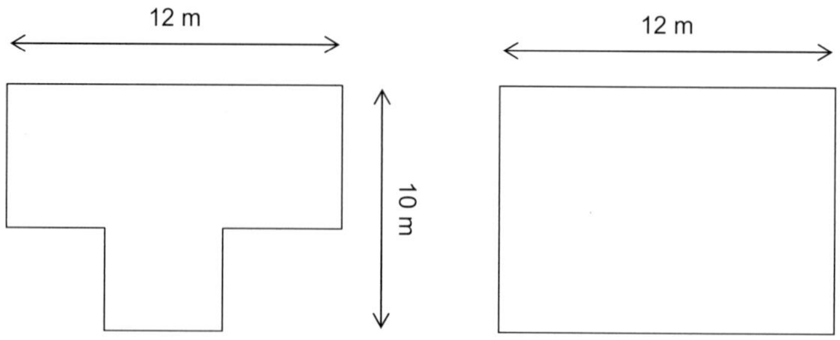
12 m
10 m
12 m

The two designs have the same perimeters even if the shapes are different.

The only time the perimeters are different and the number of external angles is more than four is when the design incorporates insets.

Example 20

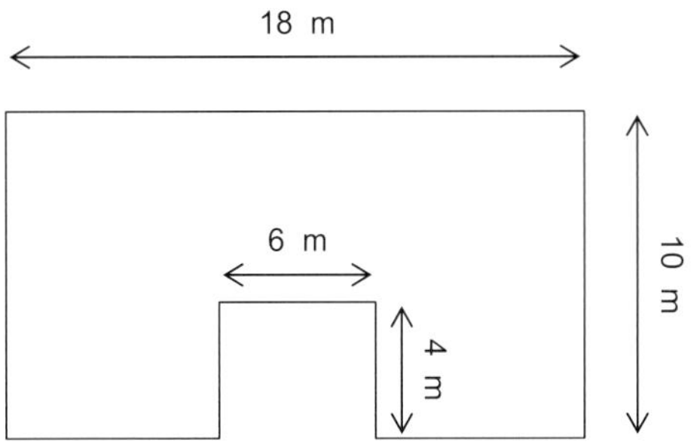
18 m
6 m
4 m
10 m

Answer 20

The perimeter is equal to that of a rectangle with sides of 18 and 10 plus the depth of each insert line.

Therefore:

$$\text{Perimeter} = (2 \times 18) + (2 \times 10) + (2 \times 4) = 64\,\text{m}$$

Method C: The centre line method

When walls have many more returns than Example 11, it can be difficult to ensure that corners are not measured twice over. Professional quantity surveyors make use of the 'centre line method' to overcome this problem when measuring up walls.

This simple but effective idea means that you first work out the total length on plan, of the centre line, of all the intersecting walls of a building having the same thickness. This figure is then used to calculate the total area of walling in one single sum (see Example 21).

The previous two examples explained how the external perimeter of a building is calculated, but it is necessary to move this line into the centre of each wall to ensure that the calculation is correct.

This is done by making a deduction for each external angle equivalent to the thickness of each wall.

If the dimensions were internal then an addition could be made for each internal angle equal to the thickness of each wall.

Example 21

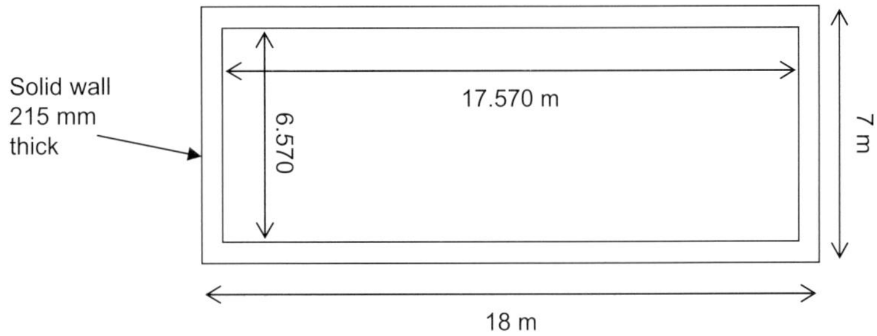
Solid wall
215 mm
thick
17.570 m
6.570
7 m
18 m

Calculate the centre line for the building with a solid brick wall.

Answer 21

Taking external dimensions:

Total external dimensions = (18 + 7) × 2 = 50 m

Deduct corners = 4 × 215 = 0.860 m

Centre line of solid brick wall = 50 − 0.860 = 49.140 m

Taking internal dimensions:

Total internal dimensions = (17.570 + 6.570) × 2 = 48.280 m

Add corners = 4 × 215 = 0.860 m

Centre line of solid brick wall = 48.280 + 0.860 = 49.140 m

The following enlarged drawing will explain in more detail why it is necessary either to add or to deduct the full thickness when adjusting for corners.

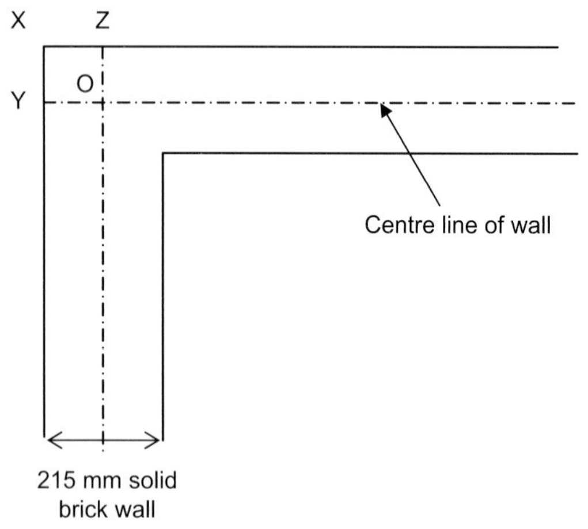
X
Z
O
Y
Centre line of wall
215 mm solid
brick wall

The drawing shows one corner of a 215 mm solid brick wall. The centre line passes through the intersection point at 'O'.

The centre lines both extend to meet the outside face of the wall at Y and Z.

This shows quite clearly that the lines XY and XZ need to be deducted to produce the correct centre line.

The two dimensions are equivalent to two half walls which together add up to a full wall thickness. There are always four corners which need either deducting when taking external dimensions or adding when taking internal dimensions.

When using the centre line method on solid walls it gives the centre line for the excavation, the foundation concrete, the damp-proof course (DPC) and the brickwork. So you can see that it is a very important calculation.

When cavity wall construction is being measured the calculation is a little more complicated but the same principle applies.

The centre line of the wall will be required for the excavation, foundation concrete and formation of the cavity and possible cavity insulation, but a deduction or addition will need to be made to calculate the centre line of the external facings and the inner blockwork.

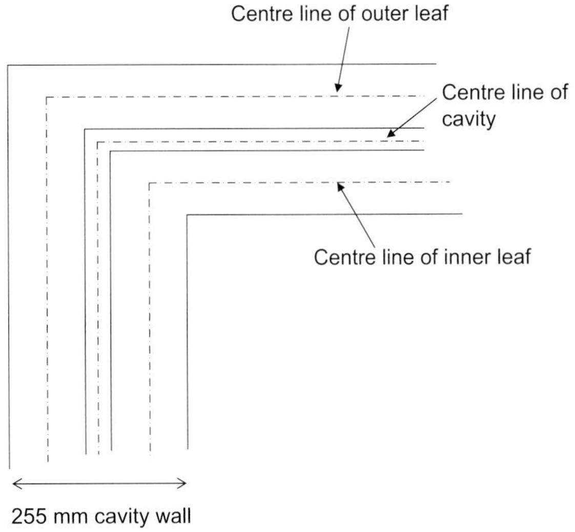
Centre line of outer leaf
Centre line of cavity
Centre line of inner leaf
255 mm cavity wall

Assume the external wall to be 102.5 mm wide, the cavity 50 mm and the inner blockwork 102.5 mm.

To arrive at the centre line of the face brickwork, taking external dimensions, you would need to deduct 4×102.5 from the overall perimeter dimensions.

To calculate the centre line of the cavity you first need to deduct the same figure (4×102.5) to arrive at the inside face of the outer leaf of face brickwork. Then deduct 4×50 to find the centre line of the cavity.

To find the centre line of the blockwork you first need to move the centre line to the inner face of the blockwork by deducting 4×50 mm cavity. Deduct 4×102.5 to find the centre line of the blockwork.

Example 22

Calculate the centre line for both the outer leaf of facing bricks and the inner leaf of blocks.

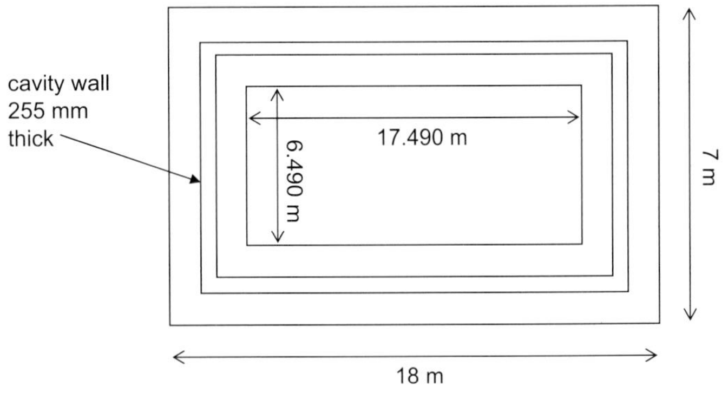
cavity wall
255 mm
thick
17.490 m
6.490 m
7 m
18 m

Answer 22a: External dimensions

Centre line of outer leaf of facing bricks:

Total external dimensions $= (18 + 7) \times 2 = 50$ m

Deduct corners $= 4 \times 102.5 = 0.410$ m

Centre line of facing brick outer leaf $= 50 - 0.410 = 49.59$ m

Centre line of inner face of facing bricks:

Centre line of outer leaf = 49.59 m

Deduct 4 × 102.5 = 0.410 m

Inner face of brick outer leaf = 49.59 − 0.410 = 49.18 m

Centre of cavity:

Centre line of inner face of facing bricks = 49.18 m

Deduct 4 × 50 = 0.200 m

Centre line of cavity = 49.18 − 0.200 = 48.98 m

Inner face of blockwork:

Centre line of cavity = 48.98 m

Deduct 4 × 50 = 0.200 m

Centre line of inner face of blockwork = 48.89 − 0.200 = 48.78 m

Centre line of blockwork:

Centre line of inner face of blockwork = 48.78 m

Deduct 4 × 102.5 = 0.410 m

Centre line of inner leaf of blockwork = 48.78 − 0.410 = 48.37 m

Answer 22b: Internal dimensions

Centre line of inner leaf of blockwork:

Total internal dimensions = (17.49 + 6.49) × 2 = 47.96 m

Add corners = 4 × 102.5 = 0.410 m

Centre line of blockwork inner leaf = 47.96 + 0.410 = 48.37 m

Centre line of inner face of blockwork:

Centre line of inner leaf = 48.37 m

Add 4 × 102.5 = 0.410 m

Inner face of blockwork inner leaf = 48.37 + 0.410 = 48.78 m

Centre of cavity:

Centre line of inner face of blockwork = 48.78 m

Add 4 × 50 = 0.200 m

Centre line of cavity = 48.78 − 0.200 = 48.98 m

Inner face of brickwork:

Centre line of cavity = 48.98 m

Add 4 × 50 = 0.200 m

Centre line of inner face of brickwork = 48.89 + 0.200 = 49.18 m

Centre line of brickwork:

Centre line of inner face of brickwork = 49.18 m

Add 4 × 102.5 = 0.410 m

Centre line of outer leaf of brickwork = 49.18 + 0.410 = 49.59 m

If you check the above answers they will be the same irrespective which method is used.

If we combine the centre line method and the material calculation we should arrive at the total materials required for any building.

Example 23

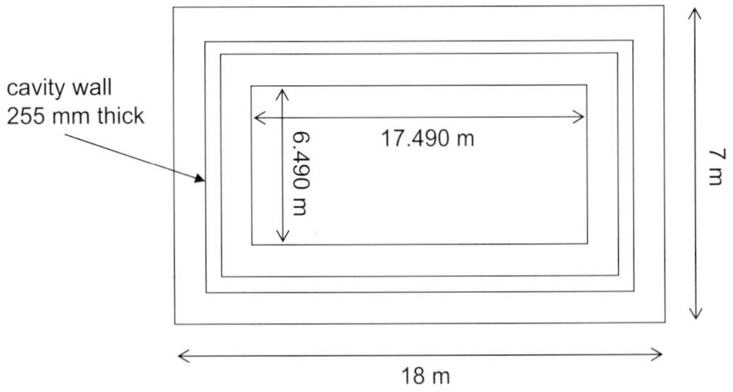
cavity wall
255 mm thick
17.490 m
6.490 m
7 m
18 m

Using the centre lines calculated in Example 22, calculate the facing bricks and blocks required for the building, assuming the walls are 2.2 m high.

Answer 23

Using 60 bricks and 10 blocks per m^2

Number of facing bricks:

Centre line of facing bricks $\times$ 2.2 = 49.59 $\times$ 2.2 = 109.098 m^2

Number of facings = 60 $\times$ 109.098 = 6545.88 bricks

Rounded off to 6546 bricks

Number of blocks:

Centre line of blockwork $\times$ 2.2 = 48.37 $\times$ 2.2 = 106.414 m^2

Number of blocks = 10 $\times$ 106.414 = 1064.14 blocks

Rounded off to 1065 blocks

Example 24

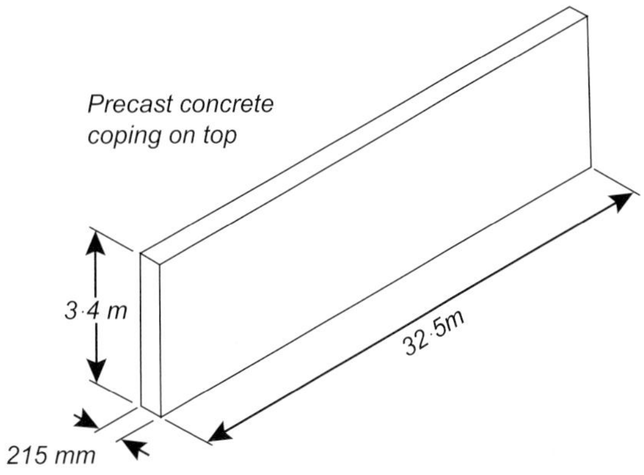
Precast concrete
coping on top
3·4 m
32·5m
215 mm

Calculate the total area for repointing the brick wall, back, front and ends.

Answer 24

Perimeter = (32.5 m) × 2 + (0.215 × 2)

= 65 m + 0.43 m

= 65.43 m

Total area of repointing = 65.43 × 3.4 = 222.462 m^2

Rounded off to 233 m^2

Example 25

Calculate the total area of raking out joints and repointing external walls of the building shown. (Ignore door and window openings for this example.)

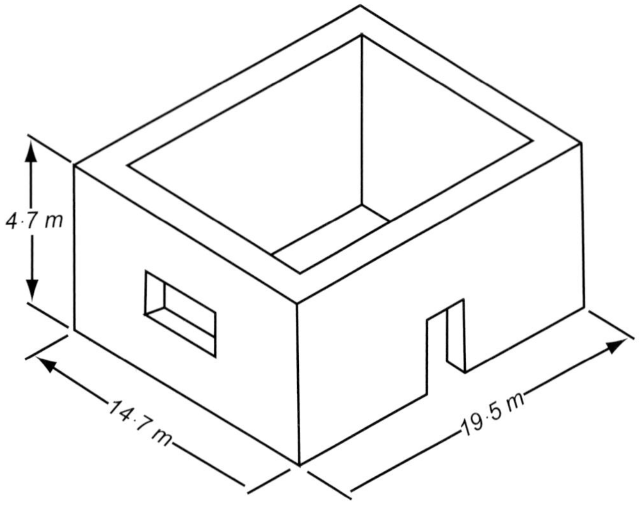
4·7 m
14·7 m
19·5 m

Answer 25

Perimeter = (19.5 m) × 2 + (14.7 × 2) = 39 m + 29.4 m = 68.4 m

Total area of repointing = External perimeter × 4.7 = 68.4 × 4.7 = 321.48 m^2

Rounded off to 322 m^2

Division of facings and commons. The outer, 102 mm thick leaf of cavity work and other half brick walling is usually built in stretcher bond, so that all the bricks required will be facings.

With 215 mm and thicker solid brickwork faced on one side only, some bricks will need to be less expensive commons to back up the facing bricks.

With 215 mm thick Flemish bond walling, 2/3 will be facings and 1/3 commons. With 215 mm English bond walling, 3/4 will be facings and 1/4 will be commons.

If 215 mm thick walls are built to header bond, then of course all bricks ordered should be facings. This information is summarized in Table 3.7.

Note

For extra practice repeat Examples 24 and 25 after changing the dimensions (invent your own).

Example 26

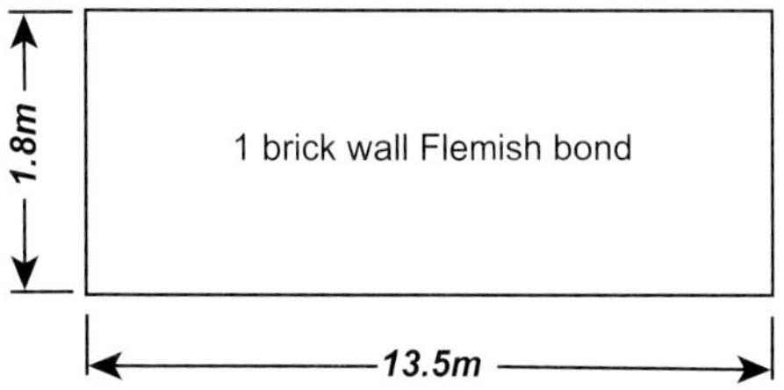
1.8m
1 brick wall Flemish bond
13.5m

Elevation

Assume that the wall shown is to be 215 mm thick Flemish bonded brickwork.
Calculate the number of facings and commons that will be required, in separate totals.

Answer 26

Taking information from Table 3.7:

80 facings per m^2

40 commons per m^2

Three lines of working method:

Surface area = 13.5 m × 1.8 m = 24.3 m^2

Number of facings = 24.3 × 80 = 1944 facing bricks

Number of commons = 24.3 × 40 = 972 common bricks

Percentage for cutting and waste

All of the foregoing examples of calculating requirements for bricks, blocks and mortar have resulted in exact or net quantities. Owing to breakages and waste when cutting and handling bricks and blocks and when using mortar, it is necessary to increase net orders by 5 per cent.

Therefore as an additional stage at the end of each calculation, 5 per cent extra should be added on, using the % button on your calculator, for the purposes of ordering.

Example 27

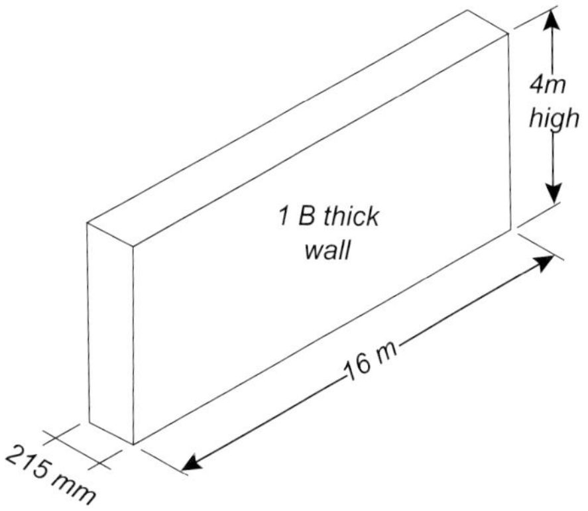
4m high
1 B thick wall
16 m
215 mm

Calculate the amount of facing bricks and common bricks for the wall shown if English bond was used.
Allow 5 per cent for wastage

Answer 27

Taking information from Table 3.7:

90 bricks per m^2

30 commons per m^2

Two lines of working method:

Surface area $= 16 \times 4 = 64\,m^2$

Number of facings $= 64 \times 90 = 5760$ facing bricks

Add 5% for wastage $= 5760 \times 5 \div 100 = 288$

Therefore, total facing bricks required $= 5760 + 288 = 6048$ facing bricks

Number of commons $= 64 \times 30 = 1920$ commons

Add 5% for wastage $= 1920 \times 5 \div 100 = 96$

Therefore total commons required $= 1920 + 96 = 2016$ common bricks

Example 28

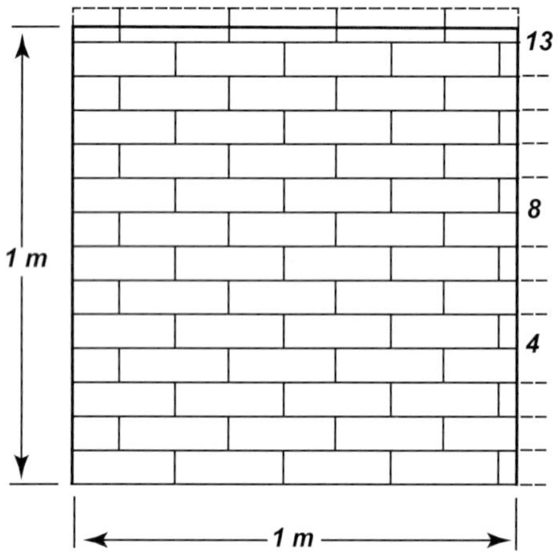

13
8
4
1 m
1 m

Calculate the area of mortar joints (say to assess the effect of mortar colour on finished walling) showing on the surface of $1\,m^2$ of stretcher bonded brickwork.
For the purposes of this calculation, assume that the elevation above is 'stack bonded'.

Answer 28

Bed joint mortar showing $= 1\ m \times 0.01 \times 13 = 0.13\,m^2$

Cross joint mortar showing $= 1\ m \times 0.01 \times 5 = 0.05\,m^2$

Total $= 0.18\,m^2$

Expressed as a percentage of $1\,m^2 = 0.18/1\,m^2 \times 100 = 18\%$ of surface viewed is mortar colour. The remaining 82% only is brick colour.

Note

Go back to Example 23 and calculate the number of bricks and blocks that should be ordered after allowing 5 per cent for cutting and waste.

ADDITIONAL APPLICATIONS OF THE CENTRE LINE METHOD

In addition to bricks, blocks and mortar, a bricklayer may be asked to work out quantities of cavity insulation, DPC, foundation concrete or even excavation to trenches.

Each such calculation is very simple, if you find out the total length of the centre line of the walling first.

Example 29

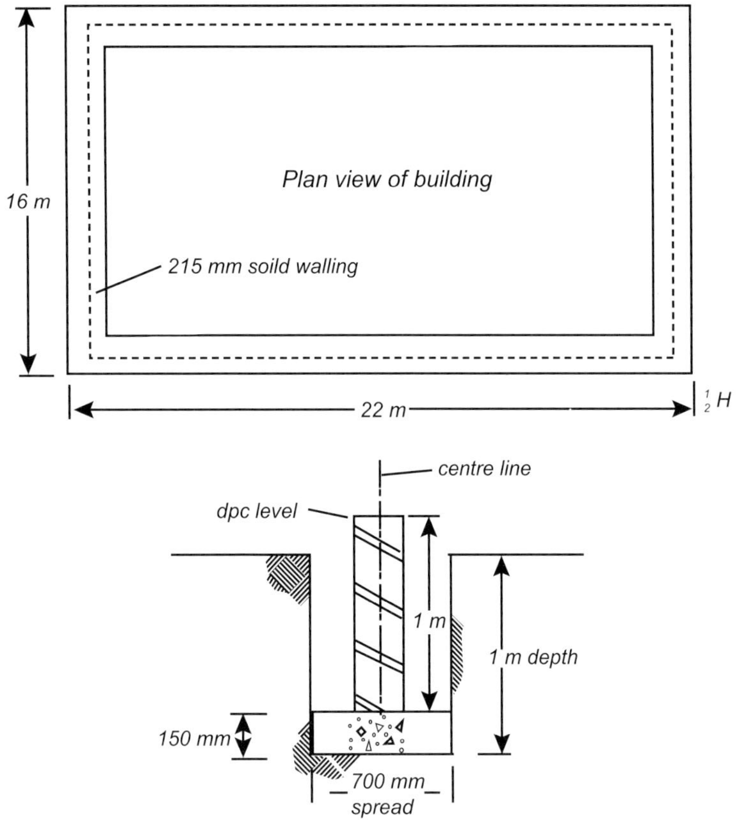
16 m
Plan view of building
215 mm soild walling
22 m
½H
centre line
dpc level
1 m
1 m depth
150 mm
700 mm
spread

Answer 29

Centre line

Total external dimensions = (22 + 16) × 2 = 76 m

Deduct corners = 4 × 215 m = 0.860 m

Centre line of solid brick wall = 76 − 0.860 = 75.14 m

The dotted centre line perimeter shown on the drawing is exactly 75.14 m long. This dotted line is the exact centre line of the trench excavation, concrete strip foundation, substructure brickwork and DPC, all with a net length of exactly 75.14 m that can be used in four separate calculations.

Trench excavation

Volume = Centre line × Width × Depth

$= 75.14\ m \times 0.70\ m \times 1\ m = 52.598\ m^3$

Rounded off: 53 m^3

Strip foundation concrete

Volume = Centre line × Spread × Depth

$= 75.14\ m \times 0.70\ m \times 0.150\ m$

$= 7.890\ m^3$

Rounded off: 8 m^3

Substructure brickwork

Area = Centre line × Height

$= 75.14\ m \times 1\ m = 75.14\ m^2$

Number of bricks = 75.14 × 120 = 901.68

Rounded off: 902 bricks

DPC

Linear measurement = Centre line + Extra for laps

= 75.14 + 5% extra for laps = 75.14 × 5 ÷ 100 = 3.757

Therefore total DPC required = 75.14 + 3.757 = 78.897

Rounded off to 79 m length of 215 mm wide DPC

Circles and triangles

In addition to those shapes of walls and buildings used as examples in this chapter so far, there are two others that the bricklayer will encounter in construction, which require measurement for setting out and calculation: the triangle and the circle, or parts of them (Figure 3.20).

Example 30

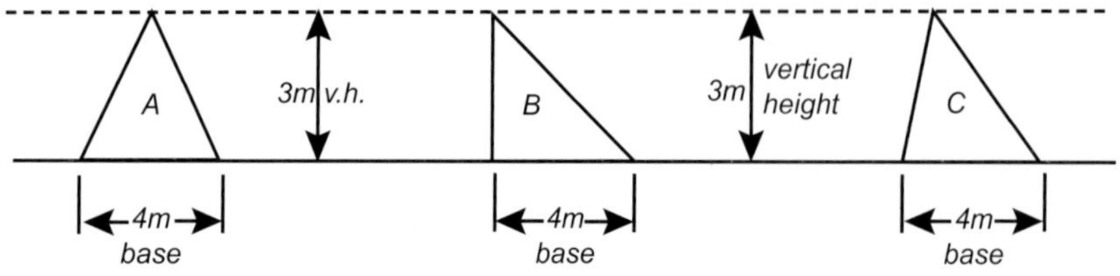
A
3m v.h.
B
3m vertical height
C
4m base
4m base
4m base

Note

Go back to Example 21 and calculate the volume of excavation to form 1 m deep by 750 mm wide foundation trenches for that example.

Warning

Rounding up

Rounding up numbers that represent quantities of building materials, when carrying out building calculations, should only be done with the final figure, or last result.

Do not round up two or three times during the working stages of a calculation, otherwise the final quantity could be excessively large, and become what is called an 'accumulated error'.

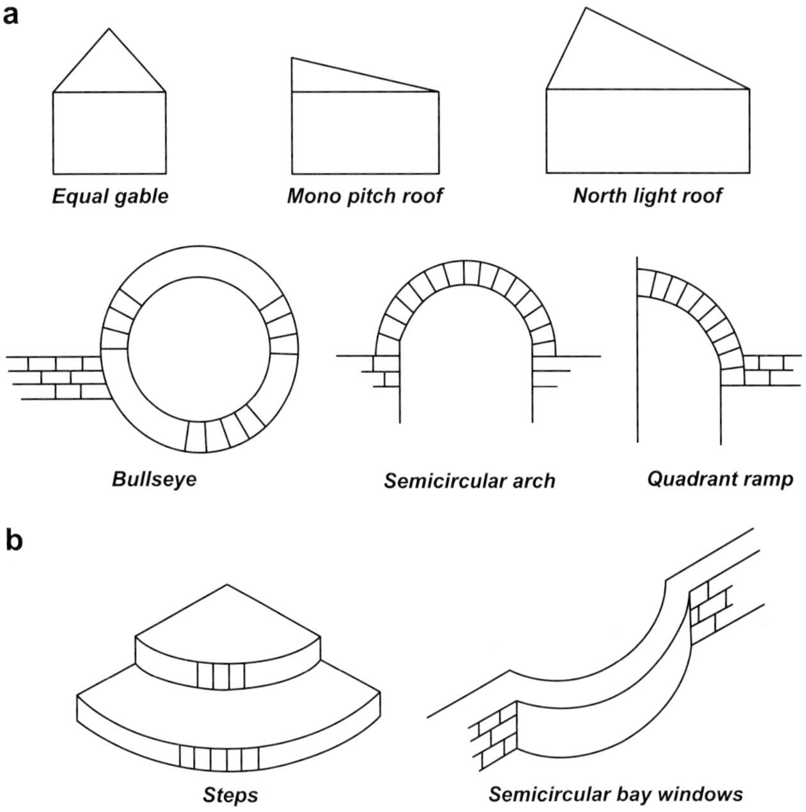
a
Equal gable
Mono pitch roof
North light roof
Bullseye
Semicircular arch
Quadrant ramp
b
Steps
Semicircular bay windows

FIGURE 3.20
Examples of (a) triangular gable end walls; (b) circular shapes or parts of a circle used in construction

The surface area of any triangle can be calculated as follows, whether shaped like A, B or C:

$$\text{Surface area} = (\text{Base} \times \text{Vertical height}) \div 2$$

$$= (4\text{ m} \times 3\text{ m}) \div 2 = 6\text{ m}^2$$

3025 The surface area of all three triangles is $6\,\text{m}^2$, despite their different shapes.

Example 31

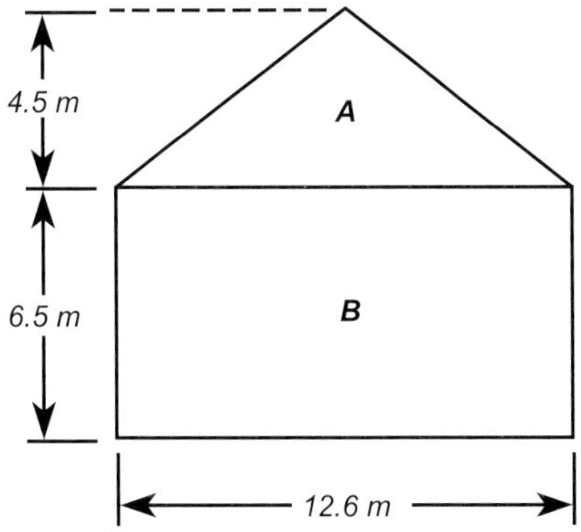
4.5 m
A
6.5 m
B
12.6 m

Calculate the surface area of the gable end wall, for repointing.

Answer 31

Total surface area = A + B

Area of A = (Base × Vertical height) ÷ 2

= (12.6 m × 4.5 m) ÷ 2

= 56.7 ÷ 2

= 28.35 m^2

Area of B = 12.6 × 6.5

= 81.9 m^2

Total surface area = 28.35 + 81.9 = 110.25 m^2

The area of any circular shape can be calculated from the simple formula:

$$\text{Area of circle} = \pi r^2$$

(pronounced pi, r, squared).

Some pocket calculators have a button marked π; if not, then use the figures 3.14 (the value for pi never changes).

The letter r in the formula stands for radius of the circle.

The perimeter of a circle is called the circumference. Another simple formula is used to calculate the circumference of a circle:

$$\text{Circumference of circle} = \pi D$$

(pronounced pi, D). π remains 3.14 and D stands for the diameter.

Example 32

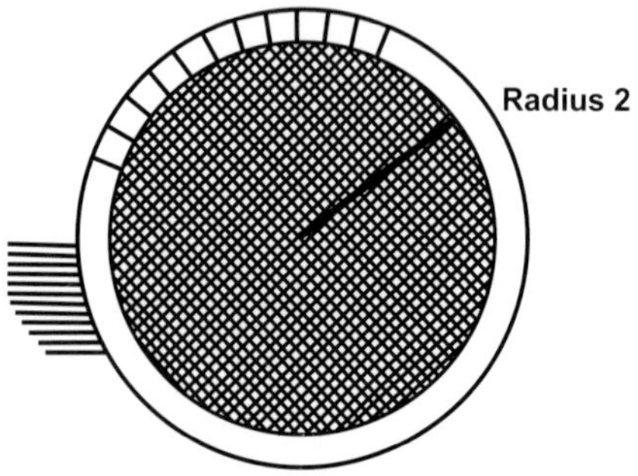
Radius 2

Calculate the surface area of the shaded part of the bullseye.

Answer 32

Surface area = πr^2

= $\pi \times r \times r$

= 3.14 × 2 × 2

= 12.56 m^2

Example 33

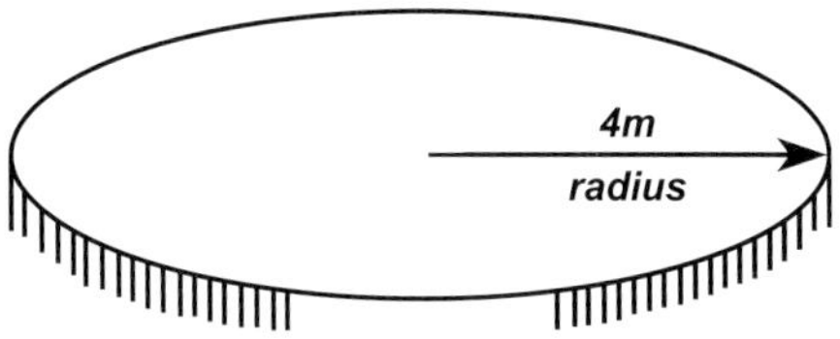
4m
radius

Calculate the surface area of the circular paved area.

Answer 33

Surface area $= \pi r^2$

$= \pi \times r \times r$

$= 3.14 \times 4 \times 4$

$= 50.24\,\text{m}^2$

Example 34

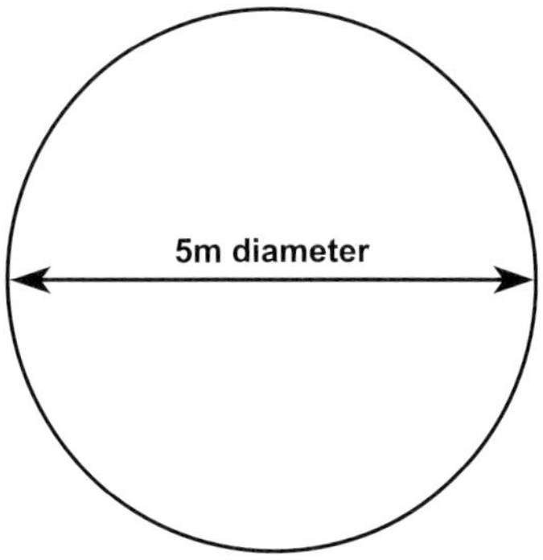
5m diameter

Calculate the circumference of the circle shown.

Answer 34

Circumference $= \pi D$

$= \pi \times$ Diameter

$= 3.14 \times 5$

$= 15.70$ m

Note

Remember that the circumference of a circle is just a plain measurement of length, in linear metres. Surface area is measured in m^2.

Estimating

Once the trainee has mastered the previous pages of calculations the knowledge gained can be used to calculate estimates for small building works.

An estimate is a price given for a certain amount of work according to the firm's set labour rates. These can change according to the firm.

On large contracts it is customary to have a bill of quantities. Under the traditional competitive tendering procedure the bill of quantities is the starting point for the main contractor tendering for the work.

The bill is drawn up by a quantity surveyor taking measurements from the final or contract drawings supplied by the architect.

From the quantities of materials shown in the bill and the nature of the finished building as shown by drawings and specifications, etc., the contractor estimates the costs for completing the contract.

The main areas included in the final estimate are:

- labour
- materials
- plant and equipment
- overheads.

The cost of labour is calculated by the application, to the total quantities, of labour rates per unit quantity, e.g. £15.50 per cubic metre for excavating, £12.60 per cubic metre for clearing away, and so on through the trades, or by the estimated costs of complete operations such as those used for bonus targets.

The cost of the material comes from expected market prices plus any extra for handling, etc. Plant and other expenses may be allowed for in the labour rates or, preferably, estimated separately.

All these, when added together, give a total figure for the contract excluding overheads and profit. Overheads are items of expenditure that cannot easily be added to each contract, e.g. telephone, stationery and main office running costs. A percentage is added to all final figures to include these overheads.

The final percentage to be added is the profit required by the firm. The final figure plus overheads and profit gives the final estimate for the work.

Variations

This estimated price and the final price for a contract can change according to certain conditions.

The client could require extra works or prefer more expensive items than originally priced for, such as floor coverings.

The Building Regulations could change and cause the cost to increase.

If the contract architect made a mistake in the drawings this could lead to extra or less work.

Profits for the builder could be affected if there were mistakes in the calculations for the estimate. It is therefore essential that all calculations are checked and double checked. A simple mistake could cause a big loss.

Relaying information

To be able to relay information gathered to the appropriate people, good communication skills are required. Good communication skills are important in every aspect of our lives; in the workplace they are essential.

You may at times be the first point of contact with the client or customer, whether you are self-employed or working for a company.

Meeting and talking to people outside your organization can be one of the more interesting aspects of any job, but it is also a challenge.

You may be talking to people on the telephone or face to face, but in each case you are acting as the representative of your company and therefore you should present a positive and friendly image.

As firms grow into very large firms it becomes even more essential to have good communication.

Methods of communication

When entering the construction industry you will have to communicate well, irrespective of your position in the company.

It may be necessary to communicate with customers, visitors and potential employers in a number of different ways, for example:

- orally – face to face
- in writing
- by telephone
- through drawings.

ORALLY – FACE TO FACE

When dealing with a customer or visitor face to face, always appear helpful and polite. Smile but do not appear to be over-friendly as this can make some people feel uncomfortable.

Never act in an aggressive or immature manner.

Remember, first impressions count. If your attitude comes across as 'couldn't care less', potential customers may think that this is your attitude towards your standard of work. Obviously this can frighten off potential customers.

Always offer as much advice and information as possible. If you cannot answer their questions or deal with their requests, find someone who can.

Verbal communication can include talking on the telephone, mobile or walkie-talkie. Verbal communication is instant and feedback can be obtained immediately.

Remember

Ask rather than tell.

Always be polite.

Keep instructions short and to the point.

Be friendly.

WRITTEN

Written communication can take many forms and one or more of the following will commonly be used:

- letters
- memos

- reports
- records
- site diary
- handbooks and manufacturer's details
- e-mails and text.

The main advantage of written communication is the evidence it provides. Written communication may take longer than verbal communication, but with e-mails and mobile texting written communication is becoming much more rapid than in the past.

LETTER WRITING

Writing a letter is an effective form of communication. We need to write letters on a number of occasions to:

- give information
- ask for details
- confirm things
- make arrangements.

Letters provide a permanent record of communication between organizations and individuals.

They can be handwritten, but formal business letters give a better impression of the organization if they are typed.

They should be written using simple, concise language. The tone should be polite and business like, even if it is a letter of complaint.

The letter must be clearly constructed, with each fresh point contained in a separate paragraph for easy understanding. When you write a letter for any reason, remember the following basic structure:

- your own address
- the recipient's address
- date
- greetings
- endings
- signature.

Report writing

A report is written to pass on information quickly and accurately to another party.

When writing a report, sufficient information should be included to allow the reader to understand fully what the report is intended to say.

A report should be divided into five parts:

- headings
- introduction to the report
- body of the report
- conclusion, with any recommendations
- any data, drawings, contracts, etc.

Site diary

This is one of the most important reports that a site manager has to produce during a contract.

It will include the following:

- the weather, morning and afternoon
- any visitors on site
- delays by subcontractors
- instructions from the clerk of works
- materials delivered
- number of staff on the site.

TELEPHONE

Telephones have a very important role to play in communications. Often, a telephone conversation may have taken place before a letter is written to confirm something. Its clear advantage over the written message is obviously the speed with which people are put in touch with one another.

A good telephone manner is essential. Remember, the person you are talking to cannot see you so you will not be able to use facial expressions and body language to help make yourself understood.

The speed, tone and volume of your voice are very important. Speak very clearly at a speed and volume that people can understand. Always be cheerful and remember that someone may be trying to write down your message.

Making notes

It is important when receiving information that some kind of record is kept. Very often information is not written down and then mistakes are made through not remembering correctly.

Instructions are very often given by word of mouth. They should be kept brief and to the point, without complicated and confusing detail. This should make it easier for the recipient to remember them.

It is always good practice to aid your memory by making notes of what is being said. In order to record and possibly pass information received to others correctly, keep a small notebook in your pocket and always use it if the oral information is complicated and has to be passed on to others.

You may not be able to write down the instructions word for word.

- Make lists.
- Use notes and sketches.
- Abbreviate notes using key words.

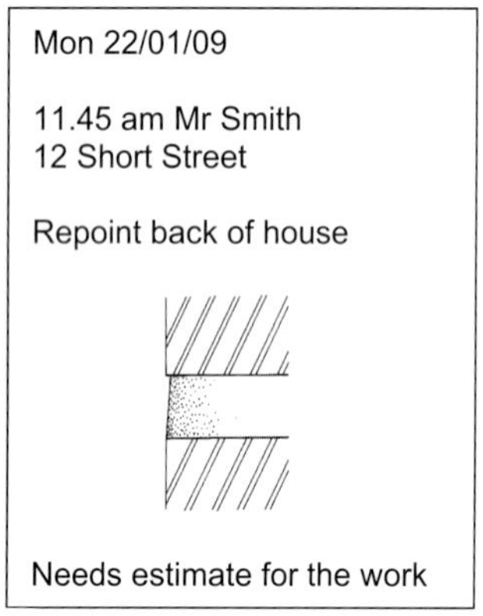
Mon 22/01/09
11.45 am Mr Smith
12 Short Street
Repoint back of house
Needs estimate for the work

Positive and negative communications

For a company to function properly and make a profit, it is essential to maintain good working relationships within the workforce.

To achieve this, various sections and members of the workforce must communicate with one another as fully as possible.

When you first join a company you are often the odd one out and therefore you should try hard to co-operate with your new work colleagues. This will mean being polite, acting on request as quickly as possible and building a good working relationship with workmates.

You may be talking to people on the telephone or face to face, but in each case you must try to be both positive and friendly, both to give a good account of yourself and to promote good, effective working relationships.

Poor communications cause negative feedback and can lead to delays and mistakes which will cause a loss of profit for the company.

Barriers to communication

In all methods of communication there can be obstacles and these are known as barriers. These are of two kinds, external and internal, depending on their origin.

External barriers are those things in the working surroundings that interrupt or suppress the process of communication, such as machine noises, telephone bells, people talking and other similar distractions.

External barriers are usually quite easy to identify and overcome, whereas internal barriers, which arise within a person, are more difficult to recognize and combat.

Lack of knowledge and experience, different responsibilities, background and levels of intellect create problems for the communicator.

It is the emotional make-up of people, however, that gives the biggest headache. Their instinctive response to circumstances and to what others expect of them results in the formation of attitudes and behaviour that are often unexpected.

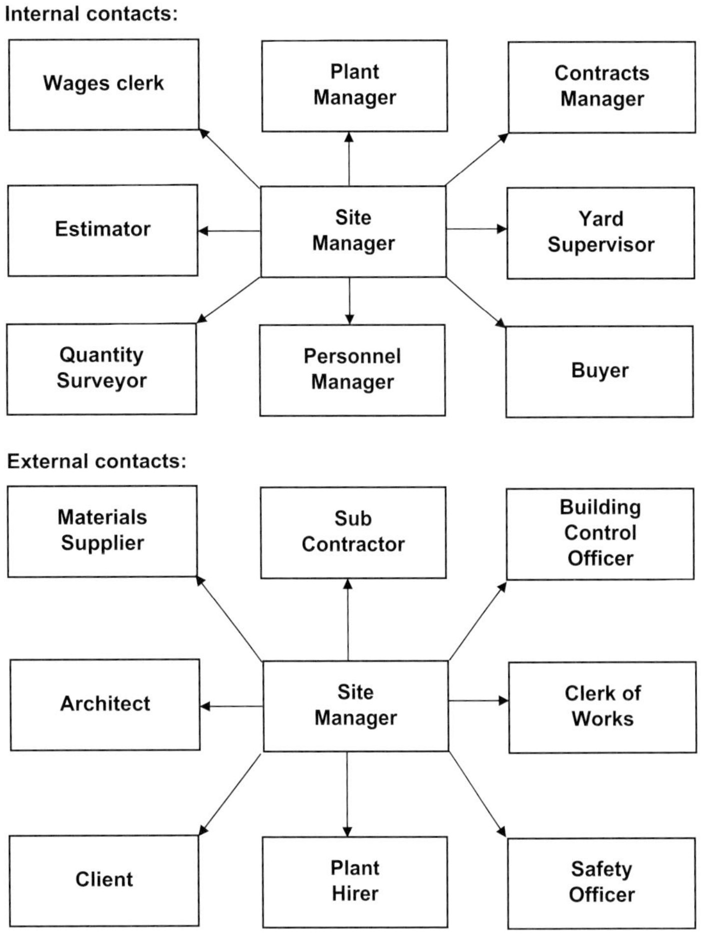
Internal contacts:
Wages clerk
Plant Manager
Contracts Manager
Estimator
Site Manager
Yard Supervisor
Quantity Surveyor
Personnel Manager
Buyer
External contacts:
Materials Supplier
Sub Contractor
Building Control Officer
Architect
Site Manager
Clerk of Works
Client
Plant Hirer
Safety Officer

FIGURE 3.21
Internal and external contacts

Communications with those outside the building team

Examples of internal and external contacts are shown in Figure 3.21.

To be beneficial to both head office and the site manager there should be two-way communications to show how work is progressing, any problems that may have developed, and whether any more information, materials, plant or labour is required to maintain progress.

Multiple-choice questions

Self-assessment

This section of the book is designed to allow you to check your level of knowledge. The section consists of revision questions for this chapter. The questions are all multiple choice and have four possible answers. The answers are to be found at the end of the book.

The main type of multiple-choice question will be the four-option multiple-choice question. This will consist of a question or statement, known as the stem, followed by a choice of four different answers, called the responses. Only one of these responses is the correct answer; the others are incorrect and are known as distracters.

You should attempt to answer the questions by choosing either (a), (b), (c) or (d).

Example

The person employed by the local authority to ensure that the Building Regulations are observed is called the:

(a) clerk of works
(b) building control officer
(c) council inspector
(d) safety officer

The correct answer is the building control officer, and therefore (b) would be the correct response.

Communication

Question 1 An assembly drawing would be most likely to be drawn to a scale of:

(a) 1:5000
(b) 1:500
(c) 1:50
(d) 1:5

Question 2 Which of the following components are best suited for a schedule?

(a) wall ties
(b) windows
(c) air bricks
(d) insulation batts

Question 3 Which of the following gives a precise description of the materials and workmanship required for a contract?

(a) specifications
(b) schedules
(c) schemes
(d) contract plans

Question 4 Which of the following formulae can be used to calculate the area of the triangle shown?

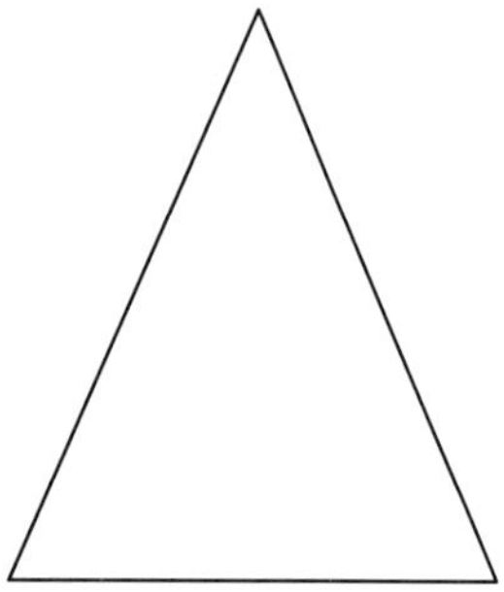

(a) base $\times$ height

(b) base $\times$ slope $\div$ 2

(c) base $\times$ height $\div$ 2

(d) base $\times$ slope

Question 5 Which of the following is the correct area for the rectangle shown?

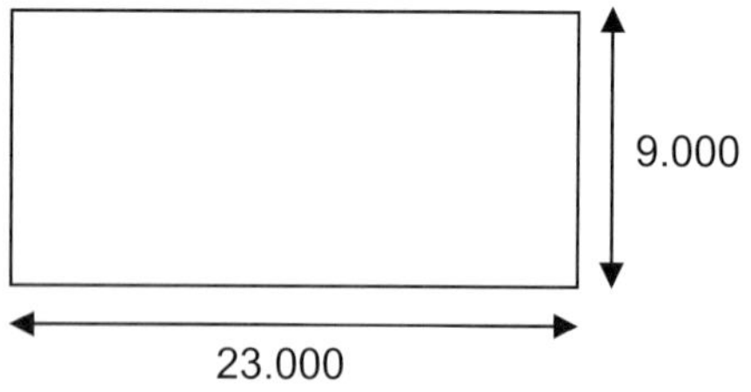
9.000
23.000

(a) 207 m^2

(b) 32 m^2

(c) 64 m^2

(d) 20.7 m^2

Question 6 Which of the following methods of communication would be written?

(a) radio

(b) letters

(c) mobile

(d) telephone

Question 7 Which method of projection is shown?

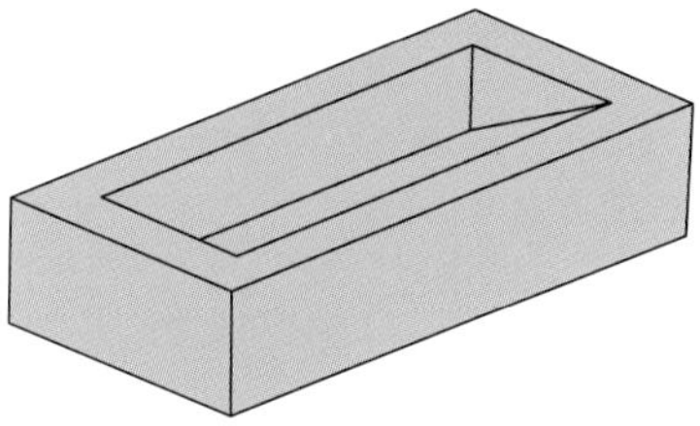

(a) oblique

(b) orthographic

(c) isometric

(d) cavalier

Question 8 Which of the following could result in the original estimate to the client being altered?

(a) rise in the cost of materials

(b) alteration to the programme

(c) rise in the cost of living

(d) alteration to the working week

CHAPTER 4

Construction Technology

This chapter will cover the following NVQ and Diploma units:

- NVQ VR02
- CC 2003K

This chapter is about:

- Interpreting building information
- Planning and carrying out productive, efficient working practices
- Working with others or as an individual

The following NVQ performance criterion will be covered:

- Performance criterion 3: Maintain records

The following Diploma outcomes will be covered:

- Know the principles behind walls, floors and roofs
- Know the principles behind internal work
- Know about materials storage and delivery of materials

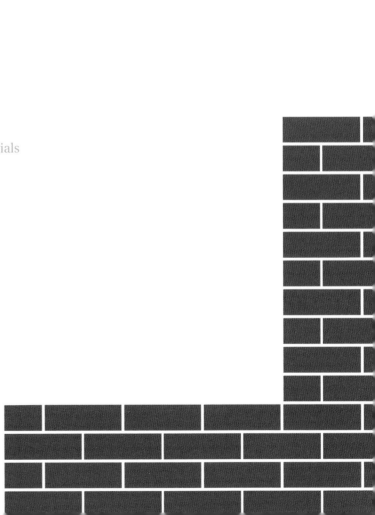

Working drawings

It has been stated before that drawings are the best way of communicating detailed information from the designer to all those involved in the works. For more details on working drawings see Chapter 3.

Because drawings are part of a legal contract between the client and designer it is essential that all drawings are precise and easy to interpret. Mistakes in either the drawing or the interpretation can be very costly.

All drawings are produced according to the guidelines produced by the British Standards Institute, BS 1192: Construction Drawing Practice. These standards ensure that all drawings are drawn to the same standard and should make it easier for everyone to interpret them in the same way.

Setting out

The person in charge of the site has to read and understand the drawings in order to be able to set out the building. It is essential that this part of the works is carried out correctly as mistakes in setting out can be costly.

Preliminary setting out of a building

Full setting-out procedures are covered in more detail in Chapter 6.

Before any setting out should be undertaken the top soil should be removed.

It is assumed that the building of a traditional semi-detached domestic dwelling is contemplated, but it should be noted that the same principles can be applied to all types of construction buildings and other more complicated forms (Figure 4.1).

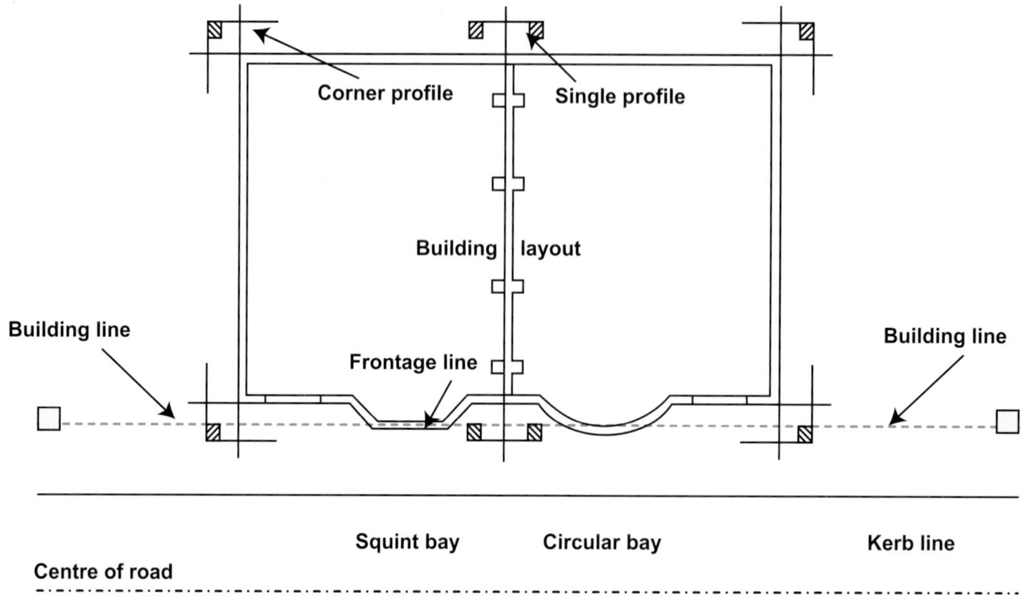
Corner profile
Single profile
Building
layout
Building line
Frontage line
Building line
Squint bay
Circular bay
Kerb line
Centre of road

FIGURE 4.1
Location of foundation profiles

When the setting-out stage is reached the site will have been previously surveyed and probably levelled to find the shape and features of the land.

Drawings showing the proposed buildings and associated site works will have been submitted to the local authority and approved.

A large development requires setting-out drawings to be prepared by the architect or engineer, but a small site would have sufficient information on the block plan or general site plan.

The most important task in setting out a building is to establish a base line to which all other setting out can be related.

This base line is very often the building line; this is an imaginary line that is established by the local authority.

You can build on the building line and any distance behind the building line, but

NEVER BUILD IN FRONT OF THE BUILDING LINE.

It is usual for the building line to be given as a distance from one of the following:

- existing buildings
- the centre line of the road
- the kerb line.

The frontage line of the proposed building must then be set out on or behind the building line, but never in front of it. In this example the frontage line is on the building line.

DEGREE OF ACCURACY

A high degree of accuracy is required when setting out and this can be achieved if a steel tape is used and supported to avoid any sag.

Checks should be made on all measurements wherever possible.

All horizontal measurements should be made horizontal. Use a level to ensure that the tape is horizontal to avoid errors (Figure 4.2).

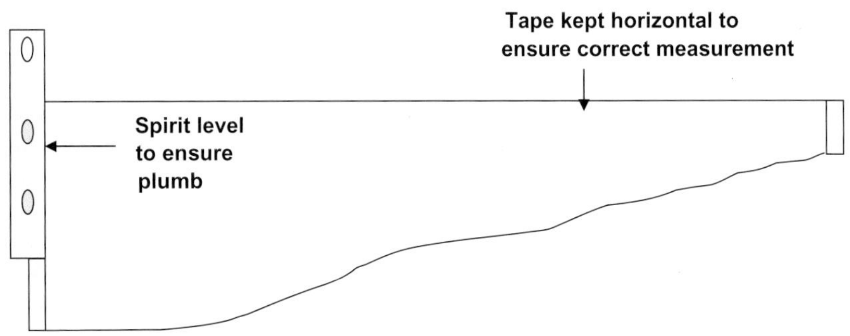
Tape kept horizontal to ensure correct measurement
Spirit level to ensure plumb

FIGURE 4.2
Accuracy with horizontal measurements

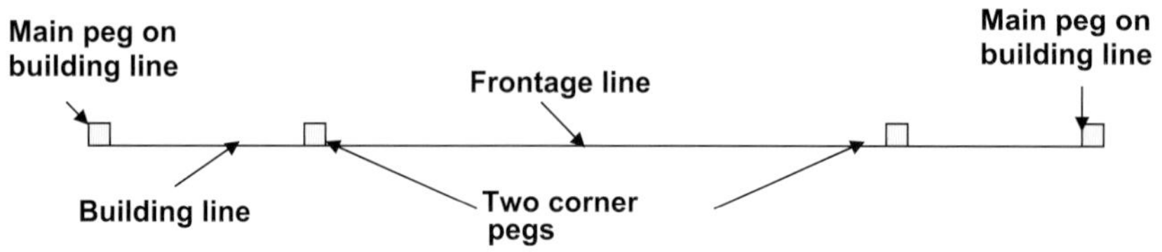
Main peg on building line
Main peg on building line
Frontage line
Building line
Two corner pegs

FIGURE 4.3
Setting out frontage line

The building line is fixed to two pegs on either side of the proposed plot, a reasonable distance away from the building.

The two corner pegs are then set out on the building line, which marks out the frontage line of the proposed building (Figure 4.3).

The side walls of the building can now be set out at right angles to the frontage line. Ensure that the pegs are placed at a distance more than the side wall length (Figure 4.4).

The right angles can be set out with a builder's square, the 3:4:5 method or an optical square.

Once the position of the side walls has been established the correct length can be measured and pegs can be knocked in. The back line can now be completed.

Once the outline has been completed it is essential that the setting out is checked for square. The two diagonals are measured and if they are equal the setting out is accurate and the building is square (Figure 4.4).

If they are not equal adjustments have to be made. Remember never to alter the frontage line as this should be correct.

The back two pegs have to be adjusted to ensure that the diagonals are equal.

When adjustments have been made always recheck the dimensions of all wall lengths.

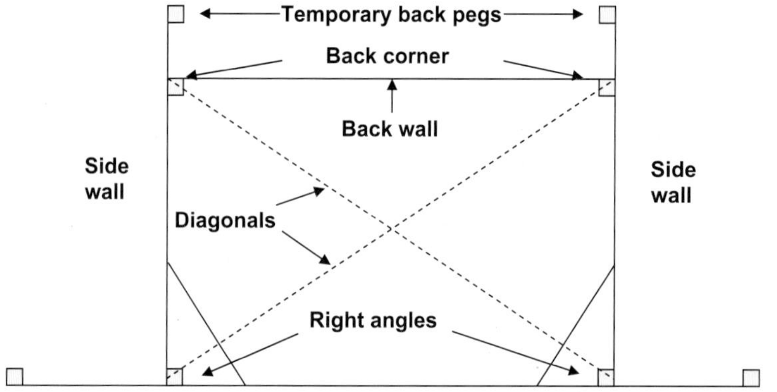
Temporary back pegs
Back corner
Back wall
Side wall
Side wall
Diagonals
Right angles

FIGURE 4.4
Checking for accuracy

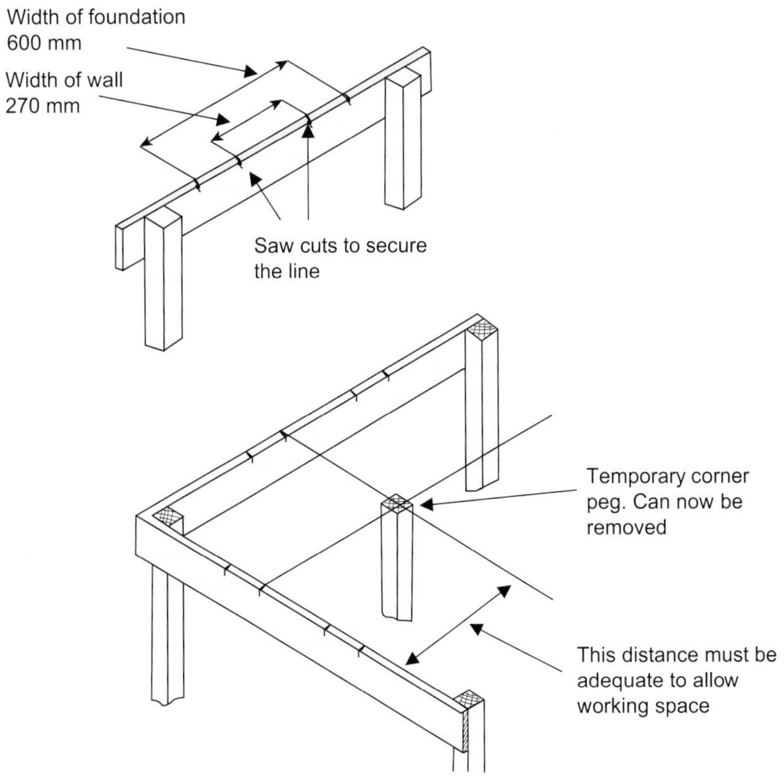
Width of foundation
600 mm
Width of wall
270 mm
Saw cuts to secure
the line
Temporary corner
peg. Can now be
removed
This distance must be
adequate to allow
working space

FIGURE 4.5
Types of profile

When the setting out has been proven correct the profiles have to be erected. At the moment the setting out pegs will be in the foundation trench so they have to be repositioned approximately 1 m away from all the wall faces. Profiles can be either corner or single type and consist of wooden pegs and rails (Figure 4.5).

Profiles are erected to enable the corner setting out pegs to be removed to allow excavation to take place without disturbing the pegs.

One of these corner profiles could be set to a given level which is known as the datum, and may relate to the finished floor level or damp-proof course (DPC). This datum peg should be protected with concrete.

The datum peg could also be positioned away from the setting pegs but close enough to be accessible and should also be protected with concrete and a small barrier of pegs and rails (Figure 4.6).

The profiles should be positioned approximately 1 m away from the face of the building to allow working space for the excavation.

The completed profiles can have foundation and wall widths marked on them in one of various methods. Saw cuts are best as nails could be accidentally removed (Figure 4.7).

Once the profiles have been accurately constructed the dimensions should all be checked again, as should both diagonals.

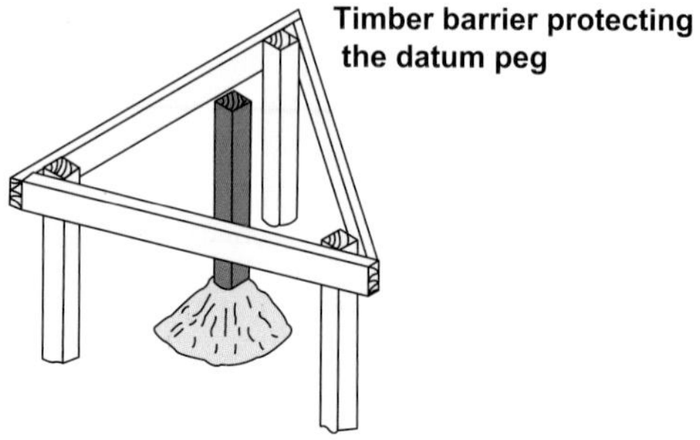
Timber barrier protecting the datum peg

FIGURE 4.6
Datum peg

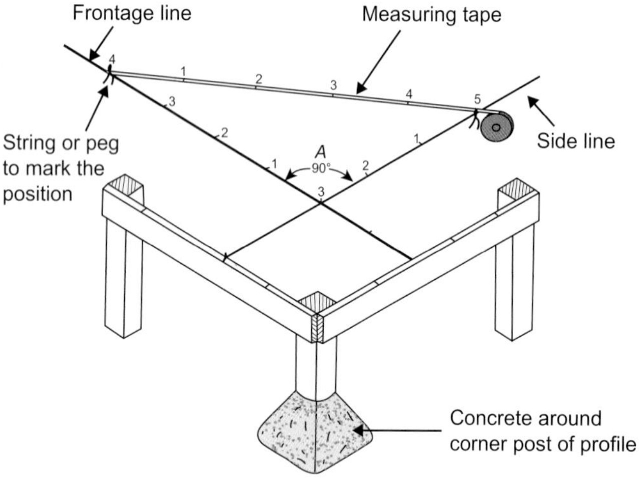
Frontage line
Measuring tape
String or peg to mark the position
Side line
A
90°
Concrete around corner post of profile

FIGURE 4.7
Checking a right angle by the 3:4:5 method

The original setting-out pegs can now be removed.

Building lines can now be fastened to the profiles and the trench marked out ready for excavation (Figure 4.8).

Sand can be used to mark out the ground ready for excavation and the lines can then be removed.

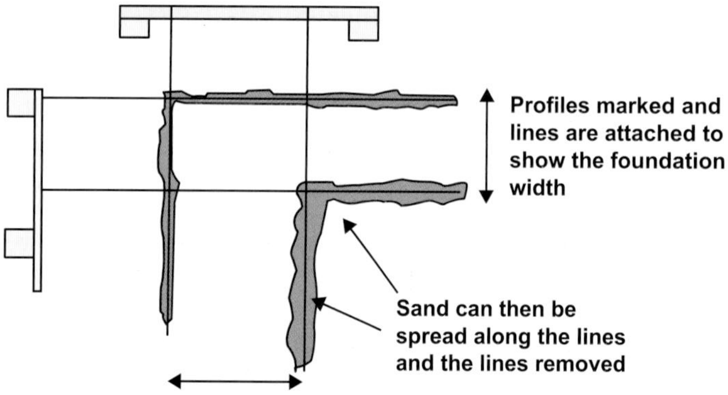
Profiles marked and lines are attached to show the foundation width
Sand can then be spread along the lines and the lines removed

FIGURE 4.8
Marking out the ground ready for excavation

Ground works

As soon as the setting out is completed the excavation can begin

Excavations

On small sites, hand-held tools such as picks, shovels and wheelbarrows are used. However, if the depth of excavation exceeds 1.2 m some other method of removing spoil from the excavation will have to be used.

On all sites mechanical methods could be used, depending on factors that are different on each site. These include the volume of soil involved, the nature of the site and any cost constraints.

Before excavation begins it is essential to check for underground cables and services and ensure that they are adequately marked and disconnected if necessary. Cable and pipe detectors are available to detect buried services.

There are national colour codes for buried services, the most common of which are:

- black – electricity
- red – high-voltage electricity
- blue – water
- yellow – gas
- grey – telecommunications
- green – cable television.

It is now common for coloured polythene tape to be placed in the backfill about 300 mm above the pipe or service cable.

The most common machine for digging site trenches is the backactor (Figure 4.9).

FIGURE 4.9
Backactor

Work below ground level

Almost all construction sites will involve some form of excavation. Every year many construction workers are involved in accidents connected to excavations. The most common problem is the collapse of trench sides or equipment or materials falling into trenches.

Any ground work must be planned and carried out correctly and safely to prevent accidents.

All subsoils vary in their abilities to remain stable during excavations.

The first operation is the removal of the top soil, which should have been completed before the building was set out. This also makes the marking out of the ground easier.

Top soils can vary in depth from approximately 150 mm to 300 mm, and they contain various organic materials, making them unsuitable to build on. They could be either removed from the site or stored for later use when landscaping.

The foundation trenches can now be excavated by either hand or machine. Hand excavation is only carried out where the site restricts the use of machinery, such as for small extensions at the rear of existing properties.

Timbering

All subsoils have a natural angle of repose which allows the subsoil to rest unless given support. As the trench is excavated temporary support is placed in the trench according to the type of subsoil.

Some simple forms for shallow trench excavations are shown in Figure 4.10.

Laying concrete foundation

After the trenches have been excavated to the required depth the concrete can be placed to form the foundation. A straight edge should be used between each steel peg to produce a level foundation on which to build (Figure 4.11).

A rough finish is required to ensure that the brickwork adheres to the concrete foundation.

The distance from the top of the datum peg to the concrete needs to be in brick courses. This is checked with a spirit level and gauge rod (Figure 4.11).

The pegs are then levelled around the trench with either a straight edge or a spirit level. On larger areas, or where more accuracy is required, there are numerous mechanical levels available such as the quickset level, the autoset level and the laser level.

Brickwork up to damp-proof-course

Once the concrete has hardened the brickwork can be built up to DPC level.

Lines are fastened to the profiles to mark out the position of the brickwork (Figure 4.12).

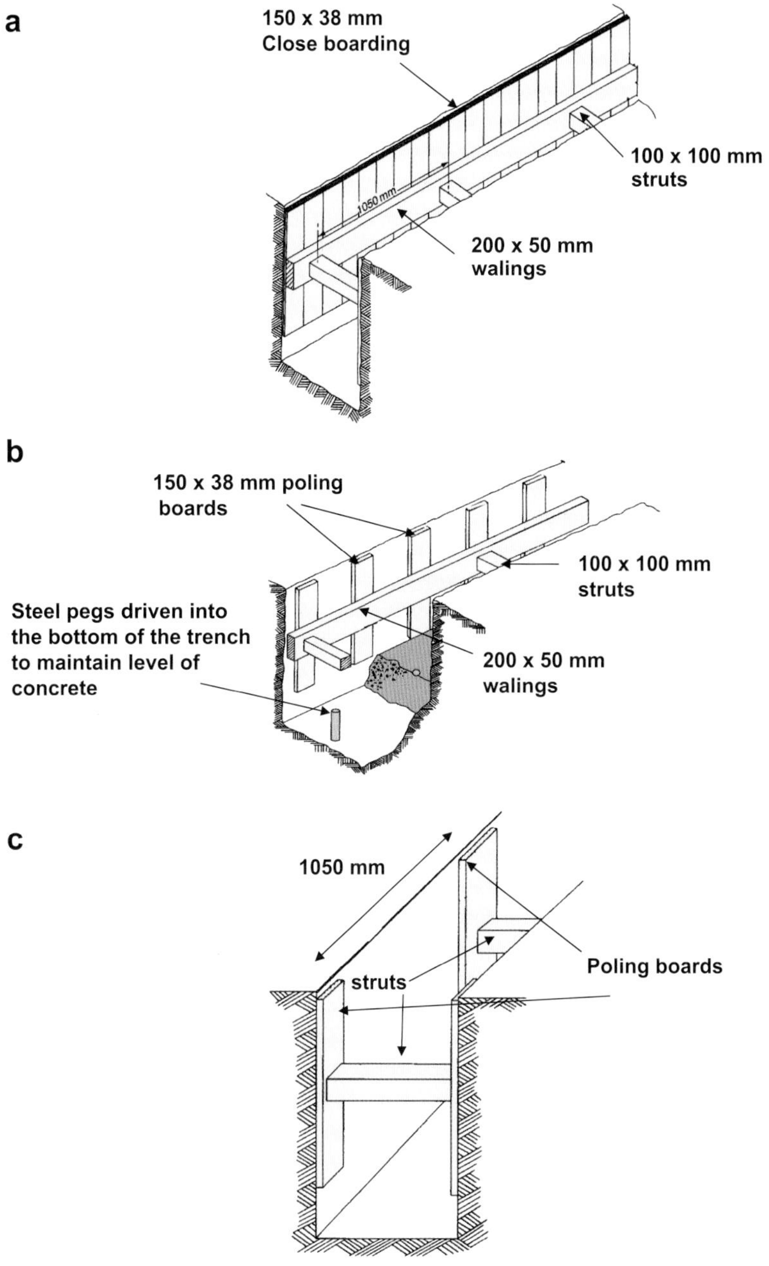
a
150 x 38 mm
Close boarding
100 x 100 mm
struts
1050 mm
200 x 50 mm
walings
b
150 x 38 mm poling
boards
100 x 100 mm
struts
Steel pegs driven into
the bottom of the trench
to maintain level of
concrete
200 x 50 mm
walings
c
1050 mm
struts
Poling boards

FIGURE 4.10
(a) Loose soil; (b) moderately firm soil; (c) firm soil

The ranging lines are set to wall marks on the profiles, and plumb lines are taken down on the main corners to the concrete foundation.

To facilitate clear marking, a mortar screed must be spread over the concrete to a thickness of approximately 3 mm at the positions where the plumb lines are being taken.

A spirit level can normally be used to ascertain plumb lines, but if the trench is too deep to allow this to be done, a 'drop bob' should be used (Figure 4.12).

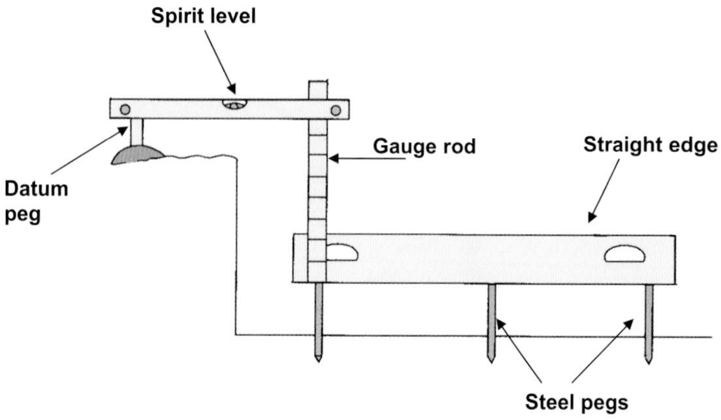
Spirit level
Gauge rod
Straight edge
Datum peg
Steel pegs

FIGURE 4.11
Placing concrete in foundations

The bricks are laid and the corners are erected by racking back until the correct height is reached.

The first course should be dry bonded to ensure that the correct bonding pattern is maintained.

The top two courses below the DPC should be facing bricks as they will be seen after the ground work has been completed.

Once the corners have been erected the main walls can be completed using lines and pins.

The procedure for working below ground level is very different to working above. The space in the trench may be very tight and all materials have to be set out on the sides of the trenches. Working below ground is often known as nose bleeding as you are mainly bending over.

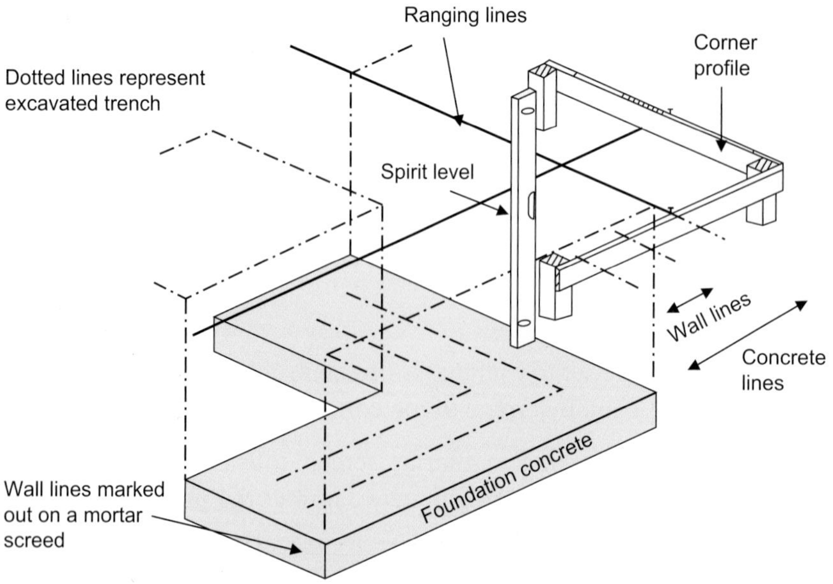
Ranging lines
Corner profile
Dotted lines represent excavated trench
Spirit level
Wall lines
Concrete lines
Foundation concrete
Wall lines marked out on a mortar screed

FIGURE 4.12
Plumbing down from ranging lines

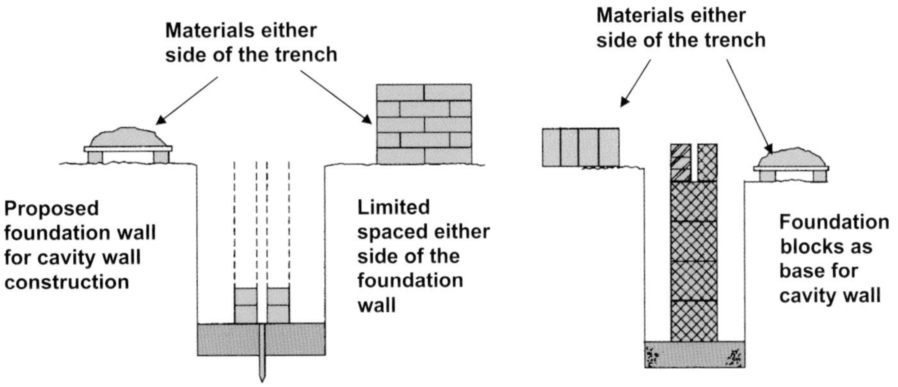
Materials either
side of the trench
Materials either
side of the trench
Proposed
foundation wall
for cavity wall
construction
Limited
spaced either
side of the
foundation
wall
Foundation
blocks as
base for
cavity wall

FIGURE 4.13
Preparation for work below ground level

It is important not to place the materials too near the side of the trenches, which may cause collapse due to too much pressure (Figure 4.13).

If the foundation brickwork is a cavity wall there will be more space for the feet while erecting the face wall. Solid brick or foundation block walls will be tight for space as most foundations are designed with a minimum 150 mm space on either side of the walls.

At this stage it is sufficient to know that the ground-floor level is approximately 225 mm above DPC level.

When this point is reached in the building of the wall, a level must be transferred from the datum which represents ground-floor level. This level will have been adhered to while constructing the wall, to ensure that the wall is horizontal at all points.

The transferred level is maintained on the wall by fixing a short length of 50 mm × 25 mm batten, termed a datum peg (Figure 4.14).

From this datum a storey rod is used to maintain correct heights and, to be of assistance to the charge hand bricklayer, it should be marked with all

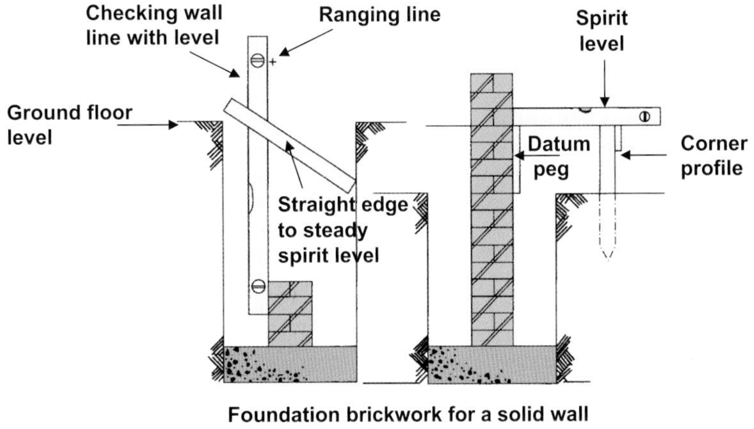
Checking wall
line with level
Ranging line
Spirit
level
Ground floor
level
Datum
peg
Corner
profile
Straight edge
to steady
spirit level
Foundation brickwork for a solid wall

FIGURE 4.14
Transferring setting out lines and datum to substructure

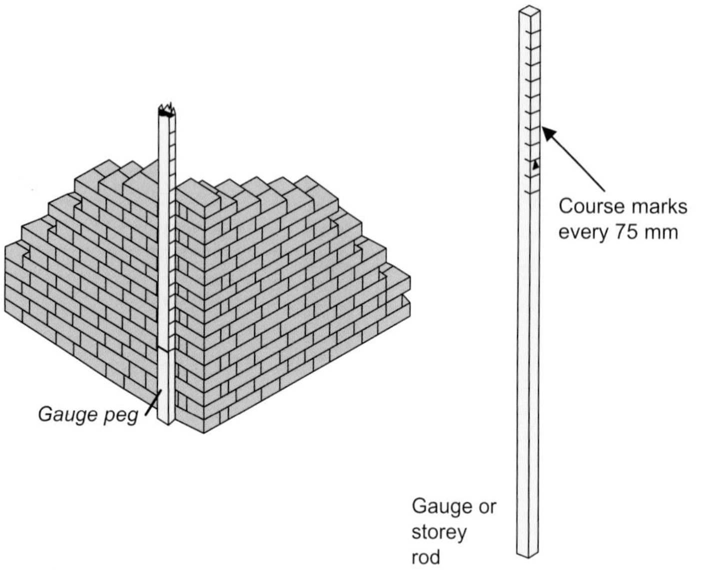
Gauge peg
Course marks
every 75 mm
Gauge or
storey
rod

FIGURE 4.15
Use of storey rod

those details appearing on the elevation of the wall, e.g. sills, arch and floor levels (Figure 4.15).

Services

During the work below ground level there are times when openings have to be left in the brickwork for services, etc. (Figure 4.16).

Sometimes the exact position is not known, so sand courses are used to allow bricks or blocks to be removed at a later date.

Foundations

The foundation of a building is that part which is in direct contact with the ground.

The current Building Regulations require that the foundations of a building shall safely sustain and transmit to the ground the combined dead and superimposed loads in such a way as not to impair the stability or cause damage to any part of the building.

The ground or subsoil on which a building rests is called the natural foundation or subfoundation, and has a definite load-bearing capacity, according to the nature of the soil (Figure 4.17).

Subsoils are of many varieties and may generally be classified as rock, compact gravel or sand, firm clay and firm sandy clay, silty sand and loose clayey sand.

It is possible to erect a wall on rock with little or no preparation, but on all soils it is necessary to place a continuous layer of in situ concrete in the trench, called the building foundation.

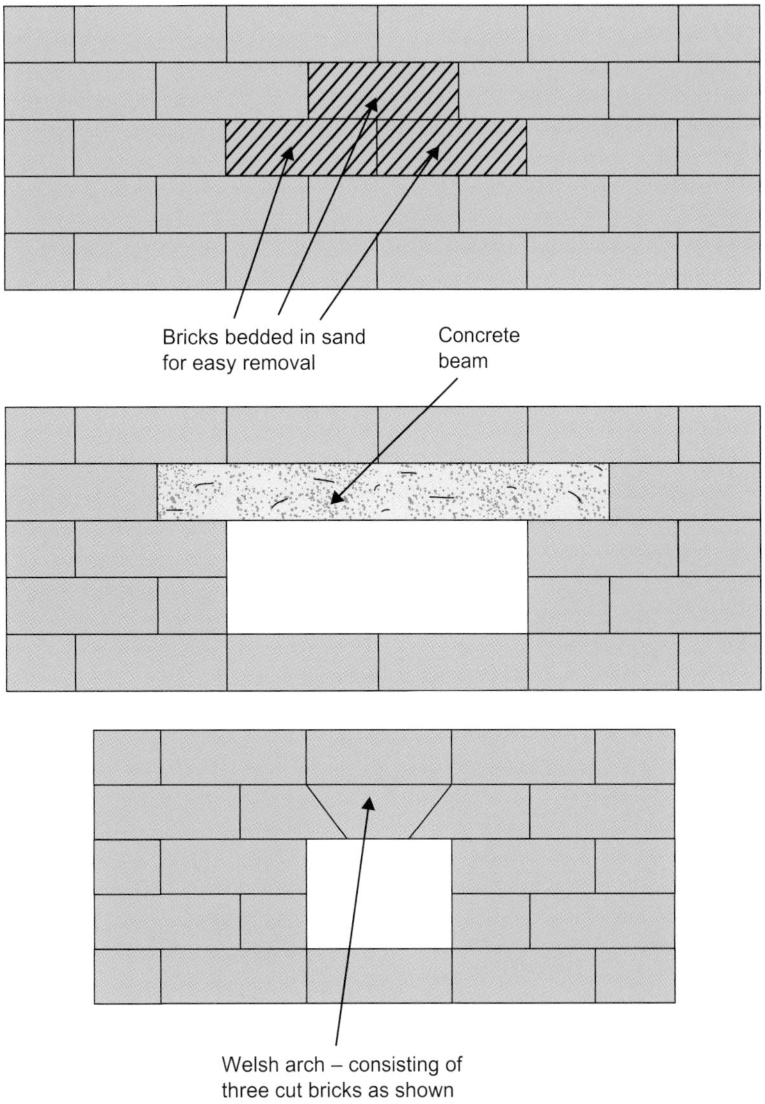
Bricks bedded in sand
for easy removal
Concrete
beam
Welsh arch – consisting of
three cut bricks as shown

FIGURE 4.16
Provision for services

The term 'in situ' means cast in place in its permanent position; unlike 'precast' which is made elsewhere, lifted and transported later to the place where it is required for use.

This cast, in situ concrete is made from Portland cement with coarse aggregate, plus sharp sand or ballast, graded from 40 mm to fine sand, and mixed in the proportion of 1:6.

Foundation types

This chapter will deal with four main types of foundations:

- strip foundations – narrow strip and wide strip foundations
- pad foundations
- raft foundations
- pile foundations.

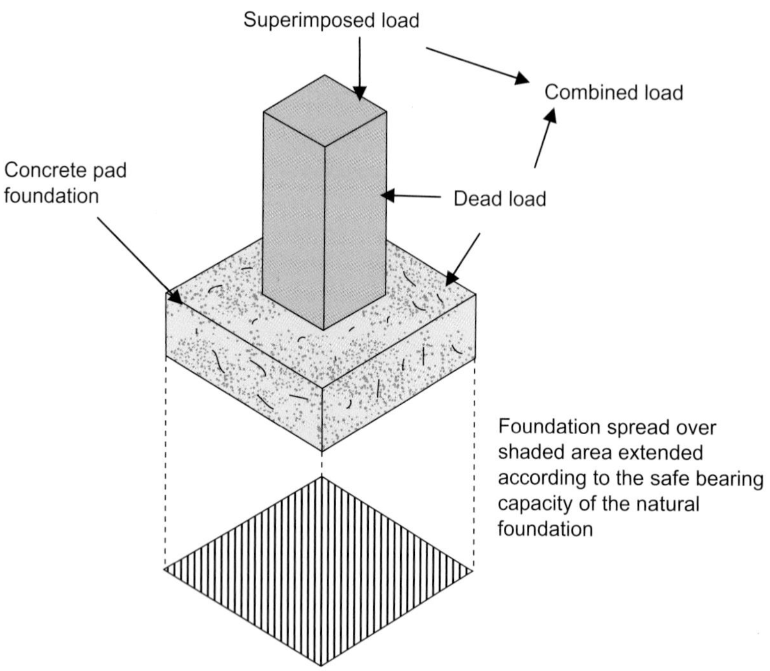
Superimposed load
Combined load
Concrete pad foundation
Dead load
Foundation spread over shaded area extended according to the safe bearing capacity of the natural foundation

FIGURE 4.17
Loads

STRIP FOUNDATIONS

When a load is placed on soil it is necessary to 'spread' or 'extend' the foundation base to ensure stability. This extended or spread foundation is referred to as a strip foundation in the case of a continuous wall structure, or a pad foundation in the case of an isolated pier.

The thickness of the concrete foundation should not be less than the projection of the strip either side of the wall, but in no case less than 150 mm. One method of ascertaining the depth of concrete is shown in Figure 4.18.

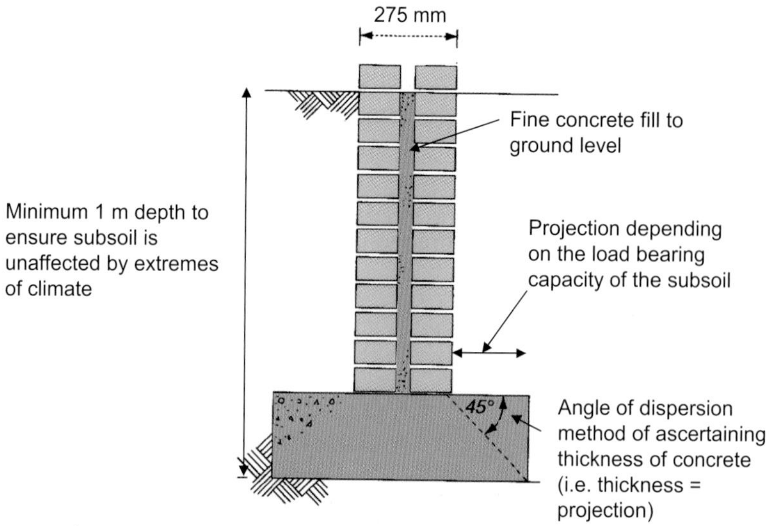
275 mm
Fine concrete fill to ground level
Minimum 1 m depth to ensure subsoil is unaffected by extremes of climate
Projection depending on the load bearing capacity of the subsoil
45°
Angle of dispersion method of ascertaining thickness of concrete (i.e. thickness = projection)

FIGURE 4.18
Strip foundation

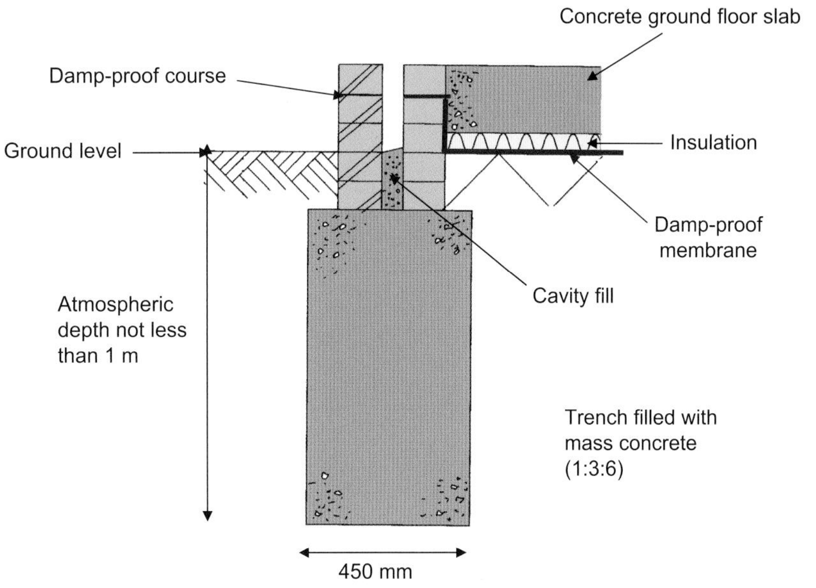
Concrete ground floor slab
Damp-proof course
Ground level
Insulation
Damp-proof membrane
Cavity fill
Atmospheric depth not less than 1 m
Trench filled with mass concrete (1:3:6)
450 mm

FIGURE 4.19
Narrow strip foundation

ATMOSPHERIC DEPTH

This is the depth below ground level to which foundations should be taken. It depends on the type of soil and is the depth at which the subfoundation ceases to be affected by the weather. This is between 600 and 1500 mm, decreasing as the proportion of gravel increases (Figures 4.18 and 4.19).

NARROW STRIP

Narrow strip, or trench fill as it is also known, is used as an economical alternative to the normal strip foundation (Figure 4.19).

A narrow strip is excavated by the mechanical excavator and backfilled with mass concrete up to a level just below the finished ground level.

A high standard of accuracy in constructing such a foundation is required. It is cheaper and quicker to fill the trenches with mass concrete than to excavate a wider trench. There is less excavated material to be removed, no brickwork below ground and backfilling is eliminated. This method also eliminates timbering to trenches.

WIDE STRIP FOUNDATIONS

Where the structural loads are very heavy or the safe bearing capacity of the soil is low, the spread of the foundation base becomes greater. This is normally referred to as a wide strip foundation. It follows that the required depth of the concrete foundation base may be considered excessive and it can be reduced by the introduction of steel reinforcement, but the foundation must always be of sufficient depth to ensure that, in combination with the steel, it will resist the stresses of tension and shear. Figure 4.20 shows a simple example of a reinforced concrete strip foundation.

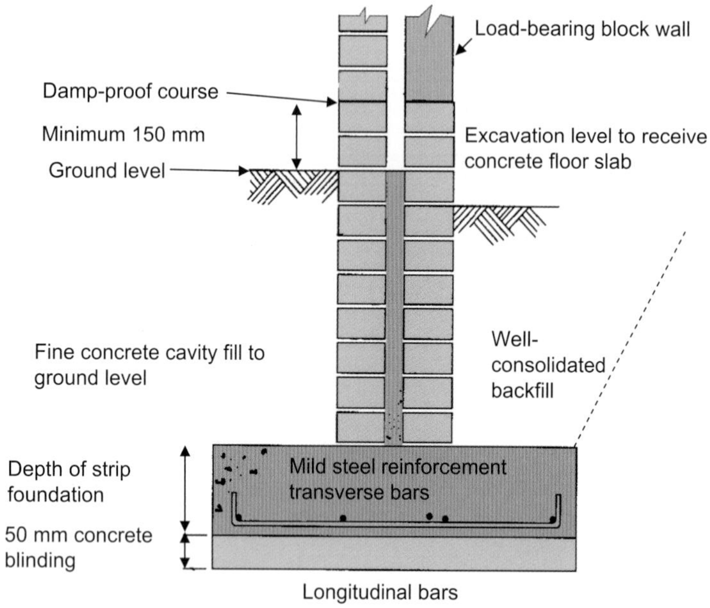
Load-bearing block wall
Damp-proof course
Minimum 150 mm
Ground level
Excavation level to receive concrete floor slab
Fine concrete cavity fill to ground level
Well-consolidated backfill
Depth of strip foundation
Mild steel reinforcement transverse bars
50 mm concrete blinding
Longitudinal bars

FIGURE 4.20
Reinforced wide strip foundation

PAD FOUNDATIONS

For single loads that are transmitted down a column, the most common foundation is a square or rectangular block of concrete of uniform thickness known as a pad foundation (Figure 4.21).

> **Note**
>
> In order to spread the load over a greater area it is necessary either to make the pad thicker or to use reinforced concrete.

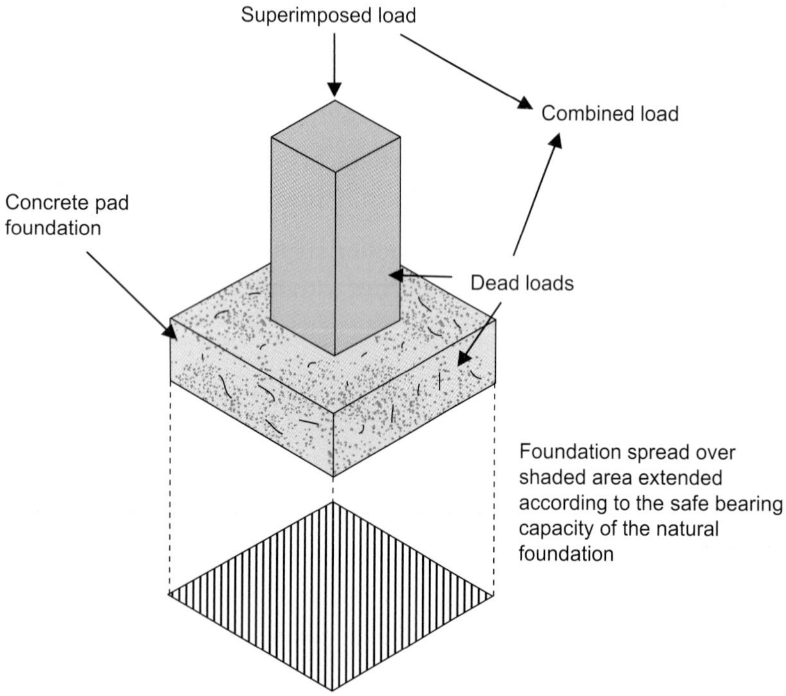
Superimposed load
Combined load
Concrete pad foundation
Dead loads
Foundation spread over shaded area extended according to the safe bearing capacity of the natural foundation

FIGURE 4.21
Pad foundation

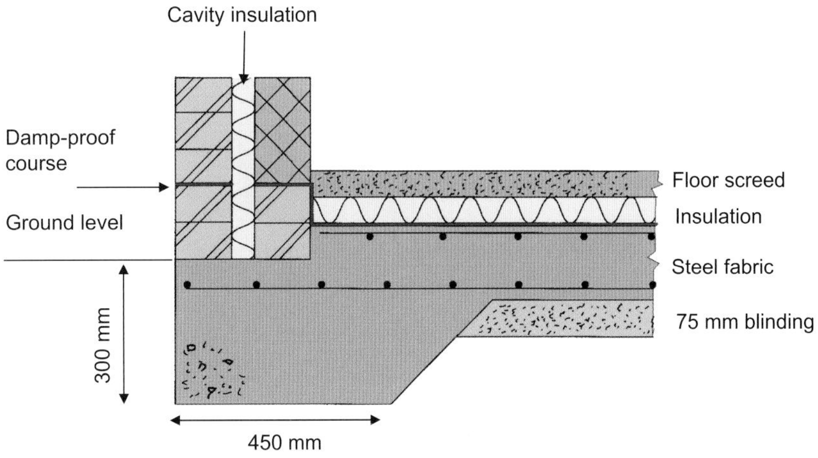
Cavity insulation
Damp-proof course
Ground level
Floor screed
Insulation
Steel fabric
75 mm blinding
300 mm
450 mm

FIGURE 4.22
Raft foundation

RAFT FOUNDATIONS

These foundations consist of a raft of reinforced concrete under the whole of the building.

Raft foundations are often used on poor subsoils for lightly loaded buildings and are considered capable of accommodating small settlements of the subsoil.

The simplest and cheapest form of raft is the thick reinforced concrete raft (Figure 4.22). Its rigidity enables it to minimize the effects of differential settlement.

SHORT BORED PILED FOUNDATIONS

If, instead of spreading the load from the wall over a wide area, it is decided to transfer it to a greater depth, an economical solution is the use of a short bored pile foundation.

Short bored piles are formed by boring circular holes 300 mm diameter to a depth of about 3 m by means of an auger. This depth is governed by the level of suitable bearing capacity ground. These are filled as soon as possible with mass concrete.

The piles are placed at the corners of the building and at intermediate positions along the walls. The piles support reinforced concrete beams which are cast in place in the ground (Figure 4.23).

Floors

Ground floors

There are two types of ground floor construction found in both domestic and communal buildings:

- solid
- hollow.

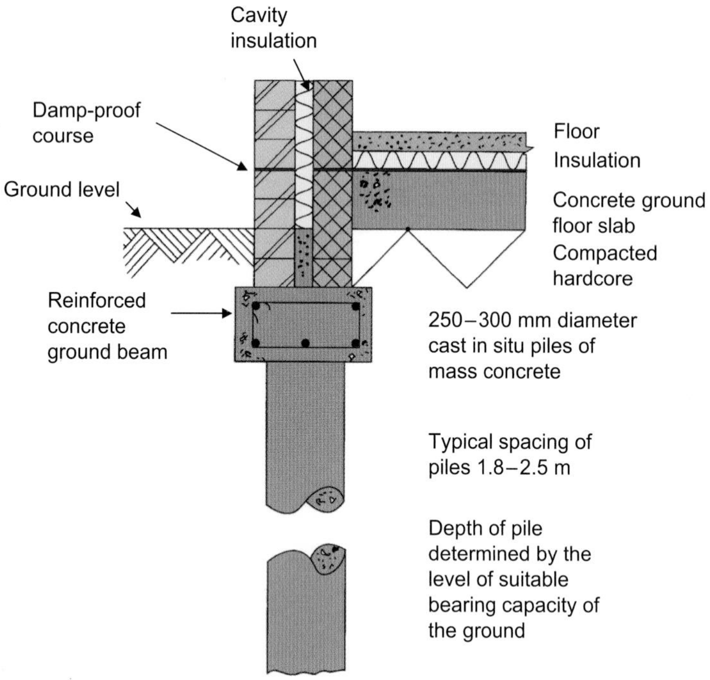
Cavity
insulation
Damp-proof
course
Ground level
Reinforced
concrete
ground beam
Floor
Insulation
Concrete ground
floor slab
Compacted
hardcore
250–300 mm diameter
cast in situ piles of
mass concrete
Typical spacing of
piles 1.8–2.5 m
Depth of pile
determined by the
level of suitable
bearing capacity of
the ground

FIGURE 4.23
Piled foundation

The functions of ground floors are to provide a level surface with sufficient strength to support all the loads, and to prevent dampness entering the building and heat loss from the building.

SOLID GROUND FLOORS

Solid ground floor consists of a compacted hardcore with sand blinding on top. This protects the damp-proof membrane (DPM). Insulation is laid on the membrane on which a slab on concrete is laid.

Details are shown in Figure 4.24.

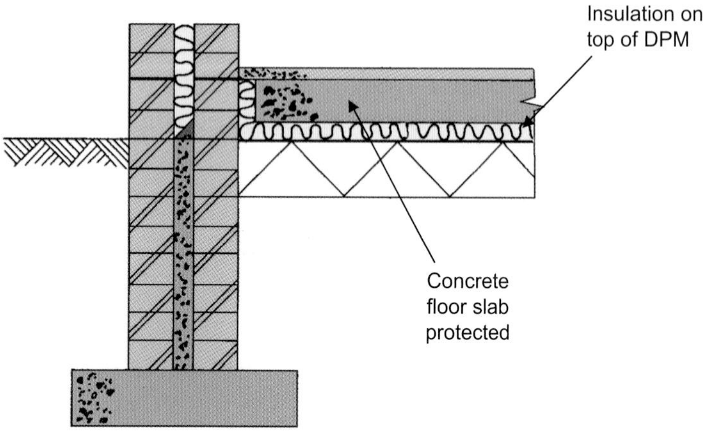
Insulation on
top of DPM
Concrete
floor slab
protected

FIGURE 4.24
Damp proofing a solid ground-floor slab with the DPM under the concrete floor slab

The horizontal DPC level must never be above the floor level.

Brick rubble or hardcore laid directly beneath the concrete floor will not only prevent settlement but, being of a porous nature, also help to prevent dampness. Its thickness should be approximately that of the concrete floor.

Solid ground floors have to be insulated, according to the current Building Regulations, to provide resistance to unacceptable heat loss through the floor.

This can be achieved in various ways but the most common is to place the insulation on top of the DPM, which is placed on a blinding layer on top of the hardcore.

This method of construction protects the concrete floor slab from any moisture or harmful salts. The only problem with this method is the risk of damage to the DPM and insulation when laying the concrete floor slab.

An alternative method is shown in Figure 4.25. This method is easier but the concrete floor slab is not protected against the ingress of moisture or harmful salts.

Remember

The DPM *must* be continuous with the horizontal DPC in the external and internal walls.

Materials suitable for floor insulation are dense resin bonded mineral or glass fibre slabs, and polystyrene and cork slabs. These should be placed above the DPM and turned up at the edges of the floor slab to prevent heat loss through the external wall.

HOLLOW GROUND FLOORS

Suspended timber floors need to have a well-ventilated space beneath the floor construction to prevent the moisture content of the timber rising above an unacceptable level, i.e. above 20 per cent, which would create the conditions for possible fungal attack.

Hardcore and oversite concrete are still required for these floors, but in this case the concrete does not require a waterproof membrane.

Hollow sleeper walls are constructed on the oversite concrete to receive the wall plate, which in turns supports the floor joists (Figure 4.26). A horizontal DPC is inserted under the wall plate to resist rising damp.

UNDERFLOOR VENTILATION AGAINST DRY ROT

Having placed the oversite, the floor must be supported, so as to allow free passage of air to prevent the floor timbers from rotting. This is achieved by

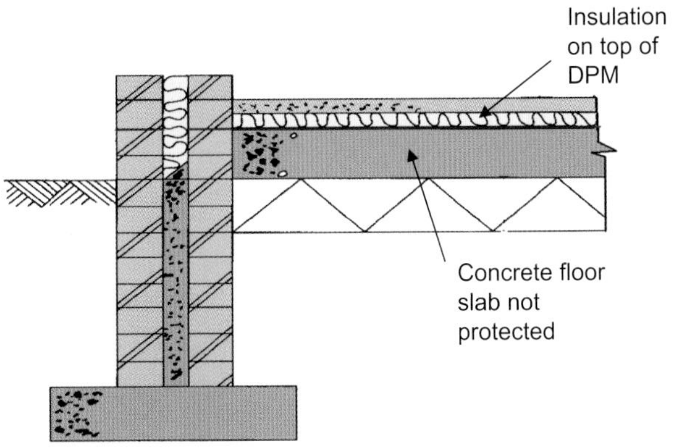
Insulation on top of DPM
Concrete floor slab not protected

FIGURE 4.25
An alternative method with the DPM on top of the concrete floor slab

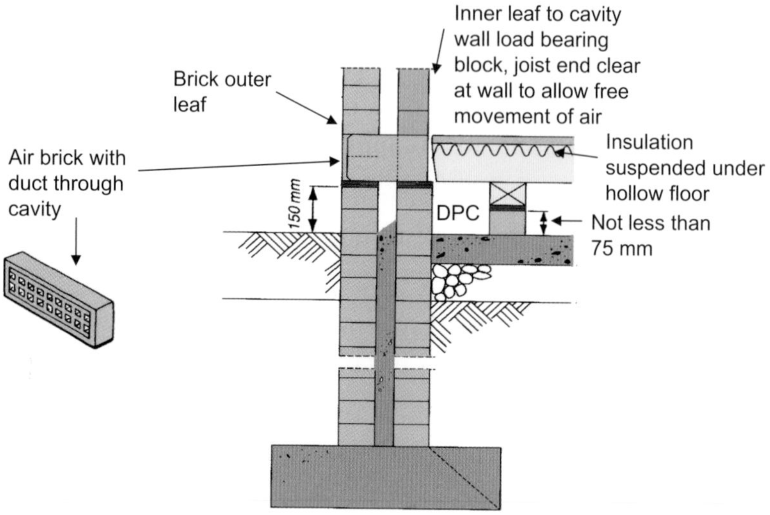
Inner leaf to cavity wall load bearing block, joist end clear at wall to allow free movement of air
Brick outer leaf
Insulation suspended under hollow floor
Air brick with duct through cavity
150 mm
DPC
Not less than 75 mm

FIGURE 4.26
Cavity wall, showing underfloor ventilation

building honeycomb sleeper walls on the oversite concrete and air bricks built into the external wall.

Two types are illustrated in Figure 4.27.

PRECAST BEAM AND POTS

Prestressed precast concrete beams were first used in commercial ground floor construction and are now common in domestic construction. They have been designed as an alternative to suspended timber floors. The same preparation is required as for suspended timber floors (Figure 4.28). They consist of prestressed beams with block inserts.

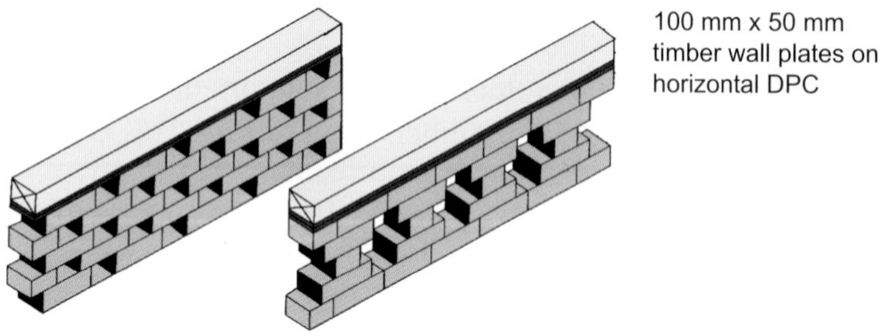
100 mm x 50 mm timber wall plates on horizontal DPC

FIGURE 4.27
Types of sleeper wall construction

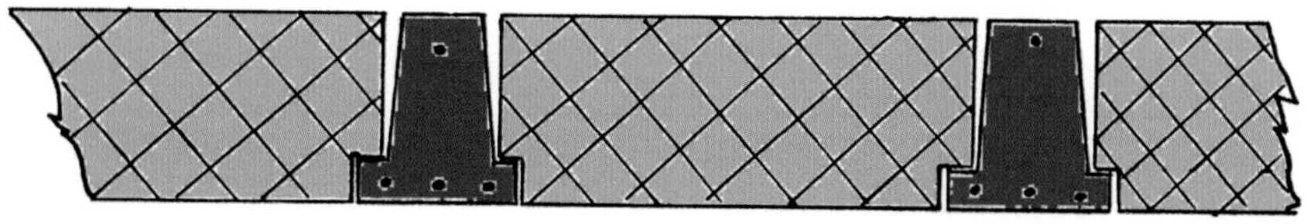

FIGURE 4.28
Precast concrete ground floors

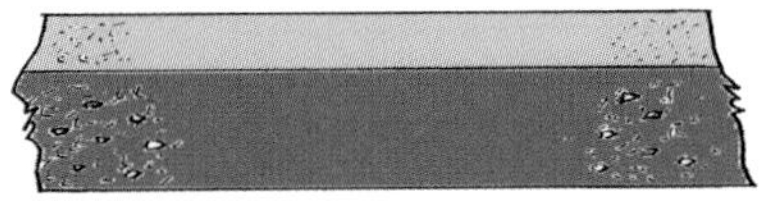

FIGURE 4.29
Cement and sand screeds to concrete floors

When precast floors have been constructed they provide a safer work area than suspended timber floors at both ground and first floor level.

Suspended timber floors require boarding out before work can proceed, whereas the complete floor area of precast floors is ready to work on and can take full loads.

FLOOR FINISHES

There are numerous floor finishes available, some of which form an integral part of the construction, such as floor boarding and screeds.

SCREEDS

These are used to give the concrete floor a durable finish to receive the final floor finish, such as tiles.

Cement and fine aggregate screed are laid up to a thickness of 75 mm directly on top of the concrete floor (Figure 4.29).

FLOOR BOARDING

Timber suspended floors can be covered with either floor boarding or sheets (Figure 4.30).

Tongue-and-groove boarding is laid at right angles to the floor joists and should be fastened using floor brads.

Sheeting is available in 600 mm × 19 mm sheets and is laid at right angles to the floor joists in a chequerboard design. Sheeting should be screwed into position to prevent movement.

FLOATING FLOORS

When quiet floors are required, such as for a library, floating floors can be constructed (Figure 4.31). They require the final floor to be separated

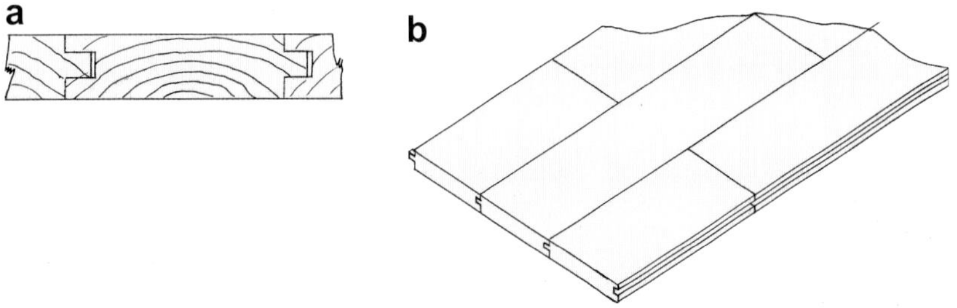
a
b

FIGURE 4.30
Covering to timber joists: (a) tongue-and-groove floor boarding; (b) tongue-and-groove sheeting

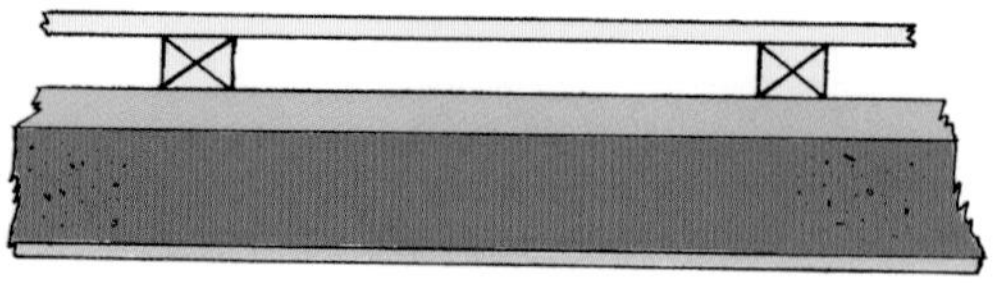
Floor boarding on 45 mm
battens on a resilient layer

FIGURE 4.31
Floating floor

from the floor structure to prevent the passage of sound up through the structure and into the room above.

Walls

External wall construction can consist of either solid or cavity walling (Figure 4.32).

Cavity walling can be further divided into masonry cavity walls (Figure 4.33) and timber-frame cavity walls (Figures 4.34 and 4.35).

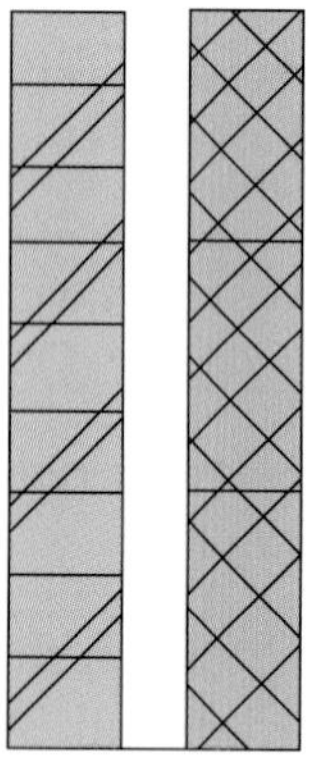

FIGURE 4.32
Solid and cavity external walls

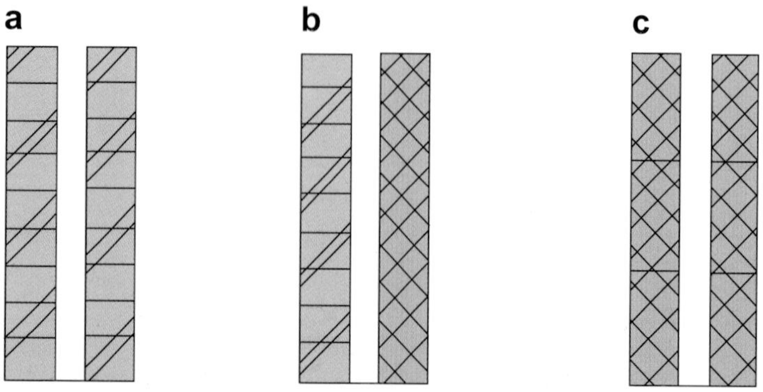
a
b
c

FIGURE 4.33
Types of cavity walling: (a) brick to brick; (b) brick to block; (c) block to block

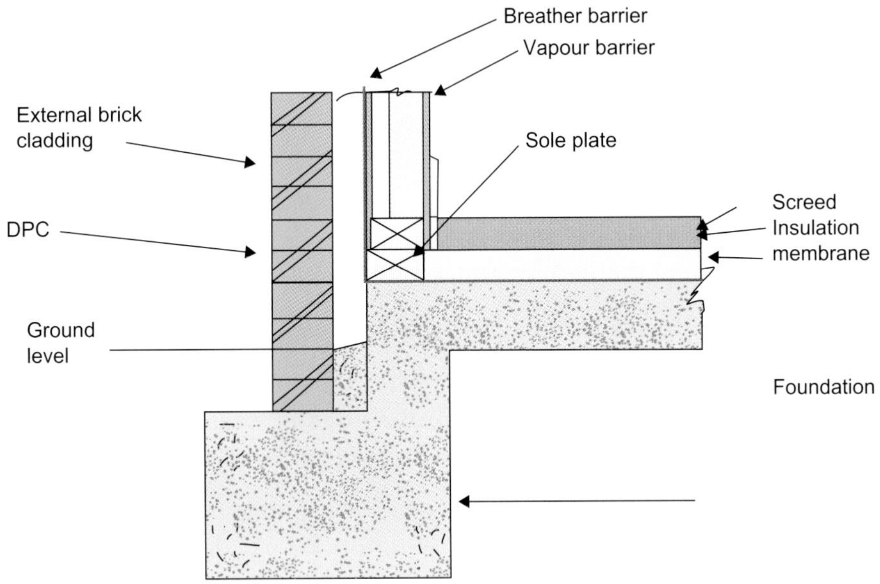
Breather barrier
Vapour barrier
External brick cladding
Sole plate
Screed
Insulation
membrane
DPC
Ground level
Foundation

FIGURE 4.34
Timber frame

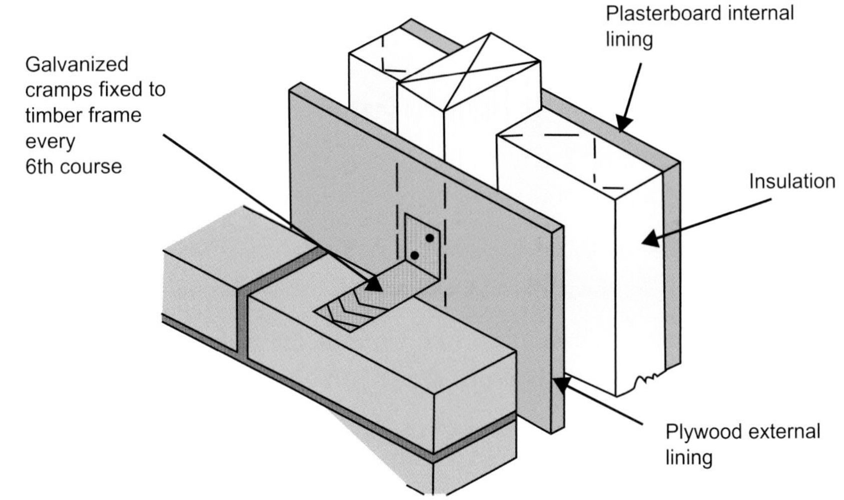
Plasterboard internal lining
Galvanized cramps fixed to timber frame every 6th course
Insulation
Plywood external lining

FIGURE 4.35
Timber frame fixing

Masonry cavity walls can be constructed in:

- internal and external skins of brick
- internal skin of block and external skin of brick
- internal and external skins of blocks
- cladded walls.

External walls to framed structures may be grouped according to the type of structural frame: steel, concrete or timber.

This chapter will deal only with cladding to timber frames.

Timber frames

Timber wall frames are erected on a foundation of conventional walls or a concrete base or raft to which a sole plate is bolted.

The frame is usually assembled as a storey-height frame, in sections around the building.

Openings for windows and doors are included in the frame and are covered with a sheathing of timber boards or plywood and a breather barrier.

The brickwork is fastened to the timber frame with galvanized steel cramps fixed every sixth course.

The standard form of construction for the external walls of domestic and communal brick buildings is called cavity walling.

This means that the bricklayer builds the two separate leaves or 'skins' of brick masonry (a general term indicating brickwork and/or blockwork), with a 50–75 mm wide space between them.

The outer skin is usually 102.5 mm thick face brickwork, but may be constructed from facing quality blocks.

The inner skin is usually 100 mm thick common blocks that are later plastered to receive internal decoration.

Both skins of brick masonry are joined together with a regular pattern of corrosion-resistant ties, so that they behave as a single wall.

The main objectives of cavity walls are to prevent rain penetration and provide a greater degree of thermal insulation than solid wall construction.

New developments in the design of cavity walls have allowed for greater insulation values to prevent heat loss through the walls.

The external walls of domestic and communal buildings can also be designed in a framed construction. Frames can be timber, steel or concrete.

One of the great advantages of framed buildings is that the main construction can be almost finished and watertight before the external skin of brickwork is added. This is often considered as a faster method of building and allows other trades to continue while the external cladding is being fixed.

When designing to meet the current Building Regulations for heat loss through walls, it is important to consider which method of construction will be used, as the cost can vary according to which method is used.

The current Building Regulations require that external walls have a maximum U-value of 0.35 $W/m^2 K$. The U-value denotes the thermal transmittance, which is the rate of heat transfer through a wall from air to air.

The most common method of insulating cavity walls, whether they are brick and block or framed, is by inserting slabs of polyurethane foam, expanded polystyrene or bonded glass fibre into the cavity space.

Figures 4.36 and 4.37 show insulation placed in the cavity of a brick and block wall and insulation to timber-framed construction, respectively.

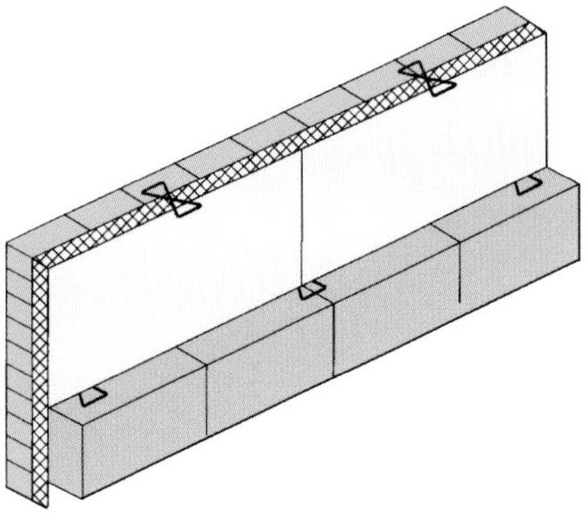

FIGURE 4.36
Insulation to cavity wall

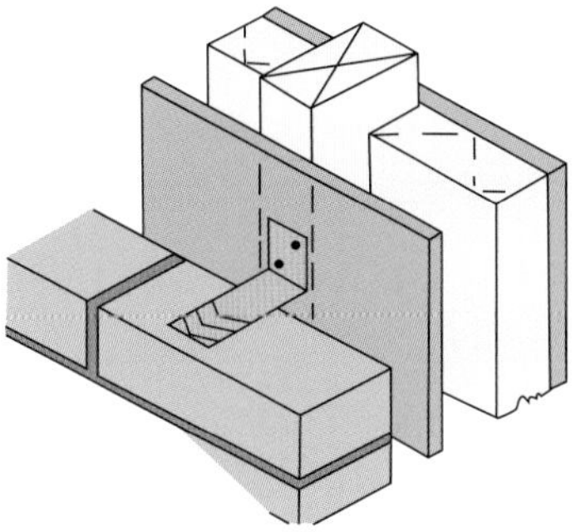

FIGURE 4.37
Insulation to timber frame cladding

External envelope

The structure of every building is divided into two parts:

- superstructure – all elements of the building above ground level (horizontal DPC)
- substructure – all elements of the building below ground level.

This envelope can be described as the external envelope and is generally a combination of walls and roofs with openings for light and access.

External walls

The external envelope should provide the building with a watertight coat which is also decorative.

Facing bricks are the most common material used in modern domestic construction, whether the design is timber framed or brick and block cavity walling.

In certain parts of the country the local material will change and other materials will be considered, such as stone.

Whatever materials or methods are chosen for the external wall they all have to satisfy the following functions:

- strength and stability
- weather exclusion
- thermal insulation
- sound insulation
- durability
- fire resistance
- appearance
- access and egress.

STRENGTH AND STABILITY

The external walls have to be strong enough to carry the loads without excessive deformation and to transfer these safely to the structural frame or foundation.

WEATHER EXCLUSION

Keeping out the weather is an extremely important consideration if the internal environment is to remain constant.

THERMAL INSULATION

Heat will flow from a high temperature inside the building to a low temperature on the outside. The external walls should be designed to reduce this heat flow to a minimum.

During the winter period there will be a constant flow of heat from the building which means that extra heating will be required to counteract the heat loss through the external envelope.

SOUND INSULATION

Noise is defined as 'unwanted sound'.

Extremely loud noises can be detrimental to the health of the users of the building, while small noises can be irritating and could cause loss of concentration.

Since the majority of people do not have control of the sounds outside the building, such as road traffic and aircraft noise, the external envelope must reduce that noise to an acceptable level.

DURABILITY

The materials from which the external walls are constructed must have sufficient resistance to the damaging effects of the climate – in the form of erosion, atmospheric pollution, frost, rain, and chemical and solar degradation – to provide a building that will be relatively maintenance free for its anticipated useful life.

FIRE RESISTANCE

The building must be able to contain any fire until the fire-fighting appliance arrives.

The external envelope should be able to contain the fire and to prevent the spread of fire to other buildings in the vicinity.

APPEARANCE

This is the aesthetic requirement which should be considered at the design stage.

The external envelope should be compatible with others in the vicinity, by either blending in with them or providing a contrast which is not too stark.

Remember that in the rural areas the envelope may be required to blend in with the landscape so that it becomes as inconspicuous as possible.

ACCESS AND EGRESS

The external envelope has to have openings formed to allow the users in and out of the building, and to allow light to enter.

Internal walls

These usually act as partitions or dividers for the internal space of the building, although occasionally they also support the upper floor and roof loads.

The most common types of internal walls (Figure 4.38) are:

- masonry partitions
- timber-frame or stud partitions
- demountable partitions
- proprietary systems (paramount).

MASONRY PARTITIONS

These can be built in either lightweight concrete or clay blocks (Figure 4.39).

Masonry partitions are one of the strongest types of internal wall and will support the loads subjected by the floors, roofs and superimposed loadings from the inhabitants and furniture.

They are built exactly as external walls, without the finish to the joints that external walls require.

The mortar is usually weaker than the material being used, to allow any shrinkage to take place in the joint rather than in the block.

If they are built on wooden upper floors they require a sole plate. If they are built on a concrete ground floor a sole plate is not necessary (Figure 4.40).

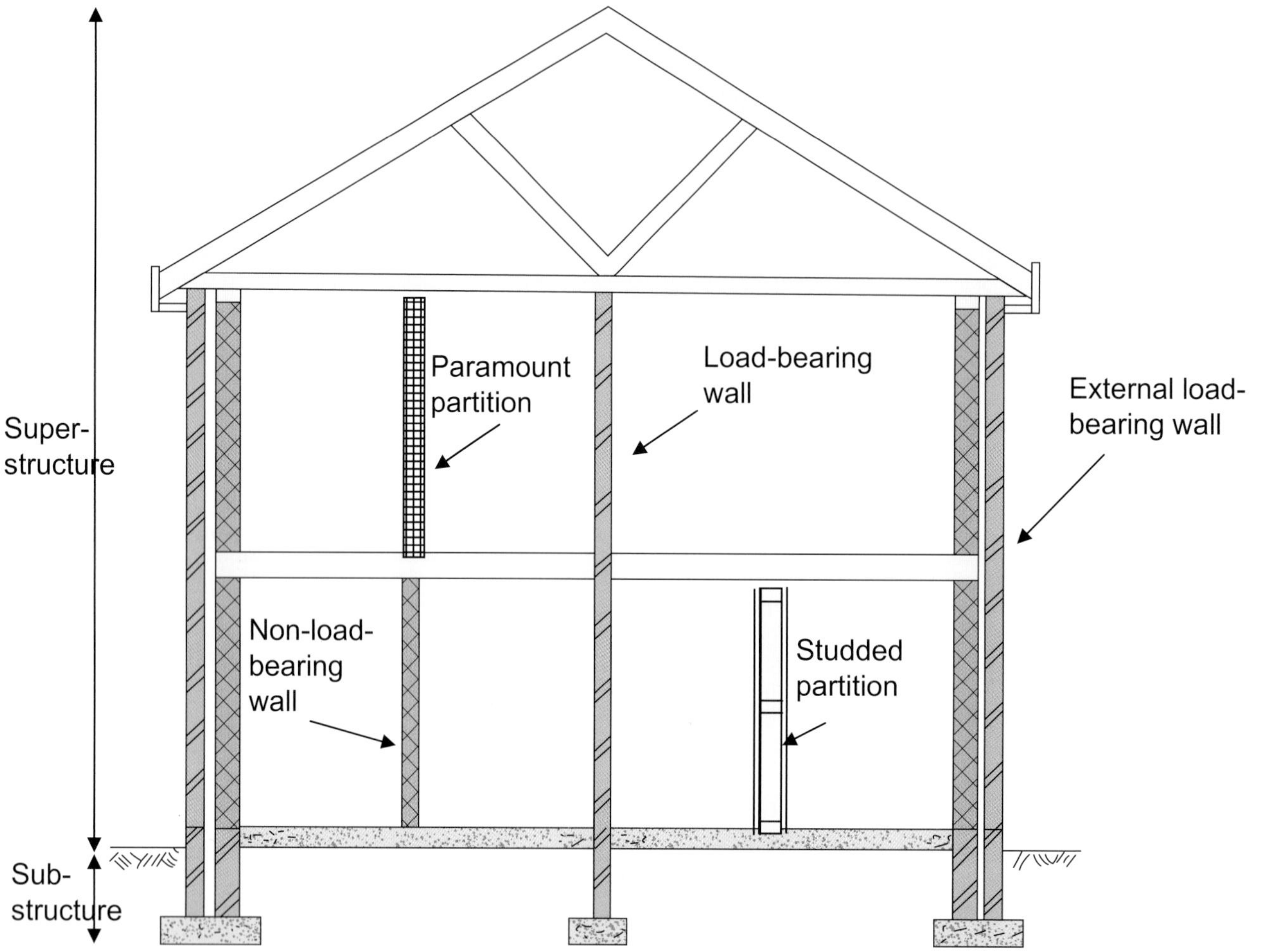
Super-
structure
Sub-
structure
Paramount
partition
Load-bearing
wall
External load-
bearing wall
Non-load-
bearing
wall
Studded
partition

FIGURE 4.38
Types of wall

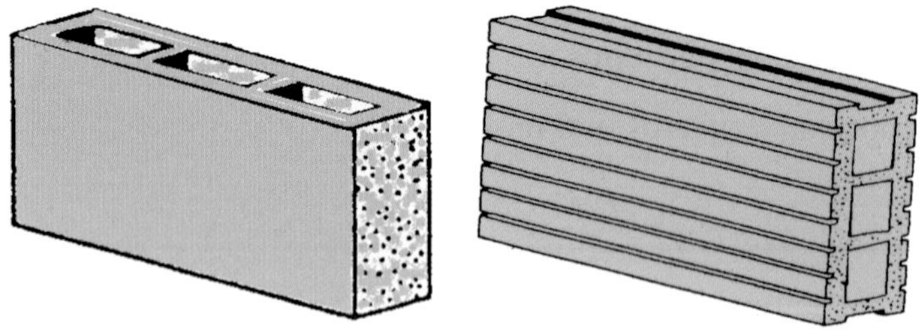

FIGURE 4.39
Types of lightweight internal block

TIMBER-FRAME OR STUD PARTITIONS

Timber studding is a non-load-bearing partition.

These partitions must be designed and constructed to carry their own weight and any fittings that may be attached to them.

They must be strong enough to withstand impact loadings on their faces and also any vibrations set up by doors being closed or slammed.

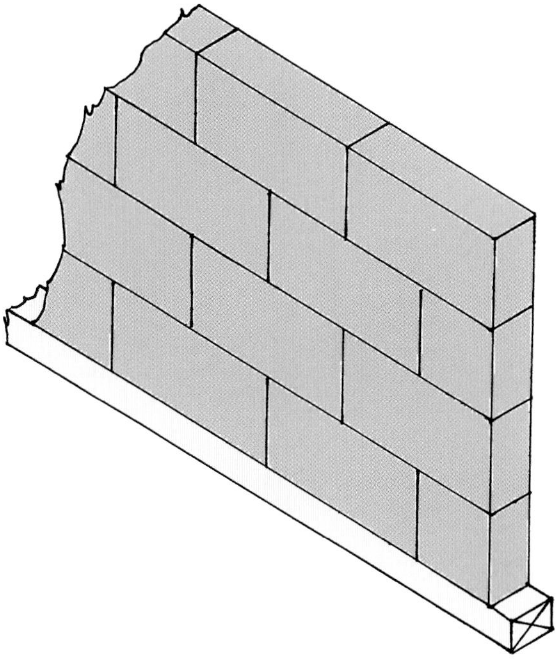

FIGURE 4.40
Lightweight block wall

Details are shown in Figure 4.41.

DEMOUNTABLE PARTITIONS

A wide range of lightweight, non-load-bearing and demountable partitions is available (Figure 4.42). The construction is usually of metal, wood or plastic sections or trim with various infill panels in a variety of materials.

Their main advantages are their versatility and non-permanent nature. They can be used to alter the floor layout of existing properties with the minimum of disruption.

PROPRIETARY SYSTEMS (PARAMOUNT)

One particular proprietary partition consists of plasterboard bonded on either side of a strong cardboard cellular core to form rigid panels. These are fixed to ceiling and wall battens and supported on a timber wall sole plate.

These units can also be provided with suitable facing for direct decoration or even with self-finish applied (Figure 4.43).

Roofs

The roof is that part of the external envelope which spans the external walls at their highest level and, being part of the envelope, it must fulfil the same functions.

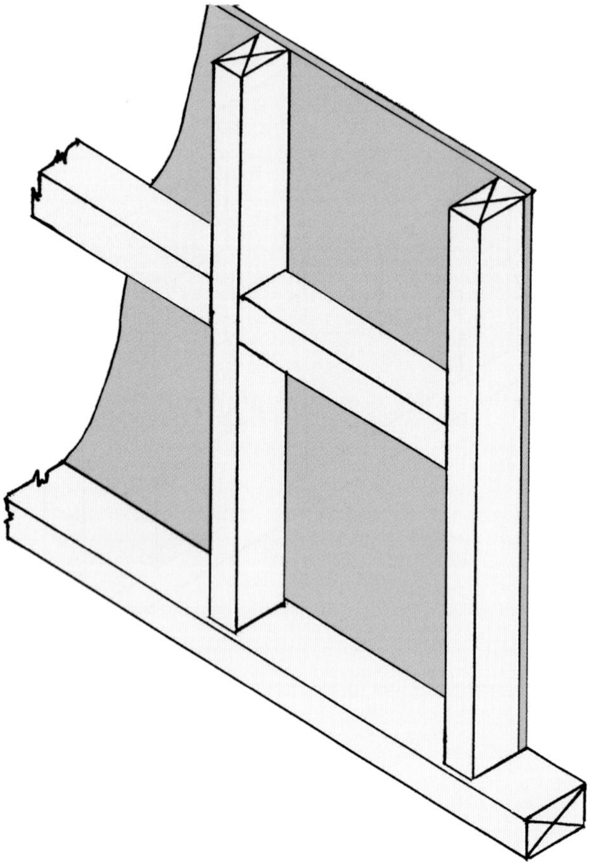

FIGURE 4.41
Timber studding

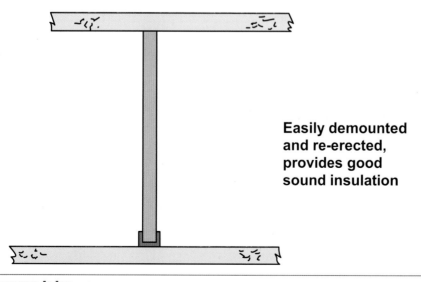

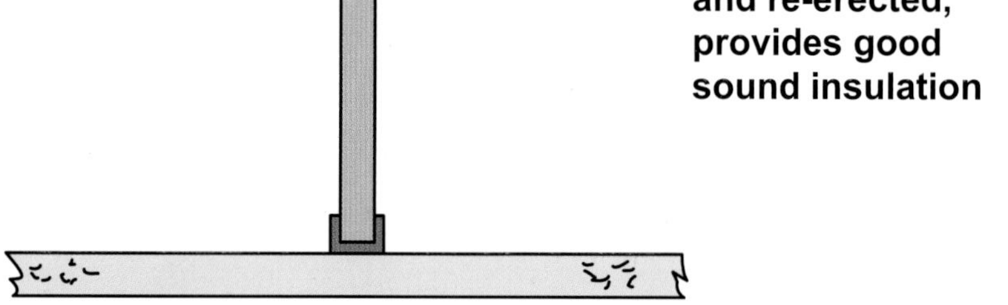
Easily demounted
and re-erected,
provides good
sound insulation

FIGURE 4.42
Demountable partition

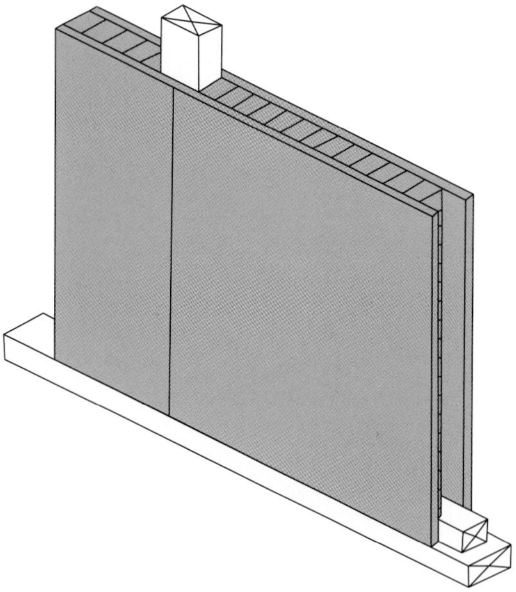

FIGURE 4.43
Paramount partitions

Basic roof forms

These may be either flat or pitched and are named according to their shape (Figure 4.44). Any roof having a sloping surface more than 10 degrees is known as a pitched roof. A roof with a slope of less than 10 degrees is known as a flat roof.

A gabled roof has a double pitched roof between two triangular brick walls known as gables. These have soffit and fascia on the front and rear only.

A hipped roof is also a double pitched roof, but the eaves continue all the way around the building.

A lean-to consists of a single pitched roof resting against a building.

ROOF COMPONENTS

The slope of the roof is known as the pitch. The edges of the roof are known as the eaves along the lowest horizontal edges and are terminated with a soffit and fascia (Figure 4.45).

The verge is found at the side and sloping edges and is finished with a barge board and soffit.

The detail of the eaves in Figure 4.45 shows tiles fixed to battens on sarking felt. This enables the roof to be watertight.

The sarking felt is continuous into the guttering fixed to the fascia.

The soffit has ventilation holes to allow the roof to be ventilated.

The roof is protected with insulation fitted between the ceiling joists.

Figure 4.46 shows roof details at the ridge. This detail shows slates fixed to battens on sarking felt on either side of the ridge of the roof.

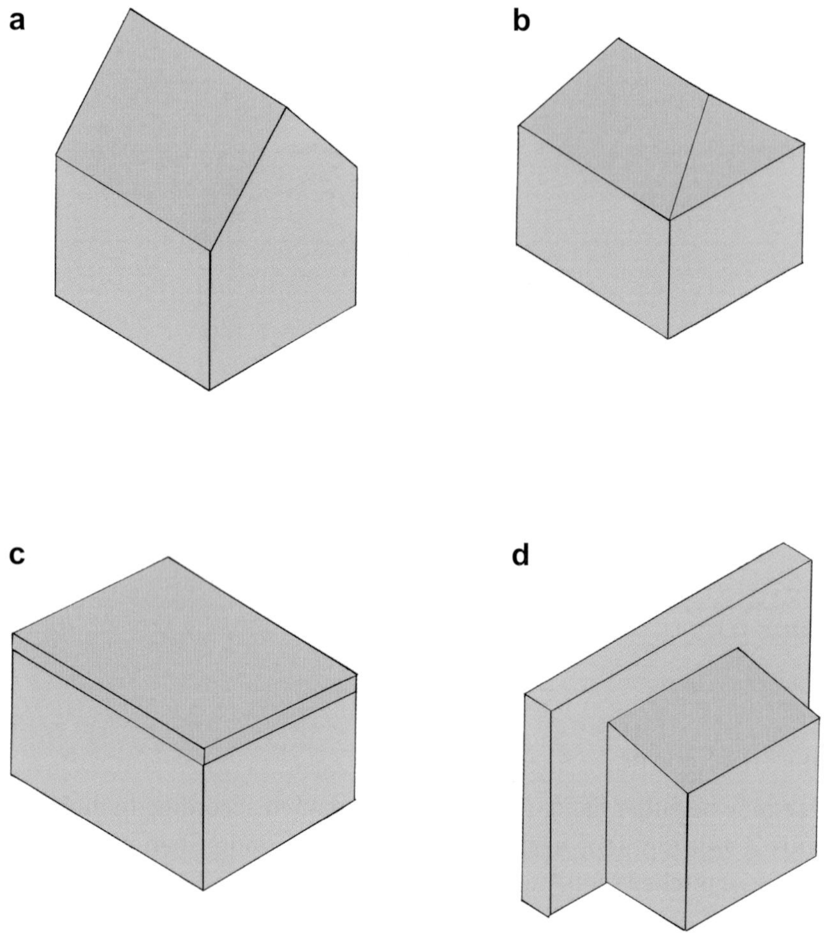
a
b
c
d

FIGURE 4.44
Basic roof forms: (a) gabled; (b) hipped; (c) flat; (d) lean-to

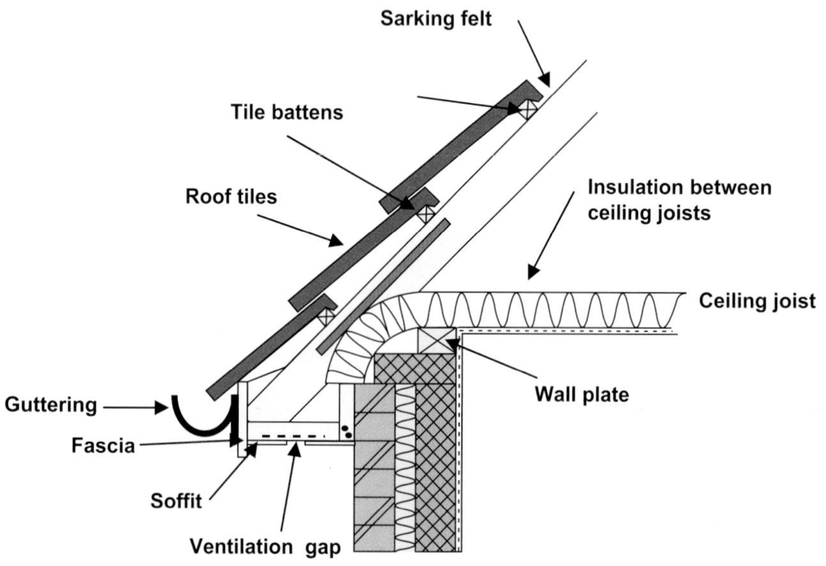
Sarking felt
Tile battens
Roof tiles
Insulation between ceiling joists
Ceiling joist
Wall plate
Guttering
Fascia
Soffit
Ventilation gap

FIGURE 4.45
Basic roof terminology at eaves

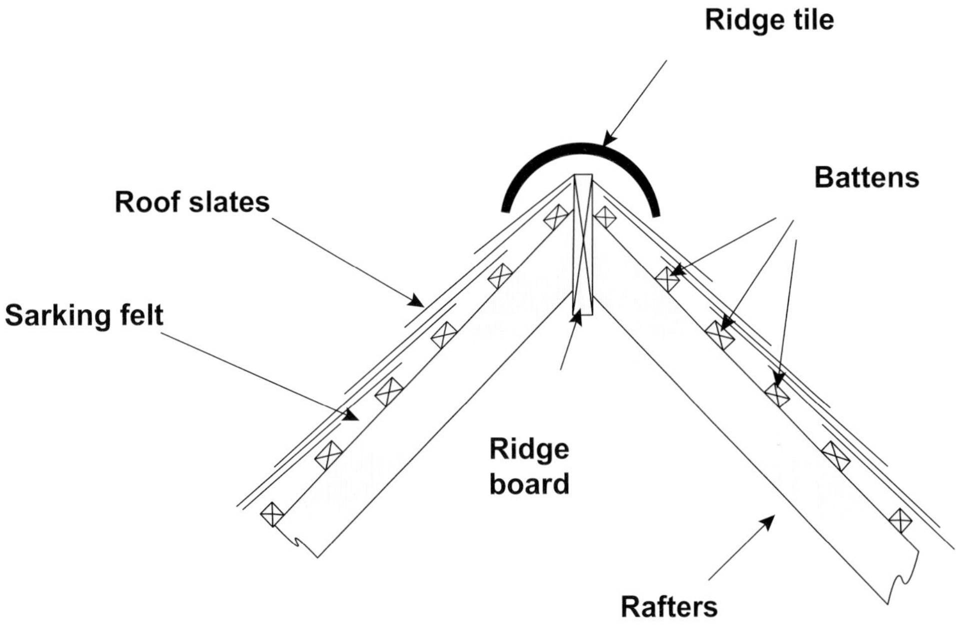
Ridge tile
Roof slates
Battens
Sarking felt
Ridge
board
Rafters

FIGURE 4.46
Basic roof terminology at ridge

Flat roofs

Flat roofs on domestic buildings are mainly of timber construction, while concrete flat roofs are used for industrial buildings.

Roofs with less than 10 degree slope are termed flat roofs. Most flat roofs have sufficient slope to prevent water standing on the roof surface.

Typical details of flat roofs are shown in Figures 4.47 and 4.48.

Where the flat roof meets the main building precautions are taken to prevent water penetrating the building below.

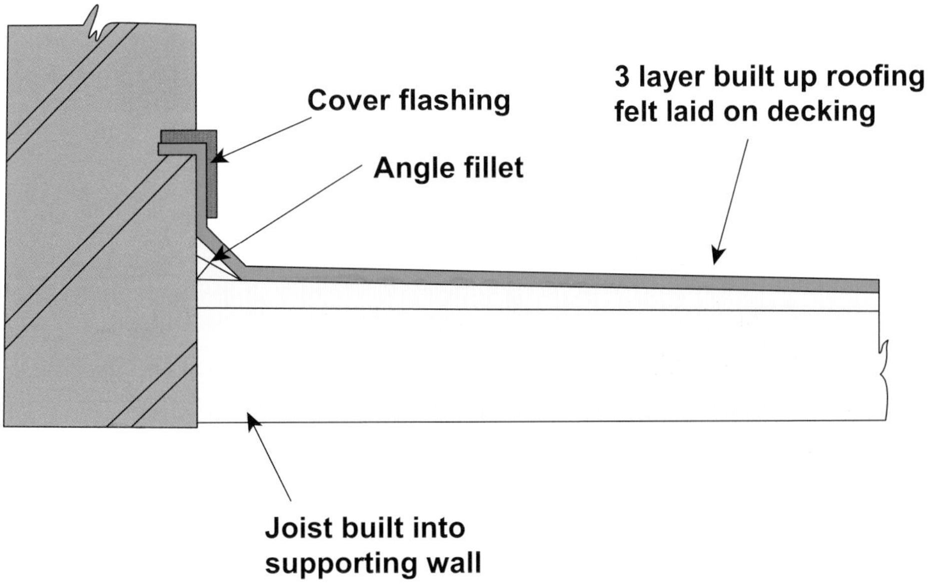
Cover flashing
3 layer built up roofing
felt laid on decking
Angle fillet
Joist built into
supporting wall

FIGURE 4.47
Junction of flat roof with main building

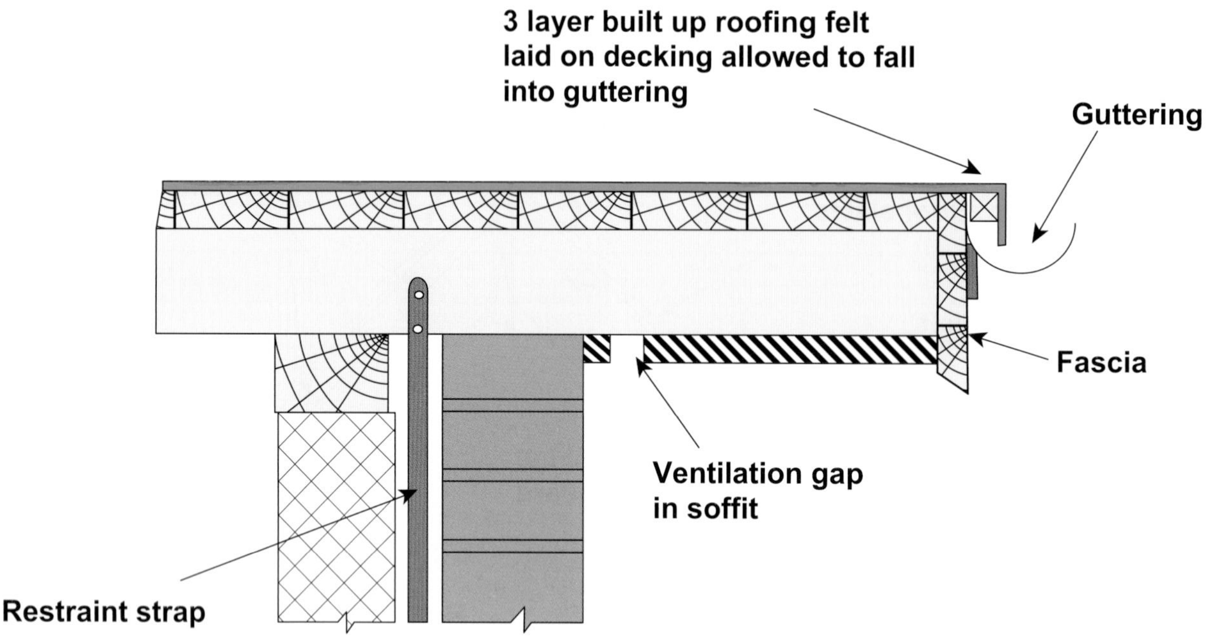
3 layer built up roofing felt
laid on decking allowed to fall
into guttering
Guttering
Fascia
Ventilation gap
in soffit
Restraint strap

FIGURE 4.48
Flat roof eaves detail

The roofing felt is turned into the brickwork and cover flashings are placed over it to complete the seal.

At the eaves the roofing felt is allowed to fall into the guttering.

Materials

A wide range of materials can be used internally and these can be either manufactured or natural.

Properties of materials

All materials used in construction must:

- be of suitable quality
- be suitable for their intended use
- be readily available at an acceptable cost
- comply with the current relevant British Standards.

Aggregates

Aggregates are granular materials that are bonded together with a binder to produce a solid mass when set.

Aggregates can be either gravel or crushed rock and are divided into fine and coarse according to the size.

By using different types of aggregate and varying the proportions of fine and coarse aggregates, various surface textures can be produced.

Cement

Cement is a mixture of clay, gypsum and either chalk or limestone.

The materials are burnt and ground into a fine powder. When water is added to cement a chemical reaction takes place, which results in the cement setting and hardening, producing great strength.

Cement is the most common binder for concrete and mortar.

There are several types of cement, but the most common is ordinary Portland cement.

Concrete

Concrete is an artificial material made from a mixture of aggregates, cement and water.

Concrete is a very versatile material and can be moulded to fit any shape. Its strength can be increased by inserting steel reinforcing bars.

Lime

Lime has been used very successfully for thousands of years as the only cementitious ingredient in mortars.

The lime used then was hydraulic, meaning that it was made from a naturally occurring chalk raw material which contains clay impurities. This clay content gave the lime a slow setting action not unlike the chemical hydration–setting–hardening process of ordinary Portland cement, but the lime mortar would take years to harden.

The raw materials used to produce building lime are either chalk or limestone. When chalk or limestone is burnt at a very high temperature it turns into quicklime. This material cannot be used for building work in this state so it undergoes treatment by the addition of water, in a process known as slaking. The end product is hydrated lime.

Mortars

It is not usually difficult to cause an argument between bricklayers – just ask them the following question: Does mortar stick bricks together or keep them apart? The answer is of course that it does both, but a lot more besides.

Mortar must stick firmly to bricks and blocks in external walling to keep the rain out. Mortar bed joints also hold bricks apart, so that the courses can be kept level and to an even vertical gauge of four courses to 300 mm, with standard metric bricks.

When walls were much thicker and Portland cement had not been invented, a mortar mix of lime and sand was used very successfully for thousands of years. The lime used then was hydraulic, as described above, and the lime mortar would take years to harden.

PURPOSE

Bricklaying mortar is the ideal material for getting bricks to rest firmly upon each other, whether these are accurately shaped class A clay engineering and calcium silicate bricks or more irregularly shaped hand-made bricks.

The mortar must remain soft enough for each brick to be pressed down to the line, before suction causes the bed to stiffen up. Not only does mortar accommodate irregularities, but it must stick firmly to each brick so as to stop rain from penetrating exposed joints.

Bricks

The study of bricks, from raw materials to delivery of finished products, is an extensive one.

Being able to recognize a brick when it appears on site, know its properties such as shape, size, weight, strength, porosity and colour – and therefore know how and where to use it correctly – is all important basic knowledge.

Brick making is a very skilful business, with many individual variations in methods of manufacture between companies and their factories.

British Standards specify a brick as a walling unit designed to be laid in mortar and not more than 337.5 mm long, 225.0 mm wide and 112.5 mm high, as distinct from a building block which is explained as a unit having one or more of these dimensions larger than those quoted for bricks.

Bricks, which are one of the most durable materials, can be described as building units which are easily handled with one hand.

There are numerous uses for bricks but the main ones are as units laid in mortar to form walls and piers and, increasingly, for brick paving.

Bricks were first made many thousands of years ago in hot climates, where a clay mixture was moulded and dried in the sun. It was found that if the clay mixture was heated to a high temperature, the bricks were much stronger.

The basic method of making bricks has not fundamentally changed.

VARIETY

Whether made from clay, sand and lime or concrete, bricks may be divided into four broad varieties: facings, commons, engineering bricks and refractories.

FACING BRICKS

These are made in a wide variety of colours and surface textures so as to be durable and attractive to look at.

COMMONS

These are for general-purpose walling that is most likely to be below ground level, externally rendered or internally plastered. They are not given particularly attractive surface features, but are hard and durable.

Common bricks have been largely displaced by lightweight building blocks for internal partition walls and the inner leaves of cavity walling.

ENGINEERING BRICKS

These are exceptionally hard, dense bricks which have a low porosity and therefore absorb very little water.

Engineering bricks are intended for walls that are heavily loaded, or very exposed to a risk of frost damage. They were originally developed by brick makers in Victorian times, in response to requests by civil engineers for a very strong brick for use in tunnels, bridges and viaducts.

REFRACTORIES

These bricks are made from specially selected clays that will withstand very high temperatures.

QUALITY

There are three qualities of brick:

- Ordinary – Ordinary quality bricks are durable enough to be used in the external face of a building. They can resist frost attack and there is no limit on their soluble salt content.
- Internal – Internal quality bricks need to be protected when used externally. There is no limit on their soluble salt content and they need not be frost resistant.
- Special – Special quality bricks must be durable enough to withstand harsh weather conditions, where they will be constantly soaked with water and attacked by frost. They must have a limited soluble salt content.

TYPE

There are five main types of brick (Figure 4.49):

- Solid – The volume of pores in a solid brick must not be greater than 25 per cent of the total volume of the brick.
- Frog – A frog is the depression formed in one or both bed faces of a brick. The volume of the frog should not be greater than 20 per cent of the total brick volume.

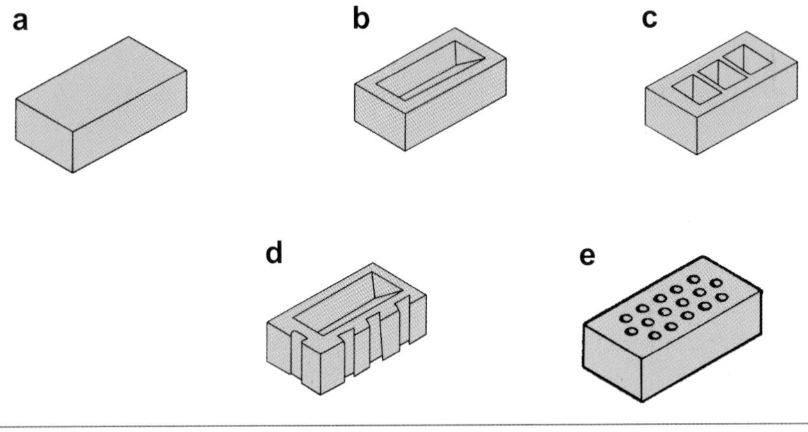
a
b
c
d
e

FIGURE 4.49
Types of brick: (a) solid; (b) frog; (c) hollow; (d) keyed; (e) perforated

- Hollow – The volume of the larger holes should be more than 25 per cent of the total brick volume.
- Keyed – Dovetail grooves in one header and one stretcher face provide a key for plaster or rendering
- Perforated – The volume of the small holes should be greater than 25 per cent of the total brick volume.

CALCIUM SILICATE BRICKS

More commonly called sand lime bricks on site, these are made from a carefully controlled mixture of 90 per cent fine silica sand plus 10 per cent lime.

This mixture is pressed into steel moulds and then steam hardened.

Coarse flint sand is also used to make flint lime bricks, which after steam curing have a higher compressive strength.

CONCRETE BRICKS

These are cast from a mixture of fine aggregate and Portland cement pressed into steel moulds. The natural setting and hardening process of cement determines the compressive strength of these bricks. They may be solid or frogged and are available as facings, commons and engineering bricks.

REFRACTORIES

More commonly referred to as firebricks, these can be considered as two sorts:

- very dense, cream-coloured, solid bricks used to contain the fire in furnace linings, cement kilns, ships' boilers and for lining steel ladles transporting molten metal in a steel works
- very lightweight bricks used to 'back up' dense refractories, or to insulate chimney shafts to prevent heat escaping from flue gases. (These are the only bricks that will float in water.)

A standard size dense firebrick of 230 × 114 × 76 mm is accompanied by a range of arch shapes, wedges and bullnose bricks made to suit circular section kilns and boilers.

Cellular blocks of this same dense refractory material are used to protect the steel deck of the brick-transporter cars used in tunnel kilns.

PURPOSE-MADE BRICKS

Many buildings have decorative features which require shaped or moulded bricks. Manufacturers keep a standard stock of bricks for this purpose, some of which are illustrated in Figure 4.50. If an unusual shape is desired, the manufacturers will make the bricks to the architect's specification.

The advantages of using a purpose-made brick are that labour is saved in cutting, waste is avoided and the natural surface is maintained.

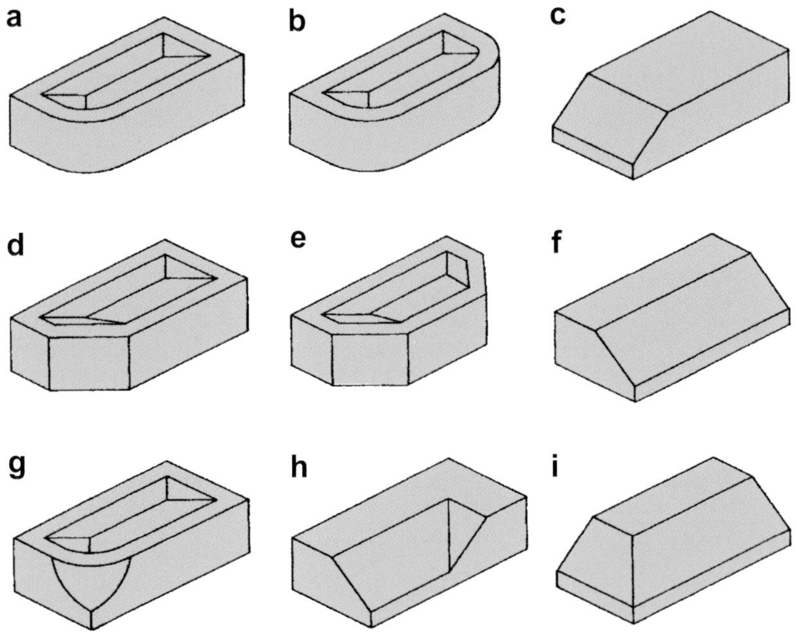
a
b
c
d
e
f
g
h
i

FIGURE 4.50
Types of purpose-made brick: (a) single bullnose; (b) double bullnose; (c) plinth header; (d) cant brick; (e) double cant; (f) plinth stretcher; (g) bullnose stop; (h) internal plinth return; (i) external plinth return

Blocks

A block is described as a walling unit exceeding the dimensions for bricks. Its height should not exceed either its length or six times its width.

Blocks are produced from clay and concrete.

CONCRETE BLOCKS

Concrete blocks are produced in a range of shapes and sizes.

The face side is usually 450 × 225 mm, the thickness varies from 37 to 225 mm and the weight varies from 6.3 to 15 kg.

They are produced in solid, hollow and multicut format. Multicuts enable a bolster cut to be made without wastage.

SPECIAL BLOCKS

Special blocks such as the return block are usually designed to stiffen walls where bonding could cause weakness.

For closing cavities the reveal block could be used, while another special is used to produce a splayed reveal.

Some manufacturers produce blocks with an insulant bonded to the outside face, while others produce hollow blocks with an insulant inserted in the voids.

Walling built with precast blocks may be divided into two main categories:

- load-bearing
- non-load-bearing.

LOAD-BEARING

These blocks are precast in moulds and compacted with the aid of vibration, or moulding machines involving the use of compressed air, or a combination of both.

These blocks are usually made of concrete comprised of Portland cement and a variety of aggregates, such as crushed stone, rock ballast or shingle.

NON-LOAD-BEARING

These can also be precast in moulds, or can be produced in slab format and cut to size when set.

These blocks are usually made with cement and a variety of lightweight materials, such as fly ash or burnt coke.

FOUNDATION BLOCKS

These are manufactured in widths from 250 to 335 mm. These blocks are used below ground level and are designed to support cavity walls. They may be dense or lightweight. The dense ones may require two people to bed them.

The current Building Regulations state that hollow blocks must have an aggregate volume of not less than 50 per cent of the total volume of the block calculated from its overall dimensions.

Hollow blocks must have a resistance to crushing of not less than $2.8\ N/mm^2$, if the blocks are to be used for the construction of a wall of a residential building having one or two storeys.

In all other circumstances blocks shall have a resistance to crushing of not less than $7\ N/mm^2$.

CLAY BLOCKS

Clay blocks are manufactured in a similar way to clay bricks, using the extrusion/wirecut method. These blocks are made from finely washed clay with certain special properties, which is forced through an extruding machine, in the process of which the blocks are cut off to length as the continuous length of clay emerges from the machine.

The thickness of the walls of the blocks is about 12 mm and this allows the blocks to dry quickly and thoroughly. The green clay blocks are then burnt at a high temperature.

The blocks are usually 300 mm long and 225 mm high and the thicknesses range from 37 to 100 mm for partition walls.

They are available with a smooth face and with dovetailed slots to provide a key for the plaster.

Clay blocks are not very easy to cut, so the manufacturers produce special half-units, enabling the bond to be formed without cutting any blocks.

Table 4.1 shows various types of block.

Table 4.1 Types of block

Type of block	Uses	Material	
Lightweight varieties	Internal non-load-bearing walls and partitions generally	Breeze or clinker; waste coke or ash and cement; burnt clay	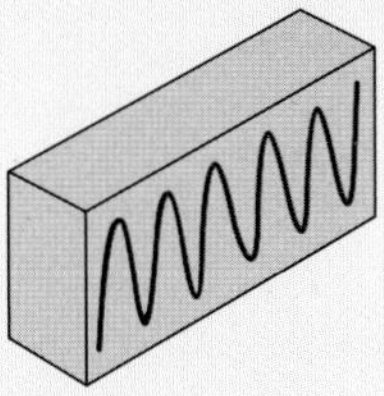
Dense heavy	Internal load-bearing walls and external walls	Usually concrete	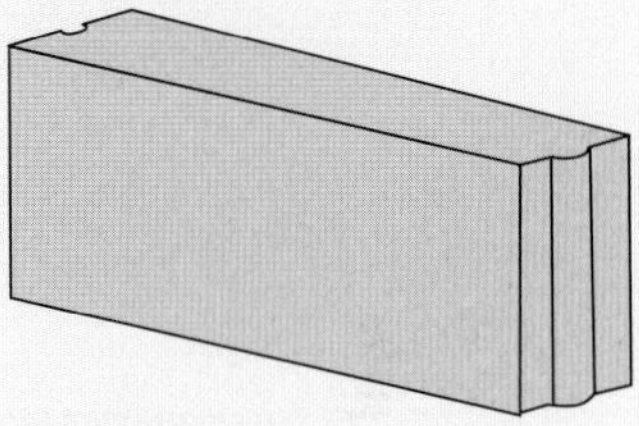
Hollow concrete	External walls, usually rendered	Concrete	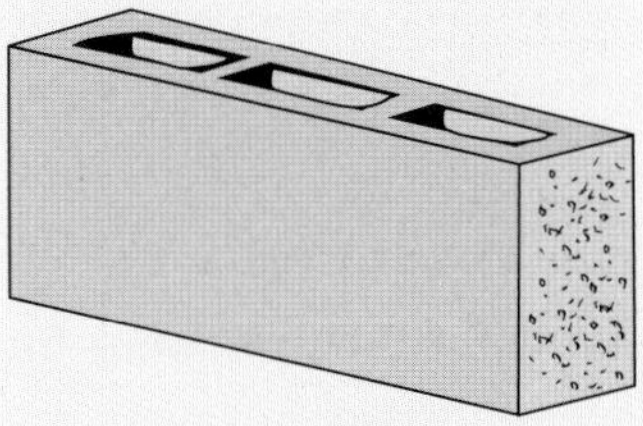
Cellular (lightweight)	Load-bearing internal walls	Burnt clay	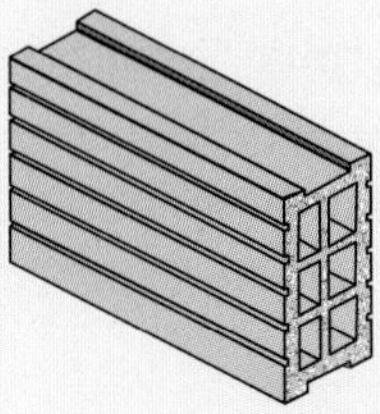

SPECIALS

Many block manufacturers provide specials cut to assist in the bonding on site and to prevent wastage from cutting (Figure 4.51).

Remember

Never use bricks and blocks in the same wall.

Stone

Stone is a natural material.

Natural stone and rocks are used in many construction products as walling, roofing, paving and floor tiling products.

Stones can be classified as:

- igneous
- sedimentary
- metamorphic.

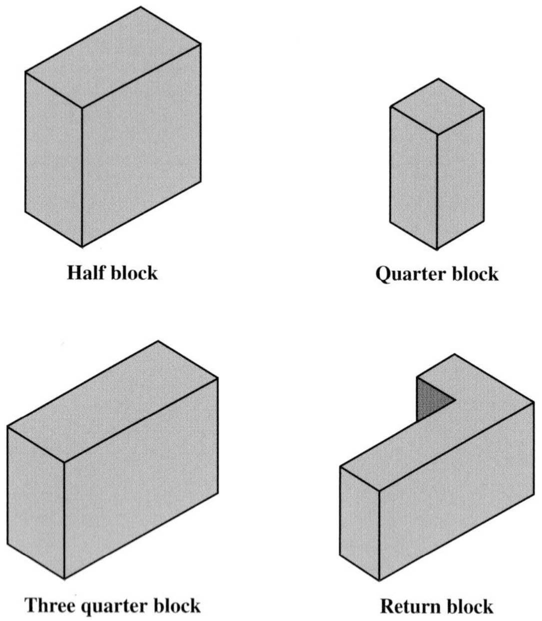
Half block
Quarter block
Three quarter block
Return block

FIGURE 4.51
Special blocks

Igneous stones were formed by the original solidification of the earth's crust. They produce a very dense material, with granite being the main type.

Sedimentary stones were formed by the settling process, when particles were deposited in layers and gradually became compacted to form the stone. Sandstone and limestone are the main types used in construction.

Metamorphic stones were formed from older stones that had been subjected to very high pressures and temperatures. This caused a structural change in the material. Marble, which was formed from limestone, is the main material in this category.

Timber

Timber is a natural material.

Although it has been in use since prehistoric times it is still a very valuable material in building. The great advantage of timber over almost all other materials is that, as trees are living, they provide a renewable source of building material.

Timber is used for many purposes in which its strength is a matter of critical importance.

Timber is low in weight, very durable and very easy to work. It is therefore one of the most versatile materials in the construction industry.

Historically, timber has been classified into:

- hardwoods
- softwoods.

There is little correlation between these terms and the actual hardness and softness of the timber.

SOFTWOODS

Softwoods are produced from coniferous trees, which are evergreens and grow chiefly in the northern temperature zones. They have cone-like leaves.

Softwoods comprise about 75 per cent of the timber used in the UK.

HARDWOODS

Hardwoods are produced from deciduous trees with broad leaves, although holly, certain oaks and the majority of tropical trees are evergreens. Their seeds are enclosed in some sort of shell.

Hardwoods include the densest, strongest and most durable timbers.

Some hardwoods contain resins and/or oils which interfere with the hardening of paints and many, such as teak and makore, include materials, e.g. silica, that make working difficult.

Hardwoods and softwoods are botanical terms and do not always relate to hardness. Thus, balsa is a hardwood, while yew, an extremely hard timber, is a softwood.

TIMBER PRODUCTS

Wood can be subject to dimensional change and distortion when used in its solid state and this often restricts the use of wood where wide or large areas have to be covered. This is the kind of work where 'manufactured' boards are mainly used.

Manufactured boards mean those sheet materials which for their greater part are composed of wood veneer, strips, particles or their combination.

Manufactured boards are available in a wide variety of sizes, shapes and surface finishes. They are easily cut and shaped to fit any area.

They fall into the following groups:

- plywoods:
  - veneer plywood
  - core plywood laminated boards
- particleboards:
  - chipboards
  - wood cement particleboards
- fibreboards:
  - insulation board
  - medium-density fibreboard (MDF)
  - hardboard.

PLYWOOD

The word plywood is usually taken to refer to sheets or boards that are made from three or more odd numbers of thin layers of wood, known as wood veneers. It is important that all veneers on each side of the core or centre veneer are balanced.

PARTICLEBOARDS

The main natural ingredients that make up particleboards are wood chippings, hence the common name of 'chipboard'.

FIBREBOARD

Fibre building boards are produced from wood that has been shredded into a fibrous state and then reassembled into a uniform sheet form.

The wood is first broken down into small chips, then steam treated to soften the lignin which binds the fibres together. Water is added (wet process) to produce a wood pulp. This is spread on to a slow-moving board-forming machine, where it is rolled out to a uniform thickness. (Most of the water has been removed before this operation.)

What follows will depend on the required density of the finished board.

Metals

Metals have to be extracted from metallic ores by smelting in a furnace. The metal obtained can then be formed into numerous shapes.

Metals are either ferrous or non-ferrous.

Ferrous metals include wrought iron, cast iron, mild steel and high-tensile steel. They will all corrode easily if not protected.

Non-ferrous metals do not contain iron, and include copper, lead and zinc. These materials corrode very slowly when exposed to the air. This process forms a protective layer which protects the metal from any further corrosion.

Mastics

Mastics are used in construction for numerous purposes, but mainly to fill gaps. A good mastic should not crack and should be water resistant. Most are applied by squeezing out of a tube with a special gun.

They are usually rubber, bitumen or plastic-based sealing compounds.

Plastics

Plastics are artificial materials made mainly from petroleum, and also coal. Plastic products are formed into their required shape while the material is still in its plastic shape.

Many building components are now produced from plastic, e.g. window and door units.

Plasters

Plaster is a finishing material mainly applied to walls and ceilings to provide a smooth, jointless and easily decorated surface.

Plaster was originally made from lime or cement but modern plasters are made from gypsum.

The normal plastering process consists of three coats, each having its own function:

- render coat – used to level off the background
- floating coat – used to produce a surface to receive the finishing coat
- finishing coat – provides a hard and smooth finish ready for decorating.

Paints

Paint is a very thin decorative finish applied to all surfaces in a liquid form.

It is applied by brush, roller or spray in three layers:

- primer – a protective coat against corrosion or moisture; provides a good base for the next coat
- undercoat – covers up the material and gives an even solid colour on which the gloss coat is applied
- gloss coat – the final colour with its required finish as gloss, matt or eggshell.

Glass

Glass is produced by melting a mixture of sand, soda ash, limestone and dolomite in a furnace. This produces molten glass, which when cooled produces a hard, clear product.

There are several uses for glass in the construction industry but the main one is for window panes.

Insulation

Numerous materials can be included into the design of buildings to restrict the passage of heat through the external walls of the property.

Basically building materials are poor insulators of heat. A good conductor of heat is still air and most insulators contain air traps to provide their insulating quality.

Many types of insulating materials are available, each having certain properties.

FIBREGLASS

Fibreglass consists of fine strands of glass woven together to form sheets or rolls. The sheet form can be used for insulation cavities and the roll form used for insulating roof spaces.

POLYSTYRENE

This is an expanded plastic resin produced in sheet form for use in cavity walls and loose granular fill for cavity walls or roof spaces.

ROCK FIBRE

This woven material is used for both cavity walls and roof spaces.

VERMICULITE

This is produced in a flaky form and sold in bags. It is used as a loose fill for roof spaces. It can also be added to concrete to produce a lightweight concrete with improved thermal insulation and fire-resisting properties.

Characteristics of materials

Each of the materials dealt with has its own characteristics, the most important of which are:

- compression
- density
- durability
- flexibility
- strength
- porosity
- conductivity
- expansion.

COMPRESSION

When materials are supporting loads they are mainly under compression, i.e. being squashed.

Concrete is very strong under compression, as are bricks.

Insulation materials are very poor under compression.

DENSITY

We are all aware that some materials are heavier than others.

A brick, for instance, is heavier than a slab of polystyrene insulation of the same size. This weight is known as its mass and is measured in kilograms (kg) or grams (g).

Density is the amount of substance in a material. The density of a material is found by dividing its mass by its volume, e.g. kg/m^3.

DURABILITY

This is a general term relating to the ability of a material to resist wear and tear. If a material takes a long time to wear away it is said to be durable.

The durability of certain materials will change depending on their position. For example, plasterboard has good durability inside a building but very poor durability on the exterior of a building.

Bricks and concrete have excellent durability.

FLEXIBILITY

Flexible materials can be bent easily without breaking.

Good examples of flexible materials are copper and lead.

A material that is not very flexible is glass in its formed state.

STRENGTH

All materials must be capable of safely supporting their own weight, dead loads and any superimposed loads, such as weather. These loads should not cause any distortion of the structure.

The strength of a material can change according to its physical properties. Some materials become weaker when damp.

When a material such as brick or concrete is being crushed this is known as being compressed or under compression. Conversely, when a material is being stretched it is said to be under tension.

Concrete is strong in compression but weak in tension. If concrete is being designed to be stronger under tension steel reinforcement bars are added, steel being strong in both tension and compression.

POROSITY

Many building materials contain water, which either was left there when the material was produced or has been drawn into the material by capillary action.

Timber has a certain amount of moisture in it when it is used, but this occurs naturally.

Concrete and bricks are examples of materials with water left in them from their production.

As this water dries out it leaves pockets of air or voids. Materials with such voids are said to be porous. When they come into contact with water they absorb it by capillarity. Capillarity is the action of the water being sucked into the material along the small voids.

This could cause permanent damage to external materials such as bricks. For example, the face could be blown off.

CONDUCTIVITY

Buildings are required to be kept at a reasonable temperature so that the occupants feel comfortable.

Heat will flow out of a building whenever the external temperature is less than the internal temperature.

Heat flows through solid materials by thermal conductivity. This is a measure of how good a material is at conducting heat.

The more air voids in a material the better the insulator, as with fibreglass quilt.

EXPANSION

Materials generally expand when heated and contract when cooled.

Large expanses of brickwork could crack under expansion, so expansion joints are built into the walls to counteract it. Concrete suffers from the same problem and again expansion joints are built in.

The actual amount of expansion varies according to the type of material and the thermal conditions. All materials have a very small amount of expansion.

Plastic guttering provides a perfect example of movement when the weather becomes warmer. You can hear the material making a noise as it moves across the brackets.

If correctly designed a building should not become dangerous due to expansion of materials.

Building defects

If a building causes problems after it has been completed it can be very costly to the client.

Poor design accounts for 7 per cent of the work of the construction industry, working to correct these faults.

Defects in the structure of the building can be caused by:

- poor design
- external effects.

Poor design

Poor design could be attributed to the wrong use of a material, poor supervision of the work or the use of substandard materials.

If materials are used in situations where they are not designed to be used, they will not last as intended. A typical example is common bricks being used in an inspection chamber when engineering-quality bricks should be used.

Poor supervision could lead to the jointing of face brickwork not being carried out correctly and dampness being allowed to penetrate the wall.

Poor storage of facing bricks could result in efflorescence at a later date.

If poorer materials are used instead of the specified quality they will fail to reach their required standard.

External effects

External effects could be attributed to weather conditions, poor supervision and substandard materials being used.

Dampness causes rapid deterioration of materials. When materials become damp inside a building this could lead to other problems such as fungal decay, insect attack, mould growth and efflorescence.

Rain could penetrate a building through cracks, joints and poor application of weatherings.

Lack of control by the site supervisor could also lead to problems with poorly fitted components and mastic seals.

DAMPNESS

It is important that the bricklayer understands the causes of dampness and the methods of damp prevention in the construction of a brick wall.

Dampness may occur even after precautions have been taken, if precautions are based on insufficient knowledge resulting in lack of care in the application of damp prevention.

Many problems with damp are cause by poor workmanship.

Figure 4.52 shows a dirty cavity with mortar droppings on the wall tie, which create a path by which dampness can enter the building.

Figure 4.53 shows garden soil above the DPC level and mortar droppings in the cavity. These form a bridge for dampness to enter the building.

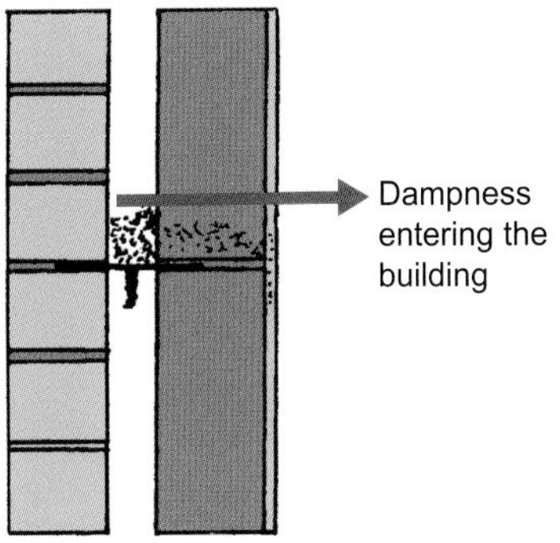
Dampness entering the building

FIGURE 4.52
Mortar droppings on the wall tie causing a bridge

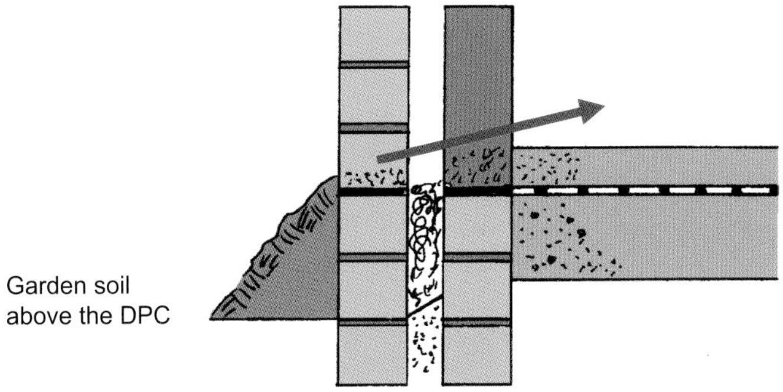
Garden soil above the DPC

FIGURE 4.53
Garden soil above the DPC level and mortar dropping in the cavity

MOISTURE

Building materials that are porous will deteriorate when wet.

This can cause defects such as swelling, warping and twisting, which cause a reduction in strength and could eventually lead to collapse.

All building materials should be correctly stored before use and protected until the building is watertight.

CHEMICAL ATTACK

When building materials are in contact with the atmosphere they may become corroded.

The atmosphere contains fine particles of dust that pollute the air, especially near industrial areas. These become diluted when mixed with rainwater and can attack the surfaces of unprotected building materials.

The corrosion of metals through rusting and the spalling of stone and brick are the main causes of chemical attack.

FUNGAL ATTACK

The most common fungal attacks occur on timber products.

There are various forms of fungus which, given suitable conditions, will attack timber until it is destroyed. These can broadly be classified into two groups:

- wet rot
- dry rot.

WET ROT

Wet rot is dark in colour and needs a source of moisture, otherwise it will die.

The affected timber is white or brown in colour and is cracked, although much less so than timber affected by dry rot. The timber becomes brittle, loses its strength and colour, and crumbles easily.

Wet rot is found in places that are constantly damp, e.g. wet timbers in leaky roofs, damp cellars, window frames, around water tanks and under ground-level floors.

Timber affected by wet rot should be cut out and replaced and the surrounding timber treated.

DRY ROT

This is the most common and serious fungus. Dry rot is off-white in colour, fluffy like cotton wool, and forms a continuous sheet over the timber. Once formed, dry rot produces its own moisture.

Timber that has been attacked by dry rot darkens in colour and forms cracks along and against the grain.

The timber becomes less dense, becomes brittle and loses its strength. It crumbles easily into a powder.

Dry rot is often found in unventilated places where humidity is greater than 20 per cent; for example, in cellars and cupboards, under floors and behind panelling.

Timber affected by dry rot must be replaced and the surrounding timber treated.

INSECT ATTACK

Most insect attacks on timber are by beetles.

A typical life cycle of a beetle occurs in four stages:

- Egg – The female beetle lays her eggs on the surface of bare timber, usually in crevices or cracks.
- Larva – The eggs hatch into larvae (grubs), which bore into and feed on the timber, leaving unwanted bore dust behind.
- Pupa – The larvae change into pupae near the surface of the timber and, during this dormant stage, develop into adult beetles.
- Adult beetle – The adult beetles emerge from the timber through a small hole known as the flight hole and fly away to mate.

COMMON FURNITURE BEETLE

The common furniture beetle is responsible for most damage caused in the UK. Commonly known as woodworm, this beetle attacks most softwoods and the sapwood of hardwoods.

It is most often found in furniture and joinery. The eggs are oval in shape and the flight holes circular with a diameter of 1.5–2 mm.

POWDER POST BEETLE

The powder post beetle attacks hardwoods and is found particularly in timber yards and sawmills.

The eggs are laid in the large pores of hardwoods and the circular flight holes are about 2 mm in diameter.

HOUSE LONGHORN BEETLE

This beetle attacks the sapwood of softwoods only and is mainly found in structural timbers. Extensive damage can be caused without its being noticed because of the long larval stage and very few flight holes. The flight holes are oval and 5–10 mm in diameter.

DEATH-WATCH BEETLE

The death-watch beetle attacks hardwoods, usually timbers of old buildings where decay has started. The flight holes are circular and about 3 mm in diameter.

PROTECTION

WOOD PRESERVATION AND PROTECTION

Except where timber and wood products are used in their natural state, for environmental, practical or economical reasons, they are generally treated with:

- paint
- water-repellent stain
- wood preservatives
- special solutions to reduce or retard the decay process.

Material records

It is generally accepted that the initial selection of materials plays a major part in achieving a satisfactory building. Unfortunately, it is less recognized that the methods adopted on site for the handling and storing these materials are equally important.

Manufacturers and suppliers go to great lengths to provide materials and components that comply with recognized standards, but much of their effort can be eroded if the same importance is not attached to the product when it is transferred to the care of the construction team.

The cost of materials used in the construction industry exceeds 50 per cent of the overall costs. The supervisor has little or even no control over the initial price of materials and every contractor usually has to pay approximately the same price. However, by employing strict methods of control and checking, efficient handling and correct use of the materials, the most efficient contractor will expend the least cost in their use.

Principles of materials handling and control

Materials handling and control can be divided into the following sectors:

- planning material deliveries
- ordering the materials
- checking deliveries on site
- storing materials on site
- processing delivery tickets and invoices
- controlling the use of materials.

PLANNING MATERIAL DELIVERIES

Planning of deliveries usually takes place in the form of charts, graphs, data sheets, diagrams or the written word. It is essential to have a record of the proposed delivery so that reference can be made to it in the future.

However, it must be stressed that these only give a guide and prompt the site manager. Adjustments should be made according to site requirements.

ORDERING THE MATERIALS

The majority of orders for the supply of building materials and components are placed by the main contractor. The main contractor is solely responsible for correct placing of material and component orders, ensuring delivery and making payment for said goods.

Before any order is placed a quotation should be obtained. In a competitive market it is wise to obtain a number of quotations for the supply of a particular item. In this way variations to the price range, delivery dates, conditions and general service can be compared between suppliers, and after careful consideration a quotation may be accepted.

When a supplier is asked to provide a quotation for a particular material or component the main contractor must provide the supplier with as much information as possible to determine the exact specification of the items.

This information can be given by any or all of the following:

- extracts from the bill of quantities
- extracts from the specification
- reference to British Standards
- detailed sketch drawings
- reference to manufacturers' literature or catalogues.

In order to provide a true price the supplier must know when the material is required. When a comprehensive picture has been built up the supplier can prepare a quotation based on reality and not an assumption.

CONTRACTOR'S ORDERS

When the quotation has been accepted the main contractor will place an order with the supplier similar to the one shown in Figure 4.54. This should include the company heading, contract name and delivery address. It should also give a full description of the materials as described in the contract documents.

CHECKING DELIVERIES ON SITE

When the materials are delivered to the site they should be accompanied by a delivery ticket. This ticket should be checked against the quantity and type of material. You should also check for any damage and inform your supervisor immediately. A typical delivery ticket is shown in Figure 4.55.

A material schedule sheet will have been completed before the site was set up, and gives the site manager the information required for controlling the materials on site.

If materials arrive on site very early they are more prone to damage or pilfering.

MATERIALS ORDER | **WEST BUILDERS LTD**

Date: ....

Deliver: ....

Contract ....

Reference Number	Description of materials	Quantity	Rate	Price

FIGURE 4.54
Order sheet

DELIVERY NOTE | **ABC BUILDING MATERIALS SUPPLIERS**

Your Ref: ....

Our Ref: ....

Order No: ....

Invoice address: ....
....
....
....

Date: ....

Delivery address : .
....
....

Description of goods	Quantity	Catalogue No

Comments:

Date and time of receiving the goods: ....

Name of recipient: .... **Signature:** ....

FIGURE 4.55
Delivery note

STORING MATERIALS ON SITE

The storage of materials must be considered during early planning.

Depending on the availability of space and the size of the contract, properly fitted-out stores, with racks and bins of varying sizes into which materials can be placed for ready issue and easy handling, should be available under the charge of a responsible, competent storeperson.

The stores should lie within an enclosed stores compound. The compound should have adequate space and provision on which to place materials that cannot be accommodated within the stores and are not affected by the weather. These stores must be kept to a minimum for security reasons.

It is important that stores are strictly controlled and only authorized employees allowed to sign for receipt. The stores should have a receiving bay and entrance door, and an entirely separate entrance for the issuing of stores, complete with a counter and register for outgoing materials. This register should indicate for which section or part of work the materials are required, and under whose authority they are issued.

Tarpaulins, plastic sheeting or protective coverings should always be available for issue.

PROCESSING DELIVERY TICKETS AND INVOICES

The site manager should receive some notification of impending deliveries. This could be an advice note from when the order was originally made.

The materials should be accompanied by a delivery note, presented by the driver as proof of delivery, quality and quantity. Meticulous checks should be made of the materials delivered for quality and quantity. Delivery notes should not be signed until the recipient is satisfied of the quality and quantity of the goods.

Each individual delivery ticket should be sent to the head office and a record of the delivery kept on site on the sheet shown in Figure 4.56.

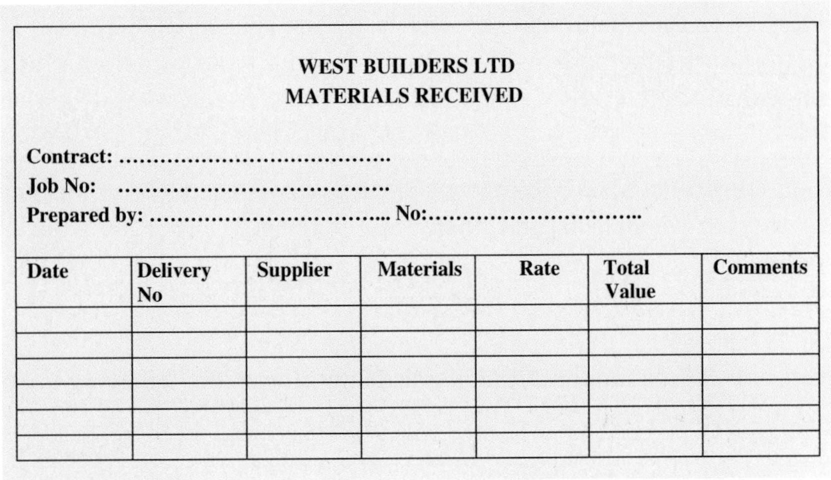
WEST BUILDERS LTD
MATERIALS RECEIVED

Contract: ....

Job No: ....

Prepared by: .... **No:**..............................

Date	Delivery No	Supplier	Materials	Rate	Total Value	Comments

FIGURE 4.56
Materials received sheet

ABC BUILDING MATERIALS SUPPLIERS

INVOICE

No: ....

Messrs: ....

Invoice date	Your order No	Dated	Sent	Code
....	....	....	....	....

Quantity	Description	Price	Items	Total

No claim can be entertained unless made three days after receipt of goods.

FIGURE 4.57
Typical invoice

Once the material has been correctly delivered the supplier will send an invoice to the building company requesting payment. A typical invoice is shown in Figure 4.57.

Material transfers

When materials need to be transferred between sites, correct documentation should be completed (Figure 4.58).

CONTROLLING THE USE OF MATERIALS

Careless off-loading and handling of materials, goods and components adds to wastage on the construction site. Each site manager should pay adequate attention to the methods used depending on the types of material being dealt with.

Secure accommodation together with an effective storage system is essential on construction sites to prevent unnecessary waste and loss of materials due to vandalism and theft, and damage caused by constant handling.

STORAGE CONTROL

Irrespective of the amount of materials being stored, the following factors need to be considered.

Deliveries

Materials received need to be checked against suppliers' delivery advice notes for correct quantities and damage.

WEST BUILDERS LTD
MATERIALS RECEIVED

No: ....
Date: ....

From site:	**To site:**
Description	**Quantity**

Issued by:	**Received by:**	**Driver's signature**

FIGURE 4.58
Materials transfer sheet

Stock rotation
Stocks of some materials (plaster, cement, paint, etc.) need to be used in the order of delivery as they have only a limited shelf-life (Figure 4.59).

Withdrawals
Materials drawn from stock by requisition need to be replaced.

Security
Storage facilities must be capable of being made secure against vandalism and theft.

Cleanliness
The store and unloading areas must be kept clean and free of debris at all times.

Safety
Materials must be handled and stacked in such a manner that injuries and damage are avoided. Hazardous materials need to be stored in separate accommodation constructed of non-flammable materials.

Minimizing waste
It is recognized that wastage of materials will occur on a construction site for many reasons; therefore estimators, when pricing for work, allow within their rates a wastage percentage ranging from as little as 1 per cent to an alarming 15 per cent.

On different construction sites, the waste of materials for similar operations varies considerably. This difference is without doubt caused by differences in the control exercised by the site supervisor.

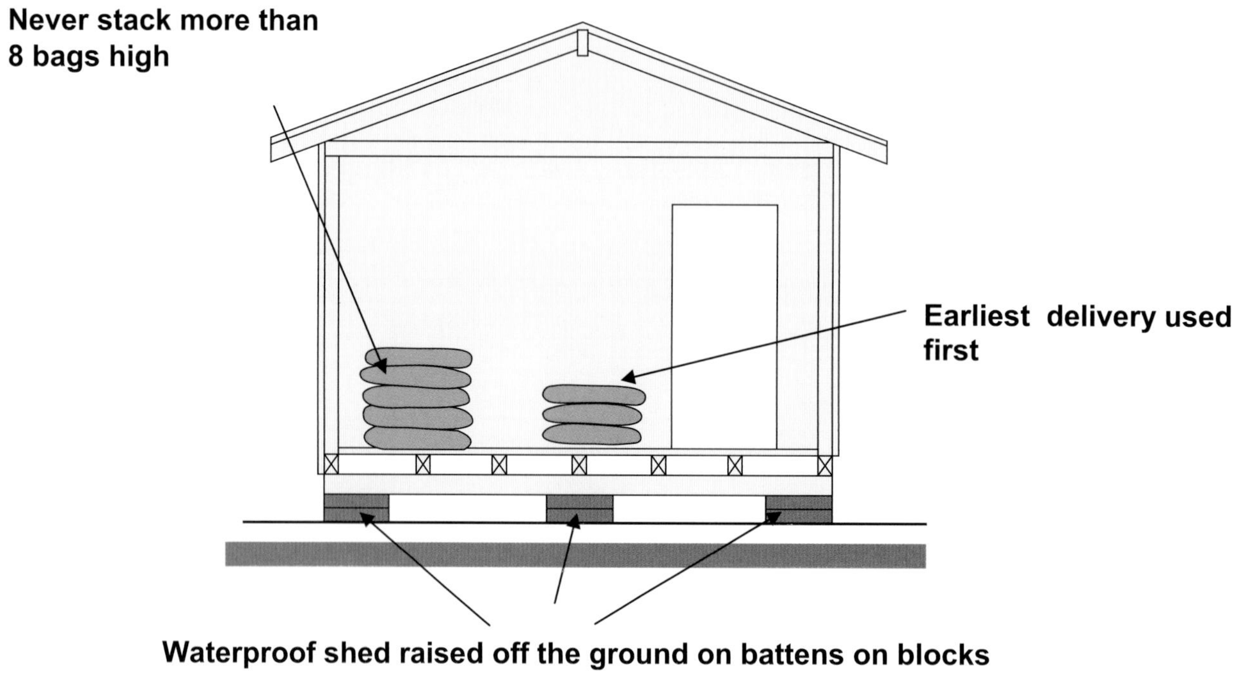
Never stack more than
8 bags high
Earliest delivery used
first
Waterproof shed raised off the ground on battens on blocks

FIGURE 4.59
Stock rotation of bagged materials

Wastage of expensive materials, in particular those in short supply, can make the difference between profit and loss.

Site management must, therefore, pay particular attention to the prevention of waste by instituting regular inspections and careful assessments of material requirements.

With timber and reinforcement the preparation of cutting lists before actual ordering takes place, combined with correct storage, can avoid unnecessary wastage.

With the growth in the industry of labour-only subcontractors, wastage should be particularly controlled. If, in the early stages of the work, it is found that wastage is taking place owing to carelessness on the part of the subcontractors, a quick reminder of a 'contract charge' is most effective.

Strict stores control can often pinpoint at an early stage who is responsible for calling up materials and then leaving them around the site.

Multiple-choice questions

Self-assessment

This section of the book is designed to allow you to check your level of knowledge. The section consists of revision questions for this chapter. The questions are all multiple choice and have four possible answers. The answers are to be found at the end of the book.

The main type of multiple-choice question will be the four-option multiple-choice question. This will consist of a question or statement, known as the stem, followed by a choice of four different answers, called the responses. Only one of these responses is the correct answer; the others are incorrect and are known as distracters.

You should attempt to answer the questions by choosing either (a), (b), (c) or (d).

Example

The person employed by the local authority to ensure that the Building Regulations are observed is called the:

(a) clerk of works

(b) building control officer

(c) council inspector

(d) safety officer

The correct answer is the building control officer, and therefore (b) would be the correct response.

Construction technology

Question 1 Which of the following should you never build in front of?

(a) the frontage line

(b) the brick line

(c) the setting-out line

(d) the building line

Question 2 Which of the following machines is best suited for trench excavations?

(a) backactor

(b) front shovel

(c) drott

(d) scraper

Question 3 The type of foundation shown is known as a:

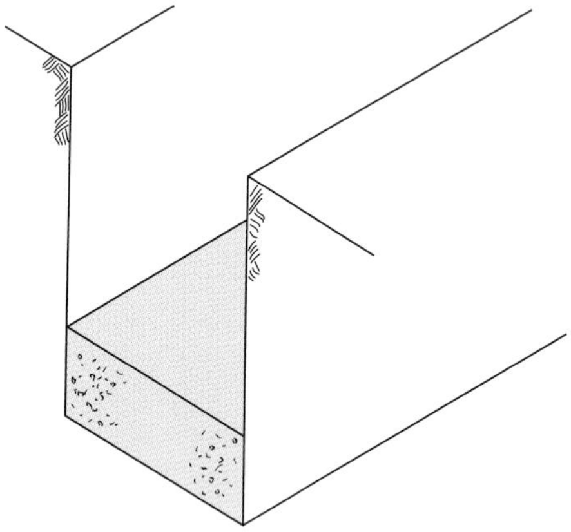

(a) raft foundation
(b) strip foundation
(c) pad foundation
(d) pile foundation

Question 4 Identify the type of insulation shown:

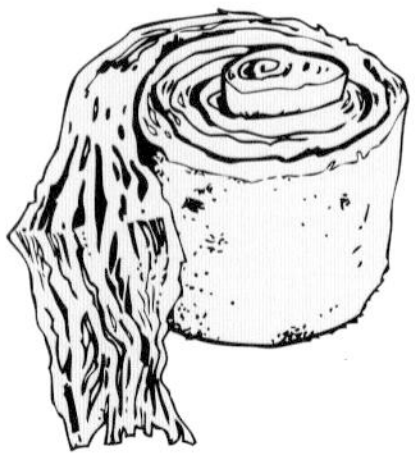

(a) expanded polystyrene
(b) fibreglass
(c) polyurethane foam
(d) vermiculite

Question 5 The type of roof form shown is known as a:

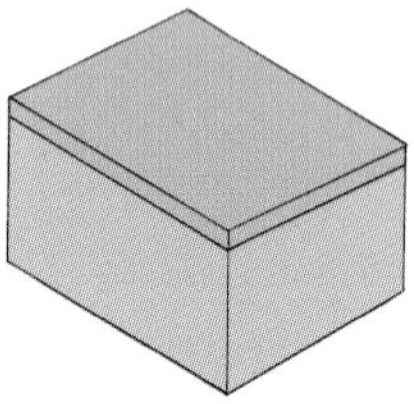

(a) flat roof
(b) lean-to roof
(c) gabled roof
(d) hipped roof

Question 6 Identify the type of brick shown:

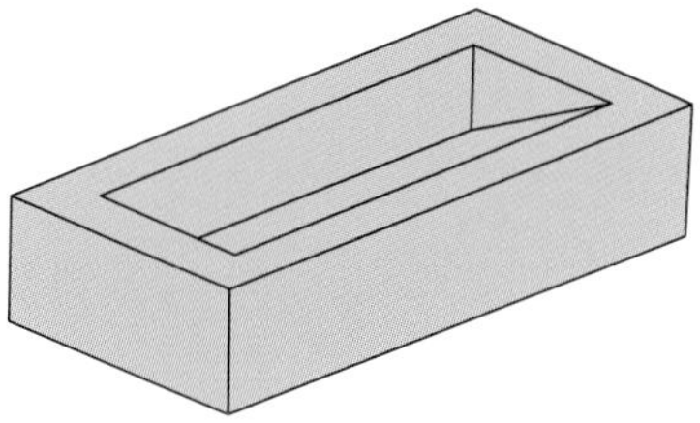

(a) engineering brick
(b) frogged brick
(c) solid brick
(d) perforated brick

Question 7 Manufactured boards made from three or more layers of wood are known as:

(a) fibreboard
(b) insulation board
(c) particleboard
(d) plywood

Question 8 The most common fungal attack on timbers is known as:

(a) woodworm
(b) efflorescence
(c) dry rot
(d) mould growth

CHAPTER 5

Moving and Handling Resources

This chapter will cover the following NVQ and Diploma units:

- NVQ VR03
- CC 2001K

This chapter is about:

- Following instructions
- Adopting safe and healthy working practices
- Selecting materials, components and equipment
- Handling, moving and storage of materials and components by manual procedures and lifting aids

The following NVQ performance criteria will be covered:

- Performance criterion 1: Comply with product information
- Performance criterion 2: Selection of resources
- Performance criterion 3: Minimize the risk of damage
- Performance criterion 4: Safe work practices

The following Diploma outcome will be covered:

- Know about safe handling of materials and equipment

Complying with product information

Material characteristics

Construction materials can be classified into the following groups:

- bulk
- hazardous
- fragile or perishable
- miscellaneous.

BULK MATERIAL

Bulk means a large amount of one type of material, e.g. bricks, sand and aggregates.

They are delivered to site in large amounts to save on delivery charges. Some material suppliers deliver loose bulk material in bags.

The following are examples of bulk materials (Figure 5.1):

- bricks, blocks, roof tiles, concrete units, drainage pipes
- aggregates – sand, shingle and granolithic chippings
- timber – hardwoods and softwoods.

HAZARDOUS MATERIALS

These materials usually have instructions written on their containers or packages explaining the handling and storing requirements.

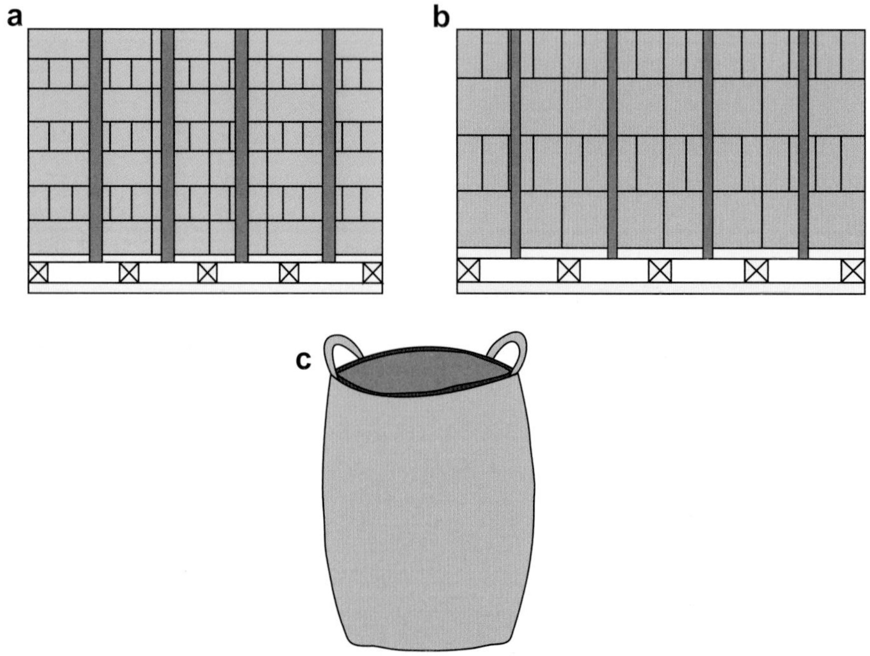
a
b
c

FIGURE 5.1
Various bulk materials: (a) pallet of bricks; (b) pallet of blocks; (c) bagged material

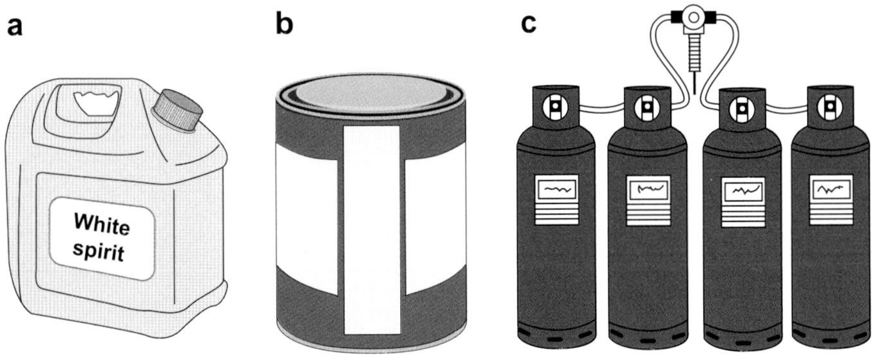
a
b
c
White spirit

FIGURE 5.2
Various hazardous materials: (a) white spirit; (b) paint; (c) gas bottles

Always ensure you read the instructions, and wear the appropriate personal protective equipment (PPE) before handling hazardous materials.

The following are examples of hazardous materials (Figure 5.2):

- paints and solvents
- mastics
- cleaning agents
- industrial gases – butane and oxyacetylene
- highly flammable liquids.

FRAGILE OR PERISHABLE MATERIALS

Many of the materials and components delivered to building sites are durable and can be stored out in the open; for example, sand, aggregates, bricks and blocks.

Other items are perishable and fragile and require special attention when handling and storing.

Durable means that a type of material or component is hard wearing and can withstand extreme weather conditions when used for its intended purpose.

Perishable items such as cement require special storage procedures to ensure the shelf-life of the material.

The following are examples of fragile or perishable materials (Figure 5.3):

- sheet materials – plywood, blockboard, chipboard, plasterboard
- glass – vitreous china – WC pans, basins, etc.
- glazed tiles
- roofing felts – damp-proof course (DPC) materials, pitched fibre and polythene products.

Several of these materials need to be unloaded, transported and stored under dry, secure conditions on building sites to prevent deterioration and loss.

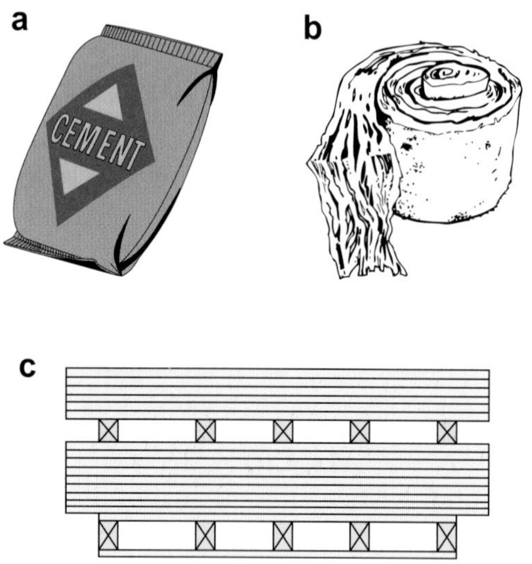
a
CEMENT
b
c

FIGURE 5.3
Fragile or perishable materials: (a) cement; (b) fibreglass; (c) plasterboard

MISCELLANEOUS

The following are examples of miscellaneous materials (Figure 5.4):

- ironmongery – nails, screws, hinges, door furniture
- doors – hardwood and softwood
- units – kitchen and bathroom
- windows – wood, metal and PVC.

All the above materials need to be stored under dry, secure conditions on building sites to prevent deterioration and loss.

This will be dealt with later in the chapter.

Material identification and limitations of use

AGGREGATES

Aggregates are particles or granules of materials that are used with a binder and water in controlled proportions to produce a solid mass when set. By the use of differing aggregates and binders, various types of materials can be produced.

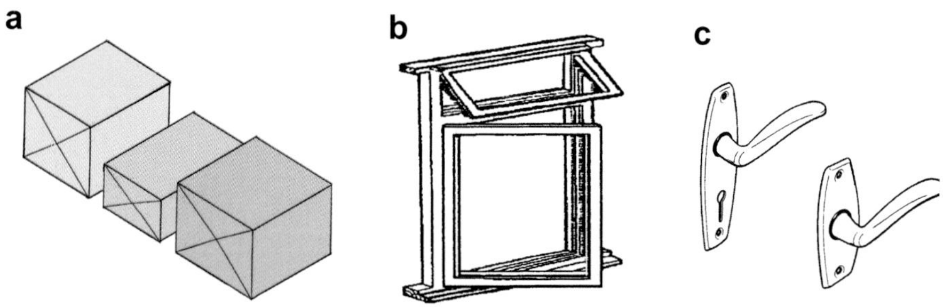
a
b
c

FIGURE 5.4
Miscellaneous materials: (a) packages; (b) window frame; (c) ironmongery

Aggregates can be divided into two further categories.

- fine aggregate or sand – used for making cement mortar
- coarse aggregate – used in the production of concrete.

Fine aggregate or sand has very small particles. It originates from either quarries or river beds.

Aggregate may arrive on site in lorries for large quantities or bagged for small quantities (Figure 5.5). It is also available in various colours.

Coarse aggregate is crushed rock or pebbles which originate from quarrying stone and are then crushed to achieve the required size of aggregate.

- Sand
  - soft sand – used in mixing mortar for bricklaying
  - sharp sand – mixed with ballast/cement to form concrete.
- Shingle
  - pea-gravel – used as a fine granular filling material, e.g. around drain pipes
  - premixed sand/shingle – sharp sand/shingle used for floor screeding and fine concreting
  - ballast – larger aggregates used with sand for mass concreting, i.e. foundations.
- Granolithic chippings – used where hard-wearing textured surface is required, e.g. garage floors, door thresholds.
- Hardcore – usually graded limestone, but broken or crushed clean brick or concrete could also be used.

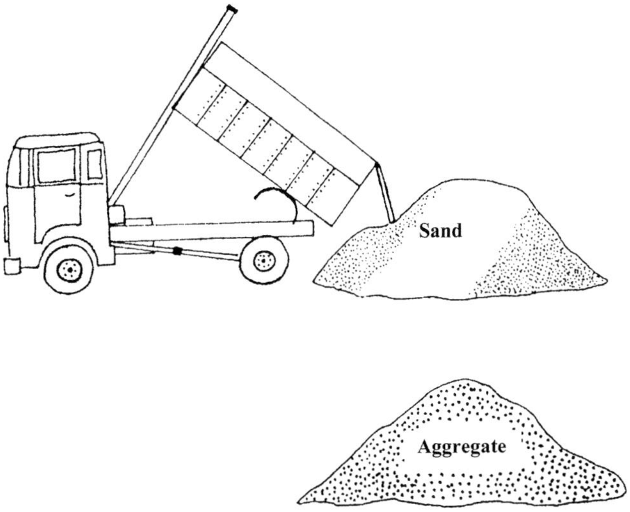
Sand
Aggregate

FIGURE 5.5
Bulk materials arriving on site by lorry

BAGGED MATERIALS

Bagged materials arrive on site in specially prepared bags. As well as taking care not to damage them when unloading, the storage is of special importance owing to their excessive deterioration through dampness.

PORTLAND CEMENT

Produced in various grades, including: ordinary, rapid hardening, super sulphate and masonry cement. They arrive on site in 25 kg bags. The bags are produced from several layers of light brown paper, one of which has been treated to prevent moisture reaching the dry cement powder.

LIME

Manufactured from limestone and used for bricklaying and plastering.

Lime arrives on site in the same way as cement and is also in bags, but white in colour.

Pallets are used on site to prevent rising damp attacking the dry powder of the lime and cement and causing it to go hard and unusable (Figure 5.6).

Plasters also arrive on site in bags and should be stored in the same way.

TIMBER MATERIALS

Timber – sheeting materials such as plywood and blockboards could be delivered either in packs or individually. Carcassing timber could also arrive either in packs or individually (Figure 5.7).

Timber is a natural material.

Timber is used for many purposes in which its strength is a matter of critical importance.

Historically, timber has been classified into:

- hardwoods
- softwoods.

There is little correlation between these terms and the actual hardness and softness of the timber.

SOFTWOODS

Softwoods are produced from coniferous trees, which are evergreens and grow chiefly in the northern temperature zones. They have cone-like leaves.

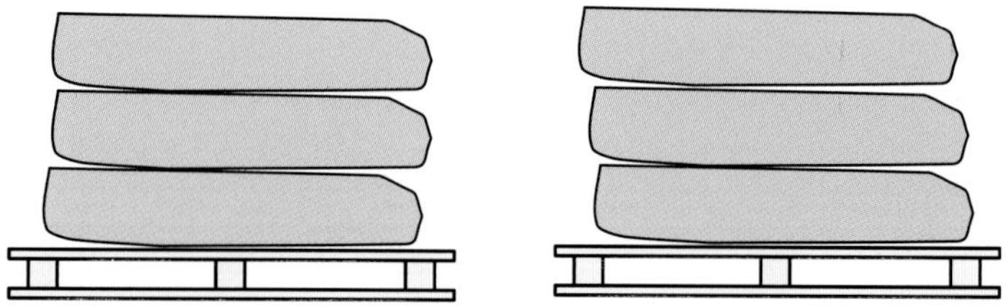

FIGURE 5.6
Bagged material on pallets

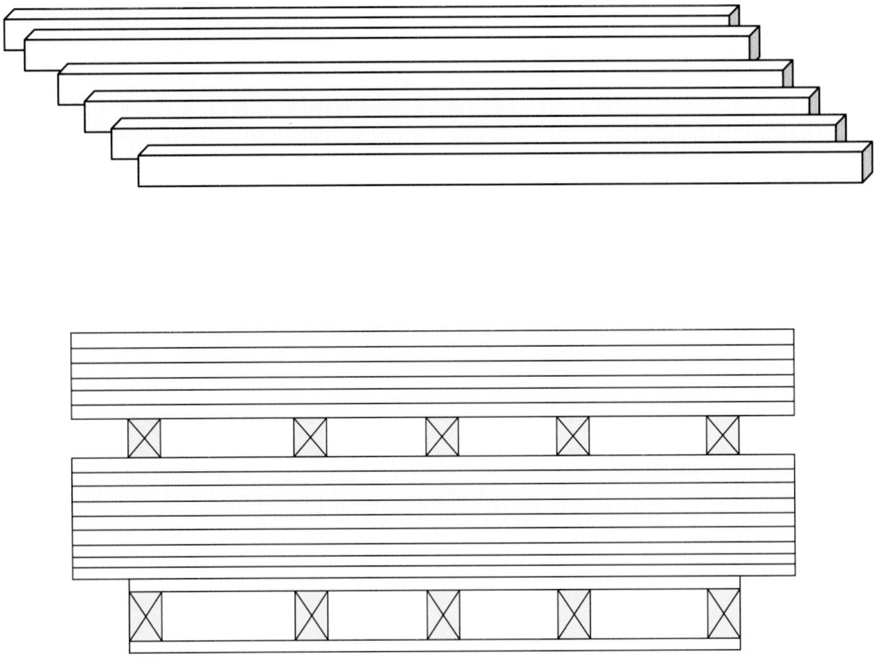

FIGURE 5.7
Timber delivered loose and in packs

HARDWOODS

Hardwoods are produced from deciduous trees with broad leaves, although holly, certain oaks and the majority of tropical trees are evergreens.

Hardwoods and softwoods are botanical terms and do not always relate to hardness. Thus, balsa is a hardwood, while yew, an extremely hard timber, is a softwood.

SHEET MATERIALS

BOARDS

In its natural form timber is one of the most useful and versatile of materials used on the construction site. It has, however, certain inherent shortcomings that tend to limit its usefulness in its solid form, i.e. as planks, boards, etc.

These shortcomings include the following:

- Width of boards – In its natural form a board is limited to the log from which it is cut.
- Stability – Timber, especially in large sizes, is likely to warp, shrink or swell.
- Strength – Timber is weak across its grain.

Artificial boards attempt to overcome these shortfalls.

PLYWOOD

Plywood is available in large, standard size sheets in a range of thicknesses from 1 mm upwards (Figure 5.8).

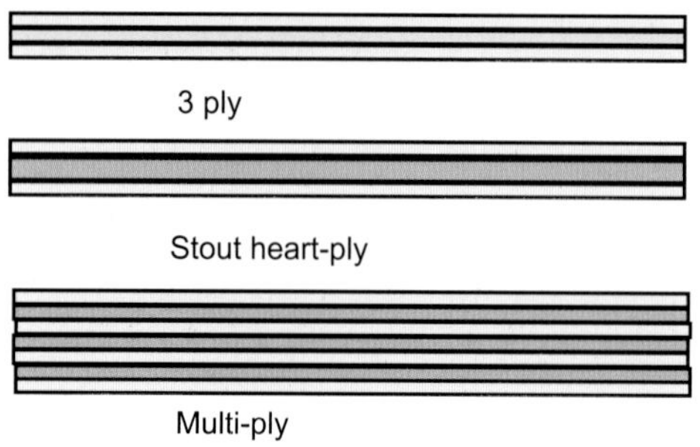
3 ply
Stout heart-ply
Multi-ply

FIGURE 5.8
Types of plywood

It comprises of a series of veneers, cemented together so that the grain of one veneer is at right angles to the next.

PARTICLEBOARDS

Generally referred to as chipboards, these are made of wood chips and flax chives, which are bonded together with a synthetic resin adhesive.

The boards are available in lengths of up to 5.3 m, widths of up to 1.7 m and thicknesses from 8 to 40 mm. There are also various densities and grades available. An example is shown in Figure 5.9.

HARDBOARD

This is available in standard size sheets from 3 to 6 mm in thickness and is used for the same purpose as plywood, being somewhat cheaper, heavier and inferior.

INSULATION BOARD

This low-density board is made from the same materials as hardboard, but is not pressed in its manufacture.

Insulation board or softboard is available in various textures and finishes to suit the wide variety of work for which it can be used.

Thicknesses range from 6 to 20 mm and it can be used in either sheet or tile form.

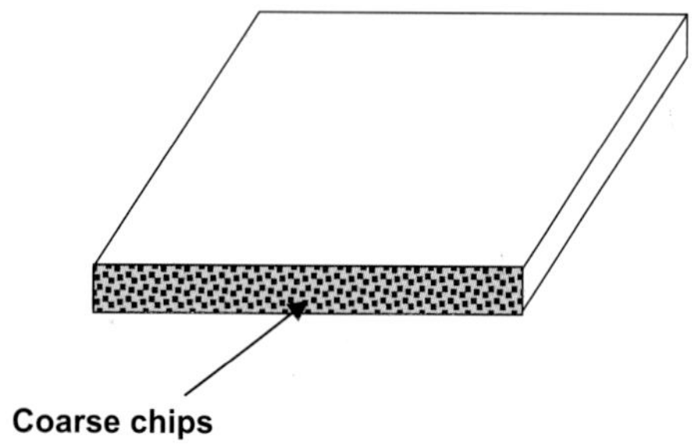
Coarse chips

FIGURE 5.9
Example of particle board

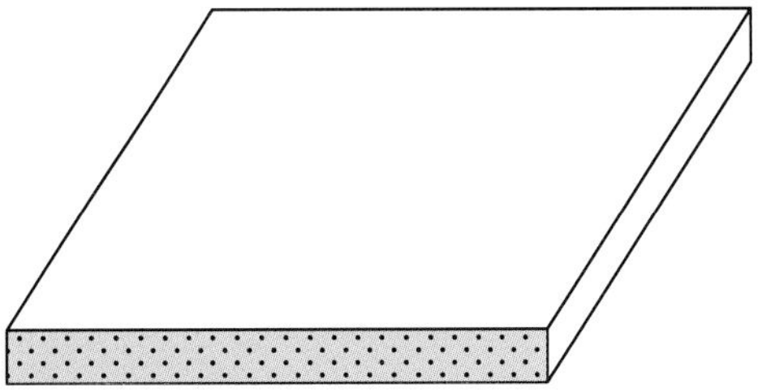

FIGURE 5.10
Example of medium-density fibreboard (MDF)

MEDIUM-DENSITY FIBREBOARD (MDF)

Fibreboard is basically a compromise between hardboard and softboard (Figure 5.10).

It has better heat insulation qualities than hardboard while retaining greater mechanical strength than softboard, and is very easy to work.

Available in standard sized sheets, it is normally 9.5 mm in thickness.

BLOCKBOARD

This is a very useful sheet material made by sandwiching a core of narrow strips of wood between two face veneers (Figure 5.11).

The best grades of blockboard have double face veneers so that the core strips and the grain of the face veneers run in the same direction.

Blockboard is available in the usual standard size sheets in thicknesses from 12 to 48 mm, and has core strips up to 25 mm wide.

LAMINBOARD

Laminboard is essentially similar in construction to blockboard, but is generally superior and more stable owing to the narrowness of the core strips, which have a maximum width of 7 mm.

Laminboard is also glued edge to edge, unlike some of the cheaper grades (Figure 5.12).

The boards are available from 12 to 25 mm in thickness.

BATTENBOARD

This material is a coarser version of blockboard and has core strips up to 75 mm wide (Figure 5.13).

It is less stable than blockboard and it is best suited for jobs where overall strength and tendency to 'ripple' are less important.

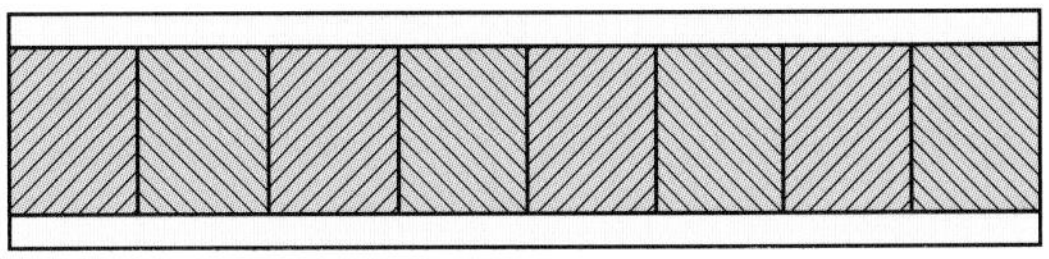

FIGURE 5.11
Example of blockboard

FIGURE 5.12
Example of laminboard

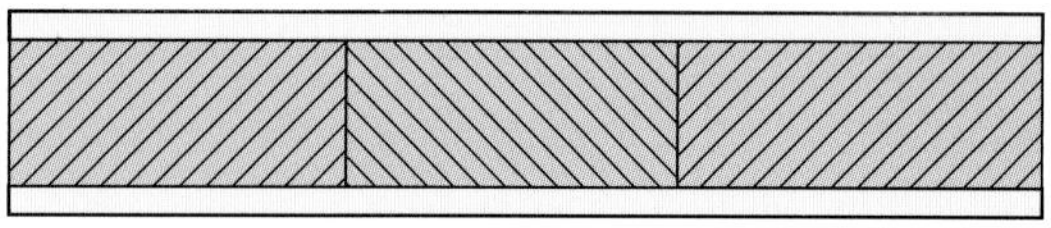

FIGURE 5.13
Example of battenboard

PLASTERBOARD

Plasterboard consists of an aerated gypsum core encased in specially prepared paper liners.

The boards are available in plain or insulating grades, the latter having a veneer or polished aluminium on one side (Figure 5.14).

There are four types of basic plasterboard:

- gypsum wallboard
- gypsum lath
- gypsum baseboard
- gypsum plank.

Plasterboards are available in many sizes, the most common being:

- 1.210 m × 0.900 m × 9.5 mm thick
- 1.829 m × 0.900 m × 12.5 mm thick.

STEEL REINFORCEMENT

Steel reinforcement arrives on the building site in large sheets or single bars.

The sheets consist of small-diameter bars welded together to form a mesh, and are used mainly for reinforcing the foundations and floors of buildings.

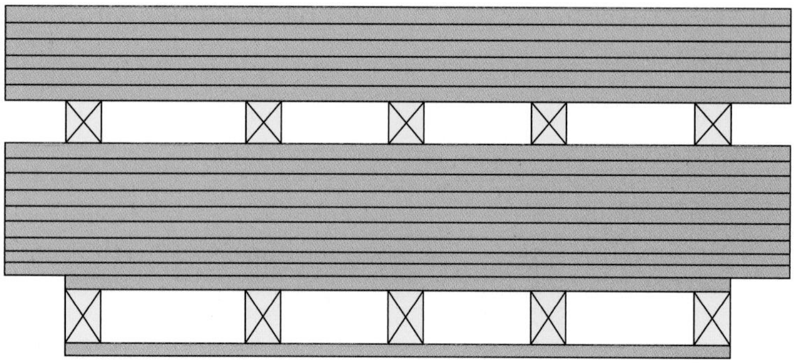

FIGURE 5.14
Example of plasterboard

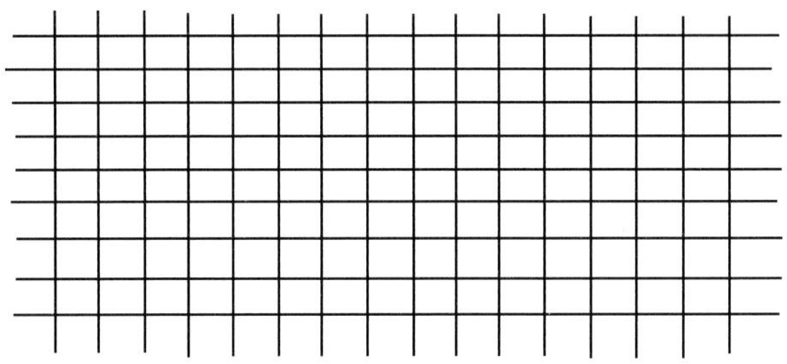

FIGURE 5.15
Example of sheet reinforcement

Sheet reinforcement is coded according to the size of the bars and the size of the mesh (Figure 5.15).

Steel reinforcing bars are available in various diameters according to the strength required, ranging from 6 to 40 mm.

There are also numerous deformed bars such as those shown in Figure 5.16. These improve the bond between the concrete and the steel reinforcement.

There will be occasions when the steel reinforcement is delivered to the site already bent and formed, ready for fixing in the appropriate foundation, column or floor, etc.

In this case it is important that you can recognize the various items of reinforcement that have been fabricated off site and are able to store them correctly.

CONTAINERIZED MATERIALS

PAINT

Paint is delivered to the normal construction site in small quantities, as and when the decorators require it, because it has a very short shelf-life.

Paint and other liquid items should be stored inside a special store and protected. Examples are shown in Figure 5.17.

There are many types of decorating materials such as emulsions, undercoat and gloss, as well as varnishes and protective coverings for timber, masonry and steel.

They should all be easily recognizable at the time of delivery.

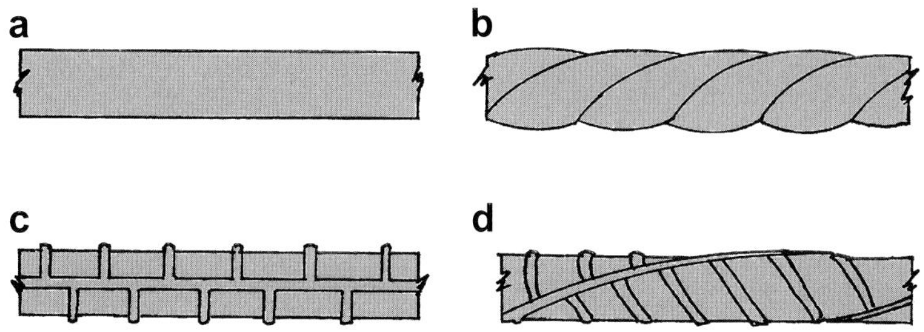
a
b
c
d

FIGURE 5.16
Examples of bar reinforcement: (a) plain round bar; (b) square twisted bar; (c) ribbed bar; (d) twisted ribbed bar

White
spirit

FIGURE 5.17
Examples of containerized materials

PETROL, DIESEL AND OIL

These items are classified as highly flammable liquids (HFLs) and should be received on site with the utmost care and attention.

They should all be kept separate and never mixed.

The canisters, barrels, etc., should be clearly marked and kept in a protected store (Figure 5.18).

The major hazard associated with these liquids is *fire* and it is essential that precautions are taken to limit the risks involved.

ROLLED MATERIAL

Examples of rolled materials are (Figure 5.19):

- DPC
- polythene (membrane sheeting)
- roofing felt

All rolled materials should be stored in an appropriate store, on end to retain their shape. They should be stored in their appropriate sizes for ease of identification.

HFL
HFL

FIGURE 5.18
Barrels of flammable liquid

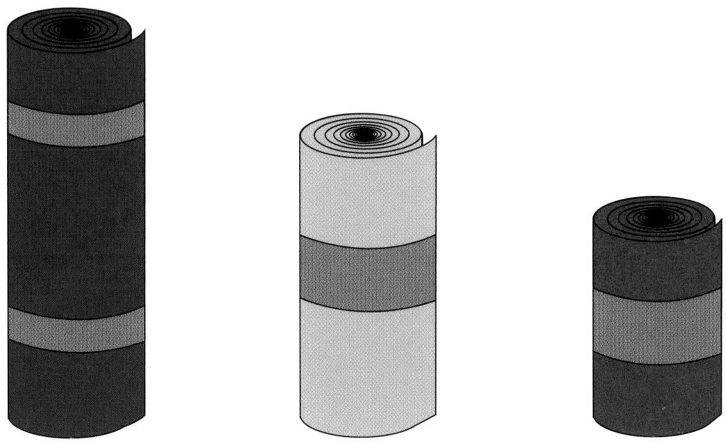

FIGURE 5.19
Examples of rolled materials

BOXED MATERIALS

Materials arriving in boxes include:

- flat packs (kitchen units)
- nails
- ironmongery
- fittings.

All of the above are delivered to site when required as they are usually some of the most valuable materials and are easily stolen.

Budget-priced units are often delivered in knock-down form, known as flat packs, ready for on-site assembly. Flat packs relate mainly to kitchen units, which consist of a range of units mass produced to standard designs by manufacturers. They should be delivered and stored in the unit where they will be fixed.

These items are easily damaged and special care should be taken when unloading and storing them. Figure 5.20 shows an assembly drawing for erecting flat-packed units.

FIXINGS

Nails, screws, etc., are delivered during the contract and are usually bought in bulk for cheapness.

There are numerous types of screw available and they are ordered according to length and gauge (thickness). There are also various types of screw heads.

Many types of nail are available; these are also ordered according to length and gauge.

IRONMONGERY

Carpenter's locks, bolts, handles, etc., are desirable items that are likely to be stolen unless stored securely under the control of a storeperson.

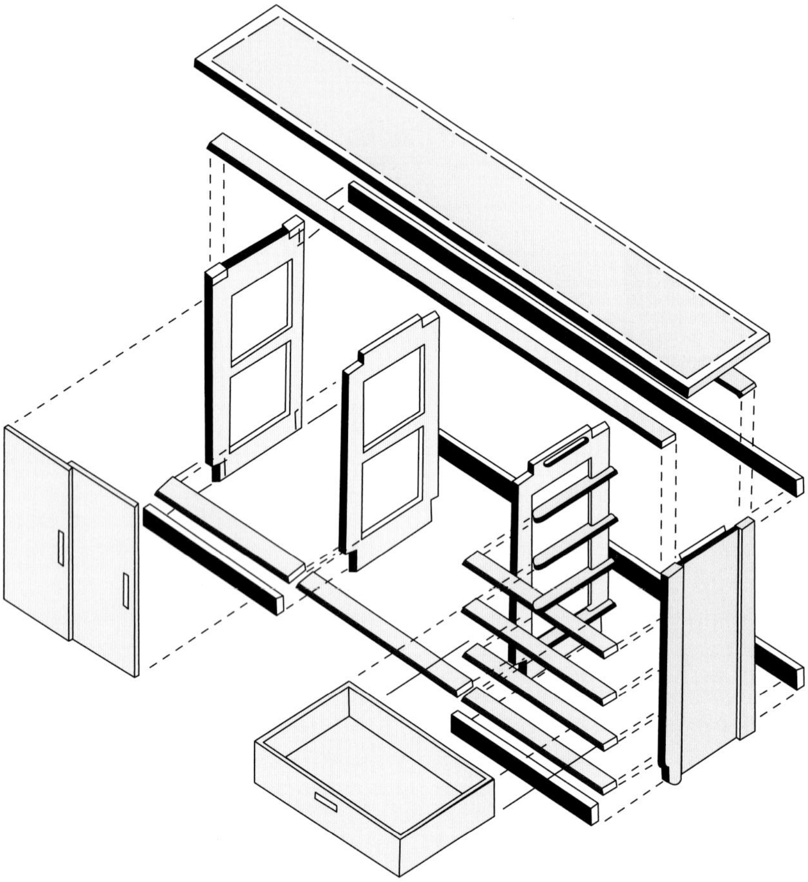

FIGURE 5.20
Flat-packed units

They are supplied as individual items, plastic bubble packed on cards or in boxed sets, e.g. 10 pairs of level handles.

Examples of typical packages are shown in Figure 5.21.

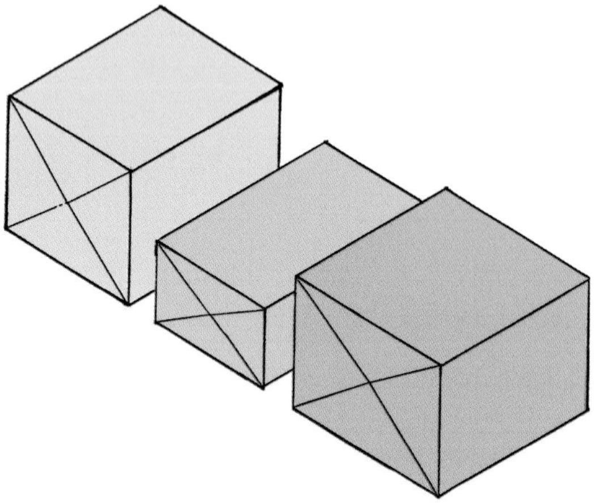

FIGURE 5.21
Examples of packaged materials

Sources of information

Information on materials can be obtained from numerous sources:

- books and technical articles
- manufacturers' literature
- specifications
- Building Regulations
- British Standards Institution
- British Standard specifications
- codes of practice
- Building Research Establishment (BRE)
- BRE digests, information papers and current papers.

Manufacturers and trade development associations:

- Brick Development Association
- Timber Research and Development Association
- Fibre Building Board Development Organisation
- Cement and Concrete Association
- Lead Development Association

and many more.

Copies of their publications are normally available for students' reference in most college and town libraries.

Many libraries also take trade periodicals which operate a reader's enquiry service whereby technical brochures and information can be obtained from various manufacturers and suppliers.

Trade periodicals:

- *Cement and Concrete Monthly*
- *The Architects' Journal*
- *Building*
- *The Brick Bulletin*

and many more.

Professional institutions and technical societies are also a source of specific information:

- Chartered Institute of Building
- Royal Institution of Chartered Surveyors.
- Institution of Civil Engineers
- Institute of Clerks of Works

and many more.

Interpreting technical information

It is important that the student understands where information can be found. It can be gathered from numerous sources.

MANUFACTURERS' TECHNICAL INFORMATION

Any efficient office should have up-to-date information regarding new and existing products from the various manufacturers who produce materials of interest to them.

Many manufacturers produce technical information which is free for the asking. These papers, brochures, leaflets, etc., should be stored in the office in an easily retrievable system.

The easiest way is to file them under each specific trade, or material.

The most elaborate system is called the SfB Classification System. To assist in the filing of literature each item is coded with appropriate SfB references in the top right-hand corner of the document (Figure 5.22).

Information can be written, oral or in drawing format.

ORAL INFORMATION

This is received from people such as supervisors and workmates by word of mouth.

Care should be taken to ensure that the person receiving the message has received the correct version of the message, although there will be no real proof of this.

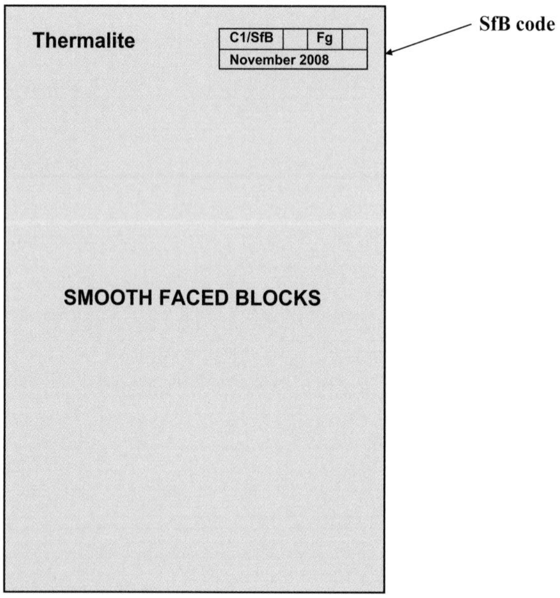
Thermalite
C1/SfB
Fg
November 2008
SfB code
SMOOTH FACED BLOCKS

FIGURE 5.22
Typical supplier's information

Noise on building sites can cause problems when oral information is being passed from person to person.

However, a quick response can be achieved with oral instructions.

WRITTEN INFORMATION

This is usually preferred to oral communication as the person receiving the message receives the correct version and has documented proof.

A permanent record can be very useful at a later date if problems occur.

However, it takes time for written information to reach to the recipient.

TABLES, GRAPHS AND CHARTS

Most construction sites have programmes to explain how a contract is progressing. Bar charts are the most common method used.

Charts can also be produced to show the financial status of the contract.

SPECIFICATION

This is a document giving a written description of materials to be used, and construction methods to be employed in the construction of the building.

SCHEDULE

Information is presented in a tabulated form for items such as:

- doors
- windows
- ironmongery
- decorative finishes
- inspection chambers (manholes).

TEXTBOOKS

There are simply thousands of textbooks available on practically every aspect of the construction industry.

They not only explain the various methods of construction but also give typical details and the reasons why particular building methods are used.

BUILDING REGULATIONS

These regulations are concerned with safeguarding the safety and health of occupants of completed buildings.

They deal with matters such as safe methods of design, standards of construction, and selection and use of materials.

BRITISH STANDARD SPECIFICATIONS

These are issued by the British Standards Institution to lay down minimum standards for materials and components (e.g. walling blocks and doors) used in construction and other industries.

BRITISH STANDARD CODES OF PRACTICE

Codes of good practice are issued by the British Standards Institution to cover workmanship in specific areas, e.g. building drainage, and brick and block masonry.

AGRÉMENT CERTIFICATE

These certificates are granted by an independent testing organization, called the British Board of Agrément, stating that the manufacturer's products have satisfactorily passed agreed tests. Subsequent to the granting of the certificate strict quality control has to be continued.

DRAWINGS

These are produced to assist the builder to produce the building exactly as the architect has designed it. Drawings have been dealt with in Chapter 3.

Refer back to Chapter 3, page 56 to reinforce your knowledge of the following types:

- sketches – quick and easy to draw and understand
- location drawings – scale drawings used to show the complete details of the building.

Component characteristics and limitations of use

Identifying and selecting components

BRICKS

See Table 5.1 and Figure 5.23.

Bricks are usually identified according to:

- place of origin – Nottingham red
- colour – buff, dapple grey

Table 5.1 Types of brick

Type	Composition and properties	Uses
Clay	Composed mainly of silica and alumina and small quantities of lime, iron and manganese	General use
Calcium silica	Made up from sand and lime or crushed flint and lime moulded under pressure and hardened by exposure to steam	On walls to reflect light
Fire brick	Produced from refractory clay having a high fusing point	Furnace work
Modular brick	Metric modular bricks available in differing sizes: 300 × 100 × 100; 200 × 100 × 100 mm	
Common brick	General bricks, which vary in quality and do not have a decorative face	Work below ground level
Facing brick	Produced with a decorative finish applied to the header and stretcher courses	Decorative work
Engineering brick	Hard and dense with a smooth texture, having a high load-bearing capacity	DPCs or supporting heavy loads
Insulation brick	Made from diatomaceous silica to give heat insulation properties	Lining chimneys
Specials	Purpose made to meet some detail of construction e.g. squint, plinth	Features

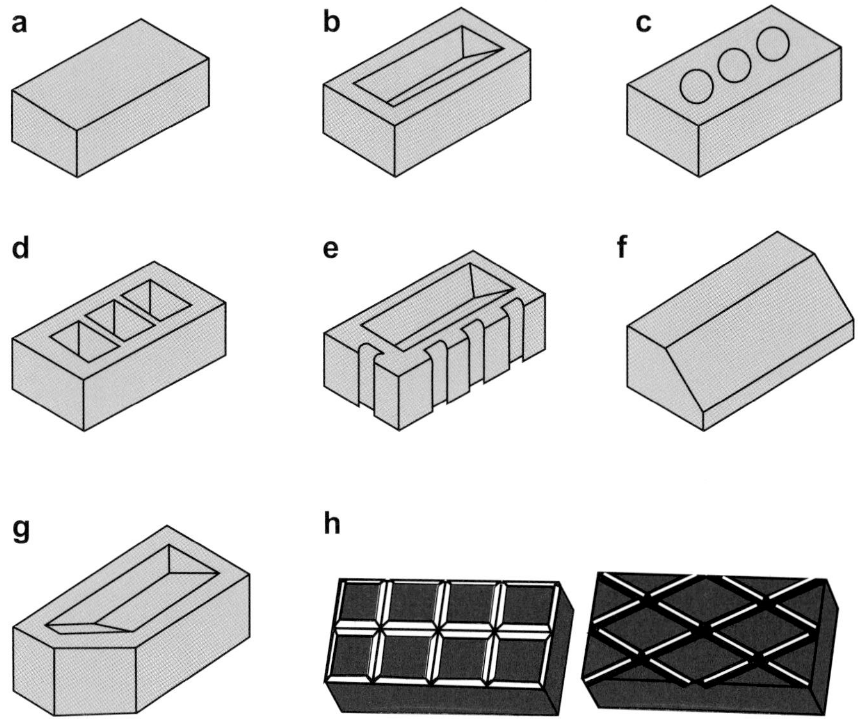
a
b
c
d
e
f
g
h

FIGURE 5.23
Types of brick: (a) solid wire-cut brick; (b) pressed or hand-moulded brick; (c) perforated wire-cut brick; (d) cellular pressed brick; (e) keyed pressed brick; (f) plinth stretcher; (g) cant brick; (h) paving bricks

- method of manufacture – pressed, wire-cut, hand-made
- surface texture – sand faced, smooth
- use – engineering brick, facing brick
- size – metric, imperial.

BLOCKS

Blocks are produced in solid, hollow or multicut form in a range of shapes and sizes (Table 5.2).

DRAINAGE PIPES

Drainage pipes can be manufactured from various materials and can be classed as rigid or flexible. A sample of pipes is shown in Figure 5.24.

Clay is the major material used for rigid drain pipes in domestic work, with cast iron as the main alternative. The usual materials for flexible drain pipes are pitch fibre and unplasticized PVC.

CONCRETE UNITS

Concrete is used for the manufacture of numerous items used in construction. Concrete kerbs, paving slabs and lintels are some the main products.

Table 5.2 Types of block

Type	Composition	Uses
Lightweight 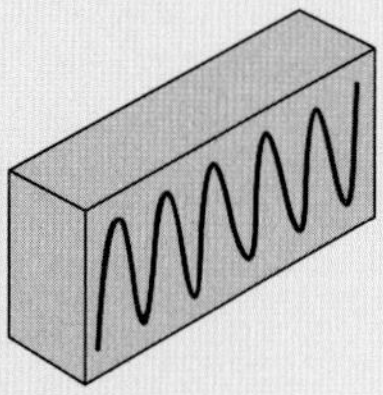	Waste coke or ash and cement; or burnt clay	Internal non-load-bearing walls
Dense/heavy 	Usually concrete with a minimum thickness of 100 mm	Internal load-bearing walls and external walls
Hollow concrete 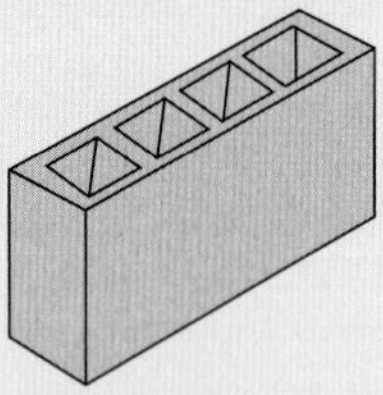	Concrete	External walls (usually rendered)
Cellular (lightweight) 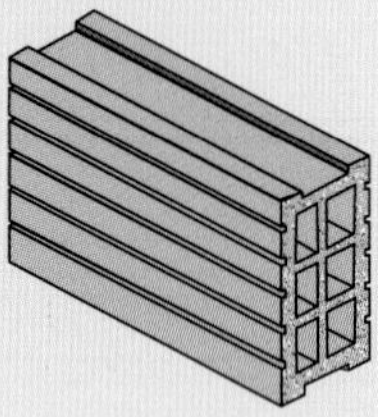	Burnt clay	Internal non-load-bearing walls

CONCRETE KERBS

Kerbs are used at the junction between the footway and the carriageway.

A kerb is a specially shaped piece of precast concrete (Figure 5.25). It serves to define the edge of the carriageway and to direct the flow of water from the pavement surface.

The kerb is bedded in concrete to give it stability.

CONCRETE PAVING SLABS

Concrete paving slabs are used on the surface of footpaths and causeways, but are used more for private footpaths than council causeways.

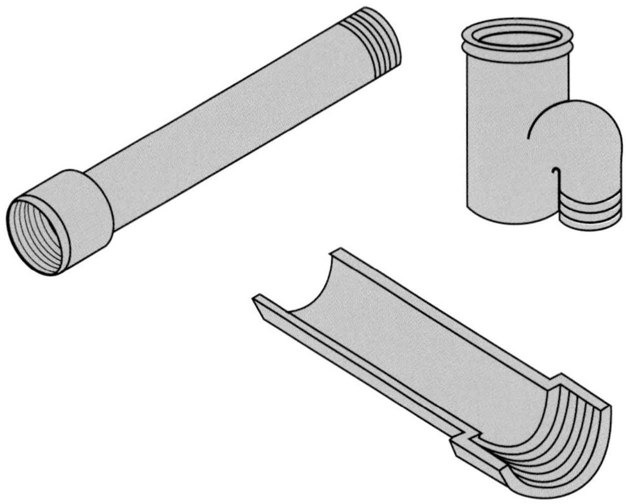

FIGURE 5.24
Types of drain pipe

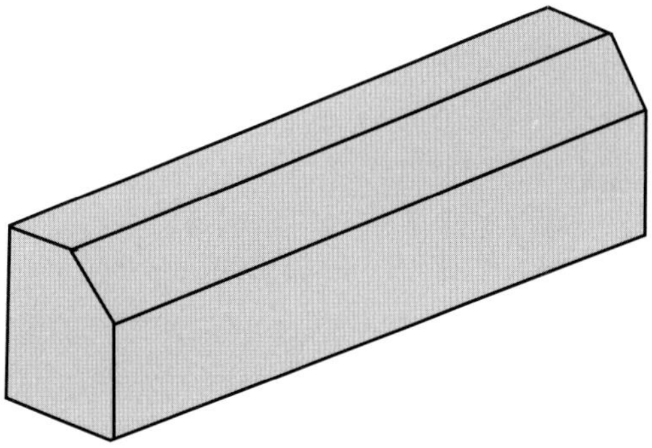

FIGURE 5.25
Concrete kerb

They are available in various shapes, sizes and colours. Typical shapes are shown in Figure 5.26.

CONCRETE LINTELS

Although other materials are more common nowadays the concrete lintel is still used on many construction sites.

Lintels are used over openings to transmit loads to the jambs at the sides of the openings. A typical concrete lintel is shown in Figure 5.27.

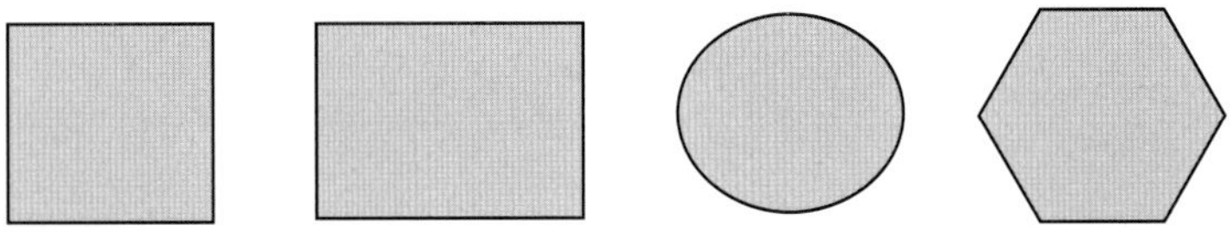

FIGURE 5.26
Concrete paving slabs

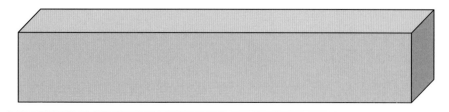

FIGURE 5.27
Concrete lintel

Roofing tiles

These are coverings for pitched roofs and can be manufactured from either clay or concrete. Various types are shown in Figure 5.28.

They are nailed to timber battens which are fixed to the rafters of the roof. The tiles are hooked over the battens by means of nibs which provide protection from the underside.

The coverings may be either double or single lap.

FLAT PACKS

Many internal fittings are delivered in flat packs, that is they are delivered unfitted, safely packaged in heavy-duty cardboard for protection. The joiner then has to assemble the units before fitting them on site.

KITCHEN UNITS

Most builders today provide a fitted kitchen as part of the dwelling. Manufacturer's details can be obtained for specific examples and quality of finish, but units fall within the following basic categories:

- base units – including provision for sinks, hobs and tables
- wall units – including upper units, cooker hoods and spot lighting
- tall units – including food cupboards, broom storage, provision for building in cookers, fridges, etc.

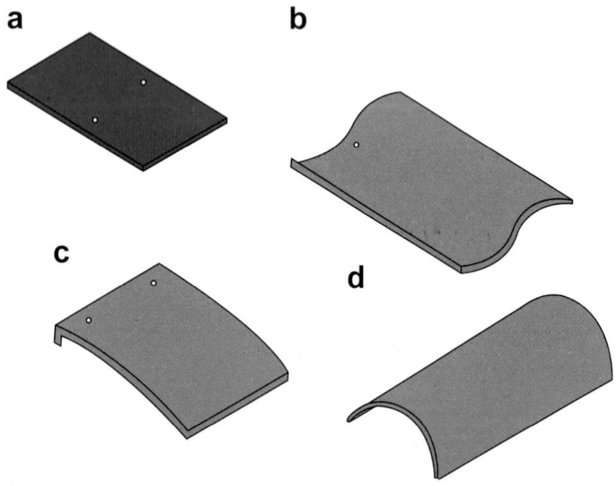
a
b
c
d

FIGURE 5.28
Roof coverings: (a) roof slate; (b) pantile; (c) standard double-lap plain tile; (d) half-round ridge tile

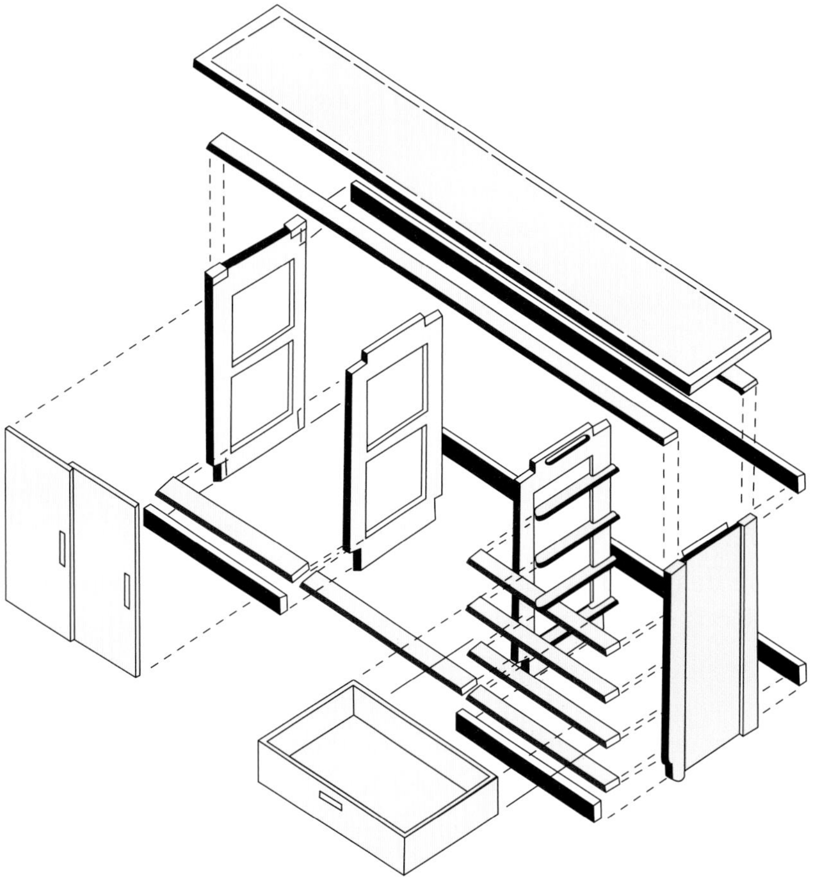

FIGURE 5.29
Flat packs

All are available in a variety of lengths based on a 100 mm module, in multiples of 300, 400, 500 and 600 mm. A typical assembly drawing is shown in Figure 5.29.

Safe work practices

The Health and Safety Commission has laid down certain regulations regarding the handling of materials to prevent people from being injured.

Summary of regulations

- No one must be employed to lift, carry or move any load as heavy as to be likely to cause injury.
- All employers must provide safe systems of work.
- Employers must ensure safety and absence of risks in the handling, storage and transportation of all types of materials.

- Employers must provide instruction and training, to ensure the health and safety of employers.
- Employees must take reasonable care for their own safety and the safety of others.

In other words, it is the duty of the employer to ensure that you as an employee have the right training as well as the right equipment, but when actually carrying out the task of lifting or handling it is you, the employee, who must take care, not only for yourself but for others.

Common injuries

TO THE BACK/MUSCLES

Strains and sprains to muscles, torn ligaments, back trouble and even hernias are all symptoms of the bad handling of materials.

They can be caused by sudden and awkward movement while handling heavy objects or by trying to lift an object that is far too heavy.

TO THE SKIN

Cuts and abrasions from rough surfaces, sharp or jagged edges, splinters and even projections, should be avoided.

TO LIMBS

Limbs (arms and legs) may be crushed by falling loads, or fingers, hands or feet may become trapped by heavy loads.

STRAINS AND TWISTS

Most of the accidents on site are a result of employees handling heavy or awkwardly shaped objects such as cement bags, concrete and steelwork. The injury is often the cumulative result of strain that has occurred over a period.

Strains are also incurred fitting wire ropes to excavators, bar bending, pick and shovel work.

Employees also injure themselves by overreaching, twisting themselves and attempting to stop loads swinging.

Kinetic handling must be taught by skilled trainers. It cannot be 'picked up'. It is important that postural errors are corrected during training before bad habits are established.

Handling materials safely

The trainee should be able to select and use appropriate safety equipment and protective clothing when handling different materials.

Appropriate equipment and aids should be selected and used to carry materials.

The trainee should also be able to demonstrate safe manual handling techniques.

Note

Always use lifting gear if it is available.

Remember

The sequence for lifting heavy and awkward loads:

1. Plan the task.
2. Bend your knees.
3. Get a good grip.
4. Lift, with your legs taking the strain.
5. Place the load down.

Lifting gear

Numerous items of small lifting equipment are available to assist with handling materials on site and in the workshop. Only use these if you are qualified to do so.

These range from the small brick lifts, slings, barrows and dumpers to mechanical fork-lift trucks. A selection is shown in Figure 5.30.

- Barrows are the most common form of equipment for moving materials on site.
- A sack truck can be used for moving bagged materials and paving slabs.
- A hod can be used for moving bricks on to higher levels such as scaffolds.
- A pallet truck can be used on hard areas for moving heavy loads

Many materials are delivered to the site on lorries equipped with mechanical off-loaders. Once the material has been off-loaded it is the builder's responsibility to move the materials to a secure place until required for use.

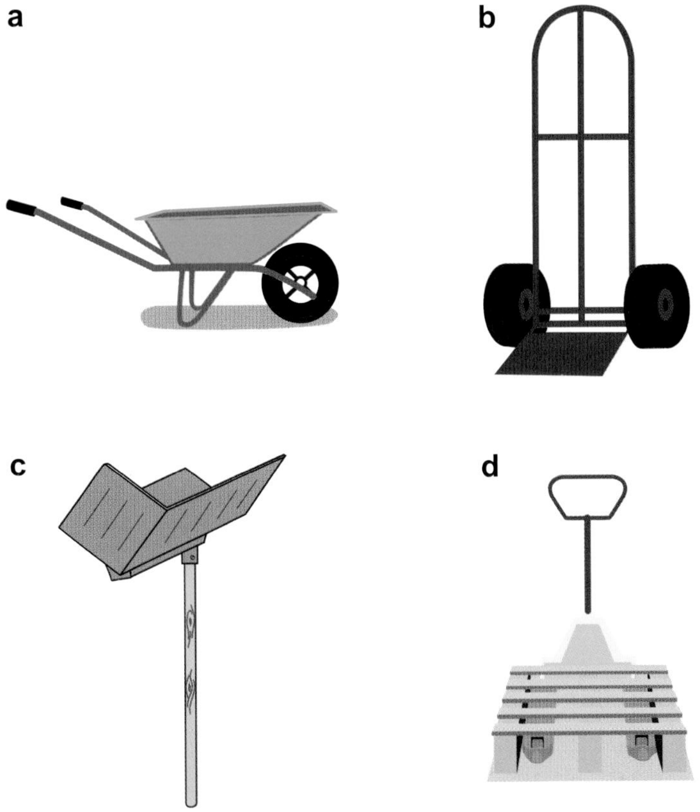
a
b
c
d

FIGURE 5.30
Various types of moving equipment: (a) barrow; (b) sack truck; (c) hod; (d) pallet truck

Personnel protective equipment

Depending on the type of workshop or site situation, the wearing of correct safety clothing and safe working practices are the best methods of avoiding accidents or injury. On some sites certain PPE is compulsory.

All construction operatives have a responsibility to safeguard themselves and others. Making provision to protect oneself often means wearing the correct protective clothing and safety equipment.

Your employer is obliged by law to provide the following:

- suitable protective clothing for working in the rain, snow, sleet, etc.
- eye protection or eye shields for dust, sparks or chipping
- respirators to avoid breathing dangerous dust and fumes
- shelter accommodation for use when sheltering from bad weather
- storage accommodation for protective clothing and equipment when not in use
- ear protectors where noise levels cannot be reduced below 90 dB(A) for 8 hours.
- adequate protective clothing when exposed to high levels of lead, lead dust fumes or paint
- safety helmets for protection against falls of materials or protruding objects
- industrial gloves for handling rough abrasives, sharp and coarse materials.

Although our skin is not proof against knocks, bumps, cuts, acid, alkalis or boiling liquids, it is waterproof. Even so, we do have to cover up at times to protect ourselves.

Workers in the construction industry are liable to injury or even death if they are not protected. Because of this, protective clothing has been developed to help prevent injury.

Various items of PPE are shown in Figure 5.31.

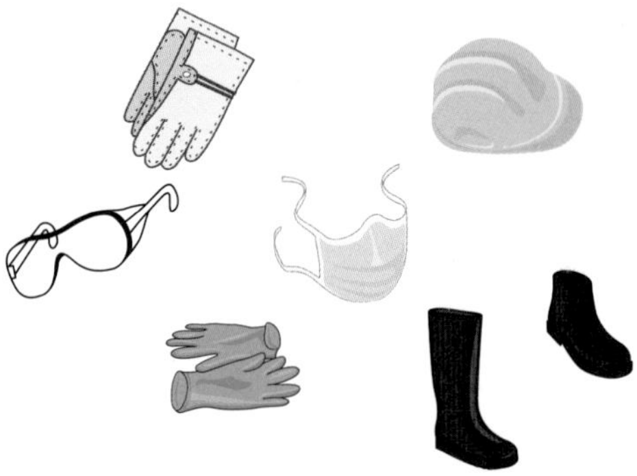

FIGURE 5.31
Selection of PPE

Storing, stacking and protecting materials and components

Objectives of storage

The objectives of storage are to enable the quick and easy location of materials, plant and equipment.

Purpose of storage

Secure accommodation together with an effective storage system is essential on the construction site to prevent unnecessary waste and loss of materials due to vandalism and theft, and damage caused by constant handling.

Storage control

Irrespective of the amount of materials being stored, the following factors need to be considered.

DELIVERIES

Materials received need to be checked against suppliers' delivery advice notes for correct quantities and damage.

STOCK ROTATION

Stocks of some materials (plaster, cement, paint, etc.) need to be used in the order of delivery as they only have a limited shelf-life.

WITHDRAWALS

Materials drawn from stock by requisition need to be replaced.

SECURITY

Storage facilities must be capable of being made secure against vandalism and theft.

CLEANLINESS

The store and unloading areas must be kept clean and free of debris at all times.

SAFETY

Materials must be handled and stacked in such a manner that injuries and damage are avoided. Hazardous materials need to be stored in separate accommodation constructed of non-flammable materials.

Site storage

Provisions for site storage will include the following.

SITE COMPOUND

A fenced area is required for storing materials and plant, providing a means of control and protection against vandalism and theft. A typical site compound is shown in Figure 5.32.

SITE STORES

Site stores are needed for storing materials that must be protected against the weather and which are easily lost or stolen. Two examples are shown in Figure 5.33.

STORAGE AREAS

Predetermined areas are required on the site where materials are to be placed before use.

Materials such as bricks and blocks could be stored in the compound or placed close to where they will be used on the site, as shown in Figure 5.32.

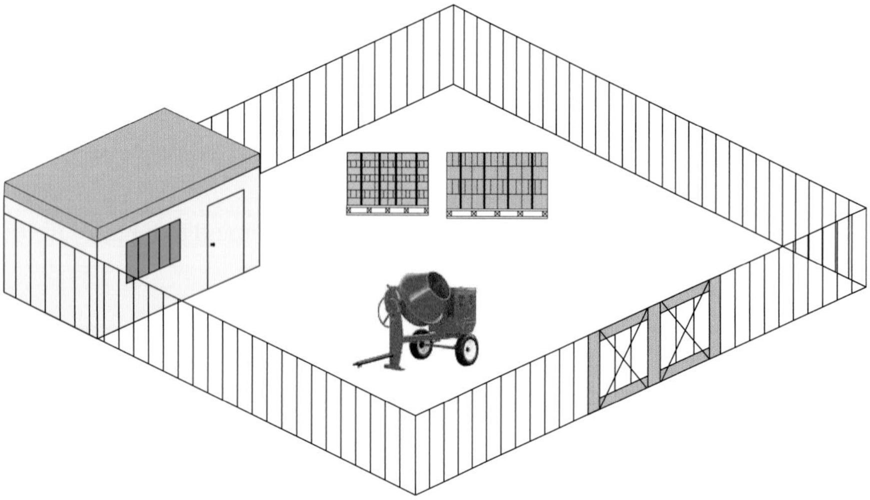

FIGURE 5.32
Typical site compound

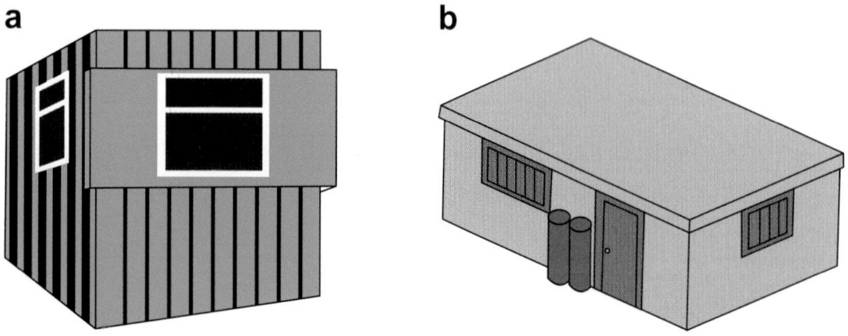
a
b

FIGURE 5.33
Types of material store: (a) metal; (b) wooden

WORKSHOP/YARD STORAGE

Convenient workshop storage should be provided for tools and equipment.

Storage under cover should be provided for timber, sheet and insulation materials.

Separate workshop accommodation may be required for the storage of paint and inflammable fuel.

A maintenance area should also be provided so that plant and equipment can be maintained and cleaned before being stored or sent out to other sites.

Material storage

The amount of storage on site depends on the size of the contract and the amount of space available for storage.

On a large site storage facilities can be very comprehensive, but on small sites the materials are often delivered as required, materials being retained at the builder's merchant until called for.

Some firms will store small items at their main depot and the site supervisor will call them forward with requisition sheets.

Even on large sites some materials are delivered as required, such as ready-mixed concrete, mortar and hardcore.

Figure 5.34 shows a typical site for a factory unit. Areas have been set out for material storage and security. The whole site has a fence around it and an internal compound.

AGGREGATES

Aggregates should always be kept in a clean condition and should not become contaminated with undesirable substances.

Not only can soil, clay, organic or deleterious materials impair the strength of the concrete, but they can also stain the mix, resulting in patchiness or an unsightly finish.

Aggregates stored in builder's yards are usually stored in specially prepared areas.

Figure 5.35 shows a typical area for storing aggregates on site. These areas are usually built with a concrete slab, sloping to the outside, and brick walls separating the various types of aggregates and sands.

On the site aggregates are very often stored directly on the ground. In this case the top soil should be removed and the pile should be as large as possible, with the bottom 500 mm never being used.

Aggregates and sands that have become contaminated should be removed from site.

In severe weather the stockpiles should be protected by sheeting materials.

Both very hot and cold materials will reduce the quality of the finished mix.

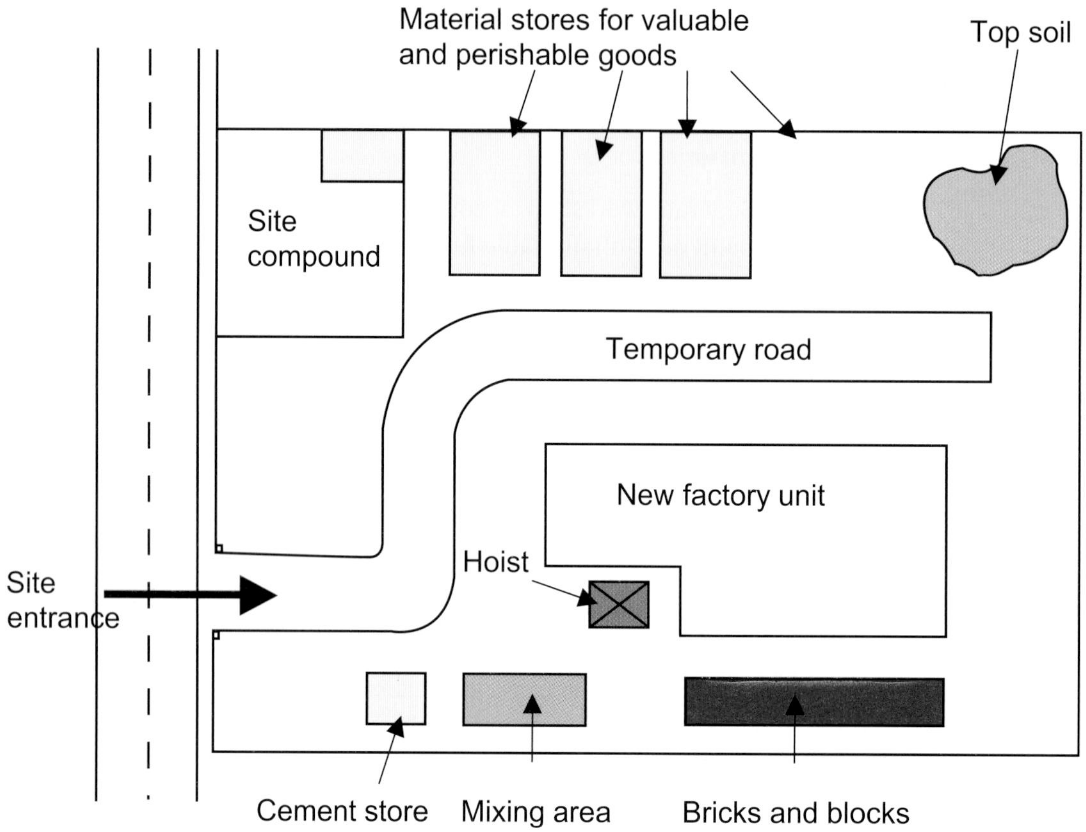
Material stores for valuable and perishable goods
Top soil
Site compound
Temporary road
New factory unit
Hoist
Site entrance
Cement store
Mixing area
Bricks and blocks

FIGURE 5.34
Typical site layout for materials

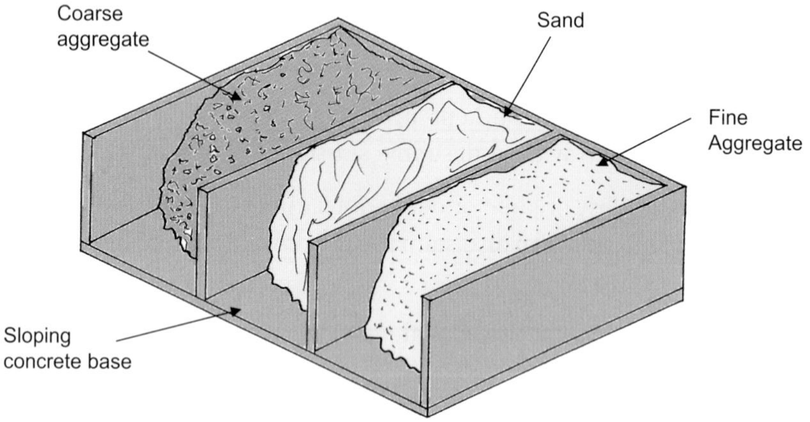
Coarse aggregate
Sand
Fine Aggregate
Sloping concrete base

FIGURE 5.35
Site storage for aggregates

BAGGED MATERIALS

These should be stored in a weatherproof shed, preferably without windows, and raised off the ground. An example is shown in Figure 5.36.

Cements, plasters and lime are the main bagged materials, all of which will deteriorate rapidly when damp.

The materials should be stored separately and clearly dated.

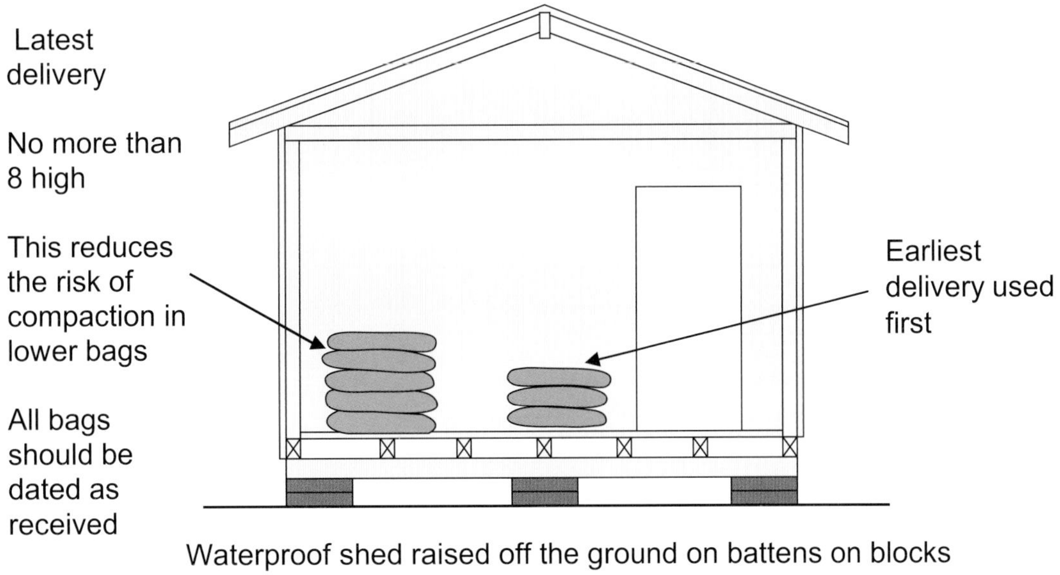
Latest delivery
No more than 8 high
This reduces the risk of compaction in lower bags
All bags should be dated as received
Earliest delivery used first
Waterproof shed raised off the ground on battens on blocks

FIGURE 5.36
Cement storage

They must be stored in such a way that they can be used on a first-in, first-out rotation. This is to minimize the storage time and prevent the bag contents becoming stale or 'air setting'.

Care should be taken when using bagged materials as the paper can cause a problem on the site.

BRICKS, BLOCKS AND STONE

Bricks, blocks and stone should be stacked clear of the ground and covered completely with tarpaulins (Figure 5.37).

Some manufacturers deliver their materials already shrink-wrapped. The wrapping should be kept on until the materials are ready for use.

DRAINAGE MATERIALS

Drain pipes should be stored in the open where they will not be accidentally damaged, on firm, level ground that has been covered with hardcore (Figure 5.38).

Because of their shape they must be restrained from rolling about and occupying a large ground area.

If the drain pipes have socketed ends they should be stacked with sockets at alternative ends for protection.

CONCRETE GOODS

Most concrete goods can be stored outside on a level, well-drained site. Examples are shown in Figure 5.39.

Paving slabs should be stored on their edge, standing on timber bearers. Different sizes, shapes and colours should be stored separately.

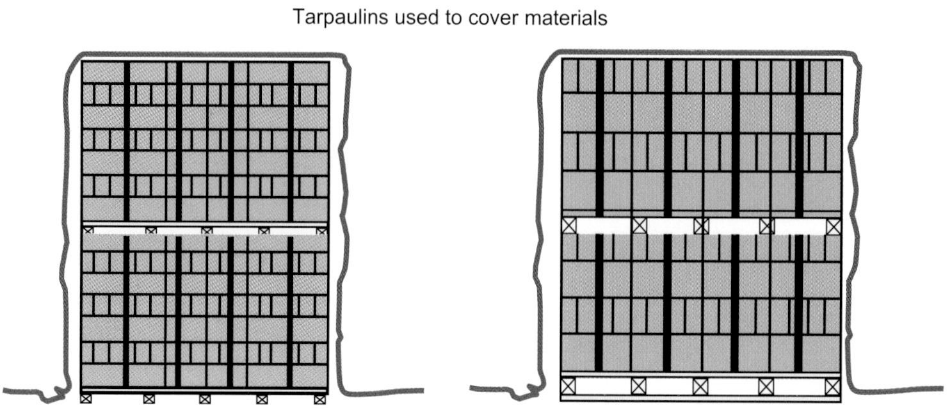
Tarpaulins used to cover materials

FIGURE 5.37
Protection of materials

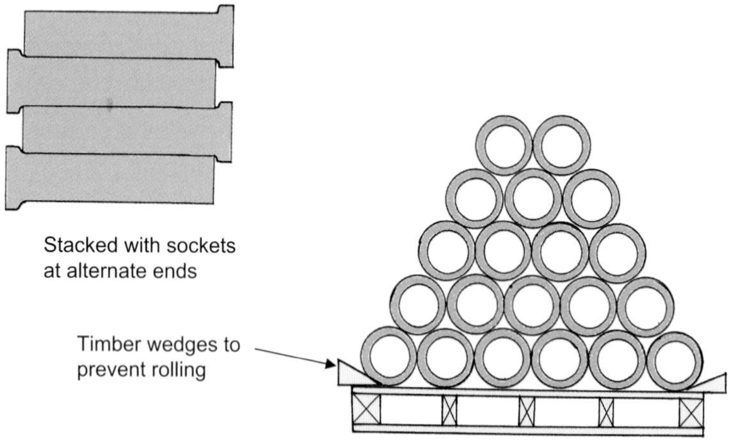
Stacked with sockets
at alternate ends
Timber wedges to
prevent rolling

FIGURE 5.38
Stacking of drain pipes

Concrete lintels should be stored on timber bearers, but it is essential to ensure that they are stored in the way they are to be fixed, i.e. top side uppermost.

Kerbs should be stored on edge on timber bearers. They may be stored in bonded layers, but never more than three high.

ROOFING TILES

Roofing tiles and slates should be stored on level, well-drained ground, possibly stoned to keep the area clean. See the examples in Figure 5.40.

Care should be taken when storing roofing tiles and slates to avoid damage.

Roofing tiles are impervious to moisture and can be stored outside. However, on a building site they should be wrapped around with a tarpaulin to protect them from splashing.

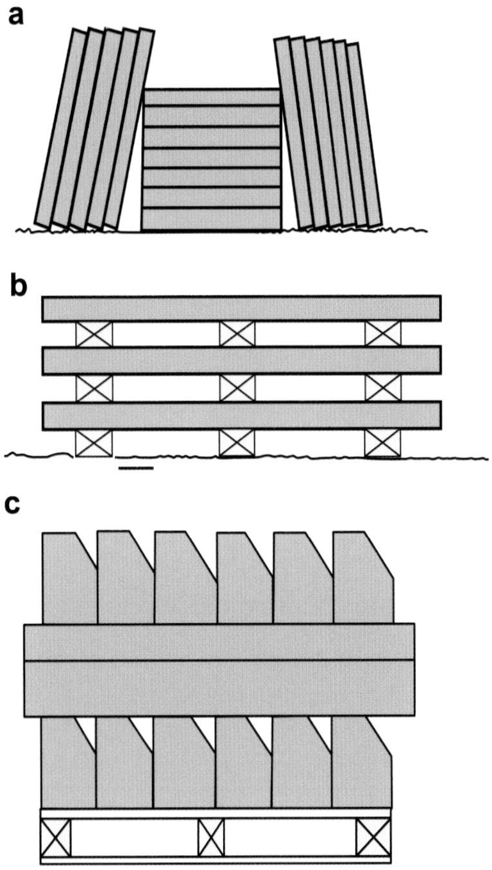
a
b
c

FIGURE 5.39
Storage of concrete goods: (a) paving slabs; (b) lintels; (c) kerbs

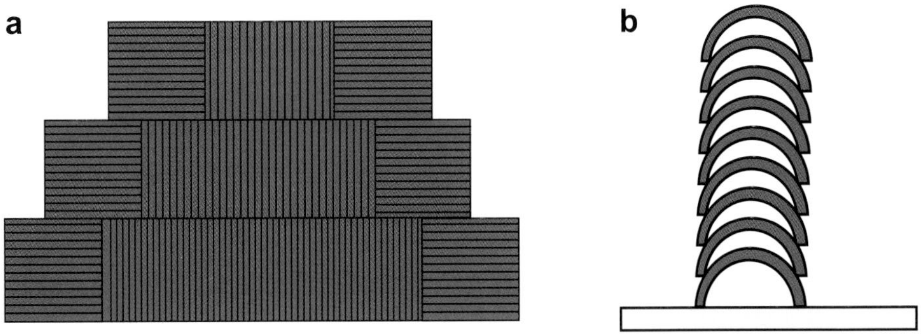
a
b

FIGURE 5.40
Stacking of roof coverings: (a) plain tiles; (b) ridge tiles

Plain tiles are stacked on end, end tiles flat and taper stacked. The tiles are placed alternately to prevent damage to the nibs. Ridge tiles are stacked as shown – never more than 10 high.

CERAMIC TILES

These are delivered to the site in boxes and after checking for damage should be stored in their boxes (Figure 5.41). They should be stored inside a store or the building where they are required.

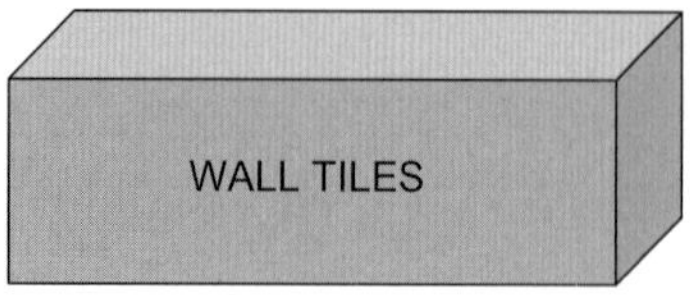
WALL TILES

FIGURE 5.41
Storage of ceramics

SHEET MATERIALS

PLYWOOD

Small quantities of plywood are best stored in racks, while large stocks can be stacked on bearers spaced at 400 mm centres to prevent distortion and covered with a tarpaulin sheet (Figure 5.42).

When handling care should be taken to avoid damage to corners and edges.

PLASTERBOARD

Plasterboard should be stored in a similar way to other sheeting materials.

Plasterboard can be unloaded directly and stored in the building where it is required, laid flat on timber bearers (Figure 5.43).

TIMBER LENGTHS

Poor storage of timber, wood products and units is a major cause of material wastage on building sites. If carcassing timber has to be stored outside then it should be stored off the ground on timber battens and protected against the weather with a tarpaulin cover (Figure 5.44).

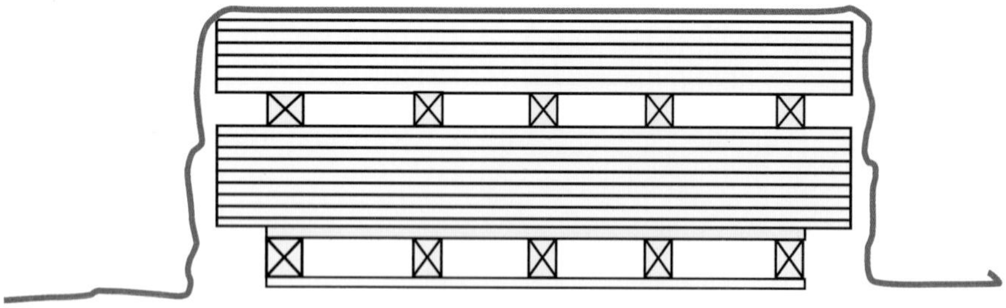

FIGURE 5.42
Storage and protection of sheet materials

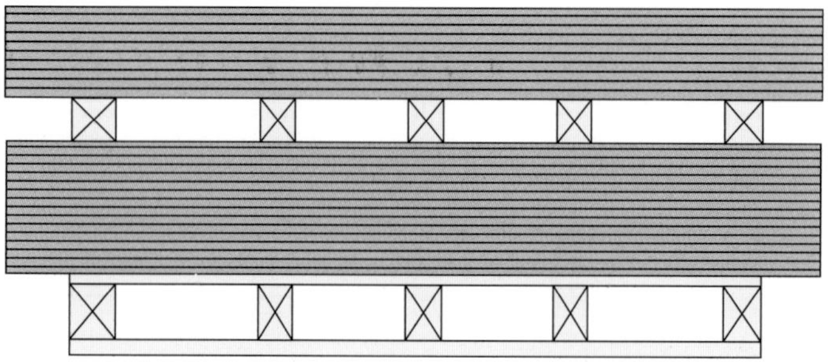

FIGURE 5.43
Storage of plasterboards

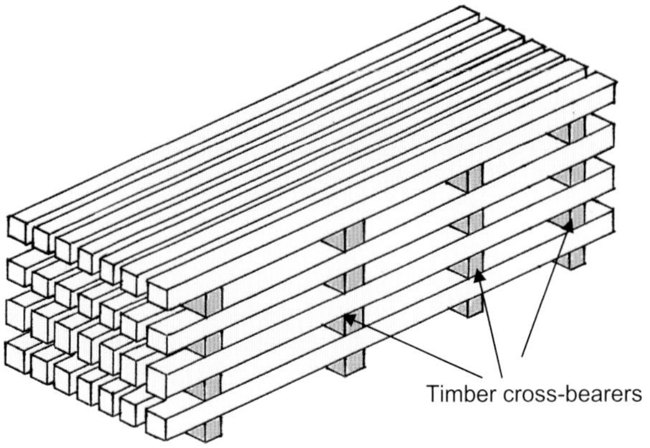
Timber cross-bearers

FIGURE 5.44
Storage of timber lengths

It is extremely important to ensure that the moisture content of the timber is kept at the right level for its use.

Timber components are usually delivered to site when required and should be stored inside a large, dry store or using part of the completed building as a store.

Always lay them flat and never lean them against a wall, etc., or they will bow, making them unusable (Figure 5.45).

PAINTS AND FLAMMABLE LIQUIDS

Paints should be stored inside a special store as shown in Figure 5.46.

Flammable liquids should be kept in securely capped cans or steel drums on which the contents are clearly marked.

Petrol, acetone, methylated spirits and other volatile liquids with flash-points below 32°C may be kept in robust metal lockers or well-ventilated non-combustible cabins, or in the open.

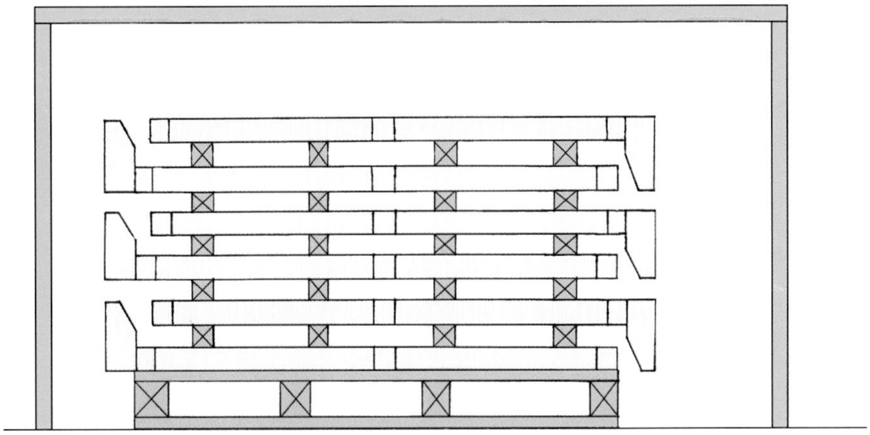

FIGURE 5.45
Storage of timber components laid flat under cover

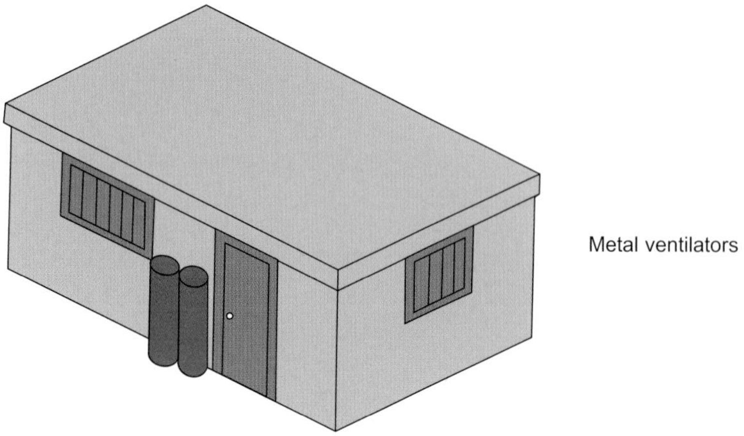
Metal ventilators

FIGURE 5.46
Storage of flammable liquids

Storage should be securely locked or fenced, and be situated at least 6 m away from any other huts, stores, site boundaries or buildings.

Smoking should be prohibited within 6 m of the store.

No other materials should be stored with flammable liquids.

CONTAINERIZED ITEMS

Any items delivered to the construction site in cardboard boxes should be ordered when required and stored in waterproof, dry conditions (Figure 5.47).

A secure store should be provided unless there are sections of the building that are complete and lockable which could be used.

Paint is delivered in tins, which also needs special arrangements for storage. Paint has a limited shelf-life so should only be ordered as and when required.

Shelving should be provided in the store so that items can be easily stored and retrieved.

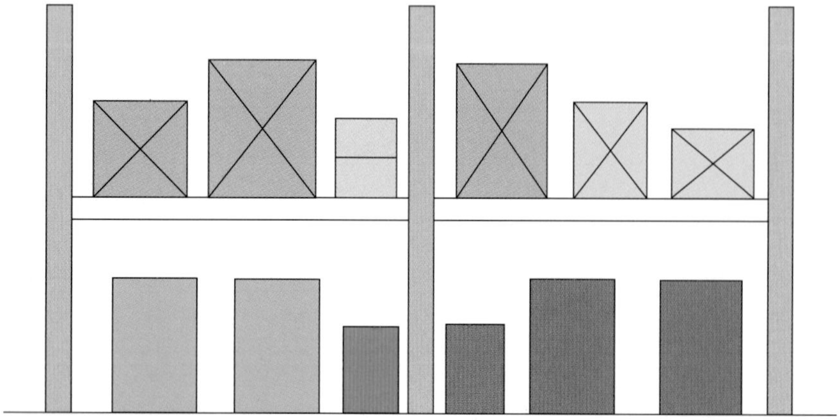

FIGURE 5.47
Storage of containerized materials

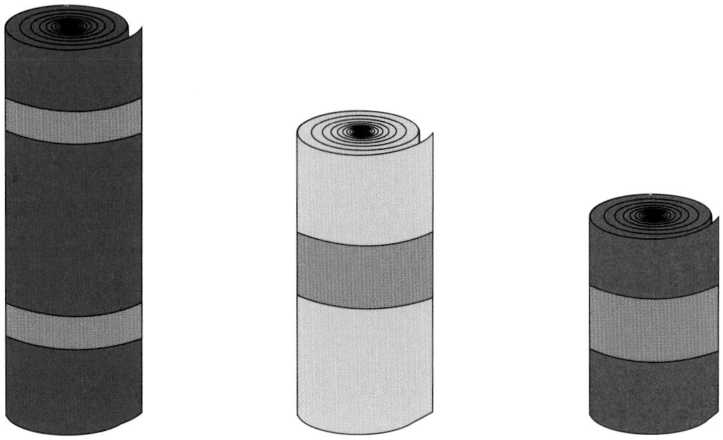

FIGURE 5.48
Storage of rolled materials

ROLLED MATERIALS

All rolled materials should be stored in an appropriate store, on end to retain their shape (Figure 5.48). They should be stored in their appropriate sizes for ease of identification.

They should be stored away from any source of heat as they could melt and become useless.

Note

NEVER stack DPCs flat as this causes the rolls to flatten and crack.

Roofing felt should be stacked in a similar fashion, on end and in special types for ease of identification.

Rolls of lead should also be stored on end in their separate codes. It is important to protect lead from damage, as damaged lead is useless. Lead should kept be in a lockable shed or building.

Minimizing damage and waste

With the present trend of material costs rising at a faster rate than labour costs, the control of waste is very important both on and off site.

The design of the building can add to the waste problem by involving cutting of materials.

Waste can be divided into:

- waste from mixing and cutting
- waste from poorly stored and protected materials
- waste from overordering material supplies
- waste from theft
- waste from the nature of the application of the material, e.g. mortar dropped on the floor by the bricklayer.

This chapter will only deal with the management of the resources on site.

As described previously, materials can be delivered to the site in various ways and the site manager is responsible for the best handling and storage to protect the materials from damage.

Materials should be ordered in exact quantities. In some cases this may not be economical owing to transport costs imposed by the supplier.

Loose materials such as building sand and ballast are ordered by the load but invoiced by the tonne.

Only order the required amount, allowing only 5 per cent for wastage, as the remainder is always left behind. A central mixing area, or using ready-mixed materials, can help to minimize wastage in this area.

Bagged materials are also a source of wastage, especially if part-open bags are left out at night and get damp.

The storage of materials on site is the direct responsibility of the site management and has a very strong influence in the control of waste. A careless regard for the value and utility of the materials on the part of management could lead to a progressive deterioration in the operatives' regard for the material.

Areas allocated on site for the storage of materials should be determined after considering the following questions:

- Will construction take place in that area? If so, when?
- Is the storage for a long or a short time?
- Can delivery vehicles safely and easily reach the area?
- Can on-site movement from storage areas to point of use be safely and economically carried out?
- Are the materials as near to their point of use as practically possible?
- If the materials have considerable value are they in the secure area?
- Will the storage area create problems in routing site transport and personnel?

The answers to the above questions will, of course, vary from site to site.

Please refer back to Figure 5.33, which shows a typical site layout with areas set out for material storage.

Disposal of materials from heights

Careful disposal of materials from heights is essential.

They should always be lowered safely and never thrown or dropped from scaffolds and window openings, etc.

Chutes can be employed as shown in Figure 5.49. These can be connected together to form any reasonable length. They should terminate in a tarpaulin-covered skip.

Waste can be kept in a skip until collected and taken to a suitable tipping area.

Remember

The protection of materials will depend on:

- susceptibility to climatic conditions
- value of materials
- size of the units
- scarcity of materials.

Hazardous waste, non-hazardous waste and debris

Waste resulting from work materials can include sand, gravel, cement, concrete, bituminous waste, sealants, bricks, stone, steel mesh, nails, timbers, fuels, oils, etc.

Waste resulting from work activities can include soils, vegetation, cuttings, inorganic materials, etc.

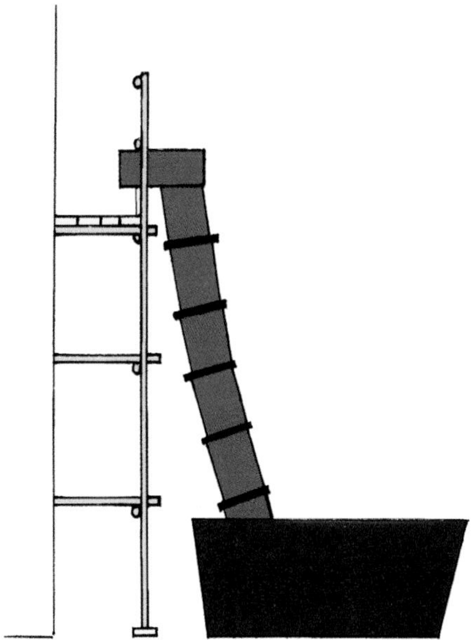

FIGURE 5.49
Removing waste from a building

Depending on the sector in which the operative works, the materials will differ. Some operatives will work in a specific trade occupation, relating to bricks or timber, while others will be involved in general construction activities, relating to excavation materials.

Most substances and materials are safe provided they are handled or worked on sensibly and with proper precautions. Almost anything can be dangerous if handled or used irresponsibly. Some materials require extra care; a few need extreme caution.

It is important to know what the hazards are, when they occur and how they can be prevented. Usually prevention will involve the use of protective clothing and equipment.

Removal of waste

Since 1 April 1992, the law on waste has included a duty of care that means you must take all reasonable steps to look after any waste you have and prevent its illegal disposal by others.

If you break the law, you could be fined an unlimited amount.

'Controlled waste' is any household, commercial or industrial waste, such as waste from a house, shop, office, factory, building site or any other business premises.

How do I know what to do?

The law requires you to complete certain paperwork and to take all reasonable steps to meet the duty of care. What is reasonable depends on who you are and on the circumstances. *The Duty of Care, A Code of Practice,* is published by HMSO (ISBN 011 75 25 57 X).

STEPS TO BE TAKEN IF THE DUTY OF CARE AFFECTS YOU

- When you have waste you have a duty to stop it escaping. Store it safely and securely.
- If you hand the waste to someone else:

  1. Secure it.
     - Most waste should be in suitable containers.
     - Loose material loaded into a vehicle or skip should be covered.
  2. Check that the person taking your waste away is legally authorized to do so.

Multiple-choice questions

Self-assessment

This section of the book is designed to allow you to check your level of knowledge. The section consists of revision questions for this chapter. The questions are all multiple choice and have four possible answers. The answers are to be found at the end of the book.

The main type of multiple-choice question will be the four-option multiple-choice question. This will consist of a question or statement, known as the stem, followed by a choice of four different answers, called the responses. Only one of these responses is the correct answer; the others are incorrect and are known as distracters.

You should attempt to answer the questions by choosing either (a), (b), (c) or (d).

Example

The person employed by the local authority to ensure that the Building Regulations are observed is called the:

(a) clerk of works
(b) building control officer
(c) council inspector
(d) safety officer

The correct answer is the building control officer, and therefore (b) would be the correct response.

Moving and handling resources

Question 1 Which of the following would be classed as bulk materials?

(a) window frames
(b) aggregates
(c) plasterboards
(d) ironmongery

Question 2 Lengths of timber should be stored:

(a) in the open
(b) leaning against a wall
(c) on timber bearers and covered with a tarpaulin
(d) on the ground and covered with a tarpaulin

Question 3 Bagged material should not be stored more than:

(a) 2 bags high
(b) 4 bags high
(c) 6 bags high
(d) 8 bags high

Question 4 Which of the following is a natural material?

(a) timber

(b) brick

(c) block

(d) steel

Question 5 Which of the following is best suited for carrying bricks up a scaffold?

(a) hod

(b) barrow

(c) pallet truck

(d) sack truck

Question 6 Which of the following is classed as perishable goods?

(a) bricks

(b) building sand

(c) plasterboards

(d) blocks

Question 7 Which of the following is the best method for storing roofing felt?

(a) laid flat in a store

(b) laid flat on ground

(c) on end on the ground

(d) on end in a store

Question 8 Identify the following type of brick:

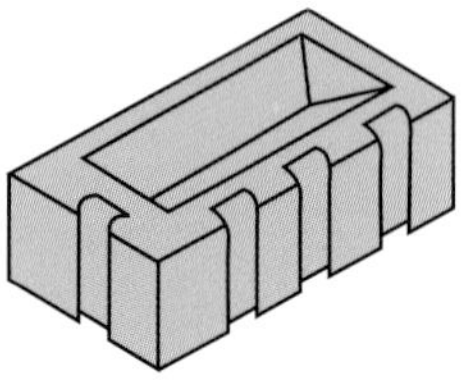

(a) solid wire-cut

(b) pressed or hand-moulded

(c) cellular

(d) keyed

CHAPTER 6

Solid Walls

This chapter will cover the following NVQ and Diploma units:

- NVQ VR40
- CC 2047K

This chapter is about

- Interpreting information
- Adopting safe and healthy working practices
- Selecting materials, components and equipment
- Preparing and erecting solid walling

The following NVQ performance criteria will be covered:

- Performance criterion 1: Interpretation of information
- Performance criterion 2: Safe work practices
- Performance criterion 3: Selection of resources
- Performance criterion 4: Minimize the risk of damage
- Performance criterion 5: Meet the contract specification
- Performance criterion 6: Allocated time

The following Diploma outcomes will be covered:

- Know how to plan and select resources for practical tasks
- Know how to erect solid walling
- Know how to erect isolated and attached piers

Interpretation of information

Sources of information

There are many items of information available to the trainee bricklayer with regard to solid walls and their application.

Many builders have numerous sources of information such as:

- manufacturers' literature
- specifications
- Building Regulations
- British Standards Institution
- British Standard specifications
- codes of practice
- Building Research Establishment (BRE)
- BRE digests, information papers and current papers.

The trainee can also visit libraries to find information from:

- books and technical articles.

Many libraries also take trade periodicals which operate a reader's enquiry service whereby technical brochures and information can be obtained from various manufacturers and suppliers.

Trade periodicals:

- *Cement and Concrete Monthly*
- *The Architects' Journal*
- *Building*
- *The Brick Bulletin*

and many more.

When information has been received it is important that the trainee can interpret it and be able to use it correctly.

Other information can be received from supervisors and workmates by oral or written methods.

Remember there is no real proof that the person has received the correct information, but a quick response can be achieved with oral instructions.

This chapter is also concerned with the selection, maintenance and safe use of power tools and equipment.

Whether dealing with powered static equipment or hand-held powered tools, or carrying out predictable and routine operations with powered tools and equipment and designated small mobile plant, identifies good practice procedures that need to be habitual to the work operator.

Before selecting the power tools for use on a contract it is essential to know all about the various tools and equipment, especially the powered ones.

Technical information can be obtained in a variety of ways and from a variety of sources. Most tools and equipment come with manufacturer's safety instructions and technical information.

Technical information can be produced in several formats:

- operating instructions – how to use the item
- safety guidelines – power supply, personal protective equipment (PPE) to be worn and recommended checks
- technical information – mechanical details and possible outputs.

Information

Normal hand tools are not usually provided with manufacturer's instructions.

Most textbooks explain hand tools in detail and the safety procedures to follow while using them.

Information on hand-held power tools may be easy to find and is usually received along with the purchase of the particular item of tool or equipment.

This may cause a problem if the item is being hired for a period, but the hirer should be able to supply technical, safety or operating instructions.

Safe work practices

Abrasive wheels/stone saws

The most common type of power tool in use by the bricklayer when building solid walls is the angle grinder or disc cutter.

These tools are used for cutting various types of materials from stone and brick to concrete and tarmac.

They are available in all types and sizes, ranging from small hand-held skill saws to bench-operated disc cutters (Figure 6.1).

Cutting with power tools

When components are cut by machine it is essential that the equipment is set up correctly and safely and the operator has had sufficient training and has acquired the required certificate.

As for hand cutting, the appropriate PPE must be worn.

The main type of hand-held equipment for cutting building components is the angle grinder or disc cutter. A table version of a disc cutter is used in workshops and building sites when a great deal of cutting has to be done.

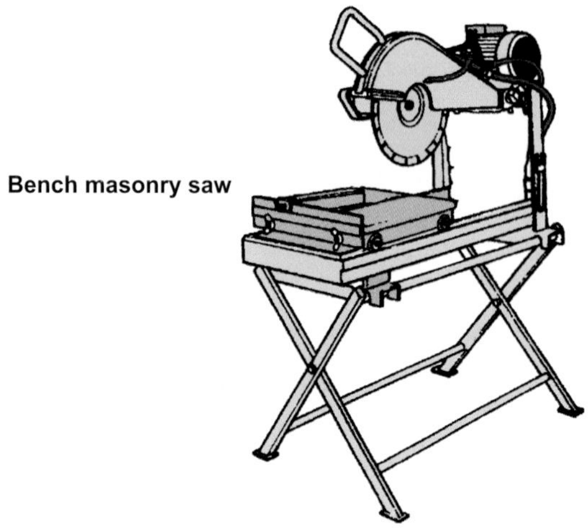
Bench masonry saw

FIGURE 6.1
Cutting equipment

Portable power tools

Numerous portable power tools are now available for all trades and it is essential that the manufacturer's instructions are read and understood before setting up and using the equipment.

Whenever power tools are being used on the construction site it is essential that the correct safety procedures are adhered to at all times.

Please refer back to Chapter 2, Health and Safety in the Construction Industry, as required.

Remember

It is the employer's responsibility to ensure that you are correctly trained to use power tools. The item of equipment provided must be maintained and safe. Figure 6.2 shows a warning sign.

Employees, in turn, have a duty to inform their employer of any work situation that presents a risk to themselves or their workmates.

Electricity

You will need a basic understanding of:

- the dangers:
  - what makes electricity so dangerous
  - what electric shock can do to the body
  - what type of work might bring you near to live electrical cables.

DO NOT CHANGE
GRINDING WHEEL
UNLESS AUTHORIZED
TO DO SO

FIGURE 6.2
Abrasive wheel warning notice

- avoiding danger:
  - how to plan your work so that you are not put at risk
  - what precautions to take so that you and others are not put at risk of electrocution
  - how to avoid injury from overhead cables
  - how to avoid injury from buried cables
  - how to recognize danger areas into which you should not go
  - the maximum safe voltage for use on site
  - how to reduce the voltage to a safe level.

ELECTRICITY SUPPLY

The supply of electricity will normally be provided by one or both of the following:

- public supply from the local electricity board
- a site generator when public supply is not practicable or uneconomic.

SITE DISTRIBUTION

The electricity supply for operatives to use is distributed around the site using a number of different units.

All supply, distribution and transformer units should be marked with the warning sign shown in Figure 6.3. An additional sign should be placed under the warning sign and should state the highest voltage likely to be present.

PLUGS, SOCKETS AND OUTLETS AND COUPLERS

To avoid plugs designed for one voltage being connected to sockets of another voltage, there are different positions for the keyway in plugs and sockets. The examples shown in Figure 6.4 conform to British Standard BS 4343.

CABLES AND WIRES

In all site offices, worksites and workshops and similar premises wiring should comply with existing Wiring Regulations.

DANGER 415 V

FIGURE 6.3
Electric safety sign

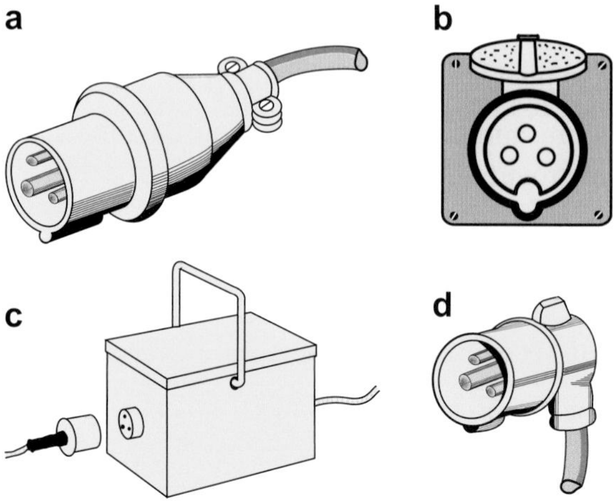
a
b
c
d

FIGURE 6.4
110 V plugs and connectors

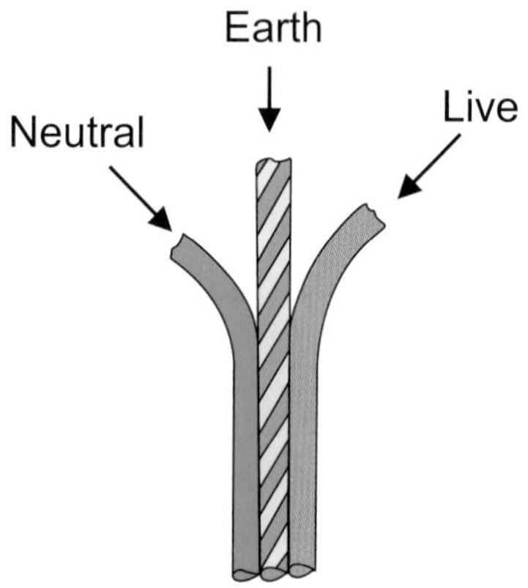
Earth
Neutral
Live

FIGURE 6.5
Wiring colours

Since 2004 all low-voltage cable colours have had to comply with the cable colours of the European Union (Figure 6.5).

These colours are:

- neutral – blue (previously black)
- earth – green/yellow (same as previously)
- phase 1 – brown (previously red)
- phase 2 – black (previously yellow)
- phase 3 – grey (previously blue).

WIRING FOR PLUGS

A wiring diagram for an electrical three-pin plug in everyday use is shown in Figure 6.6.

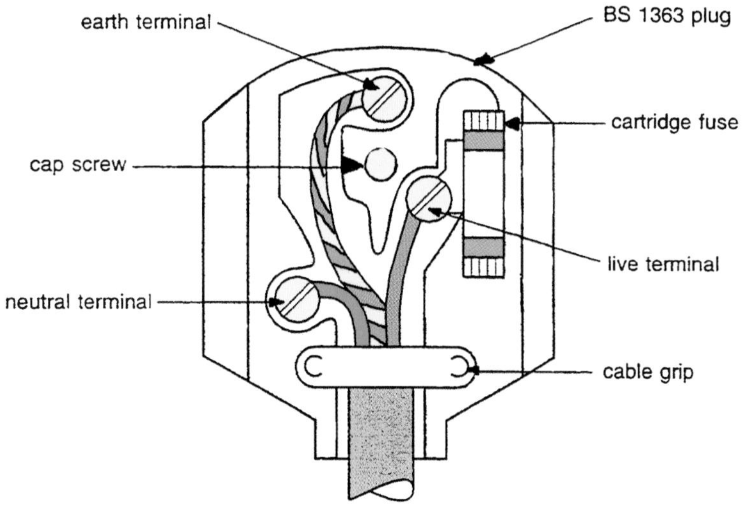
earth terminal
BS 1363 plug
cartridge fuse
cap screw
live terminal
neutral terminal
cable grip

FIGURE 6.6
Wiring diagram for an electrical three-pin plug

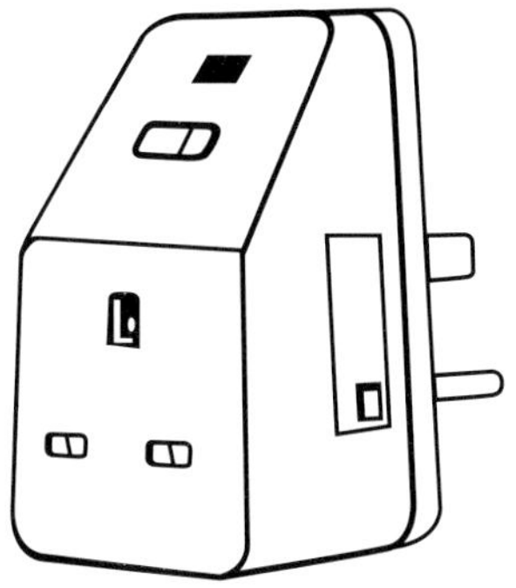

FIGURE 6.7
RCD breaker

ELECTRIC CIRCUIT BREAKERS

The residual current device (RCD) (Figure 6.7) is also known as an earth leakage circuit breaker (ELCB) or residual current circuit breaker (RCCB).

This indispensable device cuts off all current in a fraction of a second if it defects to a sudden leakage to earth due to a fault.

Always use one when working with electric power tools.

Warning

NEVER overload sockets with adapters.

NEVER use a cable that has not been fully unwound.

Selection of resources

When building solid one-brick walls the resources will simply be tools, bricks and mortar.

This chapter will assume that the work has been completed below ground level, so the first procedure will be setting out the materials at ground level.

Setting out materials

Before brickwork commences the working area must be stacked out with bricks and blocks as in Figure 6.8.

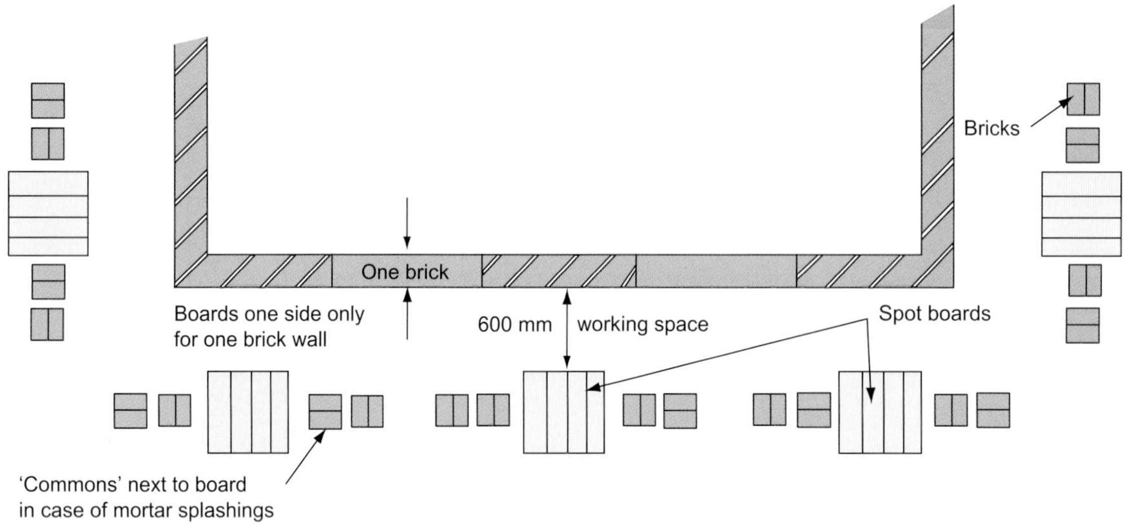
Bricks
One brick
Boards one side only
for one brick wall
600 mm
working space
Spot boards
'Commons' next to board
in case of mortar splashings

FIGURE 6.8
Setting out materials before building

Keep stacks away from edges trenches. Keep walkways and runways clear. Do not stack materials too high. Do not stack wider at the top than at the bottom.

Stack materials near the job to avoid double handling. Keep materials at working height.

Protect materials from breakages, dampness and frost. Cover bricks with tarpaulin, hessian or polythene.

Take care when opening packs of bricks. The polythene wrapping or banding should be left in place until the bricks or blocks are required. The correct way to open packs is by cutting the straps with snips. Do not chop the straps with a brick hammer or trowel or any tool that will damage the bricks. For safety reasons, take the banding to the skip immediately, or make into safe bundles and remove from the work area later.

Polythene wrapping, if removed carefully and placed to one side, could be used elsewhere for protection. If it is not required remove to the skip as soon as possible.

When loading out for the bricklayer place the bricks with the frog or perforations uppermost, for easy handling by the bricklayer.

Source the bricks from as many packs as possible, but at least three. This will prevent banding along the wall of different coloured or textured bricks. Banding of bricks can occur in the face of walls if bricks are taken from only one pack at a time (Figure 6.9).

Remember

Stack bricks and blocks safely, within easy reach of the bricklayer.

Space mortar boards at the corners of the building and not more than 3 m apart along the wall.

Level the ground off for easy walking and block up the mortar boards about 600 mm from the face of the wall.

Minimizing risks

It is important that the trainee is aware of the dangers on site and takes the necessary precautions such as wearing the appropriate PPE.

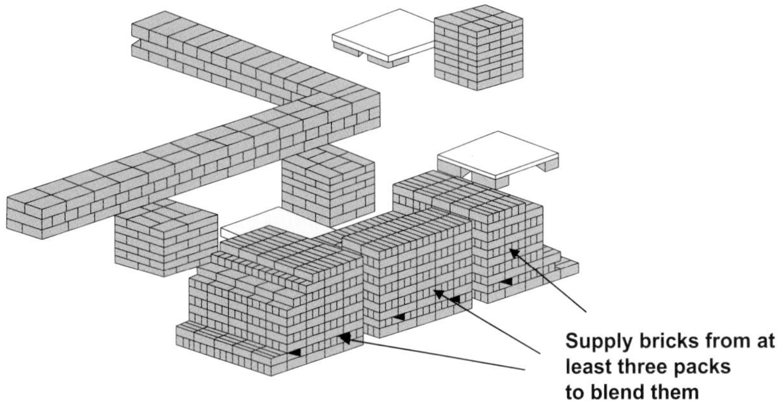
Supply bricks from at
least three packs
to blend them

FIGURE 6.9
Selecting bricks when setting out

One of the most dangerous aspects of solid brick walling, power tools, has already been covered. Another important task to watch is the movement of materials around the site.

Transporting bricks

Bricks can be carried under the arm from the stack to where they are required, but this is laborious and slow.

The use of brick clamps which can pick up six bricks on edge saves the hands and is quicker.

Barrows are much quicker, although barrow runs will have to be laid over rough ground.

Brick trolleys which carry 48 bricks held by a clamp are best used on the scaffold or on smooth ground. Bricks must be stacked on edge ready for the brick trolley.

The most common method used for moving bricks on larger sites is by fork-lift truck (Figure 6.10).

Protection of the surrounding area is important so as to minimize risks to the public.

FIGURE 6.10
Moving bricks with the fork-lift truck

The building site should be kept clean and free from as much dust and noise as possible. All waste materials should be collected in a skip, which can be moved off site when full.

The new walling should be protected with sheeting as the work proceeds, until fully set.

Meeting the contract specification

Bonding solid walls – recap on basic rules

BONDING OF BRICKWORK

The bonding of brickwork is required for its appearance as well as its strength.

- a bond used for strength – English bond
- a bond used for appearance – Flemish bond.

Bonds can also be used for their economical use of facings, such as the garden wall bonds.

Table 6.1 shows the number of facing bricks required according to the type of bond.

PRINCIPLES OF BONDING

To maintain strength, bricks must be lapped one over the other in successive courses along the wall and in its thickness.

There are two practical methods, using either a half-brick lap or a quarter-brick lap, called half-bond and quarter-bond (Figure 6.11).

If the lap is greater or smaller than these, then both appearance and strength are affected. If bricks are so placed that no lap occurs, then the cross-joints or perpends are directly over each other (Figure 6.12), and this is termed a straight joint, being either external for those appearing on the face of the wall or internal for those occurring inside the wall, and they should be avoided whenever possible.

The apprentice should note that internal straight joints will occur in some bonding problems. On the one hand, excessive cutting may solve

Table 6.1 Number of facing bricks required according to bond (per 0.84 m^2)

Type of bond	Thickness of the brick		
	50 mm	65 mm	70 mm
English bond	90	72	66
Flemish bond	80	64	59
Dutch bond	90	72	66
English garden wall bond	75	60	55
Flemish garden wall bond	69	56	51
Header bond	120	96	88
Stretcher bond	60	48	44

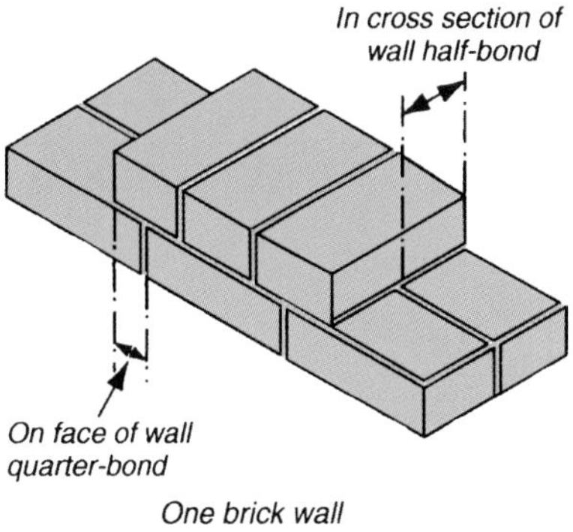
In cross section of wall half-bond
On face of wall quarter-bond
One brick wall

FIGURE 6.11
No straight joints

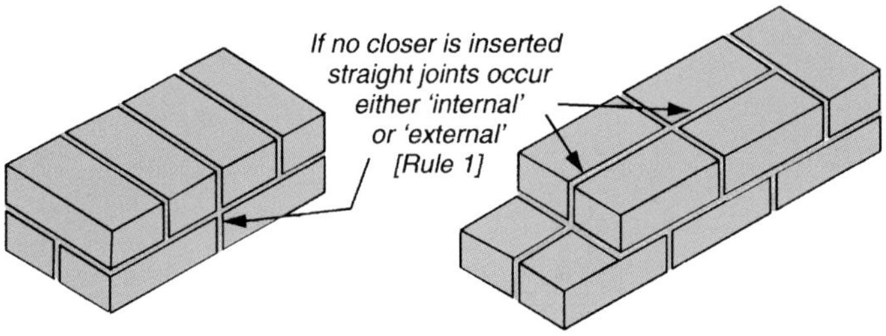
If no closer is inserted straight joints occur either 'internal' or 'external' [Rule 1]

FIGURE 6.12
Straight joints

a particular problem, but this wastes labour and materials and tends to weaken the wall. On the other hand, by introducing one or two straight internal joints, whole bricks can be used. This is a case where practice and theory must compromise.

The pattern in a brick wall is purposely arranged, has its particular use, and is called a bond.

To summarize, the two main principles of the bonding of brickwork are:

- to maintain half- or quarter-bond, avoiding external straight joints and internal straight joints wherever possible
- to show the maximum amount of specified face bond pattern possible.

To assist in maintaining these principles, rules should be remembered and applied (Table 6.2).

The apprentice should never try to remember all the problems shown as examples. Problems must be solved as they occur by the logical application of the rules. Eventually, the bonding of brickwork becomes automatic to the bricklayer.

Several bonds are in general use, but for the purpose of beginning the apprentice's bonding education, stretcher, English, and Flemish bonds

Table 6.2 Rules of bonding

Rule 1	The LAP of brickwork along the face of a wall shall be ¼ brick 56.25 mm	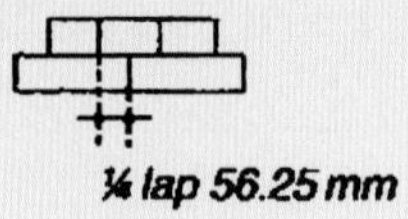¼ lap 56.25 mm
	Exception: Half-lap of 112.5 mm in stretcher bond	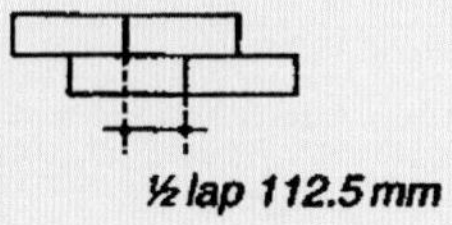½ lap 112.5 mm
Rule 2	The LAP of brickwork along the face of a wall shall be 112.5 mm	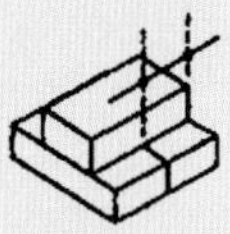
Rule 3	Brickwork should be set out on the face side and from each end of a wall or pier, so that any BROKEN BOND is centrally located Exception: Reverse bond, where end bricks do not correspond	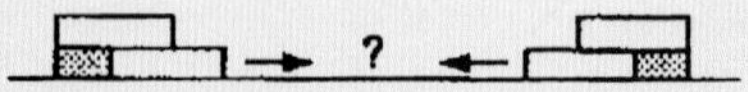?
Rule 4	In English and Flemish bonds a queen closer must be placed next to the QUOIN HEADER, in order to establish ¼ lap	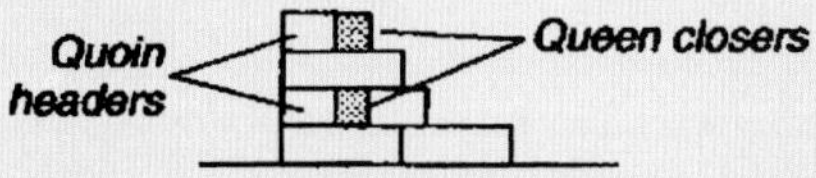Quoin headers Queen closers
	Exception: It is permissible to use a ¼ brick instead in Flemish bond, if this will avoid broken bond at centre, or is otherwise preferred	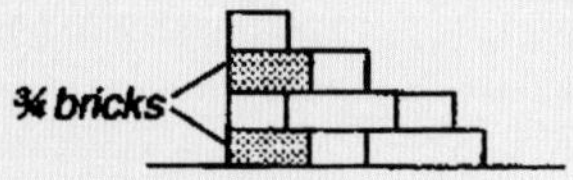¾ bricks
Rule 5	Cross-joints in consecutive courses of bricks must not coincide one above the other to create STRAIGHT JOINTS	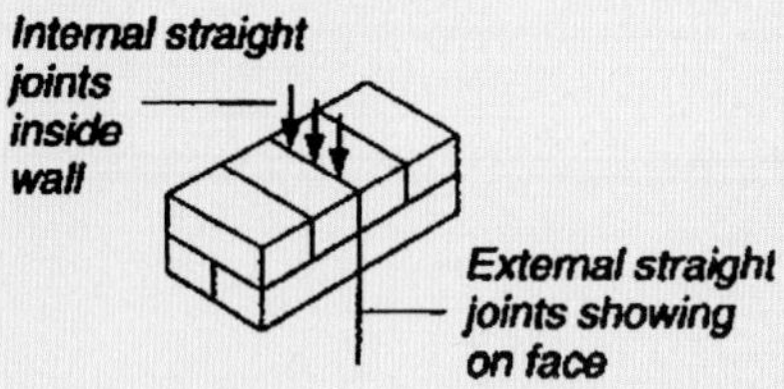Internal straight joints inside wall External straight joints showing on face
Rule 6	Cross-joints in alternate courses must coincide vertically one above the other on the face of the wall (PERPENDS)	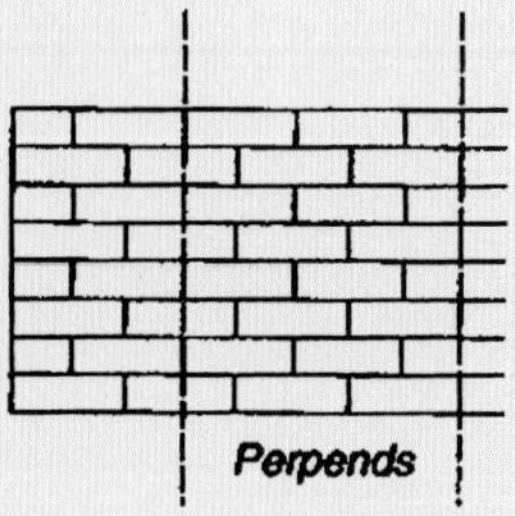Perpends
Rule 7	The TIE-BRICK between one brick thick English bond walls is always a HEADER	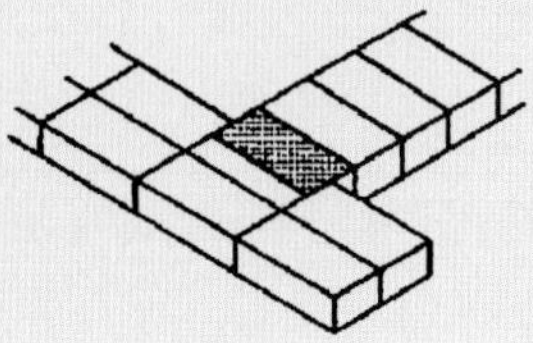

Table 6.2 Rules of bonding – Cont'd

Rule 8	The TIE-BRICK between one brick thick Flemish bond walls is always a STRETCHER	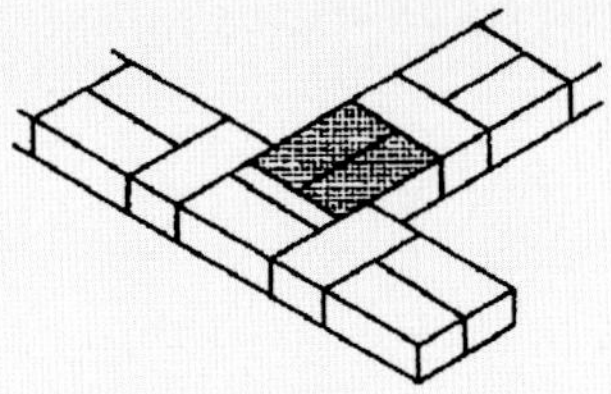
Rule 9	In English bond, when a wall changes direction, the face bond changes from headers to stretchers or vice versa in the same course	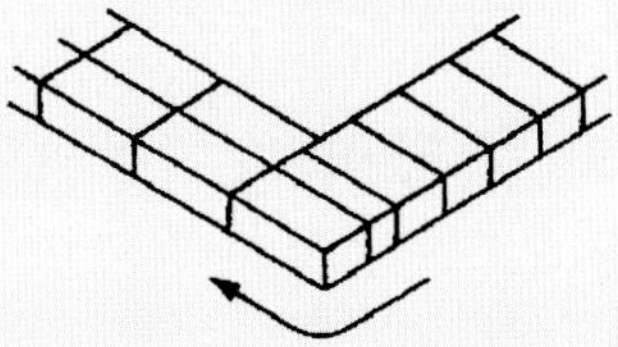
	Exception: This rule will not apply when (a) walls of different thickness interact, (b) walling curves on plan, (c) where the change of direction is 112.5 mm or less	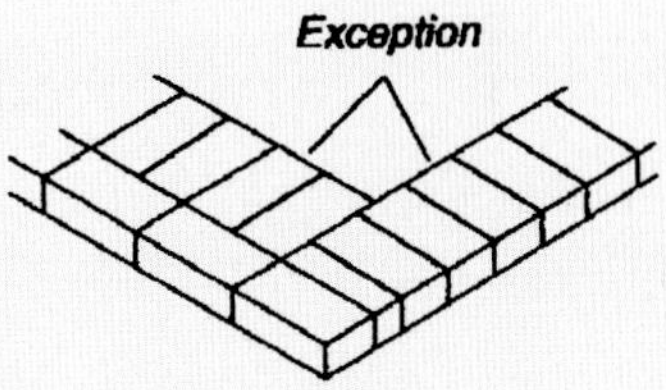 Exception
Rule 10	When bonding intersecting walls, run the bonding of each wall across the other in alternate courses	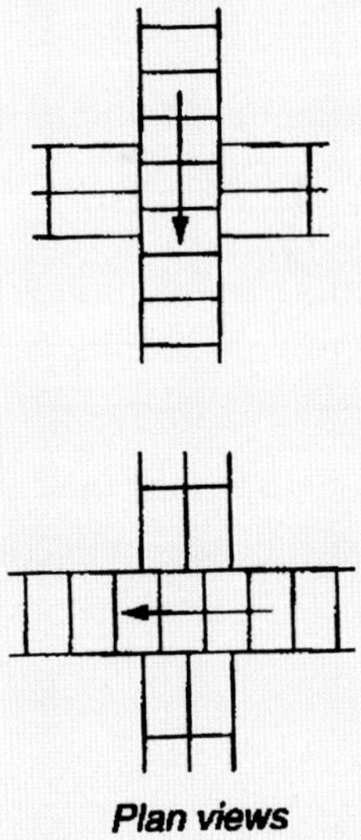 Plan views

will be explained. Problems in other bonds can be solved by the application of the same rules.

Note

Wall thicknesses are usually stated in brick sizes, e.g. the width of a brick is known as a half-brick wall; the length of a brick as a one-brick wall; the width, plus length of a brick, as a one-and-a-half brick wall; and so on.

DIMENSIONS

The bricklayer should remember the various dimensions to which bonds work without having to resort to broken or reverse bond.

Table 6.3 shows multiples of brick sizes, using normal size bricks of 215 mm, not including a 10 mm joint.

Table 6.3 Multiples of brick sizes

No. of bricks	Brick size	Joint size	Total length
1	215	–	215 [a]
1½	215 + 107.5	10	332.5
2	430	10	440
2½	537.5	20	557.5
3	645	20	665
3½	752.5	30	782.5
4	860	30	890
4½	967.5	40	1007.5
5	1075	40	1115
5½	1182.5	50	1232.5
6	1290	50	1340
6½	1397.5	60	1457.5
7	1505	60	1565
7½	1612.5	70	1682.5
8	1720	70	1790
8½	1827.5	80	1907.5
9	1935	80	2015
9½	2042.5	90	2132.5
10	2150	90	2240

[a] Not practicable.

SETTING OUT FACEWORK IN A WALL WITHOUT OPENINGS

Before setting out the bond on the face side of any wall, it is wise to make a gauge rod (Figure 6.13). This should be made of straight, smooth timber 50 × 50 mm in cross-section, and approximately 3 m long. Along one face fine saw cuts carefully made at 225 mm intervals will allow stretcher bond to be set out consistently from end to end, leaving any broken bond near the middle of the wall.

On another surface of the rod, saw cuts at 75 mm intervals serve to check regular vertical gauge when bricklaying commences and quoins are raised. Use of this dual-purpose storey rod, for checking horizontal and vertical gauge, avoids the potential risk of error when using a measuring tape.

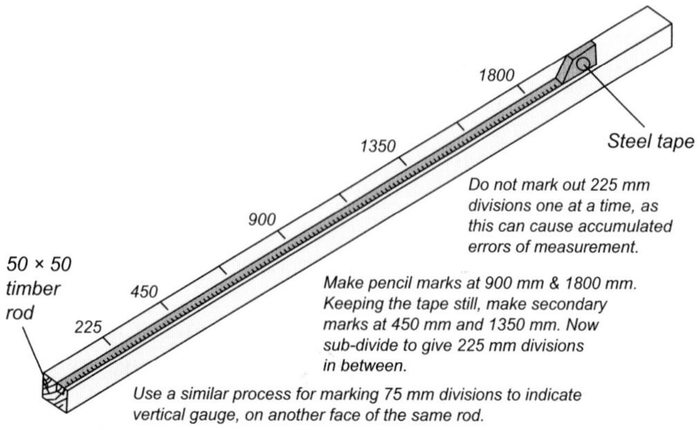
1800
1350
900
450
225
Steel tape
50 × 50
timber
rod
Do not mark out 225 mm divisions one at a time, as this can cause accumulated errors of measurement.
Make pencil marks at 900 mm & 1800 mm. Keeping the tape still, make secondary marks at 450 mm and 1350 mm. Now sub-divide to give 225 mm divisions in between.
Use a similar process for marking 75 mm divisions to indicate vertical gauge, on another face of the same rod.

FIGURE 6.13
Marking out a gauge or storey rod

SETTING OUT BONDS

When the brickwork of a building approaches ground level it is normal practice to change from common bricks to facing bricks. This changeover normally takes place two courses below finished ground level to avoid commons being shown when the building is complete.

When the facings reach ground level preparation must be made in the bond arrangement for inserting door and window openings at a future height. It is at this stage that the proposed elevation of the building must be carefully studied.

This chapter will deal only with walls without openings.

DRY BONDING

The normal practice is to dry bond the bricks from each end of the wall, placing the 'broken bond' in the centre of the wall.

Setting out facework is usually the responsibility of the supervisor who will, after consulting the architect, determine the detailed bond pattern and the location, if any, of broken or reverse bonds.

Setting out the brickwork is completely different from setting out the building, which is done before excavation begins.

It is essential that no straight joints occur.

Reveal bricks provide fixed points between which the bonding is set out.

On no account should a closer be built in the middle of the wall; it should only be placed next to the quoin header.

QUOINS

Quoins occur at the end of a straight wall, known as a stopped end (Figure 6.14).

Returns occur when a wall is being continued around the corner (Figure 6.15). The corner should be erected, continuing from the work completed up to ground level.

Remember

Change direction – change bond.

The corners should not be erected too high as it is much more economical to run in the wall with the aid of a line than to erect large corners and rake them back to form corners.

Toothing should be avoided where possible. Some clerks of works may not allow toothing at corners as it could cause a weakness if the joints are not

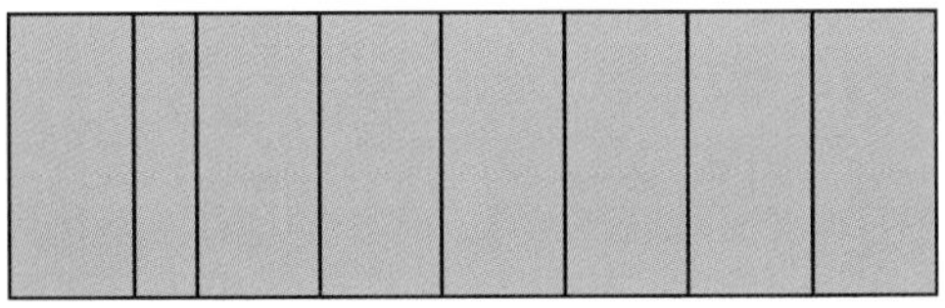

FIGURE 6.14
One brick stopped end

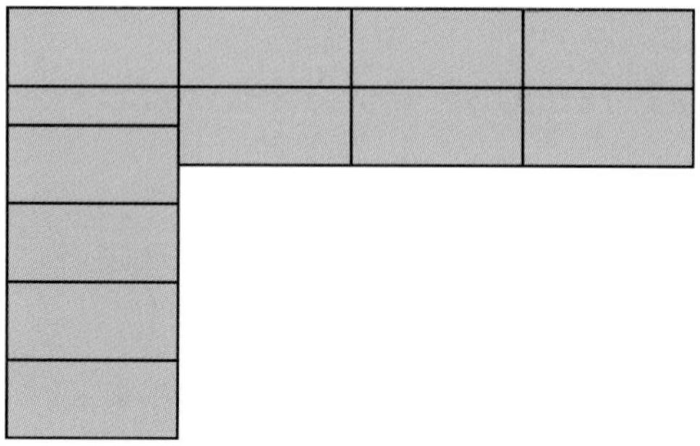

FIGURE 6.15
One brick quoin

filled correctly. See an example of raking back in Figure 6.16. Toothing the corner produces a weakness when the wall is built into it if the joints are not consolidated (Figure 6.17).

DAMP-PROOF COURSE

Brickwork above ground level usually commences with the damp-proof course (DPC) (Figure 6.18).

Engineering bricks are often used for DPCs and the common type is known as a 'blue brick' DPC. This is usually two courses laid in a mortar equal to the strength of the engineering bricks, which is 1:3.

Engineering bricks are difficult to lay as they can 'swim' in the mortar owing to their high density and lack of absorption. It is therefore essential to keep

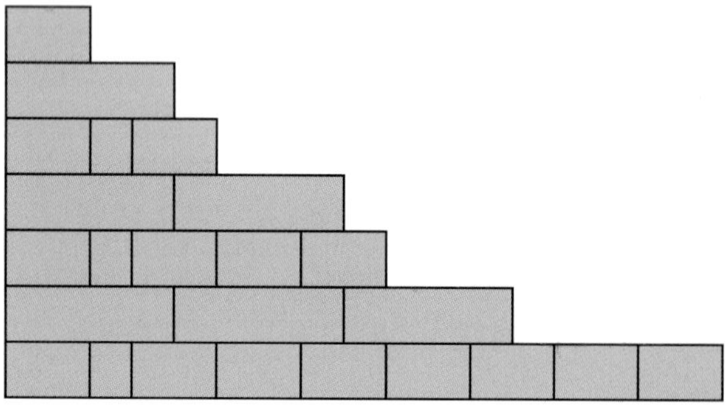

FIGURE 6.16
Raking back produces the strongest corner

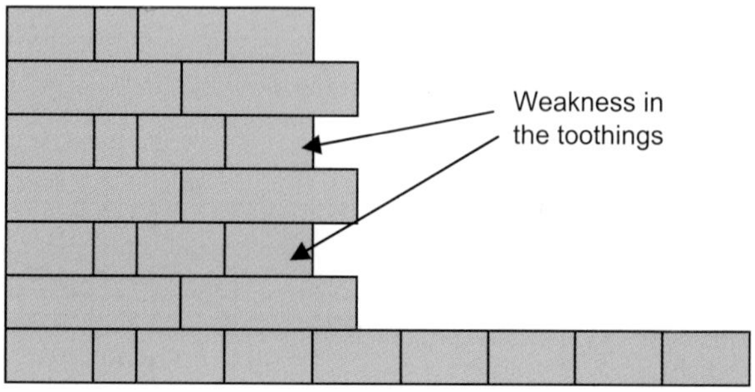
Weakness in
the toothings

FIGURE 6.17
Toothing a corner

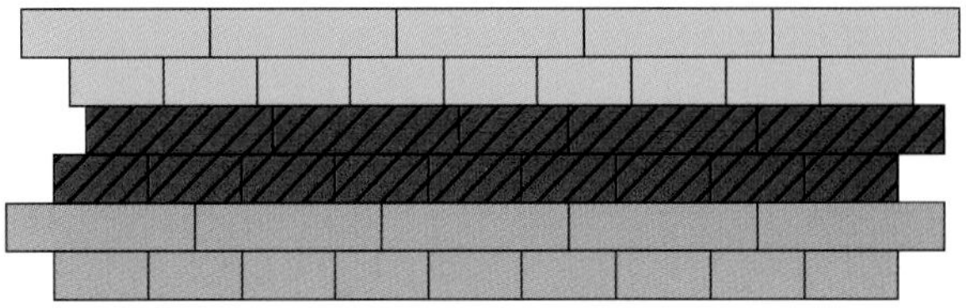
Facing bricks
Engineering bricks
Common bricks

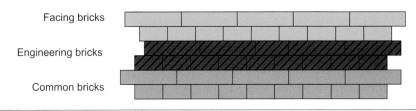

FIGURE 6.18
Damp-proof course in engineering bricks

the bricks as dry as possible and the mortar as stiff as possible to allow easy spreading.

Note

Do not attempt to joint the bricks until the mortar has hardened sufficiently.

Two courses of DPC bricks bedded in cement mortar provide a better resistance to overturning of walls than flexible DPCs.

This will be reinforced when boundary walls and piers are dealt with later.

TINGLES

When building walls longer than 1.125 m long it is always advisable to use lines and pins.

If the wall exceeds 9 m in length then it is advisable to use a tingle plate to prevent the line from sagging (Figure 6.19). The tingle brick should be as near as possible to the middle of the wall.

COLLAR JOINT

This is the joint that runs along the centre of the stretcher course when building a wall in English bond (Figure 6.20).

Different skills are required when building one-brick walls.

STRETCHER COURSE

Care should be taken when laying the stretcher course as too much mortar between the bricks will cause the face of the wall or the back of the wall to bend.

The face of the wall should be laid first to the line provided.

When the mortar is laid for backing the wall it should be pulled back from the face brick so that when the backing brick is laid it does not push out the facing brick (Figure 6.21).

HEADER COURSE

When the header course is being laid special attention should be paid to the long cross-joint (sectional joint) along the stretcher face of the brick (Figure 6.22).

The brick will have to be turned in the hand to expose the longer face of the brick.

The cross-joint should be applied in full, not top and bottom, to produce a solid joint.

The correct action is vital if full joints are to be achieved. Again, practice is necessary to achieve full joints.

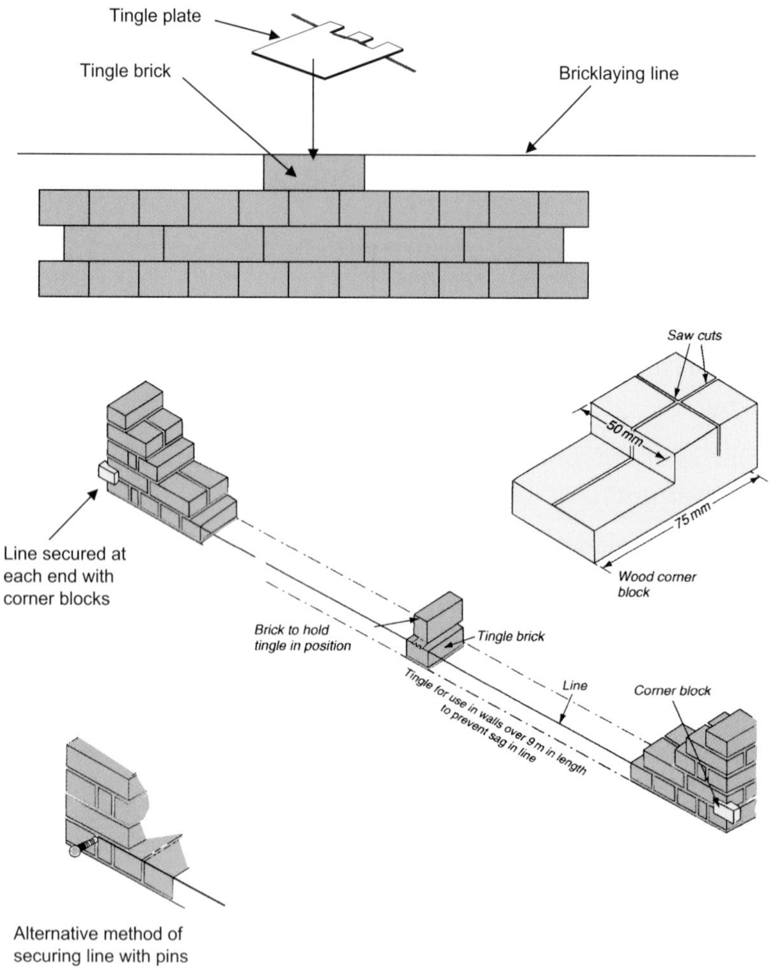
Tingle plate
Tingle brick
Bricklaying line
Saw cuts
50 mm
75 mm
Wood corner block
Line secured at each end with corner blocks
Brick to hold tingle in position
Tingle brick
Line
Corner block
Tingle for use in walls over 9 m in length to prevent sag in line
Alternative method of securing line with pins

FIGURE 6.19
Use of tingle plate to prevent line sagging

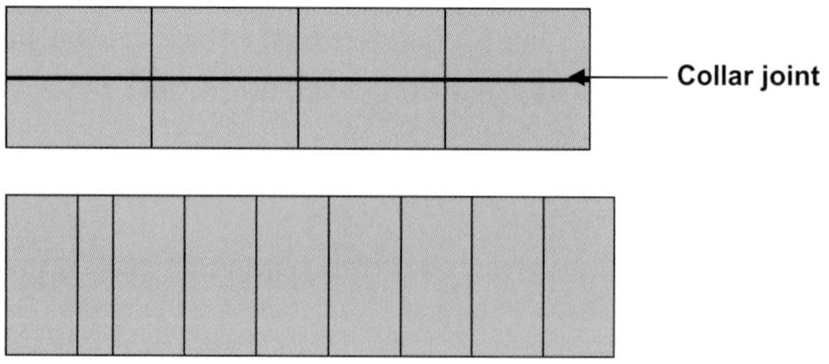
Collar joint

FIGURE 6.20
Alternate courses in English bond

Laying the back course of stretchers.

It is usual to lay the back course from the front of the wall. This is another extra skill that has to be practised to obtain a straight face to the rear of the wall.

Remember

There is no line to the back of the wall so it is essential to use the eye to line the bricks in with the course below.

PERPENDS

There are more cross-joints when building one-brick walls.

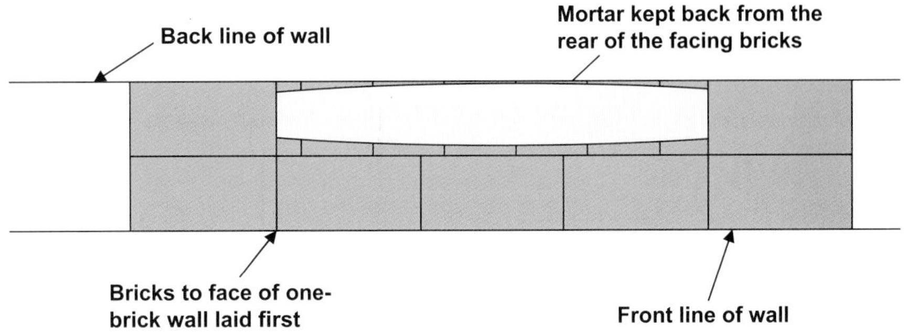
Back line of wall
Mortar kept back from the rear of the facing bricks
Bricks to face of one-brick wall laid first
Front line of wall

FIGURE 6.21
Laying back stretchers

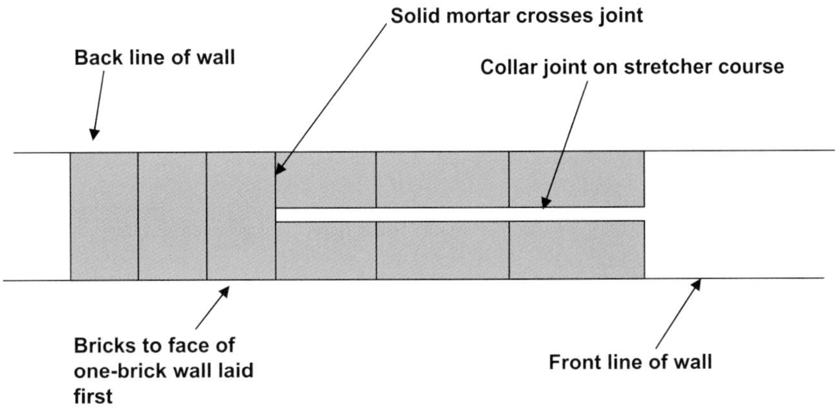
Solid mortar crosses joint
Back line of wall
Collar joint on stretcher course
Bricks to face of one-brick wall laid first
Front line of wall

FIGURE 6.22
Laying the header course

It is essential that they are controlled and that perpends are kept plumb. This can be done by checking with a spirit level or using a T square (Figure 6.23).

SOLID BRICK WALLS

A large number of bonds is available for one-brick walls and above. They have been discussed in Chapter 7, Level 1. Please refer back to this chapter for further information.

This chapter will deal with the construction of solid walls and, where appropriate, the bonding arrangements.

BASIC SKILLS FOR ONE-BRICK WALLS

Acquiring the many skills required of a bricklayer will take time and patience. What you require is dexterity: the skilful use of the hands together with eye co-ordination.

Watching your tutor or bricklayer carefully on site, and regular practice over time will help you to attain these skills.

By this point in your training you will have begun to create a grasp of the trowel, either left-handed or right-handed use, be able to pick up the mortar with an easy application and spread it on the wall to produce an economical bed of mortar.

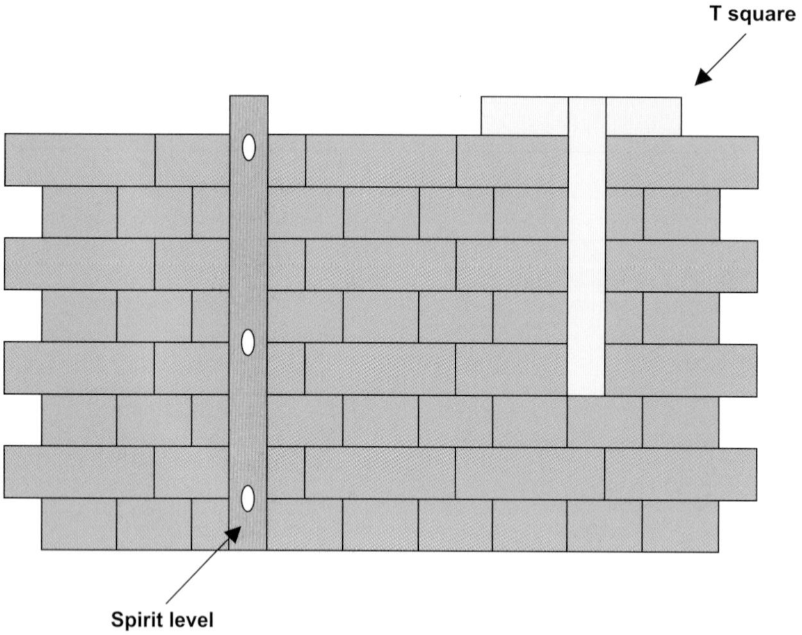
T square
Spirit level

FIGURE 6.23
Plumbing perpends

Correctly and safely handling bricks to be laid will also need continuous practice, especially when one-brick walls are being built.

The bricks may have to be turned several times in the hand, especially when laying header courses.

The building of solid walls involves some different skills from those required when building half-brick walls.

With one-brick walls the materials are all bricks, but commons could be required for the back of the wall.

ENGLISH BOND

Walls built in English bond are very strong as no straight joints occur in any part of the wall.

A unit of bond, shown hatched, of a one-brick wall in English bond, is illustrated in Figure 6.24.

FLEMISH BOND

Flemish bond is used mainly for decorative purposes as internal joints make this bond slightly weaker than English bond.

A unit of bond, shown hatched, of a one-brick wall in Flemish bond, is illustrated in Figure 6.25,

GENERAL BONDS FOR ONE-BRICK WALLS

A large number of bonds is available for one-brick walls and above. Some of the most common ones are described below.

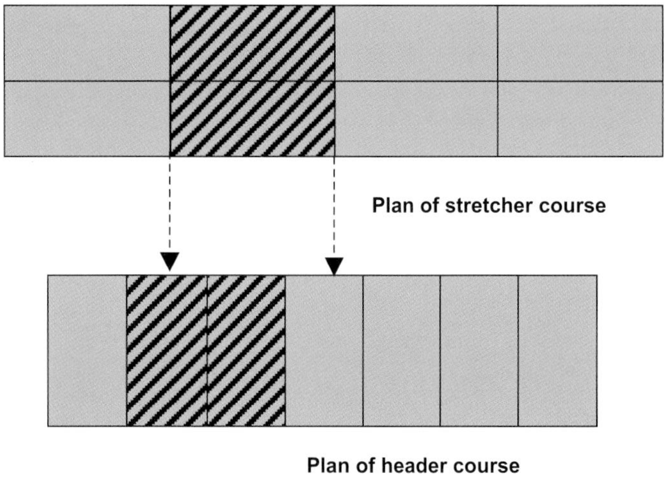
Plan of stretcher course
Plan of header course

FIGURE 6.24
Unit of bond in English bond (no straight joints)

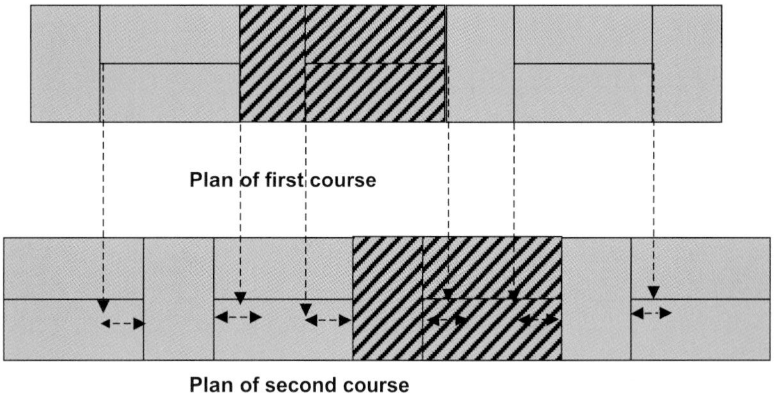
Plan of first course
Plan of second course

FIGURE 6.25
Unit of bond in Flemish bond (straight joints occur on the internal collar joint)

Header bond

Header bond is only suitable for walls one brick thick. It consists of all headers, with the bond being formed by three-quarter bats at the quoin, and is generally used in brickwork below ground or walling curved on plan (Figure 6.26).

English bond

Walls built in English bond are very strong as no straight joints occur in any part of the wall.

Alternate courses of headers and stretchers produce quarter-bonds and because of its somewhat monotonous appearance it is used where strength is preferable to appearance.

To achieve and maintain quarter-bond a queen closer must be laid next to the quoin header (Figures 6.27–6.29).

Attention to the following points helps to maintain a good standard of work.

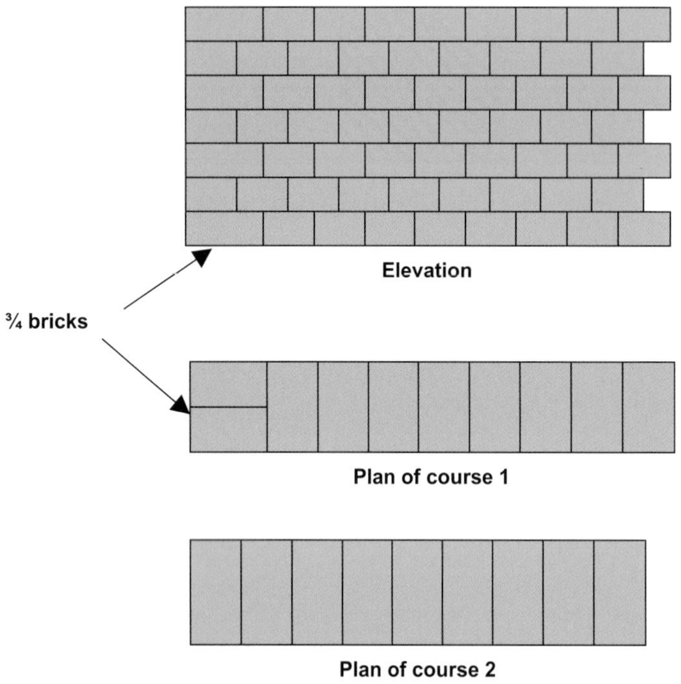
Elevation
¾ bricks
Plan of course 1
Plan of course 2

FIGURE 6.26
Header bond

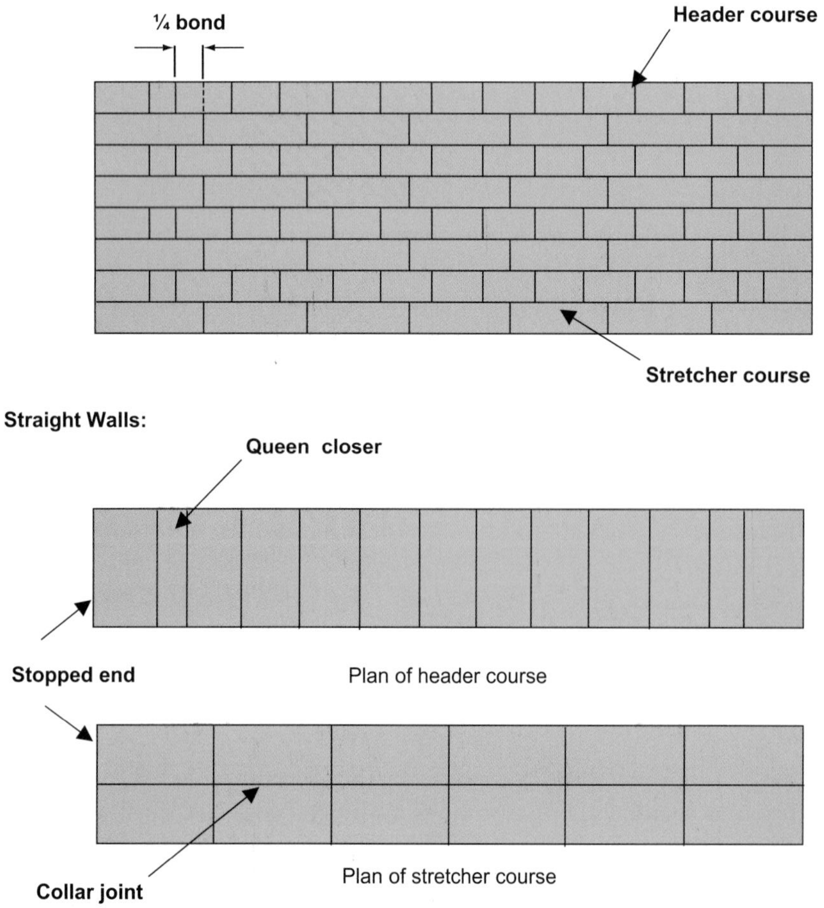
¼ bond
Header course
Stretcher course
Straight Walls:
Queen closer
Stopped end
Plan of header course
Collar joint
Plan of stretcher course

FIGURE 6.27
Bonding arrangements for English bond

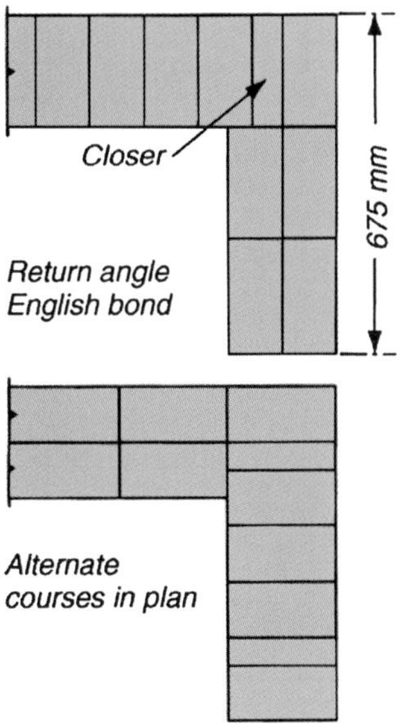
Closer
675 mm
Return angle
English bond
Stopped end
450 mm
Tee junction
English bond
Alternate
courses in plan

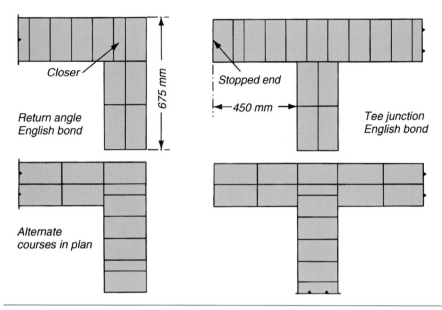

FIGURE 6.28
Return quoins and junctions in English bond

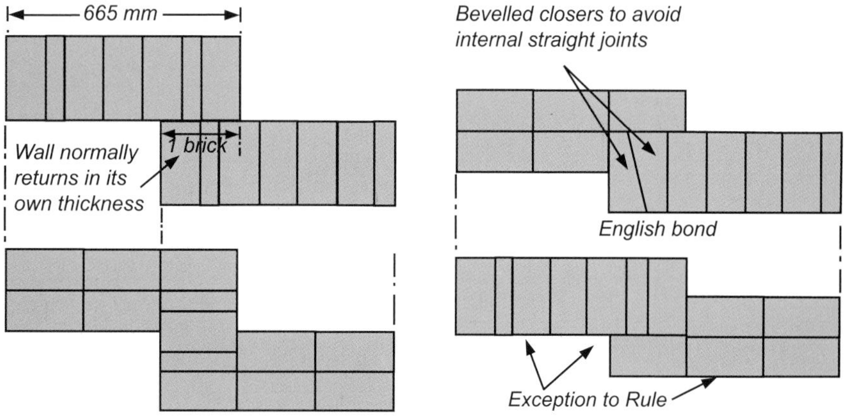
665 mm
1 brick
Wall normally
returns in its
own thickness
Bevelled closers to avoid
internal straight joints
English bond
Exception to Rule

FIGURE 6.29
Double returns in English bond

- Cut queen closers neatly and keep them regular in size.
- Keep perpends uniform and plumb as large cross-joints can soon cause you to lose quarter-bond and can bring straight joints on to the face.
- Remove mortar from the back of the bricks against the collar joint as this could prevent the backing up from being laid level.
- When backing up avoid the use of too much mortar near the collar joint, since the backing up could cause the face bricks to move out in front of the line (see Figure 6.21).

Straight walls: Flemish bond

Flemish bond consists of courses of alternate headers and stretchers with the headers in one course placed centrally over the stretcher in the course below (Figures 6.30–6.32).

A closer is placed next to the quoin header to form the correct quarter-bond.

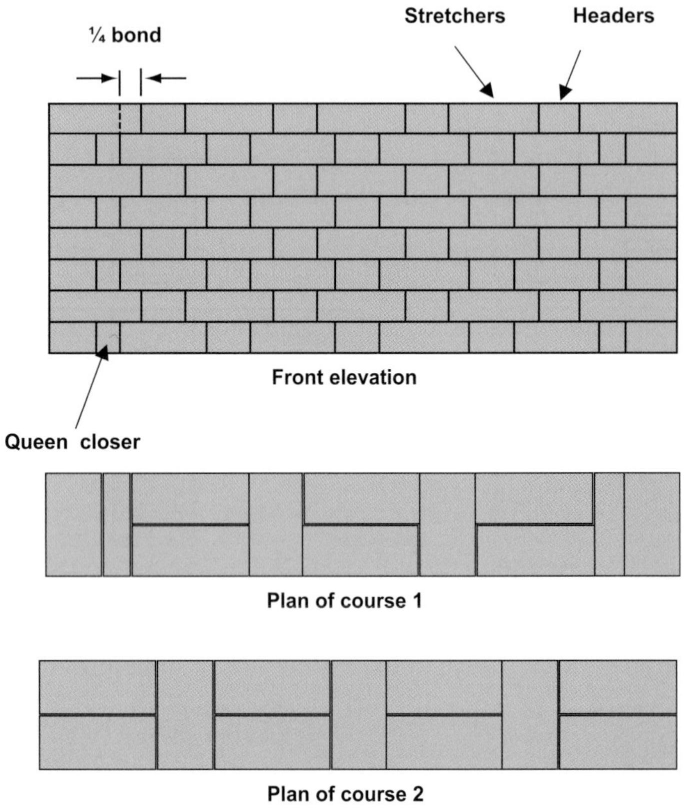
¼ bond
Stretchers
Headers
Front elevation
Queen closer
Plan of course 1
Plan of course 2

FIGURE 6.30
Bonding arrangements for Flemish bond

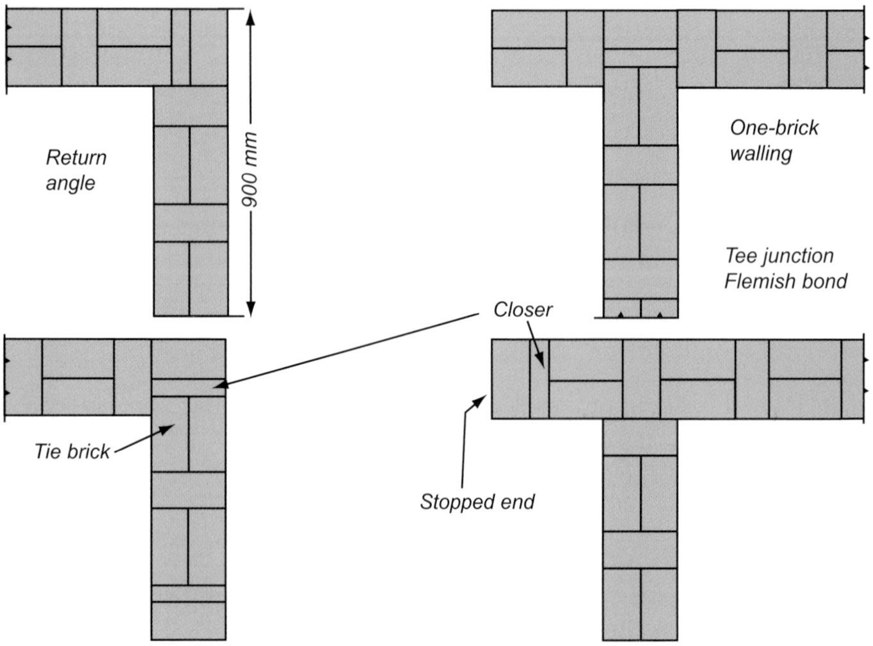
Return angle
900 mm
One-brick walling
Tee junction Flemish bond
Closer
Tie brick
Stopped end

FIGURE 6.31
Return quoins and junctions in Flemish bond

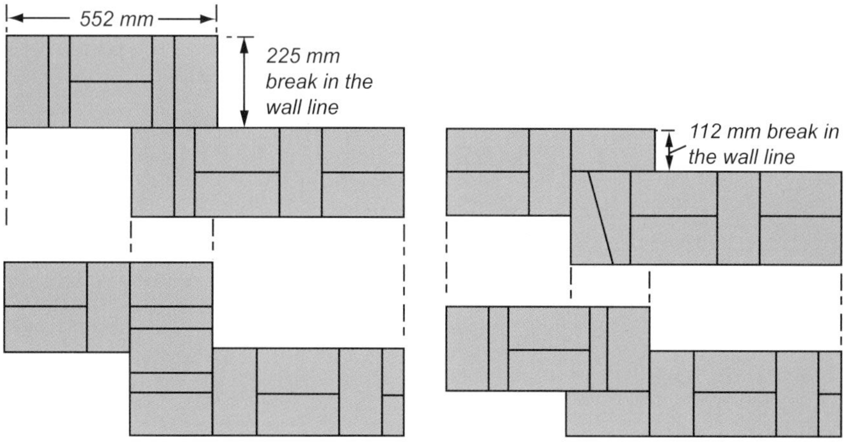
552 mm
225 mm break in the wall line
112 mm break in the wall line

FIGURE 6.32
Double returns in Flemish bond

Straight walls: Dutch bond

Dutch bond is similar to English bond in that it consists of alternate header and stretcher courses, but there are no closers in the header course, and the bond is formed by starting each stretcher course with a three-quarter bat (Figure 6.33).

In addition, the stretcher courses are laid half-bond to each other. This is achieved by placing a header on alternate stretcher courses next to the three-quarter bat.

The perpends in this bond follow each other diagonally across the wall in an unbroken line.

Garden wall bonds

This is a facing bond in which many more stretchers than headers are used. Several internal straight joints occur but a fair face can be achieved on both faces of the wall.

English garden wall consists of three courses of stretchers to one course of headers. The stretcher courses are laid half-bond with the header course being quarter-bond.

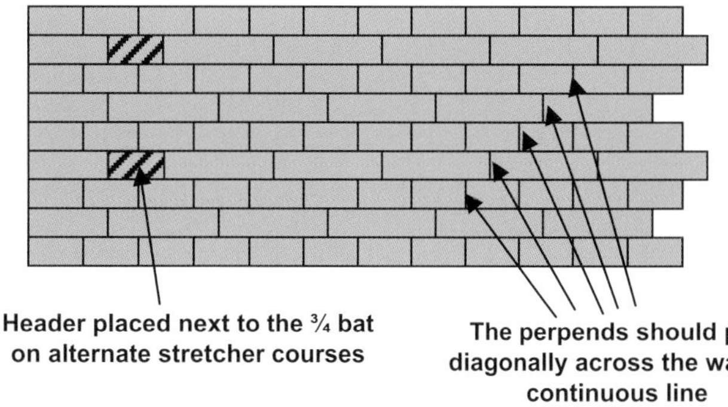
Header placed next to the ¾ bat on alternate stretcher courses
The perpends should pass diagonally across the wall in a continuous line

FIGURE 6.33
Dutch bond

Other versions have up to five courses of stretcher to one course of headers to save bricks.

English garden wall bond is sometimes known as Sussex bond. This is very popular in the USA, where six courses of stretchers are used, and is known there as 6 and 1 bond.

Flemish garden wall consists of three stretchers to one header in each course. To maintain the correct bond, the header in one course must be in the centre of the middle stretcher in the courses above and below.

Flemish garden wall bond is stronger than English garden wall bond as the headers are more evenly distributed.

Although garden wall bonds are occasionally used as face bonds, their main aim is in one-brick-thick walls where a face side of neat brickwork is required on both sides of the wall.

The varying brick lengths make it difficult to keep a fair face on both sides of the wall unless the number of headers is greatly reduced.

If the strength of the wall is of secondary importance, then the number of stretcher courses to each header course in English garden wall bond may be increased to five instead of three.

Similarly, in Flemish garden wall bond, the number of stretchers to one header in each course may be increased to five.

English garden wall bond

English garden wall bond consists of between three and five courses of stretchers to one course of headers.

The stretcher courses are laid half-bond to each other (Figure 6.34).

Flemish garden wall bond

Flemish garden wall bond consists of three or sometimes five stretchers to one header in each course.

The header in one course is laid centrally over the middle stretcher in the course immediately below (Figure 6.35).

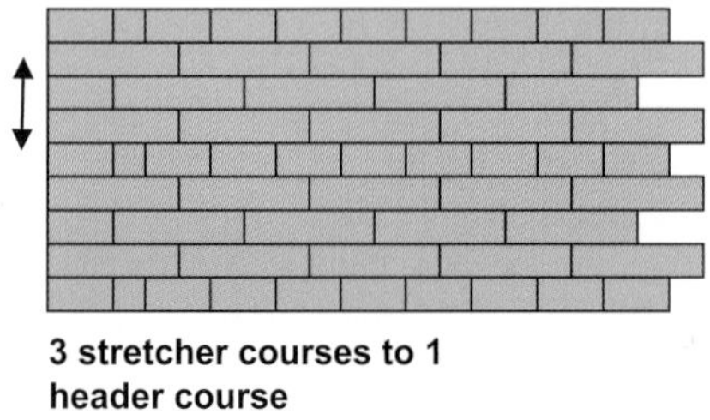
3 stretcher courses to 1 header course

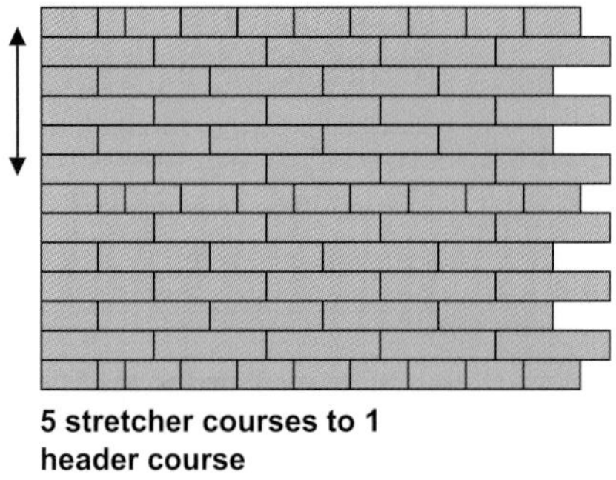
5 stretcher courses to 1 header course

FIGURE 6.34
Variations of English garden wall bond

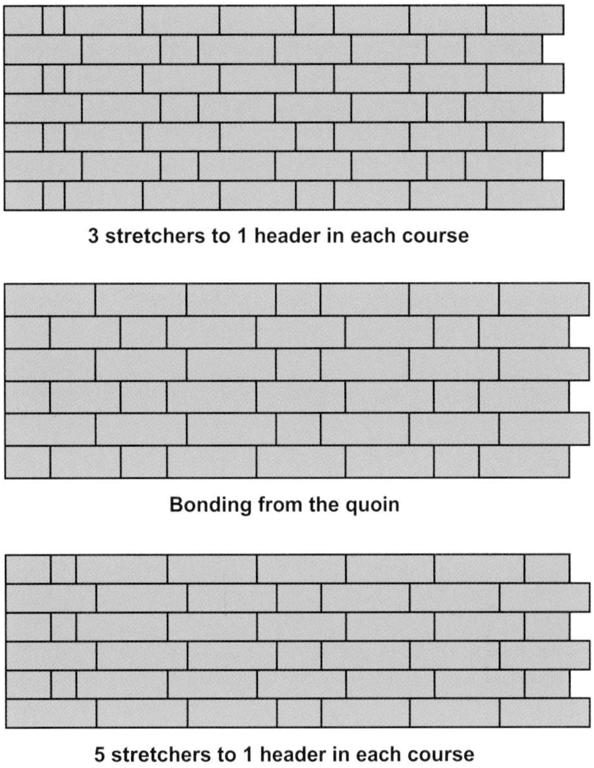
3 stretchers to 1 header in each course
Bonding from the quoin
5 stretchers to 1 header in each course

FIGURE 6.35
Variations of Flemish garden wall bond

SETTING OUT BONDS

It is essential that the bricklayer understands fully the reason why walls should be set out before laying any bricks.

The basic principles have been dealt with, but there will be situations when it will not be possible to adhere to the rules, at quoins, junctions and stopped ends.

When a wall is to be built which is not a standard brick length then broken bond or reverse bond may have to be used.

Remember

Any rule concerning bonding in general can only be applied as far as is practicable.

Remember one of the rules of bonding – change bond when you change direction.

If there are stretchers on one face then headers should be on the adjoining face. However, it may not always be possible to adhere to this rule.

Broken bond

This can happen in short lengths of brickwork between windows, doors and openings where the walls are not in multiple brick lengths.

Broken bond can sometimes be avoided by 'tightening' or 'opening' the joints. In doing so, bricklayers should work to the brick size plus a 10 mm joint = 225 mm.

It if far better if broken bonds can be avoided altogether, but where openings and piers do not allow for brick sizes, the broken bonds should be carefully set out. Once their position has been decided, it should be maintained throughout the height of the wall.

On no account should a closer be built in the middle of the wall; it should only be placed next to the quoin header.

If a wall length is 56 mm less than brick sizes a half-bat and three-quarters should be used in the centre of the wall.

Stretcher bond

When the wall length in stretcher bond is not in brick sizes broken bond can be included, but should be positioned in the centre of the wall.

Three variations of broken bond are shown in Figures 6.36–6.38.

Remember

The cut brick should not be less than half a brick.

English bond

The example in Figure 6.39 shows a wall in English bond which, on the first header course, works out correctly.

But when setting out the alternate courses, working from each corner it leaves a header in the centre of the wall, i.e. broken bond.

A three-quarter bat may be used in both the header and stretcher courses (Figure 6.40).

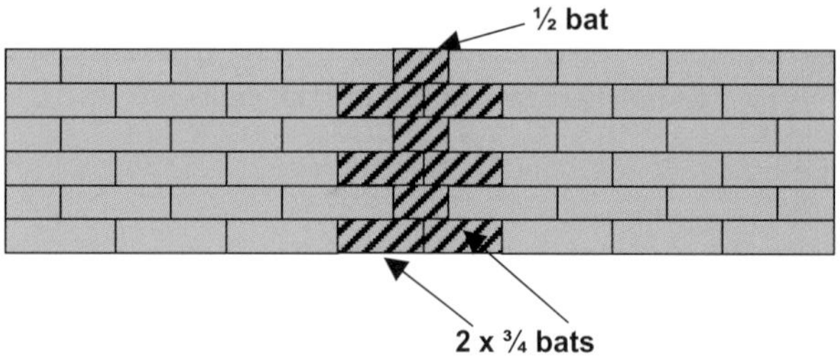
½ bat
2 x ¾ bats

FIGURE 6.36
Broken bond in stretcher bond: two three-quarter-bats in one course and a header in the next

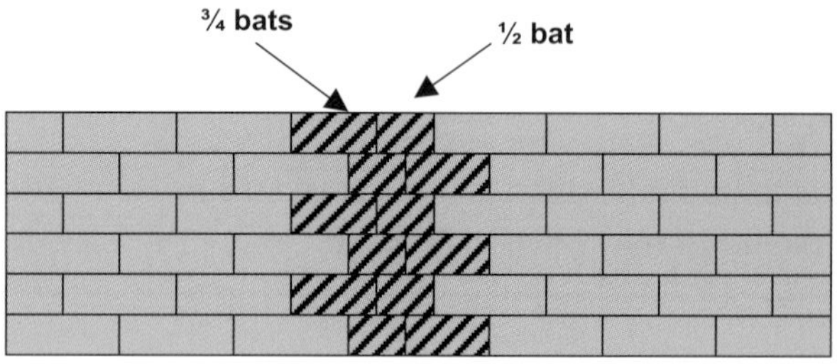
¾ bats
½ bat

FIGURE 6.37
Broken bond in stretcher bond: a three-quarter-bat and a header in each course

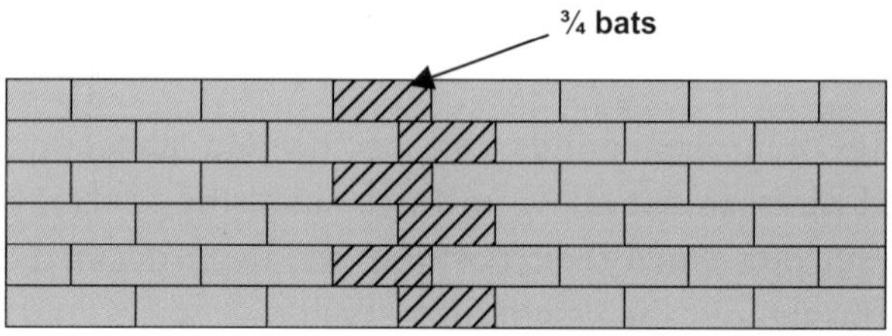
¾ bats

FIGURE 6.38
Broken bond in stretcher bond: a three-quarter-bat in each course

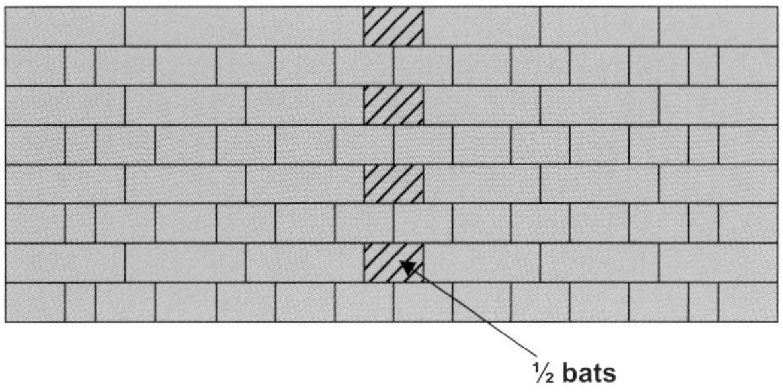
½ bats

FIGURE 6.39
Broken bond in English bond: half-bats in stretcher courses

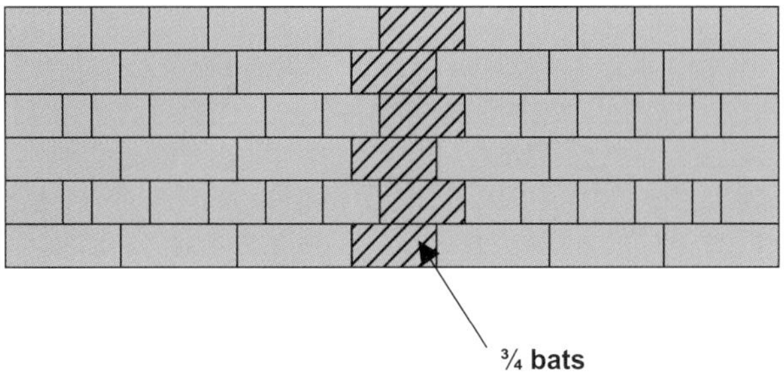
¾ bats

FIGURE 6.40
Broken bond in English bond: three-quarter-bats in header and stretcher courses

Flemish bond

When setting out Flemish bonds the stretchers can be shortened to three-quarters to accommodate the broken bond (Figure 6.41). In the next example (Figure 6.42) an extra header is inserted into each course, showing two headers together, to maintain bond.

A header and three-quarter bat have been inserted to maintain bond when the wall dimension is 56 mm less than brick sizes (Figure 6.43).

Two headers and a three-quarter bat have been inserted into one course and a three-quarter bat and an extra stretcher in the alternate course, when the wall dimension is 56 mm more than brick sizes (Figure 6.44).

Garden wall bonds

Broken bond can also occur in garden wall bonds.

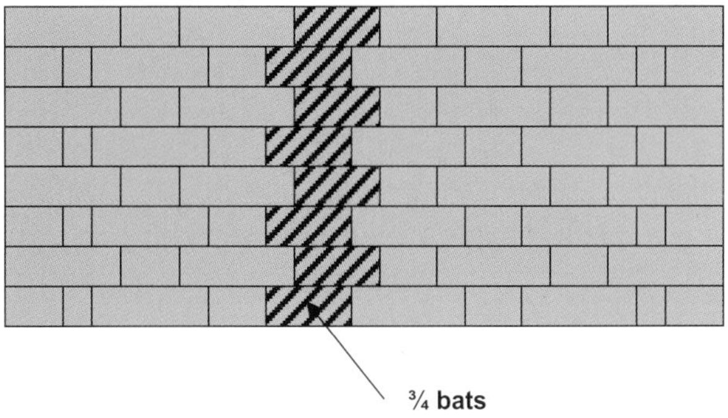
¾ bats

FIGURE 6.41
Broken bond in Flemish bond: a three-quarter-bat in each course

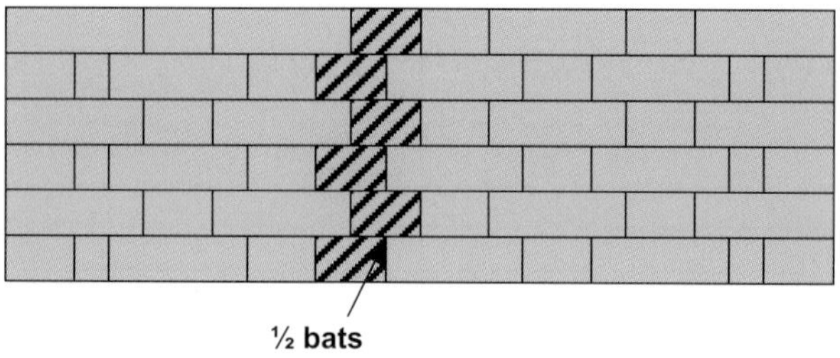
½ bats

FIGURE 6.42
Broken bond in Flemish bond: an extra header in each course

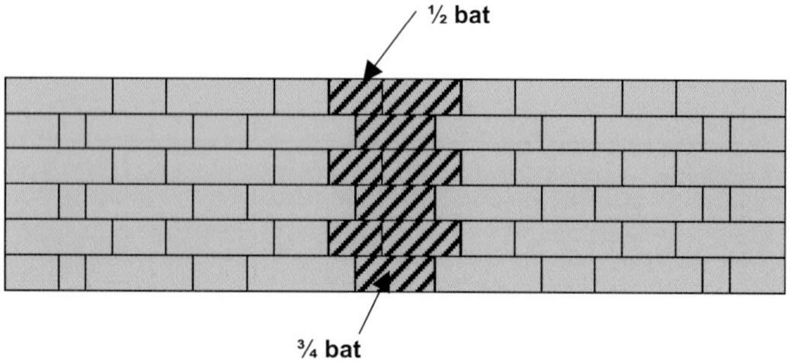
½ bat
¾ bat

FIGURE 6.43
Broken bond in Flemish bond (wall dimension 56 mm shorter than brick sizes)

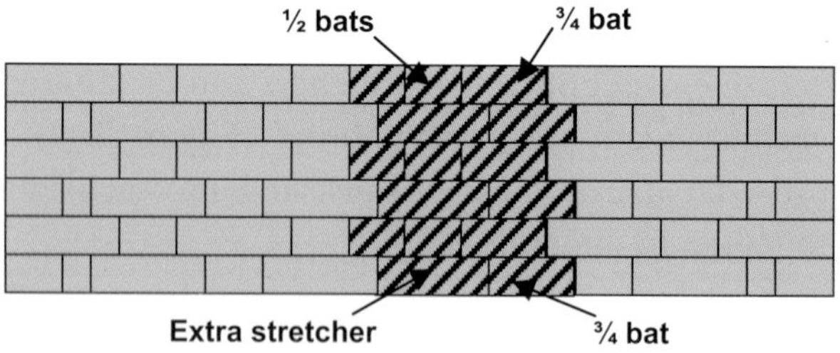
½ bats
¾ bat
Extra stretcher
¾ bat

FIGURE 6.44
Broken bond in Flemish bond (wall dimension 56 mm longer than brick sizes)

English garden wall bond

When the wall is 56 mm longer than brick sizes, a three-quarter bat can be inserted into the header course, and a header and a three-quarter bat can be inserted into the stretcher courses (Figure 6.45).

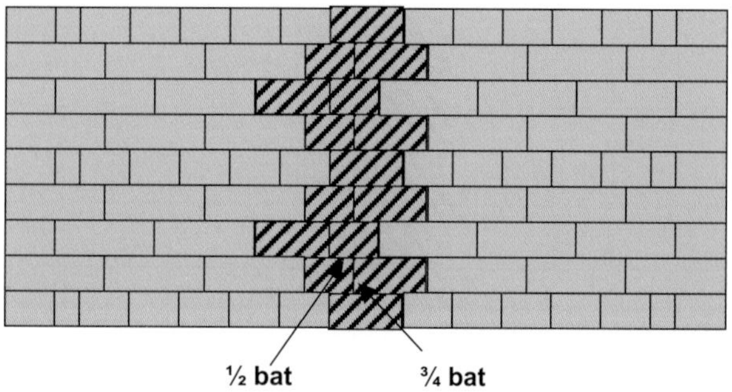
½ bat
¾ bat

FIGURE 6.45
Broken bond in English garden wall bond (wall dimension 56 mm longer than brick sizes)

When the wall is 56 mm shorter than brick sizes, a three-quarter bat can be inserted into the header course instead of a header, and a three-quarter bat can be inserted into the stretcher courses instead of a stretcher (Figure 6.46).

Flemish garden wall bond

When the wall is 56 mm longer than brick sizes, a three-quarter bat and header can be inserted into one course instead of a stretcher, and a three-quarter bat can be inserted into the stretcher courses instead of a header (Figure 6.47).

Another method of using three-quarter bats when the wall is longer than brick sizes is shown in Figure 6.48.

Reverse bond

It is normal in bonding to start both ends of the wall the same way.

If you have a stretcher at one end on the first course you should have a stretcher at the other end (Figure 6.49).

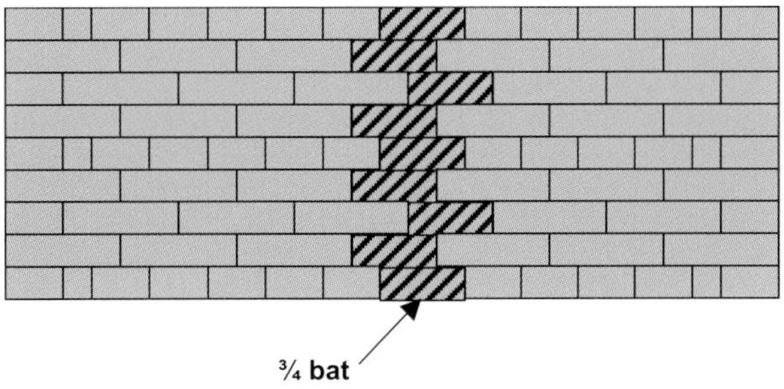
¾ bat

FIGURE 6.46
Broken bond in English garden wall bond (wall dimension 56 mm shorter than brick sizes)

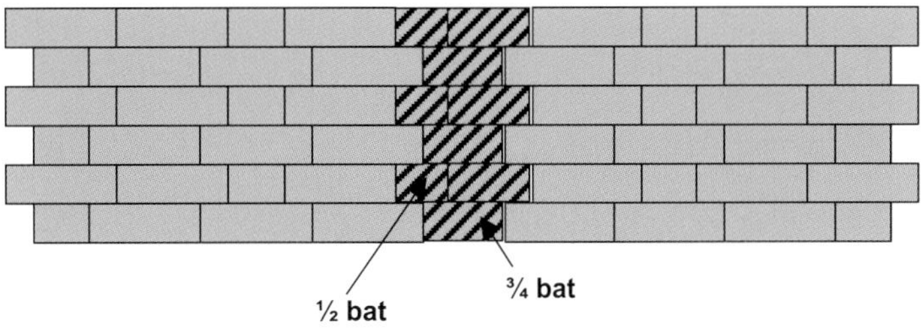
½ bat
¾ bat

FIGURE 6.47
Broken bond in Flemish garden wall bond (wall dimension 56 mm longer than brick sizes)

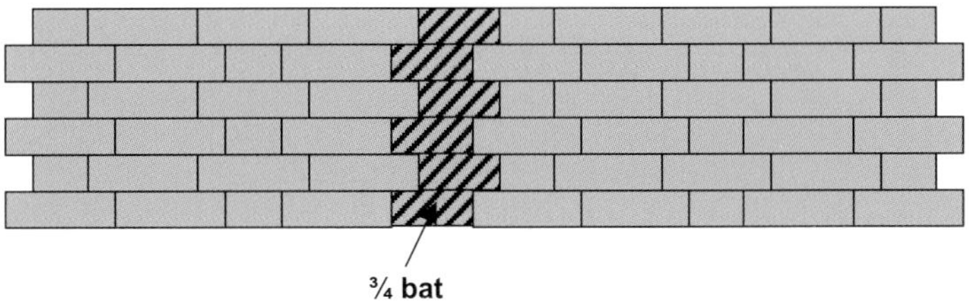
¾ bat

FIGURE 6.48
Broken bond in Flemish garden wall bond (wall dimension 56 mm longer than brick sizes)

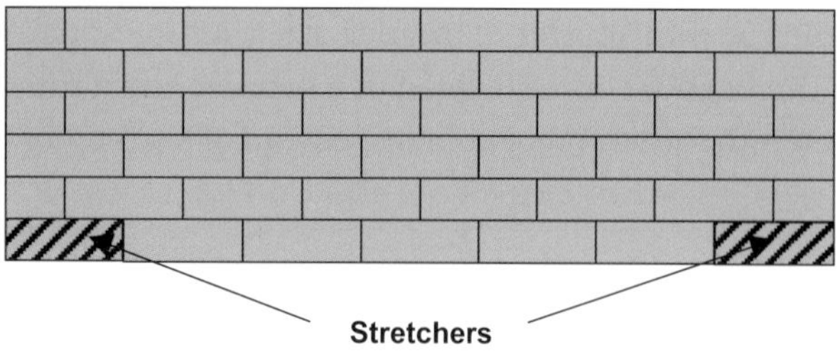
Stretchers

FIGURE 6.49
Normal stretcher bond

On occasions it is possible to change the bonding at the ends of the wall. This would be done to avoid cutting in the centre of the wall. For instance, if you bond the wall dry and you had a half-brick in the centre you would take it to one end.

This is known as reverse bond (Figure 6.50).

English bond
When the dimension of a wall is not multiples of bricks the end bricks can be reversed to avoid cut bricks in the centre of the wall.

The example shown in Figure 6.51 avoids the use of a three-quarter brick in the centre of the wall.

Flemish bond
The example in Figure 6.52 again avoids the use of a three-quarter brick in the centre of the wall.

BOUNDARY WALLS

These are described as 'free standing' because the brickwork receives no support from floors or roof structure, as do the walls in a building.

For this reason local authority building control department's approval is usually required if a wall higher than 1 m is to be built next to a public

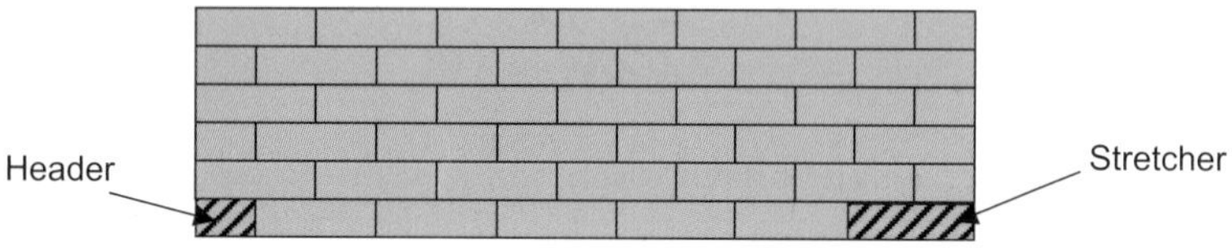
Header
Stretcher

FIGURE 6.50
Reverse stretcher bond

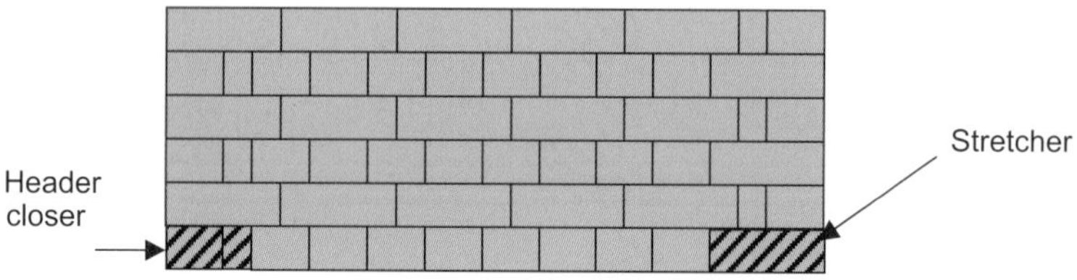
Header
closer
Stretcher

FIGURE 6.51
Reverse English bond

Header
closer
Stretcher

FIGURE 6.52
Reverse Flemish bond

road or 2 m high elsewhere, to check that the design is stable. Boundary walls include attached piers, detached piers, gates and raking cuts. Attached piers must be properly bonded to the wall

The DPC for boundary walls is best provided by using two courses of black or red class A engineering bricks, bedded and jointed in cement mortar. The brickwork above and below is solidly bonded to these DPC bricks. The result bonds better than if separated by a flexible DPC, which in a free-standing boundary wall creates a plane of weakness.

DETACHED PIERS

Solid free-standing (detached) piers are capable of transferring considerable compressive loads through their structure down to the concrete foundation. Although this type of wall can withstand heavy loads it also has to resist other forces. It has to withstand soil movements, pressure of soil placed behind the wall, inclement weather applied to all faces of the wall and side load pressure.

There must always be adequate foundations provided at the correct depth to resist frost. If the foundations are not deep enough problems with soil movement will be substantial. Cracks will appear and cause irreversible structural damage.

Many piers and boundary walls have soil placed against them which not only applies pressure to the wall but also causes dampness which eventually will cause damage to the brickwork.

The pier has four sides a top and base, and these faces are all subjected to the weather, i.e. rain, wind, snow, etc. The whole wall is constantly wet or dry throughout the year and this can cause considerable damage.

Piers used to support gates, fences, etc., can also have side pressure placed on them. If a side load is placed on the pier it creates a totally different set of stresses which cannot be carried by the brickwork.

Remember

Brickwork is strong in compression but weak in tension.

Possible problems with piers

Inclement weather can affect the pier on all sides, including the base if a DPC course is not included (Figure 6.53).

A free-standing brick pier or isolated pier is a pillar of brickwork which stands alone. It can be used to carry the ends of beams over a large span or to support garden gates, etc.

Piers can vary considerably in size and bond and can be very difficult to build because of the plumbing points.

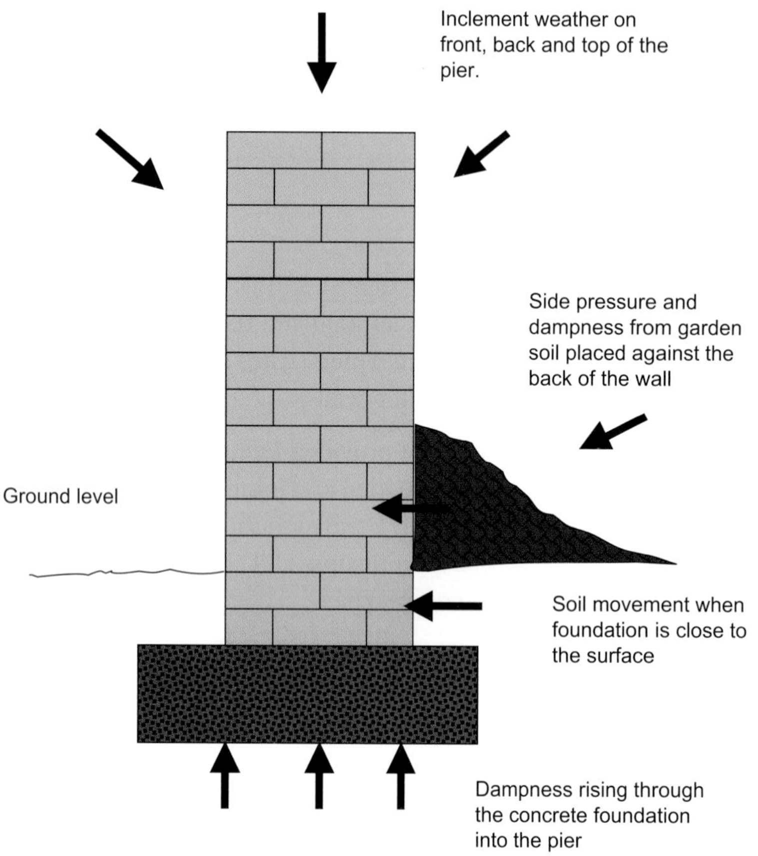
Inclement weather on front, back and top of the pier.
Side pressure and dampness from garden soil placed against the back of the wall
Ground level
Soil movement when foundation is close to the surface
Dampness rising through the concrete foundation into the pier

FIGURE 6.53
Problems with detached piers

The bond specified for a particular pier may have to be amended slightly to avoid the excessive cutting required by some bonds. Excessive cutting is expensive not only on materials but also on time. When there are many cuts the bricklayer has to take time to ensure that all the cuts are the same so as to avoid any irregularity in the perpends or plumbing.

Small-section piers are often the most difficult to build owing to the limited number of perpends.

If several piers are in line, the piers at either end should be built first and then a line attached to either end pier so that the remainder of piers can all be built in line, both front and back.

Always set the piers out square and to brick sizes.

Hollow piers

Piers are often constructed in stretcher bond with a void in the middle (Figure 6.54).

Construction

All piers are difficult to build and require extra skill. Hollow piers are easier than solid piers because of the void in the centre.

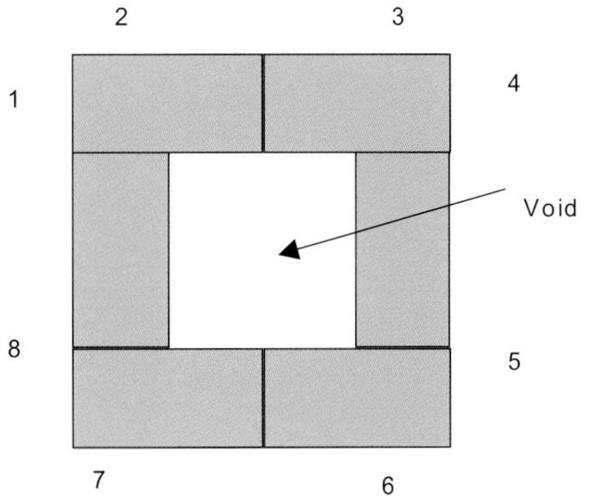
2
3
1
4
Void
8
5
7
6

FIGURE 6.54
Two-brick pier in stretcher bond

Once the setting out has been completed the pier can be built up to DPC level. All detached piers have eight plumbing points, as shown in Figure 6.54.

Each one must be accurate. It is essential that other corners are not pushed out while plumbing. Never leave the plumbing too late; always plumb before the mortar sets. DPCs should always be engineering bricks as these provide better resistance to overturning of the pier than flexible DPCs.

Overcoming problems

The DPC should be built in engineering bricks. Ensure that soil is not placed at the rear of the wall. Ensure the correct thickness and depth for concrete foundations. The top of wall should be protected by coping stones (Figure 6.55).

Solid English bond piers

Piers built in English bond are solid and usually 1½ bricks square or larger. See the examples in Figure 6.56 showing 1½-brick and 2-brick piers.

Flemish bond piers

Piers built in Flemish bond are also solid and usually 1½ bricks square or larger. See the examples in Figure 6.57 showing 1½ × 1-brick and 2 × 2-brick piers.

Construction of English bond piers

Solid piers are more difficult to build than hollow piers.

There are also eight corners to plumb, but because the pier is more solid there is no room for internal movement.

Always lay the corner bricks first and ensure that they are plumb, level and gauge.

When laying the corner bricks ensure they are immediately gauged and plumb (Figures 6.58 and 6.59).

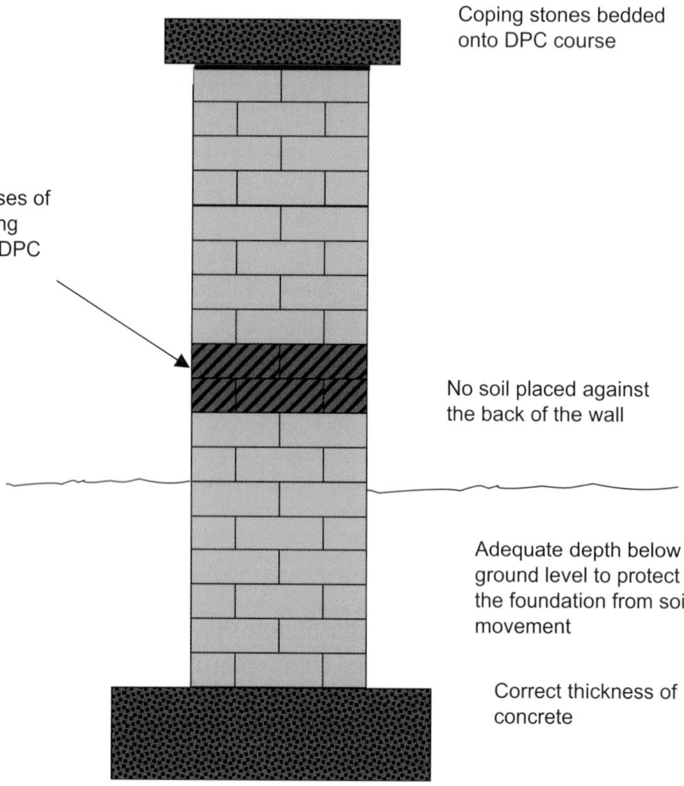
Coping stones bedded onto DPC course
Two courses of engineering bricks as DPC course
No soil placed against the back of the wall
Adequate depth below ground level to protect the foundation from soil movement
Correct thickness of concrete

FIGURE 6.55
Detached pier with problems cured

When the corner bricks have been laid, plumbed, levelled and gauged the rest of the bricks can be filled in (Figure 6.60). These are laid with caution so as not to disturb the corner bricks.

There is no correct method for filling in the headers and queen closers. You can lay the two headers as shown above, making sure they are levelled and ranged in. Then lay the four queen closers.

> **Remember**
>
> It is essential that the corner bricks are not disturbed while filling in the remainder of the course.
>
> Complete one full course at a time.

Attached piers

Purely for economic reasons, most boundary walls are constructed in thin walling in stretcher bond and strengthened with attached piers (Figure 6.61).

The attached piers help to prevent the boundary walls from overturning with pressure. Attached piers should not be less than 100 mm thick and must be bonded into the wall (Figure 6.62).

Attached piers could project on one or both sides. They should project by at least half a brick to make the bonding easier.

The DPC should continue through all piers.

CONSTRUCTION OF BOUNDARY WALLS

Stretcher bond

Attached piers in stretcher bond are the most common type of pier.

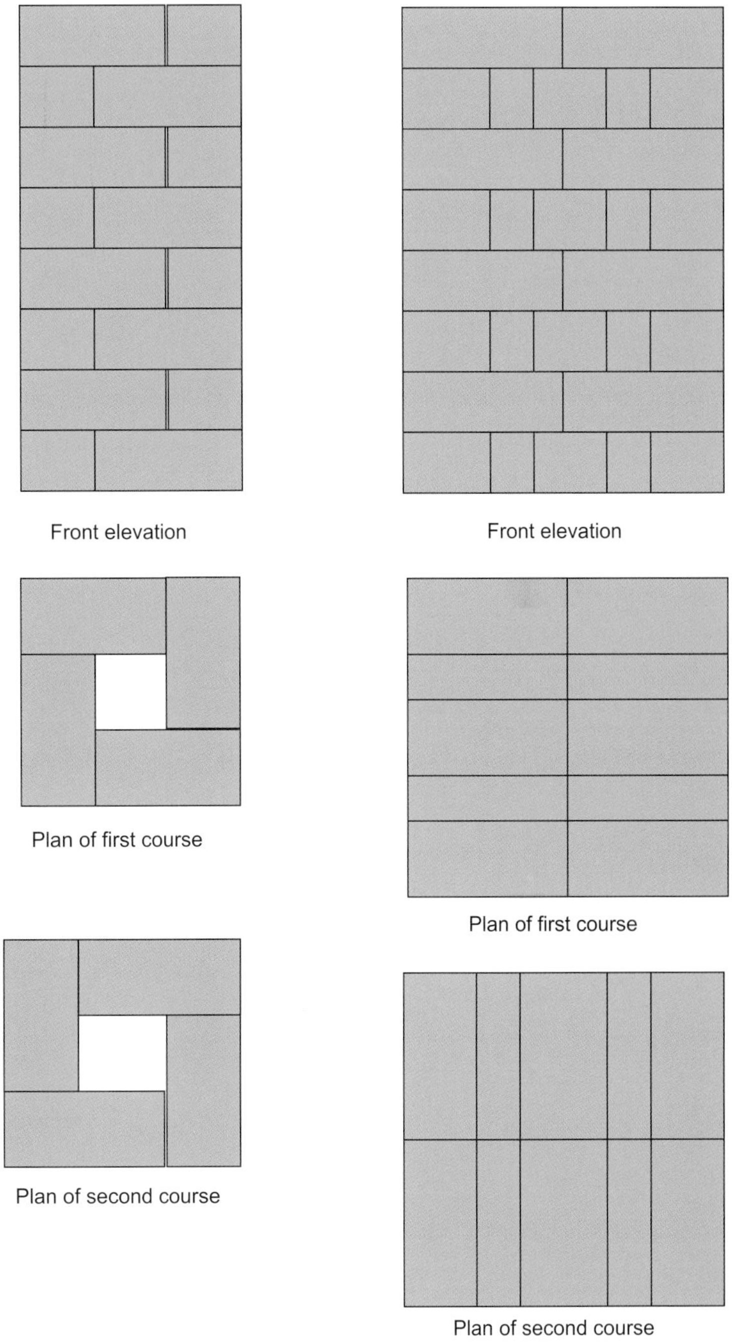
Front elevation
Front elevation
Plan of first course
Plan of first course
Plan of second course
Plan of second course

FIGURE 6.56
Examples of bonding to piers

The most important consideration with this type of pier is the appearance of the face bond. If maintaining the correct bond on the face of the wall is essential then the example shown in Figure 6.63 would be one solution.

The pier can be strengthened by bedding reinforcement over the pier on alternate courses.

Solid brick walls

Sometimes garden walls can be built one brick wide with two brick attached piers (Figure 6.64).

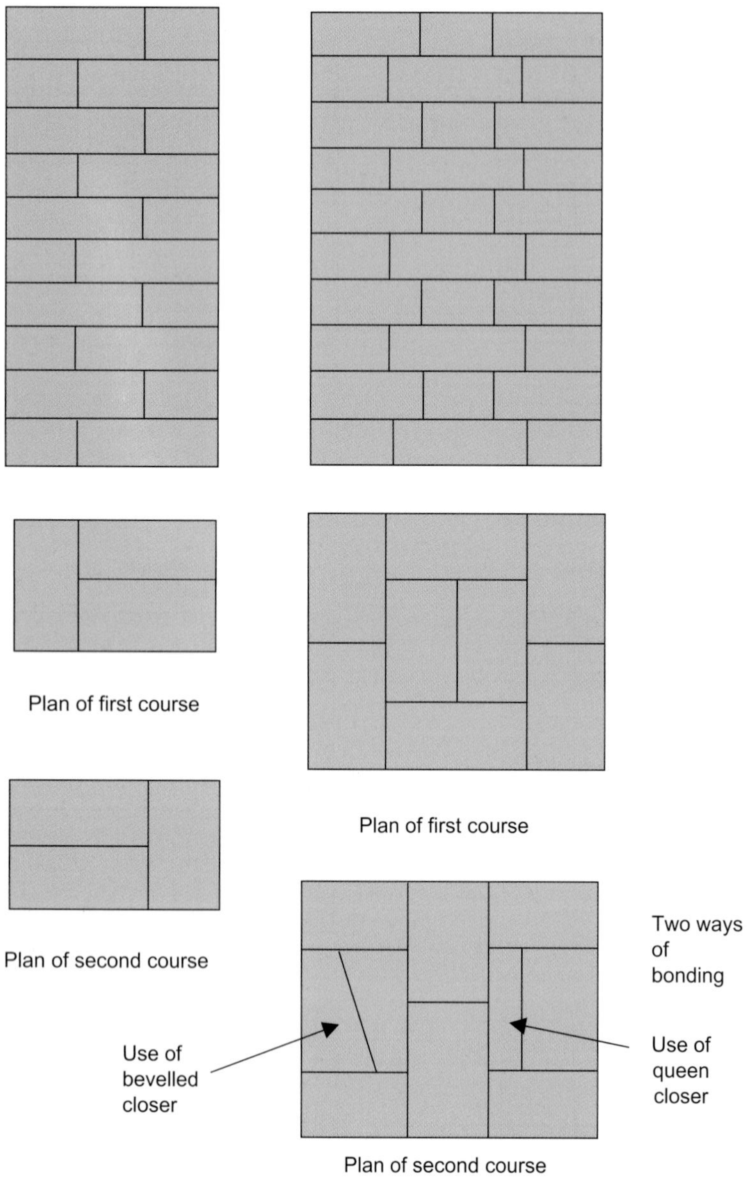
Plan of first course
Plan of second course
Plan of first course
Two ways of bonding
Use of bevelled closer
Use of queen closer
Plan of second course

FIGURE 6.57
Examples of bonding to piers

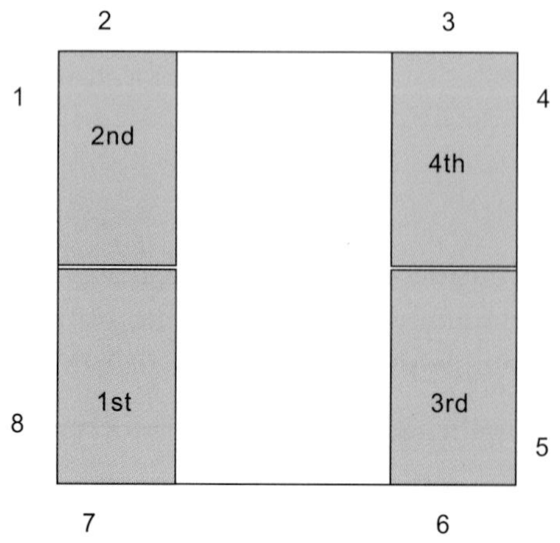
2
3
1
4
2nd
4th
1st
3rd
8
5
7
6

FIGURE 6.58
Building piers

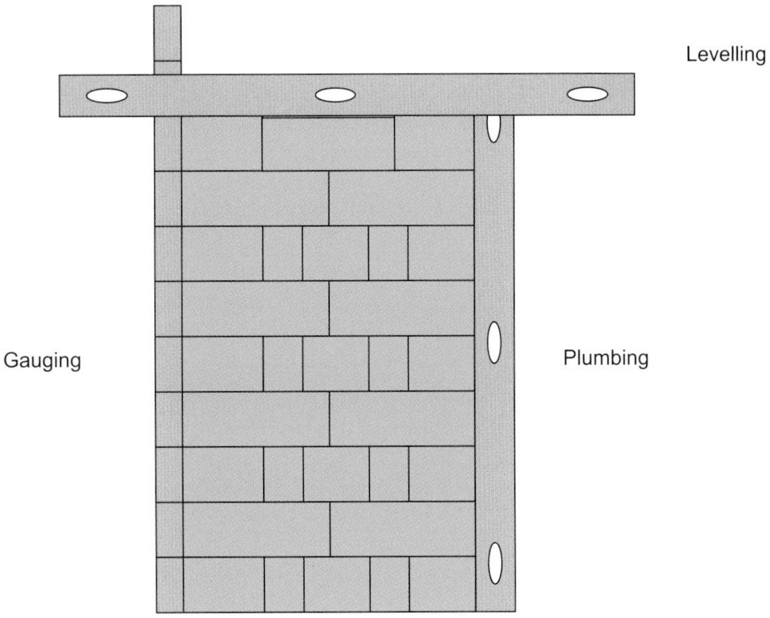
Levelling
Gauging
Plumbing

FIGURE 6.59
Checking piers

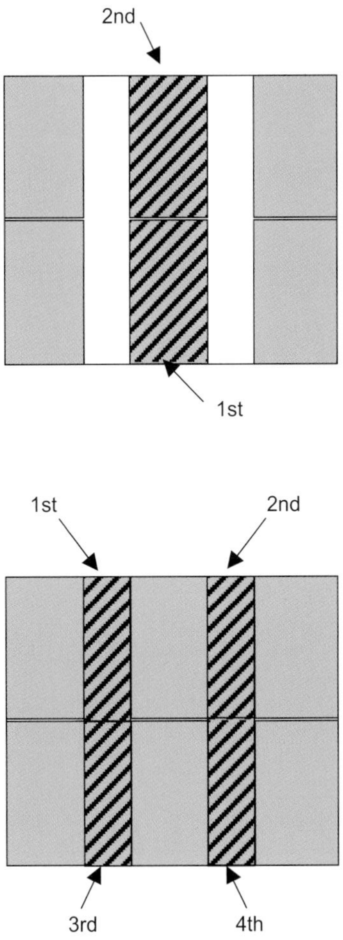
2nd
1st
1st
2nd
3rd
4th

FIGURE 6.60
Building piers

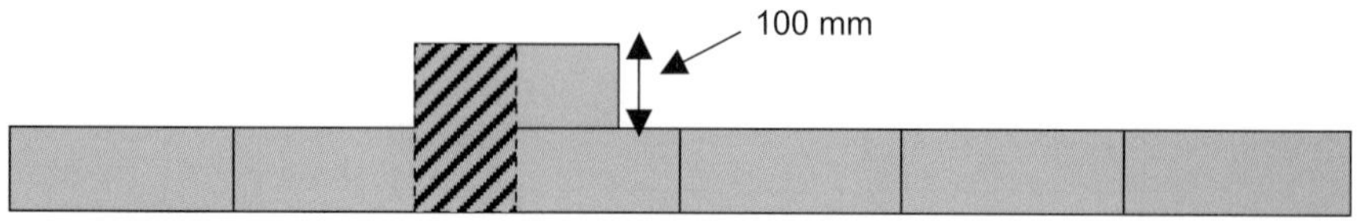
100 mm

FIGURE 6.61
Bonding single attached piers (dotted bricks show second course)

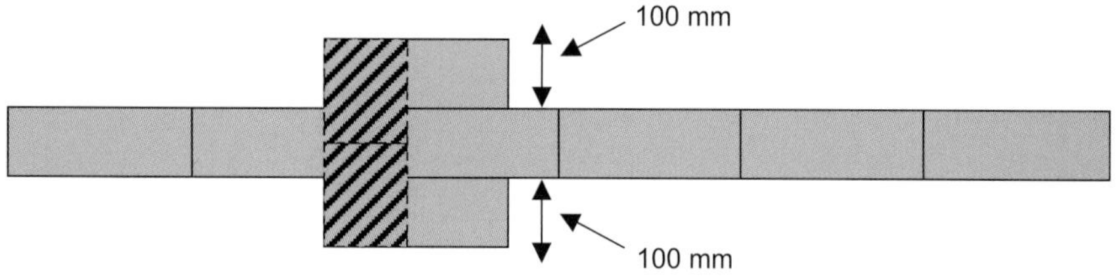
100 mm
100 mm

FIGURE 6.62
Bonding double attached piers (dotted bricks show second course)

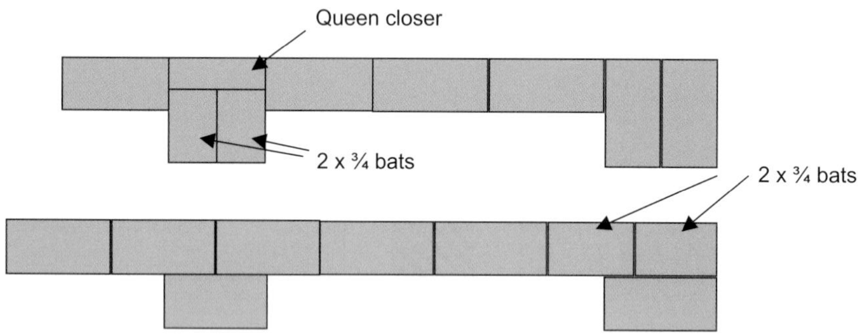
Queen closer
2 x ¾ bats
2 x ¾ bats

FIGURE 6.63
Bonding attached piers in stretcher bond

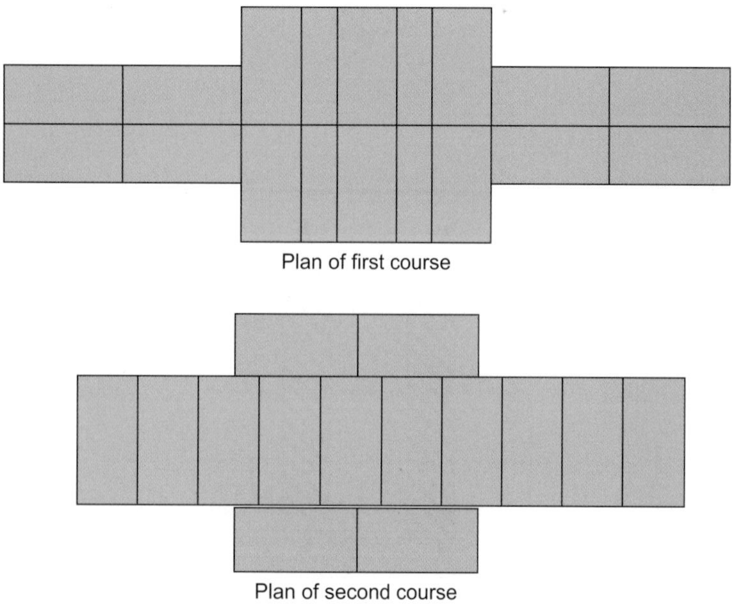
Plan of first course
Plan of second course

FIGURE 6.64
Bonding attached piers in English bond

Attached piers can be used on other brick walls to provide stability, e.g. to long or high walls. The important part is the bonding of the attached piers to the main wall.

The building of the attached piers needs care to ensure plumb and level. The gauge should not need checking as the pier should be built along with the main wall to a line.

As with detached piers, take care not to push the back of the pier when plumbing the front, and vice versa.

Ramps

Ramps are decorative features that could be added to boundary walls, and which the bricklayer must be able to construct.

Ramps are usually next to the pier when gates are being fixed, to allow the pier to be built higher than the boundary wall to support the gates.

Straight ramps can be set out with the use of bricklayer's lines and pins to the required angle (Figure 6.65).

Cutting to rake

A raking cut is required on walls that have to be finished with a sloping end, generally on gable ends, wall ramps and stair strings.

On boundary walls the brickwork is raised to the full height of the raking cut by raking back and toothing the wall, finishing just short of the required line of cut.

Each brick to be cut should be bedded onto the wall in its correct position and then marked in pencil ready to be cut.

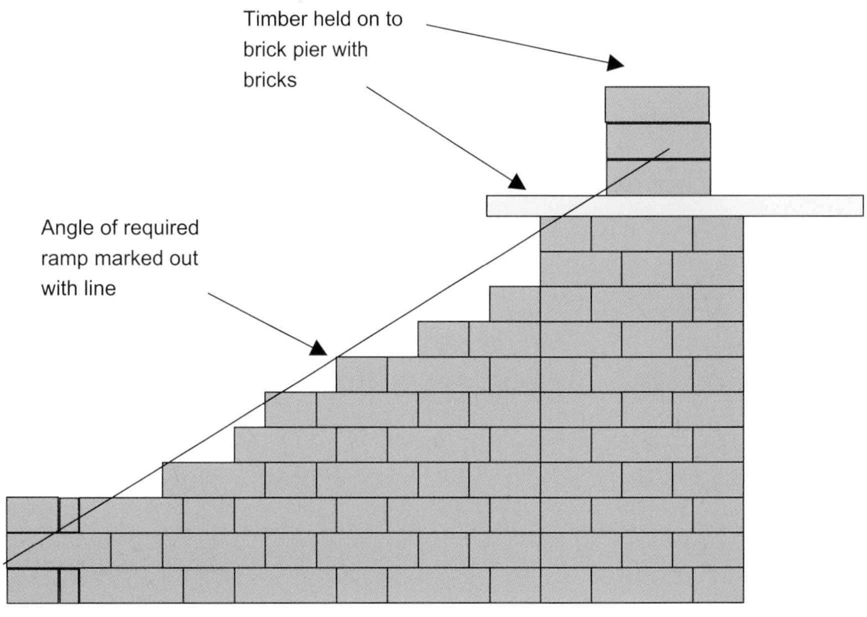
Timber held on to
brick pier with
bricks
Angle of required
ramp marked out
with line

FIGURE 6.65
Setting out cutting to rake

Use softing to prevent damage and cut with a hammer and bolster or a mechanical saw.

Deadman

Deadman is the term given to a temporary brick pier bedded to gauge and plumbed to the face of the wall. This acts as a corner and allows a horizontal line to be extended beyond the line of the raking cut (Figure 6.66). A profile could also be used (Figure 6.67).

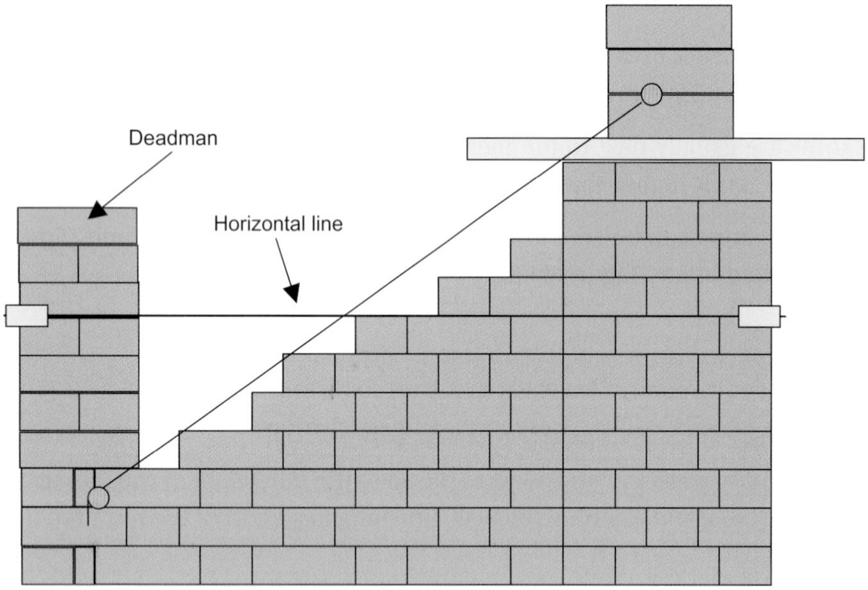
Deadman
Horizontal line

FIGURE 6.66
Use of a deadman to support the horizontal line

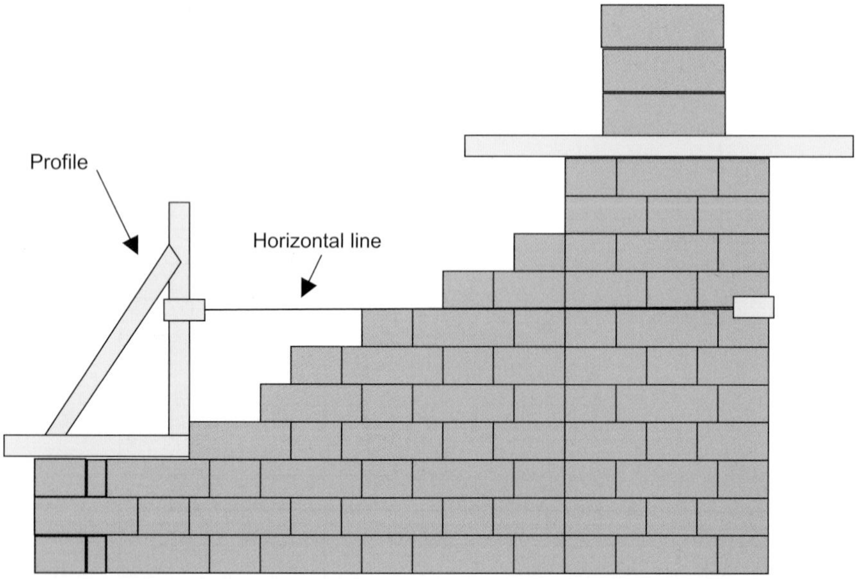
Profile
Horizontal line

FIGURE 6.67
Use of a profile to support the horizontal line

Procedure for marking and cutting bricks to rake

1. Set the sliding bevel at the necessary angle of the raking cutting.
2. Measure the top edge of the cut brick required.
3. Deduct 10 mm for one cross-joint.
4. Mark in pencil the remaining measurement on the brick to be cut, back and front.
5. Draw in pencil the angle of the sloping (raking) cut, using the sliding bevel, also back and front, taking care to match the slope on the face (Figure 6.68).
6. Cut the brick with hammer and bolster, or a mechanical masonry saw if available.
7. Trim the cut face with a scutch or comb hammer if necessary to avoid any projections.
8. Carefully bed the cut brick to line and level. See examples of marking in Figure 6.69.

GATE PILLARS

As mentioned in the previous sections on detached and attached piers, gates and ornamental ironwork are very often built into the piers.

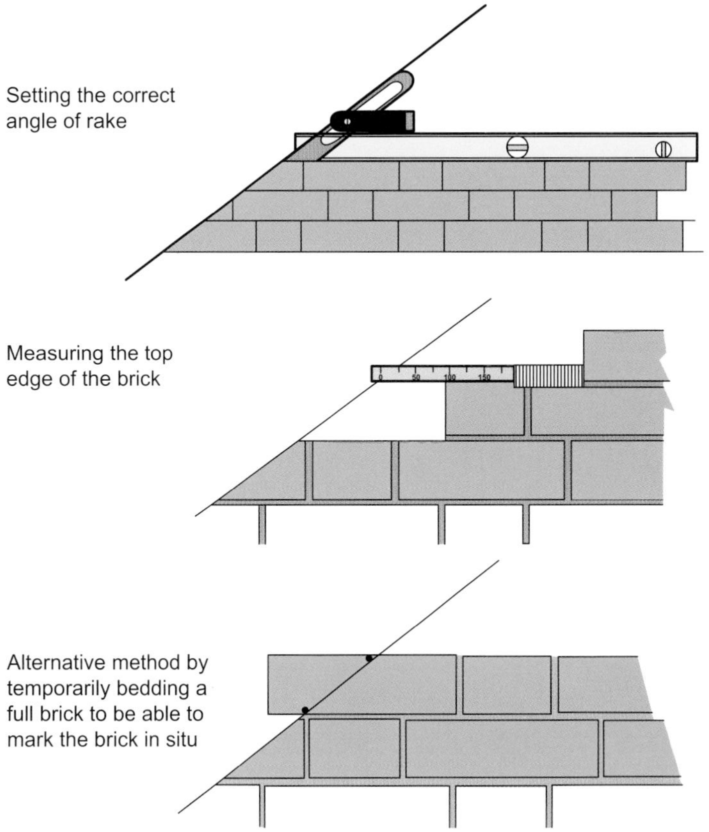
Setting the correct angle of rake
Measuring the top edge of the brick
Alternative method by temporarily bedding a full brick to be able to mark the brick in situ

FIGURE 6.68
Marking the cut bricks

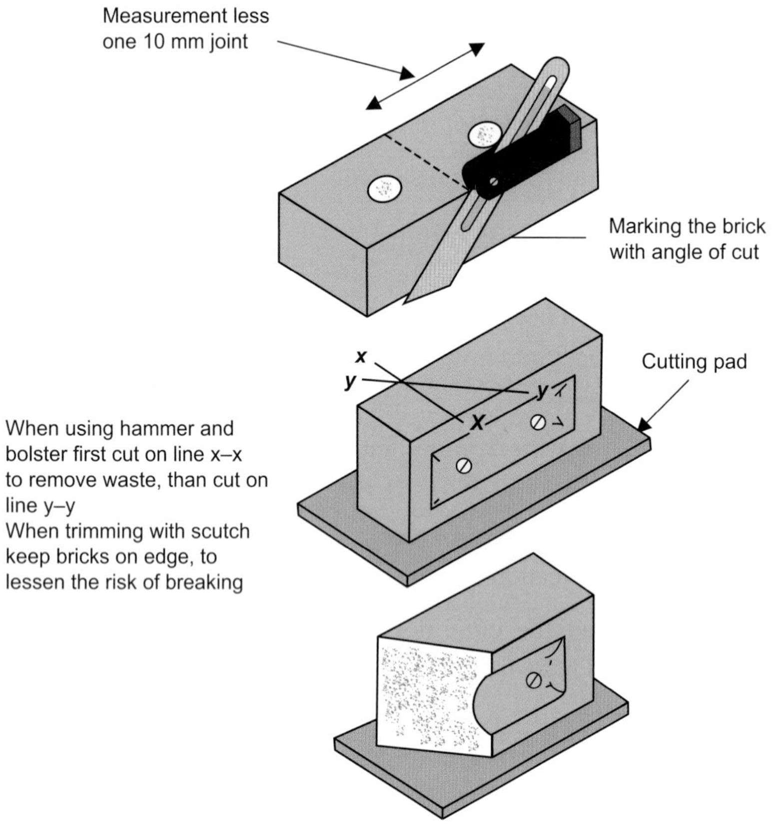
Measurement less
one 10 mm joint
Marking the brick
with angle of cut
Cutting pad
x
y
y
x
When using hammer and
bolster first cut on line x–x
to remove waste, than cut on
line y–y
When trimming with scutch
keep bricks on edge, to
lessen the risk of breaking

FIGURE 6.69
Marking and cutting the bricks

It is essential that the boundary wall is thickened at any point where gates are being fixed so that it is strong enough to withstand the extra pressures applied.

Occasionally the pier supporting a gate will be a detached pier, but more often it is attached to the boundary wall (Figure 6.70).

Because of the stresses placed on the piers by the gates the piers have to be extra strong. One method of achieving this is to construct the piers hollow and infill with concrete.

The concrete can also be reinforced and even attached to the foundation for extra resistance to side movement. Alternatively, the piers may be built of solid brickwork (Figure 6.71).

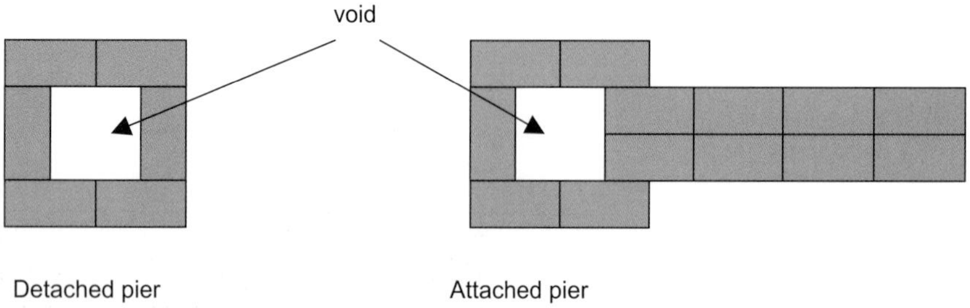
void
Detached pier
Attached pier

FIGURE 6.70
Hollow gate pillars

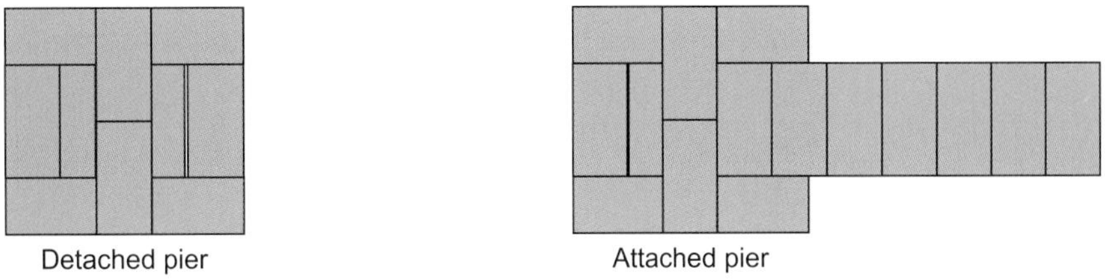
Detached pier
Attached pier

FIGURE 6.71
Solid gate pillars

Hollow piers

Detached and attached piers can be strengthened by the introduction of reinforced concrete into the void (Figure 6.72).

The brick pier is constructed as a hollow pier, therefore allowing reinforced concrete to be placed into the pier to provide additional strength.

It is important to allow the brick pier to set hard before pouring the concrete, or damage may be done to the pier.

The concrete will require compacting into the pier to achieve maximum strength.

Steel reinforcement should be placed into the concrete foundation when it is being laid. The pier will then have to built around it.

It is essential to ensure that the void in the pier is kept free from mortar droppings, which would affect the finished strength of the pier.

Fixing for gates

When piers are being built to receive gates it is essential to plan beforehand.

Gates can be fixed after the brickwork has hardened or built in as the work proceeds (Figure 6.73).

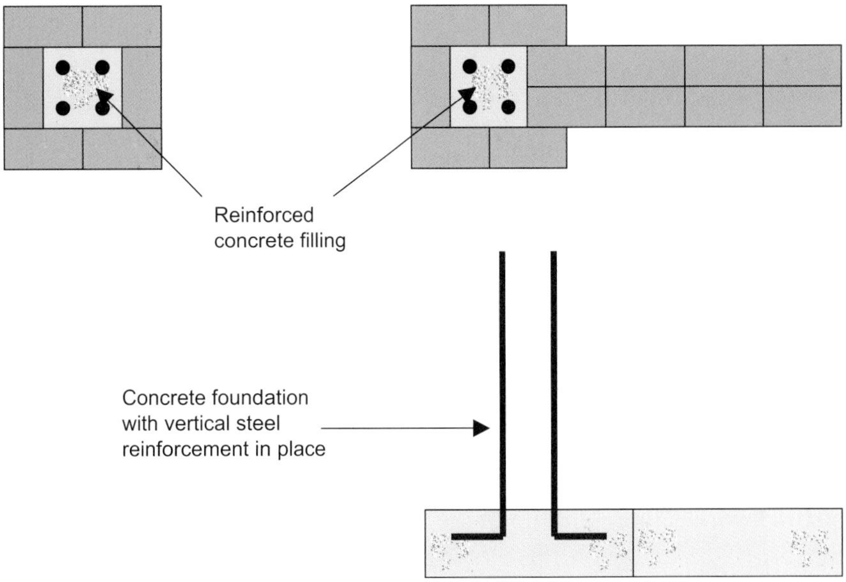
Reinforced
concrete filling
Concrete foundation
with vertical steel
reinforcement in place

FIGURE 6.72
Reinforcing gate pillars

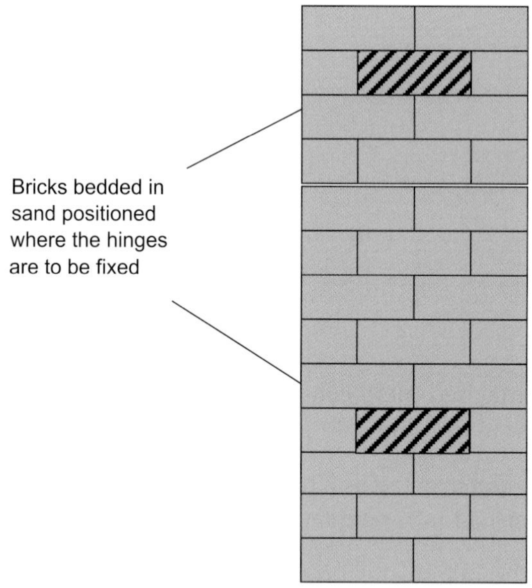
Bricks bedded in sand positioned where the hinges are to be fixed

FIGURE 6.73
Building in sand courses

Fixings for gates can be fitted after the piers have been built by either:

- building in sand courses or
- drilling and fixing hinges.

To fix the brackets after the brickwork has set the sand courses have to be removed and all the sand must be brushed out to leave a clean recess.

The recess should be dampened to ensure adhesion of the mortar.

The gates could be erected in position to identify the correct position of the brackets. These can then be fitted and the brickwork replaced as solidly as possible to ensure that there is no movement when the gate swings.

Fixing as the work proceeds

When gates are being fixed as the work proceeds they can be erected in position and supported in their final position by timbers (Figure 6.74). The brackets can then be built in as the work proceeds and could also be allowed to pass into the void so they would be further attached to the concrete infill. The piers must be constructed absolutely plumb, otherwise the gates will not swing correctly. Once the gate has been placed in its final position on timbers it can be plumbed and supported with raking struts.

The brickwork can be built up to the first bracket; this may include the DPC.

The bracket is then built in, allowing it to extend into the void, and could be fixed to the vertical reinforcement.

The brickwork is then built up to the next bracket, which fitted in the same way.

Double gates are fitted in exactly the same way (Figure 6.75). Care again should be taken to set the gates in their required position on battens and

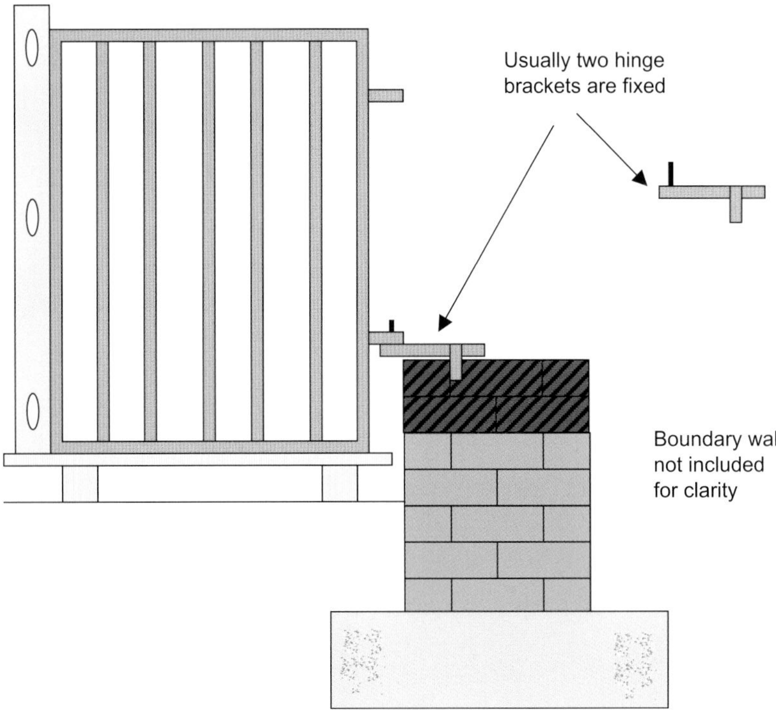
Usually two hinge brackets are fixed
Boundary wall not included for clarity

FIGURE 6.74
Fixing gates to pillars

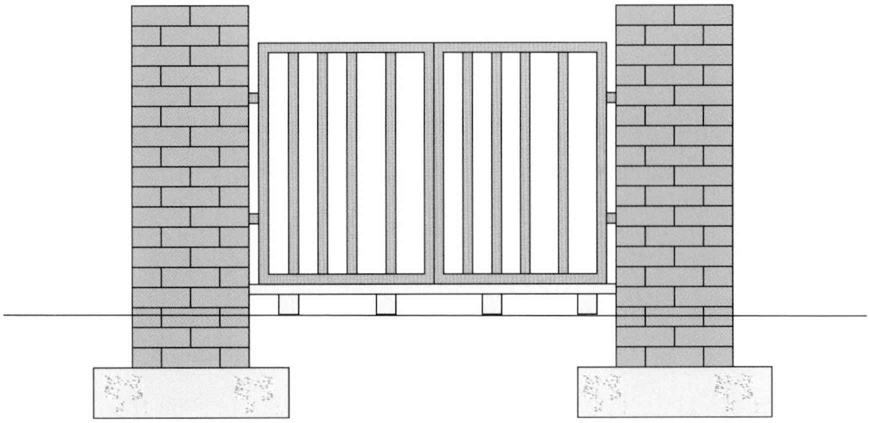

FIGURE 6.75
Fixing gates to pillars

support them in an upright position with raking struts. Alternatively, sand courses could be built in and the brackets fixed afterwards.

COPINGS

Copings are used to protect the tops of walls and piers from moisture penetration; for example, the tops of boundary walls and parapet walls (Figure 6.76).

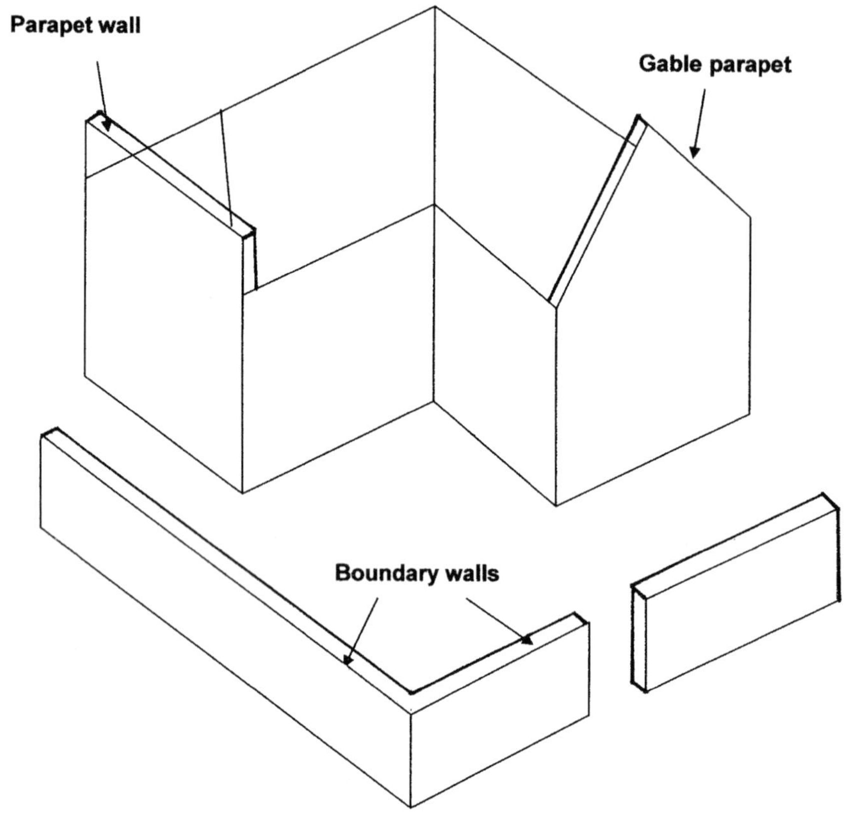
Parapet wall
Gable parapet
Boundary walls

FIGURE 6.76
Positions of copings

Materials

Precast concrete copings or standard bricks may be used. The use of engineering bricks is advisable in exposed situations.

Bricks

Standard well-burnt bricks or engineering class bricks may be used as copings.

Precast concrete copings

These are available in standard shapes and lengths or may be specially produced (Figure 6.77). Concrete copings are available in various shapes. Each provides excellent protection if fitted correctly (Figure 6.78). Flexible DPCs should be provided below the copings.

- Saddleback coping (Figure 6.78a) – The most effective method is to cap the wall with an overhanging concrete coping.
- Feather edge coping (Figure 6.78b) – This needs to be placed so the water runs off onto your own property. A good overhang is essential to run the water clear of the brickwork.
- Detail (Figure 6.78c) – The drip, a groove along the underside of the coping, prevents the passage of water from the top of the coping, the weathering, to the face of the wall.

Remember

Flexible DPCs should be sandwiched between mortar beds.

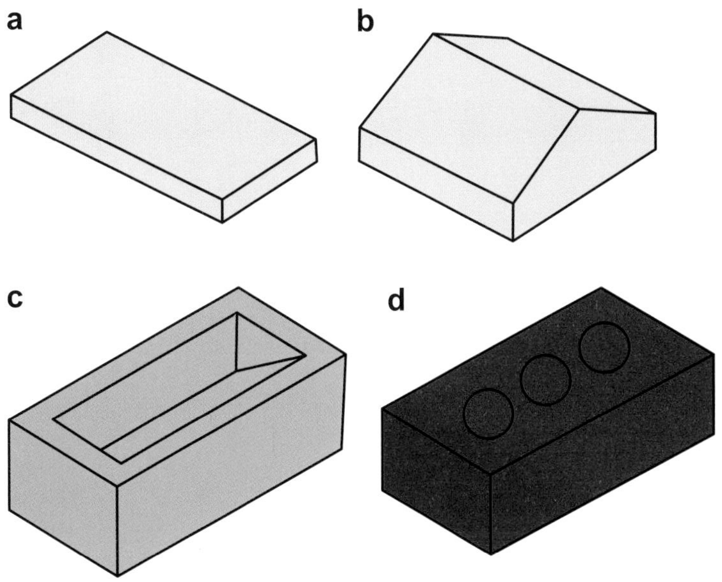
a
b
c
d

FIGURE 6.77
Coping materials: (a) flat coping; (b) saddleback coping; (c) standard brick; (d) engineering quality brick

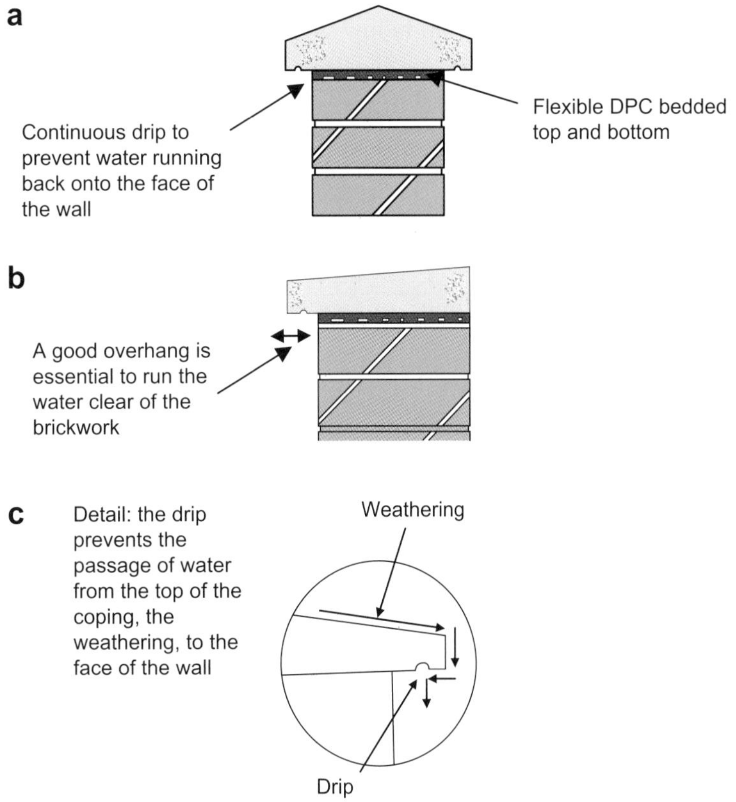
a
Continuous drip to prevent water running back onto the face of the wall
Flexible DPC bedded top and bottom
b
A good overhang is essential to run the water clear of the brickwork
c
Detail: the drip prevents the passage of water from the top of the coping, the weathering, to the face of the wall
Weathering
Drip

FIGURE 6.78
Various coping details: (a) overhanging concrete coping; (b) feather edge coping; (c) detail

Lime staining

Lime staining is caused when free lime leaches from mortar or concrete products under wet conditions (Figure 6.79). The lime accumulates on the face of the brickwork, leaving unsightly white patches along the path of the water that carries it.

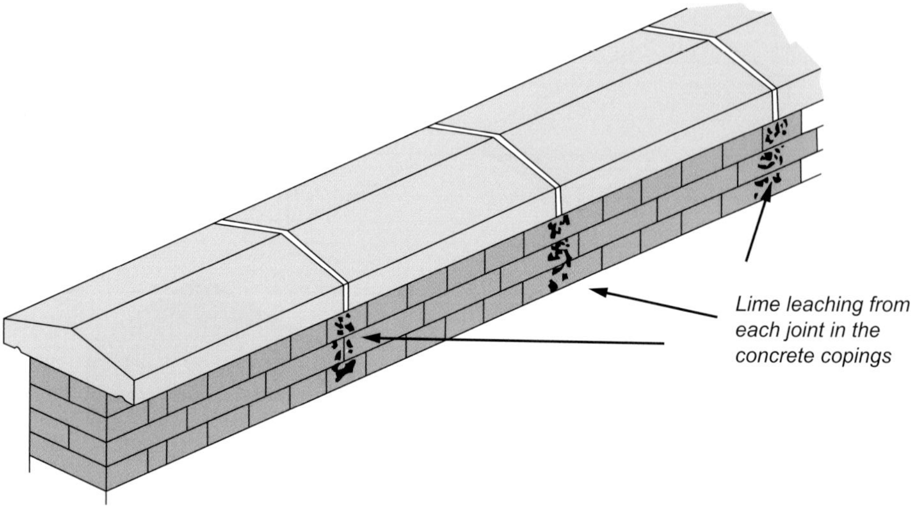
Lime leaching from each joint in the concrete copings

FIGURE 6.79
Lime leaching

Lime staining is very often found below the cross-joints of concrete coping stones as a result of the drip being blocked.

When cross-joints are applied take special care not to block the drip groove as this will cause moisture to run back onto the face of the wall, where lime staining will occur.

Positioning copings

Bed and position the end copings and ensure that they are level.

Attach building lines to front and back edge if wide copings are being laid. For narrow copings a front line is sufficient.

The line is positioned to the eyeline, which is either the upper or lower edge of the coping, depending on its height in relation to where it is viewed from.

The eyeline is determined by which edge is seen by the eye. As a general rule, copings above hand height have the eyeline to the lower edge. See the example in Figure 6.80.

To lay copings, spread a bed joint and butter the joint on the previously laid coping.

Position the coping, making any necessary adjustments by light tapping with a rubber-headed hammer or onto a piece of timber to protect the coping from damage.

Brick capping

Protection can be provided using a brick on edge and tile creasing. These are not as effective as concrete copings owing to the increase in cross-joints.

Bricks should be class A or B engineering quality, bedded in a strong cement–sand mix.

Remember

Lime leaching will not wear off. It has to be removed by cleaning with a diluted solution of hydrochloric acid.

Always wear the correct protective clothing and equipment when cleaning brickwork with acids.

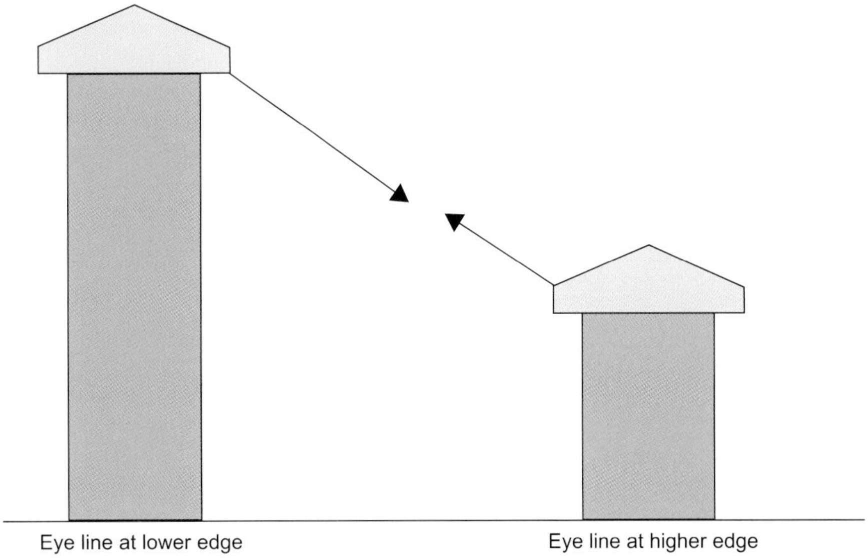
Eye line at lower edge
Eye line at higher edge

FIGURE 6.80
Eyeline of copings

Walls can be capped with a brick on edge without an oversailing course. This is not a very satisfactory method as it allows the water to penetrate the top surface and then run down the face of the wall (Figures 6.81 and 6.82).

Note

When engineering class bricks are used as a brick on edge, the flexible DPC may be omitted.

Construction

When using a brick on edge, with or without tile creasing, the flexible DPC should be correctly bedded onto the boundary wall.

The tile used is usually the nibless type, but with care the bottom course could have nibs to form a decorative effect.

Lay the first course to line with the nibs projecting downwards. This course should be lined in along the lower edge.

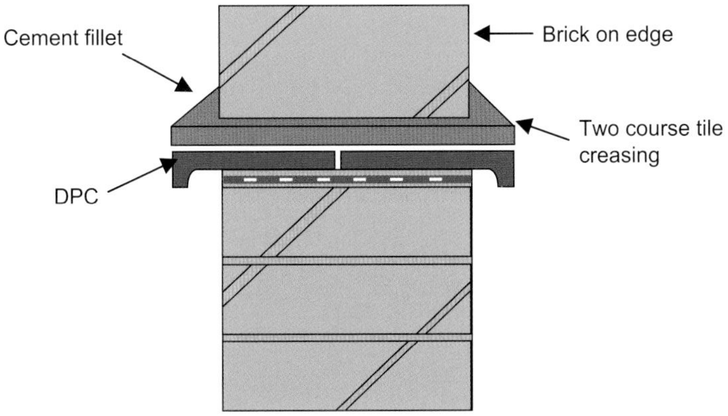
Cement fillet
Brick on edge
Two course tile creasing
DPC

FIGURE 6.81
Brick on edge with tile creasing

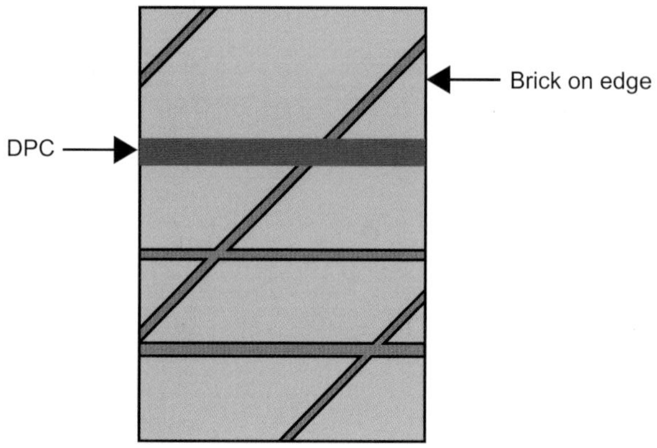
Brick on edge
DPC

FIGURE 6.82
Brick on edge

The second course should have the nibs removed and be laid half-bond and lined in along the upper edge.

Prepare three bricks at each end of the wall, unless special stop ends are being used (Figure 6.83).

The remainder of the bricks should be laid to line, taking care to check the gauge (Figure 6.84), and ensuring that the vertical joints are fully buttered and any frogs filled.

A regular joint size must be maintained on the brick-on-edge course and no cuts are permitted.

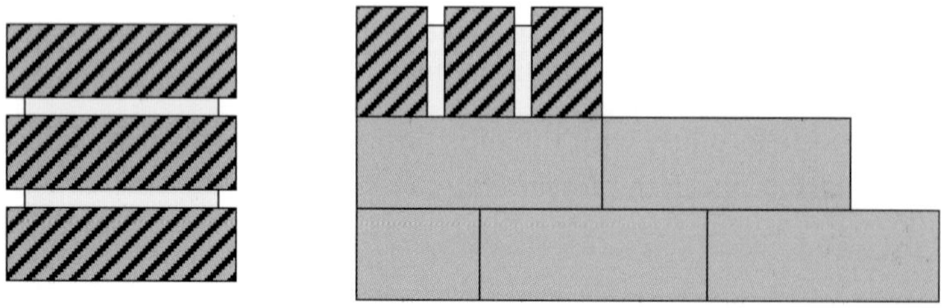

FIGURE 6.83
End block of three bricks bedded to gauge on floor

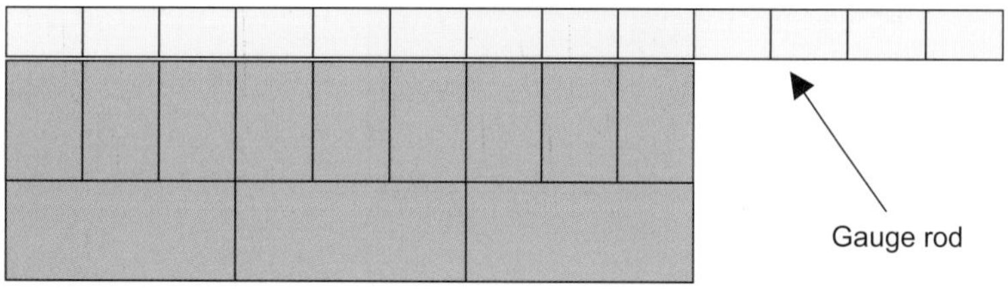
Gauge rod

FIGURE 6.84
Bricks laid to horizontal gauge

If the length of a wall to receive the brick on edge is composed of whole stretchers in any course, then each stretcher will accommodate three bricks on edge together with their joints.

At stopped ends cramps or retainers can be used and are bedded in below the brick-on-edge course (Figure 6.85). These help to protect the vulnerable corner bricks.

DECORATIVE FEATURES

A bricklayer will have to construct many decorative features during work activities; for example:

- string courses
- oversailing courses.

String courses

These are horizontal courses built into the face of a wall to form a decorative feature.

A soldier course is a form of string course, when it continues around a building. An example is shown in Figure 6.86.

Construction

Because soldier courses are classed as decorative features, good selection of the bricks is essential.

Ensure that they are all the same length, accuracy of shape, colour and texture. Reject any bricks with bows, twists, chips, etc.

The course the bricks sit on should be gauged and marked out accordingly to receive the appropriate number of soldiers. A pencil should be used to mark the gauge on the bricks.

All the bricks should have buttered, slightly furrowed joints.

Take care not to smudge the face of the bricks.

Tap the brick into position and check immediately for plumb with a small boat level. Because most soldier courses are built at eye level it is essential

Remember

If the opening does not work then gauge; tighten the joints rather than open them.

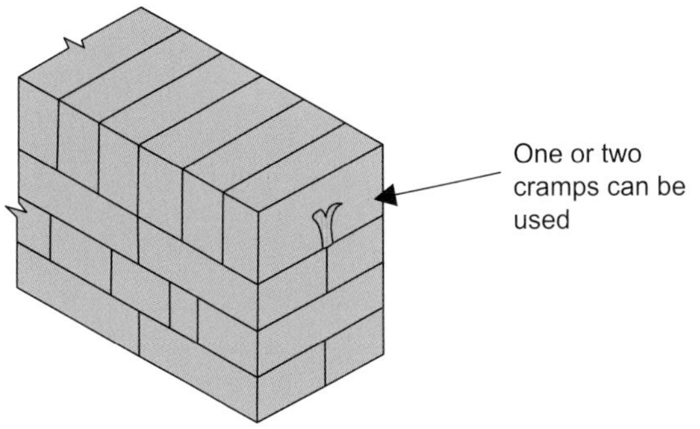
One or two cramps can be used

FIGURE 6.85
Protection to end bricks

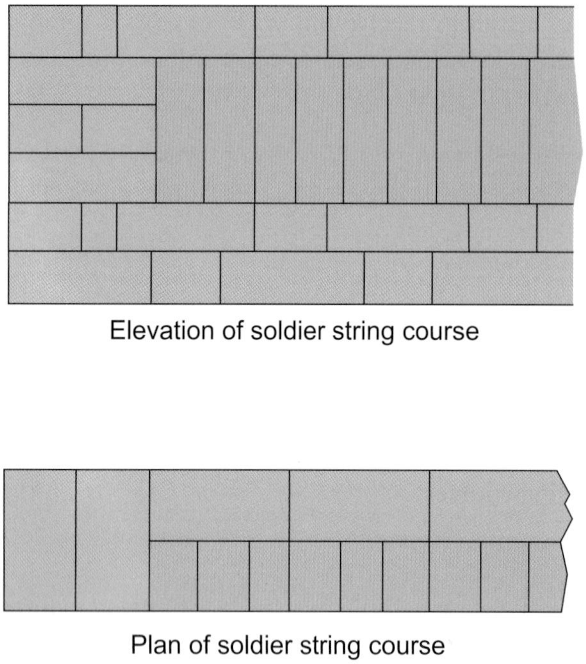
Elevation of soldier string course
Plan of soldier string course

FIGURE 6.86
Soldier string course

to plumb every brick, otherwise a very poor effect will be produced and the bricks will appear to be leaning.

Lay the bricks from the ends to the centre either side to ensure even weight distribution. Take care when placing the last brick, ensure that it is fully buttered on both sides and gently slide it into position, checking that there is no build-up of mortar on the head of the opening (Figure 6.87).

Oversailing courses

Oversailing courses or corbels is the term given to a brick or several bricks which project from the face of a wall (Figure 6.88).

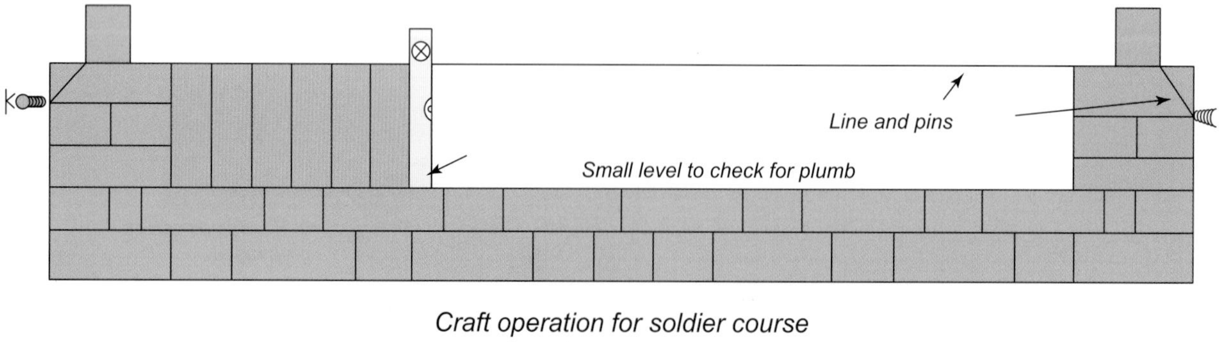
Line and pins
Small level to check for plumb
Craft operation for soldier course

FIGURE 6.87
Constructing a soldier string course

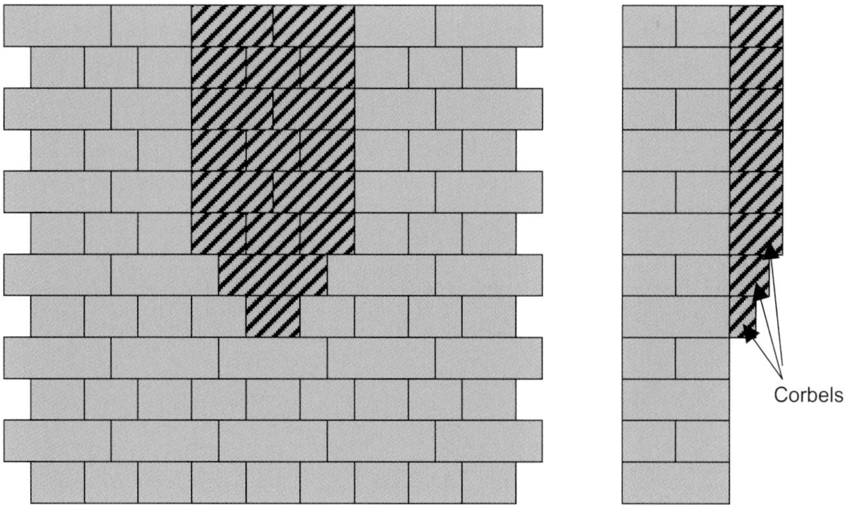
Corbels

FIGURE 6.88
A simple form of corbel to form a decorative feature

Oversailing or corbelling work may be used to thicken a wall or form a decorative feature.

When this type of work is undertaken it is important that the underside arris of each course is level because it is this line that is seen by the eye.

It is good practice to use headers for corbelling out and restrict the projection to a quarter of a brick.

The Building Regulations state that the extent to which any part of a wall overhangs the wall below shall not be such as to impair the stability of the wall or any part of it.

To prevent this happening, under no circumstances should a wall overhang more than the thickness of the wall below the corbel.

REINFORCEMENT

Brickwork under normal loading conditions is usually strong enough without reinforcement.

When brickwork is put under pressure which is trying to crush the wall, reinforcement can be added to withstand these pressures and increase the strength of the brickwork.

Reinforcement can be placed either vertically or horizontally.

We have already dealt with vertical reinforcement to gate pillars in Figure 6.72. Figure 6.89 shows pillars to a boundary wall being reinforced in a similar manner. Vertical reinforcing rods are placed into the foundation concrete and extend the height of the hollow pillars. The hollow pillar is then filled with concrete to produce reinforced brick pillars.

Brick walls can be reinforced by building in horizontal mesh as the work proceeds to strengthen the wall from side pressure (Figure 6.90).

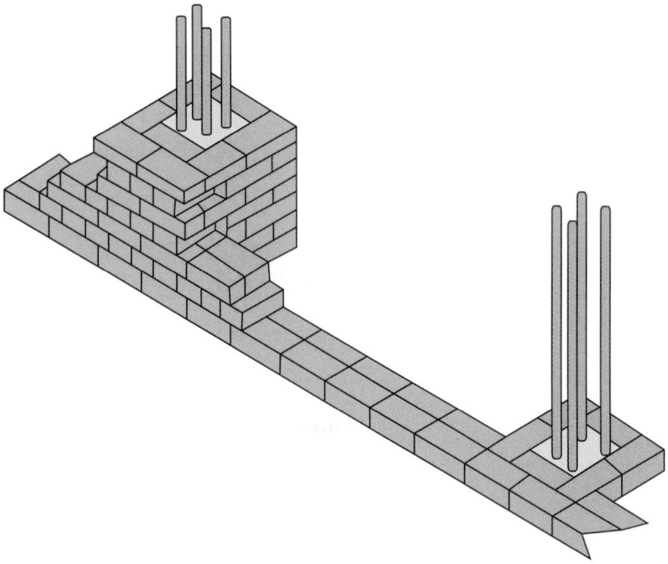

FIGURE 6.89
Vertical reinforcement

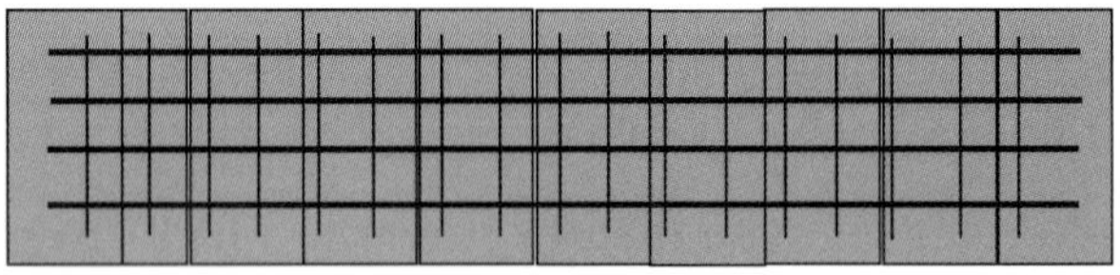

FIGURE 6.90
Horizontal reinforcement

The wire mesh is bedded into the mortar every third or fifth course as required.

VERTICAL MOVEMENT JOINTS

The outer leaf of face brickwork in a wall will expand and contract owing to changes in temperature and moisture content.

Allowance must also be made for sideways expansion and contraction, however, so that the outer leaf of brickwork does not cause damage to itself or the structure.

A typical vertical movement joint, with expanded plastic foam joint filler projecting, can be seen in Figure 6.91. This shows a 102 mm thick brick wall.

Vertical movement joints should be provided at intervals between 9 and 12 m for clay brickwork; not exceeding 7–9 m for sand lime (calcium silicate) bricks; and at 6 m intervals for concrete blockwork.

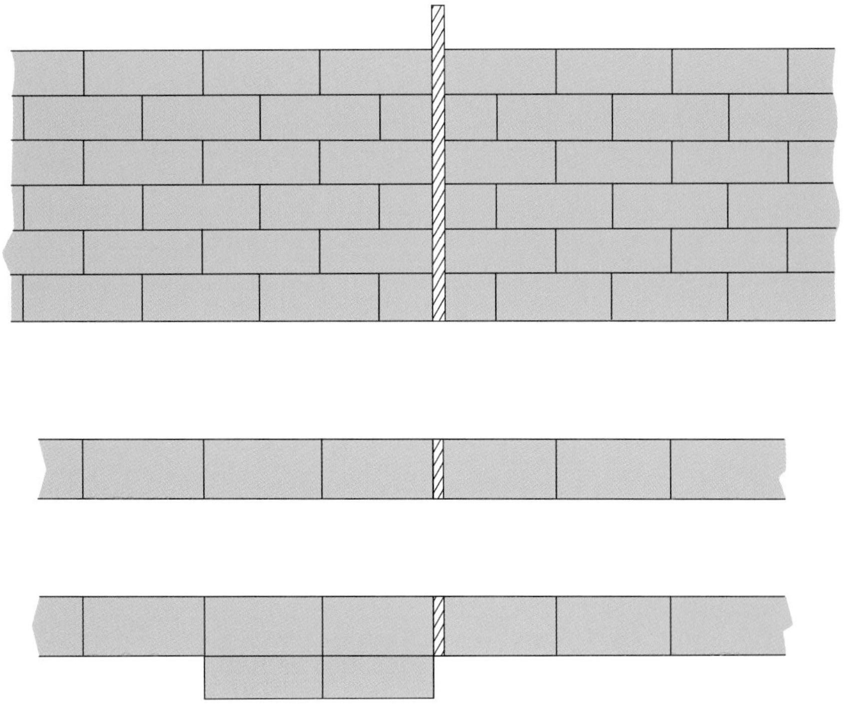

FIGURE 6.91
Vertical movement joints

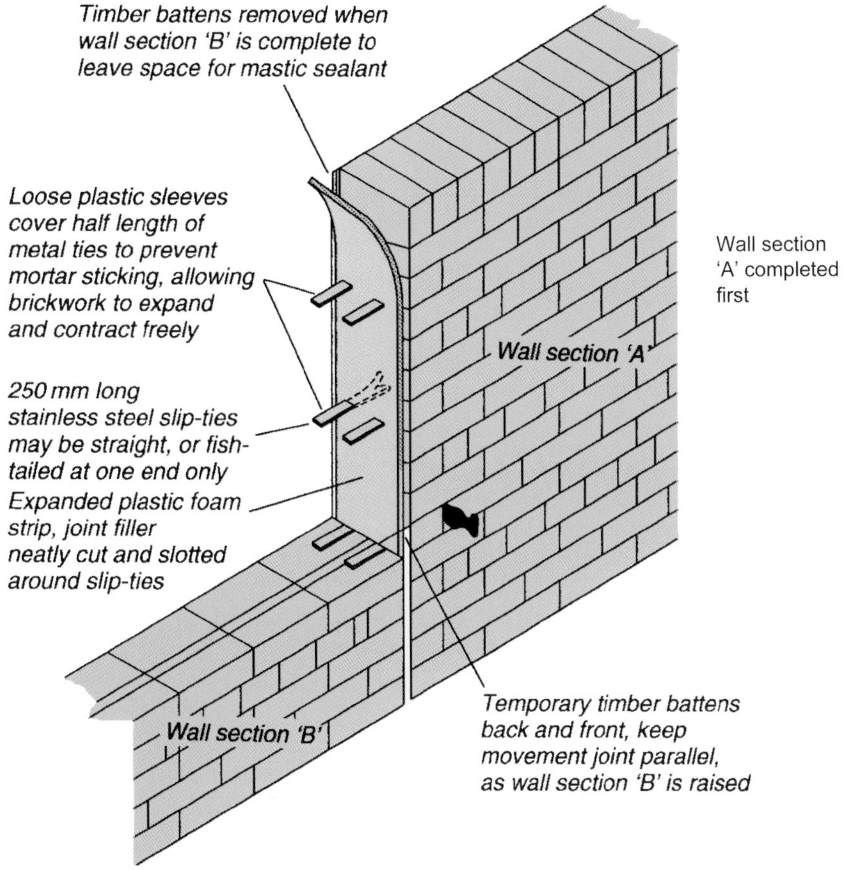
Timber battens removed when wall section 'B' is complete to leave space for mastic sealant
Loose plastic sleeves cover half length of metal ties to prevent mortar sticking, allowing brickwork to expand and contract freely
250 mm long stainless steel slip-ties may be straight, or fish-tailed at one end only
Expanded plastic foam strip, joint filler neatly cut and slotted around slip-ties
Wall section 'A'
Wall section 'A' completed first
Wall section 'B'
Temporary timber battens back and front, keep movement joint parallel, as wall section 'B' is raised

FIGURE 6.92
Building in a vertical movement joint

Clearly, vertical movement joints form a straight joint weakness in a wall, and so must be strengthened with stainless steel slip-ties every fourth course (Figure 6.92).

The expanded plastic foam filler is kept back 12 mm from both faces so that a sealant mastic can be applied to seal these movement joints against rain penetration.

Multiple-choice questions

Self-assessment

This section of the book is designed to allow you to check your level of knowledge. The section consists of revision questions for this chapter. The questions are all multiple choice and have four possible answers. The answers are to be found at the end of the book.

The main type of multiple-choice question will be the four-option multiple-choice question. This will consist of a question or statement, known as the stem, followed by a choice of four different answers, called the responses. Only one of these responses is the correct answer; the others are incorrect and are known as distracters.

You should attempt to answer the questions by choosing either (a), (b), (c) or (d).

Example

The person employed by the local authority to ensure that the Building Regulations are observed is called the:

(a) clerk of works
(b) building control officer
(c) council inspector
(d) safety officer

The correct answer is the building control officer, and therefore (b) would be the correct response.

Solid walls

Question 1 Identify the following, which could be used to finish off the top of a boundary wall:

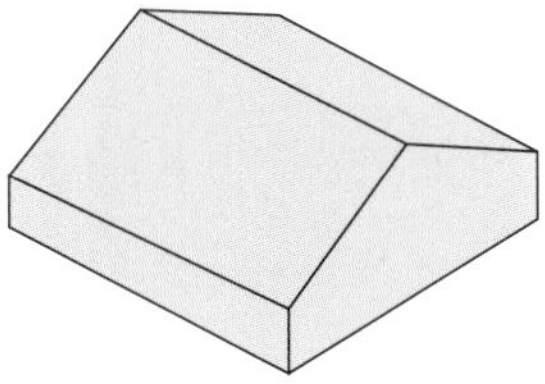

(a) feather edge coping
(b) saddle back coping
(c) gabled coping
(d) pitched coping

Question 2 Which of the following is the correct definition of English garden wall bond?

(a) alternate courses of headers and stretchers

(b) alternate headers and stretchers on each course

(c) three stretchers to one header in each course

(d) three courses of stretchers to one course of headers.

Question 3 Identify the following bond:

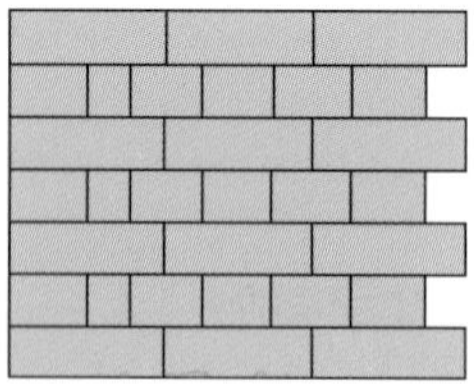

(a) English bond

(b) Flemish bond

(c) English garden wall bond

(d) Flemish garden wall bond

Question 4 What is a tingle plate used for?

(a) to fasten the lines to the stopped end

(b) to keep the wall plumb

(c) to support the line in the centre of a long wall

(d) to keep the wall level

Question 5 Why is it necessary to load out bricks for a wall from several different packs?

(a) to prevent banding on the walls

(b) to select the best bricks for the wall

(c) to use the bricks as quickly as possible

(d) to ensure the correct number of bricks

Question 6 Identify the following bond:

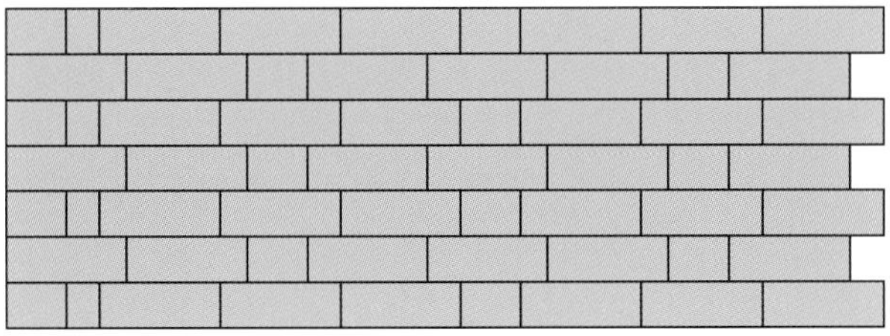

(a) English bond
(b) Flemish bond
(c) English garden wall bond
(d) Flemish garden wall bond

Question 7 When a cut brick has to be placed in the centre of a wall it is known as:

(a) broken bond
(b) reverse bond
(c) Dutch bond
(d) garden wall bond

Question 8 What is the purpose of the cramp built into the end of a brick on edge course as shown in the drawing?

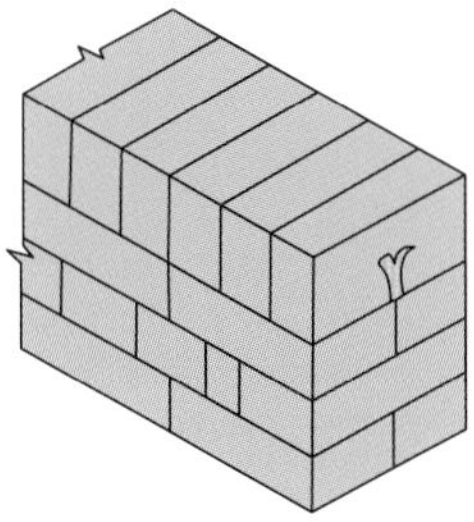

(a) for decorative purposes
(b) to keep the end of the wall plumb
(c) to protect the end of the wall
(d) to keep the brick on edge level

CHAPTER 7

Cavity Walls

This chapter will cover the following NVQ and Diploma units:

- NVQ VR40
- CC 2048K

This chapter is about:

- Interpreting building information
- Adopting safe and healthy working practices
- Selecting materials, components and equipment
- Preparing and erecting brickwork and blockwork structures

The following NVQ performance criteria will be covered:

- Performance criterion 1: Interpretation of information
- Performance criterion 2: Safe work practices
- Performance criterion 3: Selection of resources
- Performance criterion 4: Minimize the risk of damage
- Performance criterion 5: Meet the contract specification
- Performance criterion 6: Allocated time

The following Diploma outcomes will be covered:

- Know how to plan and select resources for practical tasks
- Know how to erect cavity walling
- Know how to form openings in cavity walls

Building information

Definition of cavity walls

The standard form of construction for the external walls of brick buildings is called cavity walling. This means that the bricklayer builds the two separate 'leaves' or 'skins' of brick masonry (a general term indicating brickwork and or blockwork), with a 50–75 mm wide space between.

The outer skin is usually 102.5 mm thick face brickwork, but may be constructed from facing quality blocks. The inner skin is usually 100 mm thick common blocks that are later plastered to receive internal decoration (Figure 7.1).

Both skins of brick masonry are joined together with a regular pattern of corrosion-resistant ties, so that they behave as one single wall.

A continuous space termed a cavity, which is approximately 50 mm wide, is produced and the walls are supported by specially prepared ties of metal called wall ties.

The prevention of the passage of moisture from the outer to the inner walls is of the utmost importance in cavity wall construction, and to achieve this, damp-preventing materials must be inserted in the structure at special points, such as window and door reveals, and window and door heads.

Purpose of cavity walls

Cavity walls are very useful for damp and exposed positions to ensure a dry interior to the building. The reasons for this are:

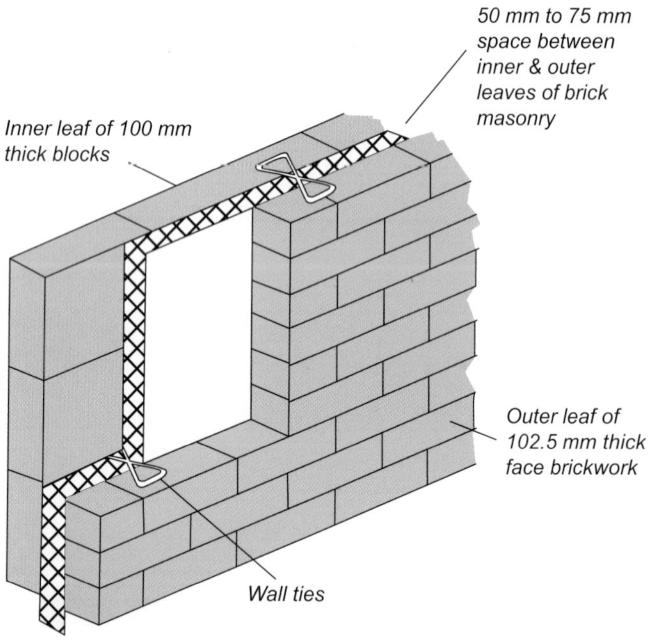
50 mm to 75 mm space between inner & outer leaves of brick masonry
Inner leaf of 100 mm thick blocks
Outer leaf of 102.5 mm thick face brickwork
Wall ties

FIGURE 7.1
Typical insulated cavity wall

- The cavity provides a break between the outer wall of the building, which may damp, and the inner wall.
- To achieve a better balanced temperature inside the building. The cavity keeps the inside cool in the summer and helps to retain heat in the winter.

Where proper care is exercised in the building of this type of wall, it will prevent the passage of moisture from the outer or protective wall to the inner wall.

Cavity wall construction began to be widely used from the 1920s, as a way of preventing dampness from soaking through the outer walls of buildings.

The 50–75 mm wide gap stopped rain penetrating from the outer surface to the plastered inner surface of external walls by capillary action (when water is drawn through hairline channels within a porous structure, e.g. a brick, by the action of surface tension). This was possible when outer walls were commonly 215 mm thick solid brickwork.

As a secondary advantage, this space between the inner and outer skins also provides thermal insulation for modern buildings, because heat energy mainly escapes by conduction through solid material. Air is a poor conductor of heat energy, therefore the rate of heat loss is very much slower than was the case when buildings had solid outer walls (Figure 7.2).

Building Regulations

The basic advantages of cavity wall construction, over solid 215 mm thick brickwork, for the outer walls of a building have been incorporated in the current Building Regulations.

In order to satisfy requirements of the current Building Regulations and enable planning permission to be obtained, cavity wall construction is usually specified whether the structure is low rise or multistorey.

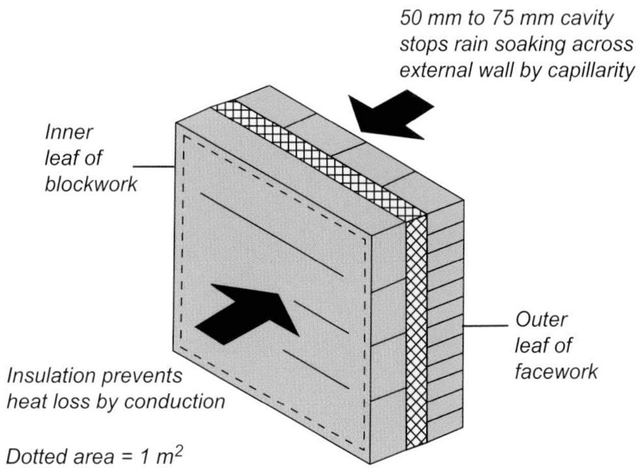
50 mm to 75 mm cavity stops rain soaking across external wall by capillarity
Inner leaf of blockwork
Outer leaf of facework
Insulation prevents heat loss by conduction
Dotted area = 1 m²

FIGURE 7.2
Function of typical cavity walling

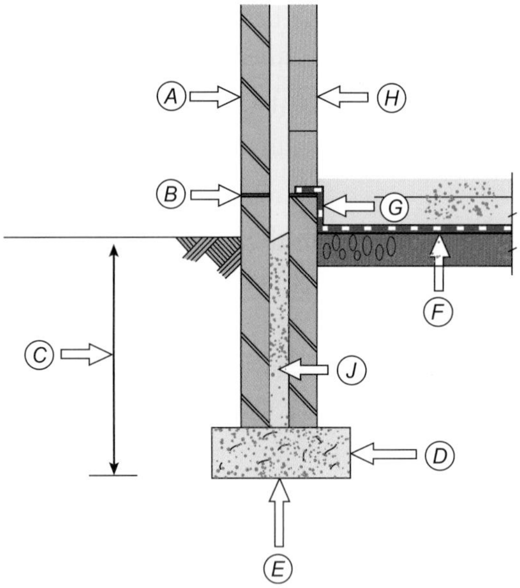
A
B
C
D
E
F
G
H
J

FIGURE 7.3
Cavity walling on standard strip foundation.

Figure 7.3 indicates how the basic requirements of the Building Regulations are satisfied, where standard strip foundations are specified with a solid ground floor slab. Figure 7.4 shows a trench-fill foundation, associated with a suspended ground floor construction of prestressed, precast concrete floor beams supporting standard size concrete blocks.

The following labels are used in Figures 7.3 and 7.4:

A: external cavity wall, providing resistance to through-penetration of rain

B: horizontal damp-proof course (DPC) in both leaves, not less than 150 mm above ground level, to prevent dampness rising from the soil

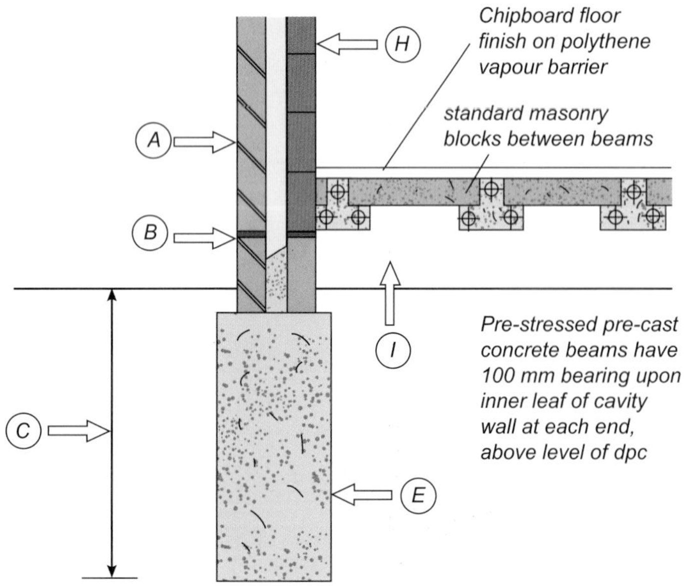
H
Chipboard floor finish on polythene vapour barrier
A
standard masonry blocks between beams
B
C
I
Pre-stressed pre-cast concrete beams have 100 mm bearing upon inner leaf of cavity wall at each end, above level of dpc
E

FIGURE 7.4
Cavity walling on trench fill foundation with suspended concrete floor

C: a minimum distance of 1 m between ground level and the underside of the concrete foundation, as a protection against frost heave in winter and drying shrinkage of clay subsoils in summer

D: a minimum 150 mm thickness of foundation concrete to transfer the building load adequately on to the natural foundation of the subsoil

E: sulphate-resisting cement in the foundation concrete and substructure brickwork, necessary where soluble sulphates are present in subsoil water

F: continuous damp-proof membrane (DPM) across the ground floor areas to prevent dampness rising from the soil

G: DPM and DPC, lapped and joined within the floor thickness around the perimeter of rooms

H: external walls built of lightweight blocks and other thermal-insulating material to give a U value of 0.35 (in other words, heat energy must not escape through the outer walls at a rate greater than 0.35 W/m^2/hour, per degree difference in temperature internally and externally)

I: ventilated air space separating the suspended floor from damp soil

J: weak cavity fill to prevent collapse of walls owing to pressure from the ground.

Function of cavity walls

Figure 7.5 shows that it is the inner leaf of cavity walling that largely supports the load from floors and roof in a low-rise building of load-bearing wall construction. Common building blocks have totally replaced common bricks for this inner leaf, owing to improved thermal insulation values and

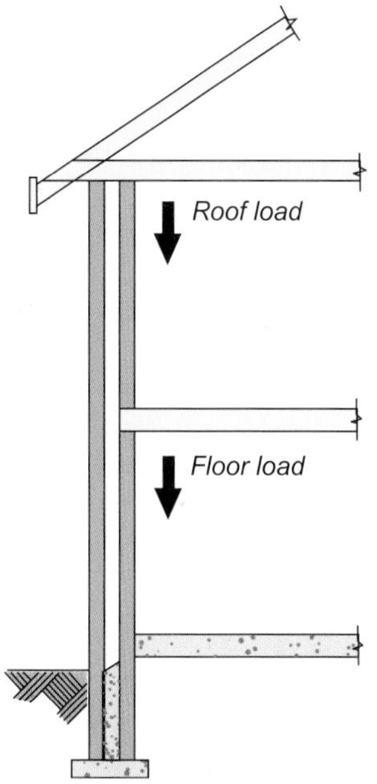
Roof load
Floor load

FIGURE 7.5
Loads on cavity walls

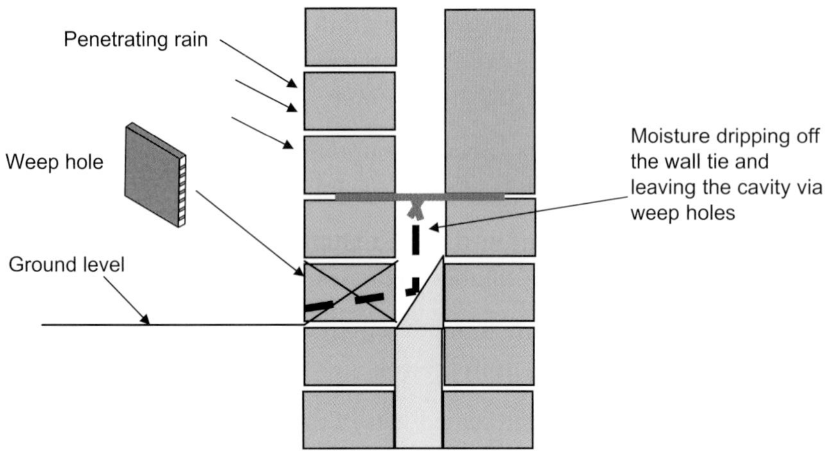
Penetrating rain
Weep hole
Ground level
Moisture dripping off the wall tie and leaving the cavity via weep holes

FIGURE 7.6
Use of drip on wall tie

bricklayer output (one standard size 100 mm thick block is equal to six bricks).

The purpose of the outer skin of facework is to give the building a weather-resistant and pleasant appearance, by selection from the wide range of colours and surface textures of bricks and facing blocks available to the designer.

Although cavity walling is constructed with properly finished, solidly filled mortar joints, it is expected that this outer skin will let rain water soak through as far as the cavity. This is because mortar, bricks and blocks are porous to varying degrees. This through-penetration will be highest on those elevations of a building exposed to prevailing wet winds (Figure 7.6).

In the early days of cavity walling it was considered good practice to ventilate the cavity as the flow of air around the cavity would keep it dry. During the 1960s, this theory changed with the increase in fuel prices. There was more emphasis on retaining heat in buildings and the cavities were then sealed to increase the resistance to heat loss through the cavity wall.

This has now changed again as the current Building Regulations are asking for even less passage of heat through the cavities. To achieve this reduction in heat loss cavities are now either partially or totally filled with insulation.

Constructional requirements

Bricks or blocks should be properly bonded and solidly built with mortar not weaker than 1:2:9.

The leaves should be securely tied together with wall ties.

The cavity should not be less than 50 mm or more 75 mm in width at any level.

The leaves should not be less than 100 mm thick at any level.

The overall thickness should not be less than 190 mm.

The wall should not exceed 3.5 m in height and 12 m in length, or 9 m in height and 9 m in length.

The inner leaf cannot be less than 75 mm thick if the walls form part of a private dwelling house.

The current Building Regulations specify that the external wall must have a DPC at a height of not less than 150 mm above the finished surface of the adjoining ground.

Advantages of cavity wall construction

- Able to withstand driving rain.
- Gives good thermal insulation.
- Gives good sound insulation.

Disadvantages of cavity wall construction

- Requires a high standard of design and workmanship and supervision.
- Vertical DPC at all openings.
- More expensive than solid walls.

Adopting safe and healthy working practices

Safety legislation has already been covered in Chapter 2. Please refer back to this chapter if necessary.

Particular areas to consider when building cavity walls are:

- personal protective equipment (PPE)
- manual handling of materials
- mechanical tools and equipment
- working at heights.

Personal protective equipment

Depending on the type of workshop or site situation, the wearing of correct safety clothing and safe working practices are the best methods of avoiding accidents or injury. On some sites certain PPE is compulsory.

All construction operatives have a responsibility to safeguard themselves and others. Protecting oneself often means wearing the correct protective clothing and safety equipment.

Manual handling of materials

More than a quarter of all reported accidents are associated with manual handling. Manual handling is defined as the transporting or supporting of loads by hand or bodily force such as human effort.

The Manual Handling Operations Regulations outline how to deal with risks to the safety and health of construction site operatives.

The site operative should be able to select and use appropriate safety equipment and protective clothing when handling different materials.

If the item to be moved is an awkward size or shape the site operative should be able to select and use appropriate equipment or aids to carry materials.

They should also be able to demonstrate safe manual handling techniques.

MANUAL HANDLING OPERATIONS REGULATIONS 1992

These regulations came into force on 1 January 1993 and apply to the manual handling of loads. They help to prevent injury, not only to the back, but to any part of the body.

The regulations take into account the physical properties of loads that could affect the grip and cause injury by slipping, roughness, sharp edges and extreme temperatures.

The employer should take the following measures when there is a possibility of risk from manual handling:

- Avoid hazardous manual handling operations so far as is reasonably practicable.
- Assess any hazardous manual handling operations that cannot be avoided.
- Reduce the risk of injury so far as is reasonably practicable.

It is a requirement of the Health and Safety at Work Act 1974 that employers provide their employees with health and safety information and training.

Remember

Always report any hazardous materials or unsafe situations to the site supervisor immediately.

Mechanical tools and equipment

These have exposed moving parts such as grinding wheels, sanders, drills, chipping hammers, portable saws, rotary wire brushes and air compressors.

There are special regulations concerning grinding wheels and portable saws, and people must be authorized to use them.

- Never use any mechanical equipment that is unfamiliar.
- First read the manufacturer's instructions or seek advice.
- Never lay a tool down while it is still rotating.
- Never wear loose-fitting clothes when using tools with fast moving parts.

Most accidents are caused by lack of knowledge, misuse, makeshift repairs or using faulty tools and equipment.

All brickwork apprentices have to achieve the Abrasive Wheels Certificate as part of their programme

Working with crane-handled or mechanically handled loads

Mechanical handling is when a crane, fork-lift truck, etc., is used.

Remember that not all manual handling is eliminated by using mechanical handling. Very often the materials or components to be mechanically moved have to be manually handled into position first.

The load being mechanically moved may also require support by the hands or any other part of the body, e.g. the shoulders.

MOVEMENT OF MATERIALS

Materials can be moved around the site either by hand or by machine.

Most materials are delivered to site on pallets and possibly shrink wrapped. These can then be stored on site, possibly in the compound, until required. They can be moved around the site with a fork-lift truck (Figure 7.7).

Although mechanical lifting methods are becoming more common on larger sites, on small sites materials may still be moved with wheelbarrows.

Working at heights

Working platforms have already been explained in Chapter 2.

The Work at Height Regulations 2005 apply to all work at height where there is a risk of a fall liable to cause personal injury.

A scaffold is a temporary staging to assist bricklayers and other tradespeople to construct a building.

The scaffold must be spacious and strong enough to support people and materials during construction.

FIGURE 7.7
Equipment for moving materials

Many accidents are due to simple faults such as misuse of tools, untied ladders, a missing toeboard, etc.

Unfortunately, a number of accidents are also due to ignoring the Construction (Health, Safety and Welfare) Regulations, which specify basic scaffolding requirements.

The three basic requirements for scaffolds are:

- They should be suitable for the purpose.
- They should be safe.
- They should comply with the regulations.

GENERAL SAFETY

No scaffold should be erected or be substantially added to or altered or be dismantled except under the immediate supervision of a *competent person* and, so far as possible, by *competent workers* possessing adequate experience of such work.

When work cannot be safely carried out from ground level or from part of a building or other permanent structure, there shall be provided either scaffolds or where appropriate ladders or other means of support, all of which shall be suitable for the purpose.

ACCESS

It is important that you are familiar with a wide range of access equipment.

It is usual to access a bricklayer's scaffold from a ladder, which must be positioned so that this can be done easily and safely.

In the case of access to a scaffold, the ladder should be securely attached to the scaffold, preferably inside the structure (Figure 7.8).

Remember

The Work at Heights Regulations 2005 require that ladders should only be considered where a risk assessment has shown that the use of other more suitable work equipment is not appropriate because of the low risk, and the short duration of the task or considerations of where the work is located.

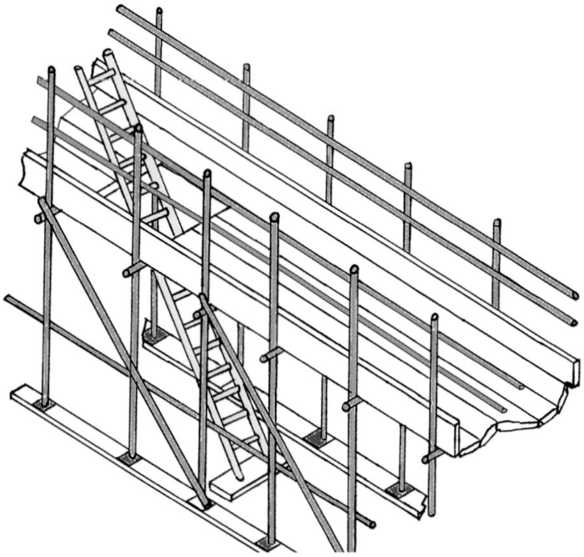

FIGURE 7.8
Access to working platform

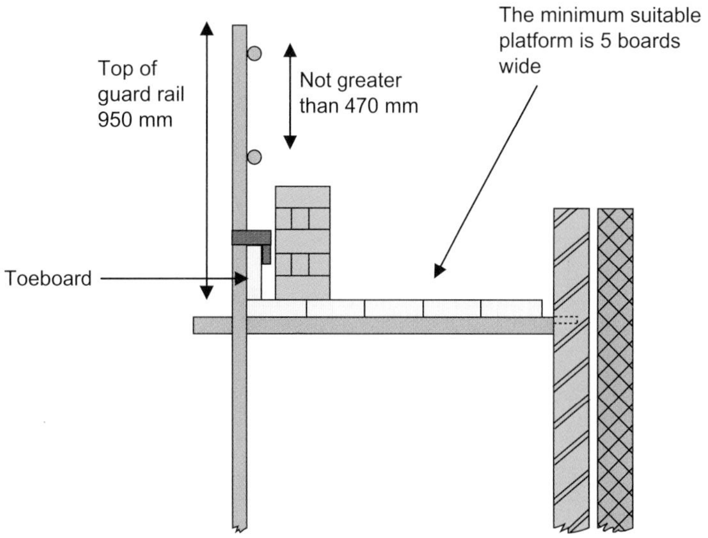
The minimum suitable platform is 5 boards wide
Top of guard rail 950 mm
Not greater than 470 mm
Toeboard

FIGURE 7.9
Bricklayer's working platform

Any surface on which the ladder rests should be stable and of sufficient strength to support the ladder so that its rungs remain horizontal and it can support any loading placed on it.

The final rung of the ladder from which the operative steps onto the platform should ideally be just above the surface of the platform.

The ladder, which should be secured at both the top and bottom, should extend at least 1.05 m (approximately five rungs) above the platform.

Note

There is no need for a centre guardrail when brickguards are being used

Bricklayers require a scaffold wide enough to allow:

- the stacking of materials
- working space for the bricklayer
- room for the passage of people and materials (Figure 7.9).

GUARDRAILS AND TOEBOARDS

Access platforms more than 2 m high must have guardrails and toeboards. The top guardrail should be at least 950 mm above the platform with intermediate guardrails fixed so that any gap between the rail and any other means of protection is not greater than 470 mm.

The risk of falling materials causing injury should be minimized by keeping platforms clear of loose materials.

In addition, materials or other objects must be prevented from rolling, or being kicked, off the edges of working platforms. This can be achieved by fixing toeboards, solid barriers, brick guards or similar at open edges.

At the end of the working day it is important to turn back the batten nearest the new brick wall to prevent any splashing if it rains (Figure 7.10).

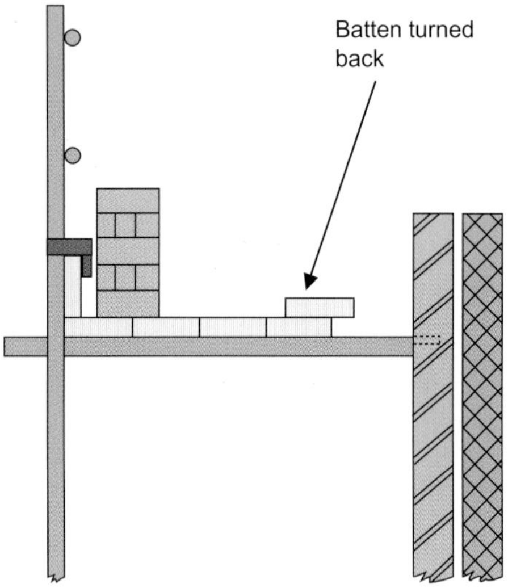
Batten turned back

FIGURE 7.10
Inner batten turned back at the end of the working day

Selecting materials, components and equipment

Cavity walls are constructed using bricks and blocks and other components as described in the following paragraphs.

Bricks and blocks

Normal construction consists of facing bricks on the outside wall and lightweight blocks on the inside (Figure 7.11). Variations of this are found with blocks on both walls.

Older versions of cavity walls can be found with common bricks on inside walls.

In response to the requirements of approved Document L, many manufacturers of blocks have produced blocks that help to meet the thermal requirement (Figure 7.12).

Future developments will see the U value fall even more and could result in manufacturers again revisiting the design of blocks.

Blocks for inner walls of cavity walls are available in widths greater than 100 mm, to give greater insulation value

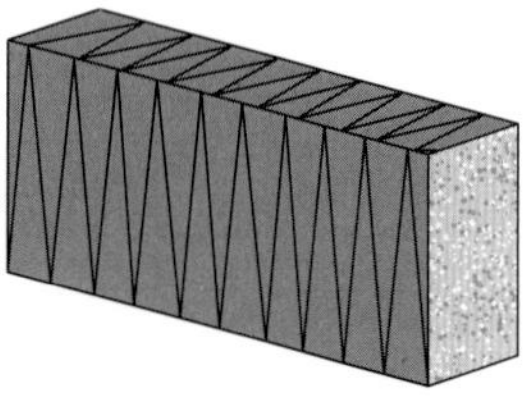

FIGURE 7.11
Bricks and blocks

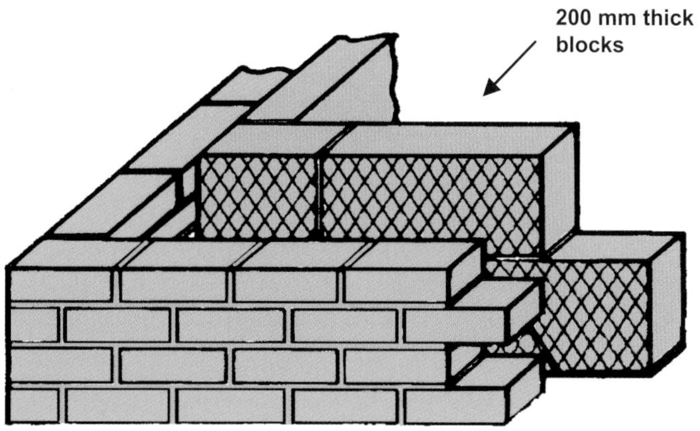
200 mm thick
blocks

FIGURE 7.12
Typical cavity wall quoin

Wall ties

Wall ties are used to give stability to cavity walls by tying the inner and outer leaves together.

A range of proprietary ties has been produced for this job of tying together inner and outer skins of brick masonry, as 'substitute headers' so that both leaves behave as one wall. These wall ties are made from stainless steel, galvanized steel or polypropylene, so they do not provide a passage for moisture.

Ties are shaped so as to form a drip at their centre and the ends should be designed to be able to be bedded securely in both leaves. Examples are shown in Figure 7.13.

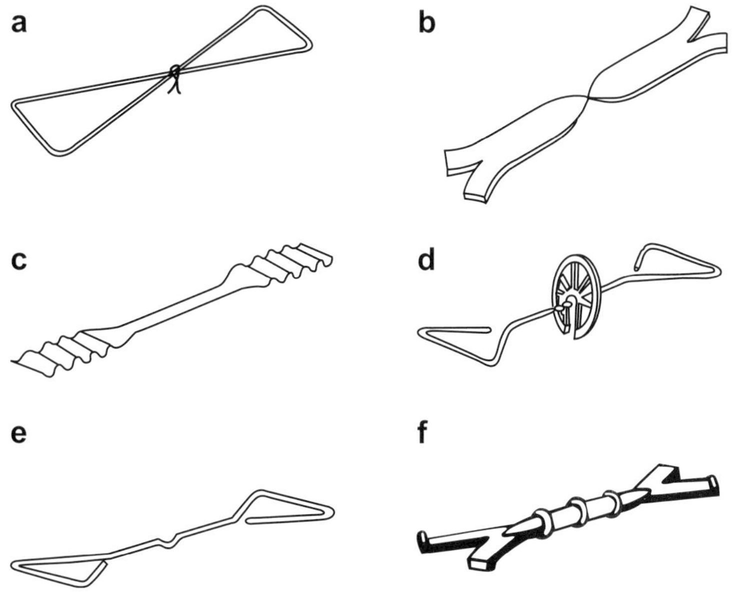
a
b
c
d
e
f

FIGURE 7.13
Types of wall tie: (a) butterfly; (b) galvanized mild steel wall tie; (c) stainless steel tie; (d) double triangle with plastic clips; (e) double triangle; (f) plastic

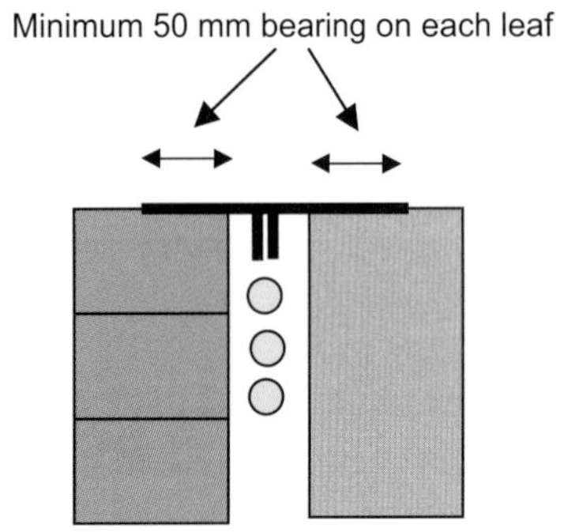
Minimum 50 mm bearing on each leaf

FIGURE 7.14
Position of wall ties

Wall ties should be bedded level or slope slightly towards the outer leaf with the drip positioned downwards in the centre of the cavity (Figure 7.14).

The standard maximum spacing for wall ties is at intervals of 900 mm horizontally and every sixth course vertically.

Each horizontal layer should be offset. For the purposes of estimating quantities of wall ties required, this works out at approximately 2.5 per square metre.

Air bricks

Ventilation into cavities or through cavities is maintained by building air bricks into the external wall. These are available in normal brick sizes and colours to match the facing bricks (Figure 7.15).

Air bricks are built into the external walls along with air ducts. These are built into the external walls at approximately 2 m centres (Figure 7.16).

Thermal insulation

Cavities can be partially filled or total filled with various types of insulation. The insulation is made from various materials and available in various thicknesses such as expanded polystyrene bead board and extruded expanded polystyrene.

Insulation slabs are usually 450 mm high and 1200 mm long.

For partially filled cavities solid insulation is required so that it can be fixed back to the internal wall with special wall ties (Figure 7.17).

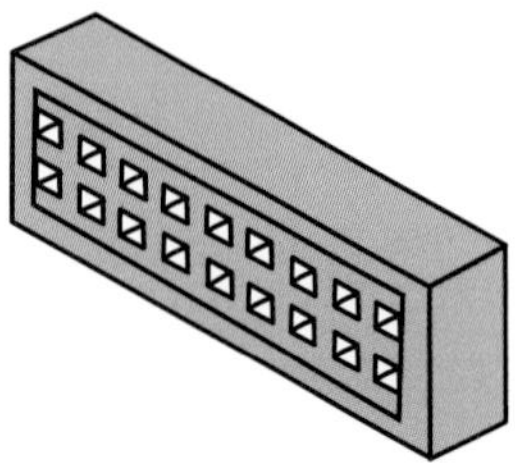

FIGURE 7.15
Air brick

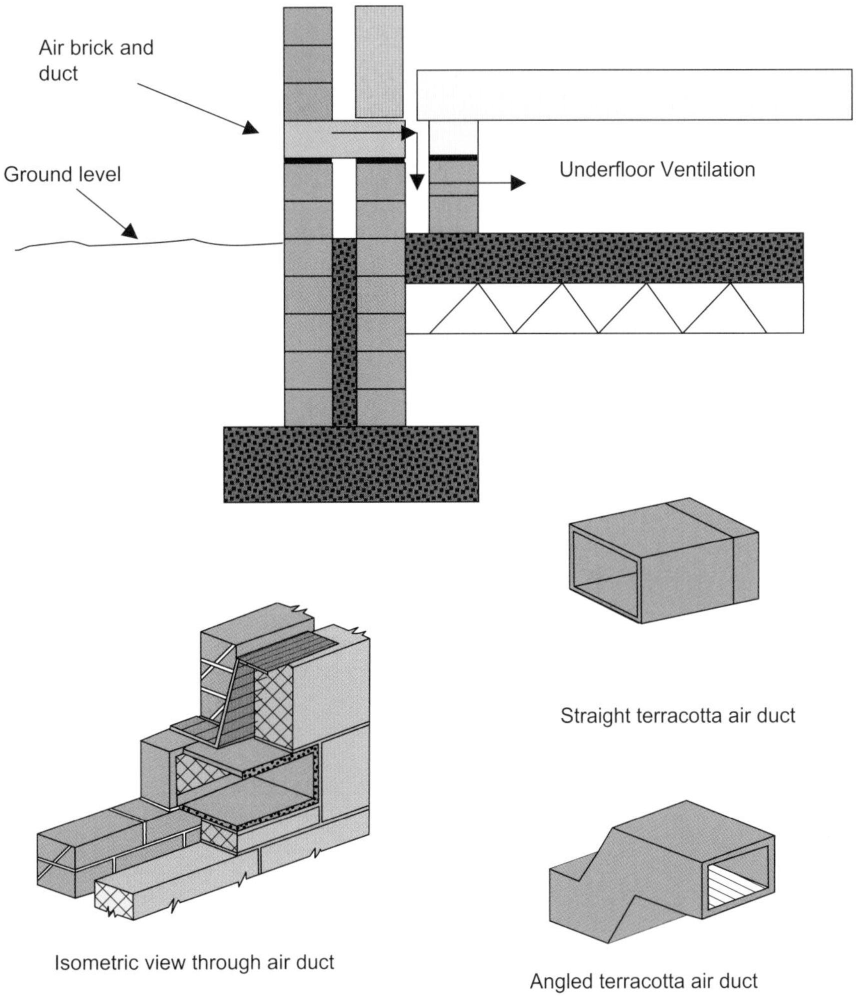
Air brick and duct
Ground level
Underfloor Ventilation
Straight terracotta air duct
Isometric view through air duct
Angled terracotta air duct

FIGURE 7.16
Air brick details

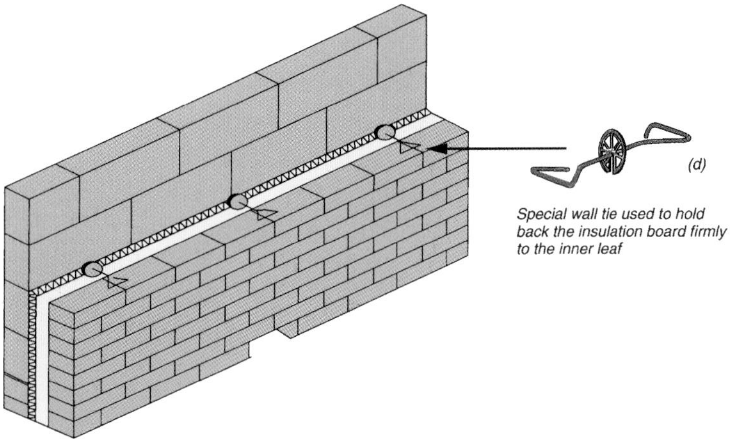
(d)
Special wall tie used to hold back the insulation board firmly to the inner leaf

FIGURE 7.17
Partially filled cavity

When cavities are totally filled solid or flexible insulation can be used (Figure 7.18).

Door and window frames

Door and window frames are required to be built into cavity walls. They can be made of various materials including wood and PVC.

Door frames are usually standard sizes, but windows are available in numerous shapes and sizes. A sample of door and window frames is shown in Figure 7.19.

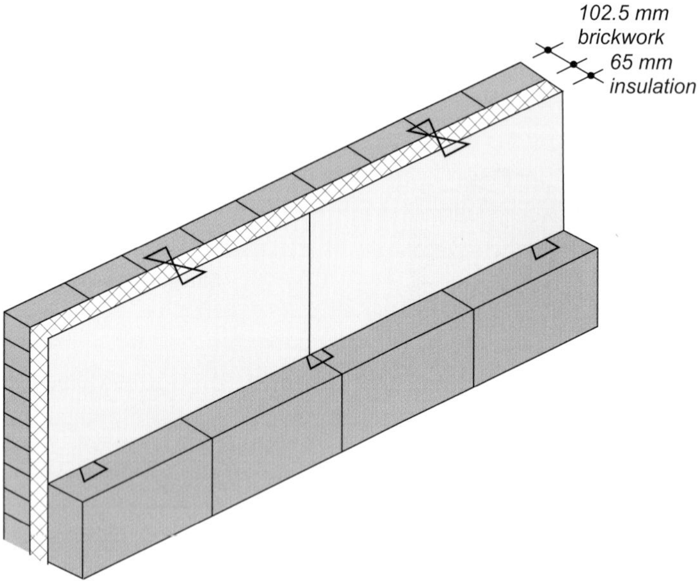
102.5 mm
brickwork
65 mm
insulation

FIGURE 7.18
Totally filled cavity

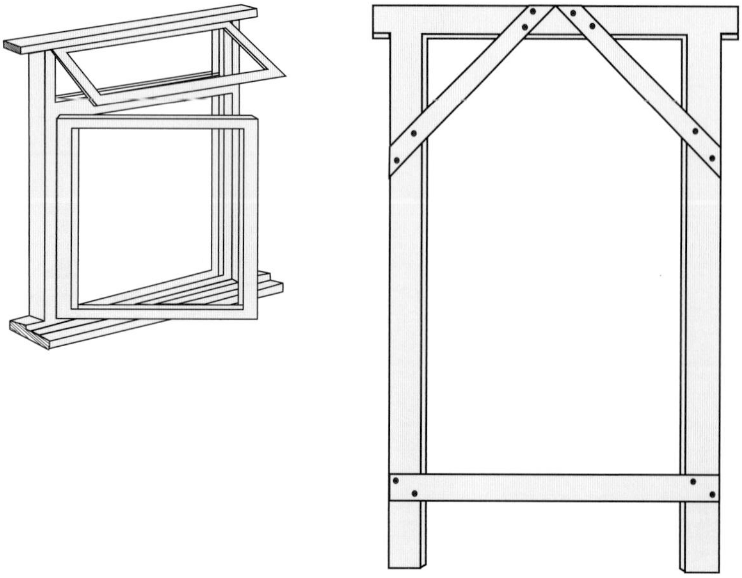

FIGURE 7.19
Door and window frames

Construction of cavity walls

Cavity walls commence below ground level unless the work is on a raft construction. Work below ground level has been covered in Level 1, Chapter 10. Please refer back to this chapter if required.

Level 2 deals with cavity walls above ground.

As with all brick and block walls it is essential to set the work out correctly before starting to lay the materials.

The facing bricks should be taken from several packs to avoid banding, but unlike one-brick walls the materials are required for both sides of the wall (Figure 7.20).

Remember the setting-out details from Chapter 6. Good practice means lifting up the mortar board on blocks to make it more convenient to reach the mortar. Place the mortar boards 600 mm from the face of the wall to allow for adequate working space.

Setting out bonds in cavity walls

When the brickwork of a building approaches ground level it is normal practice to change from common bricks to facing bricks. This changeover normally takes place two courses below finished ground level to avoid commons being shown when the building is complete.

When the facings reach ground level preparation must be made in the bond arrangement for inserting door and window openings at a future height. It is

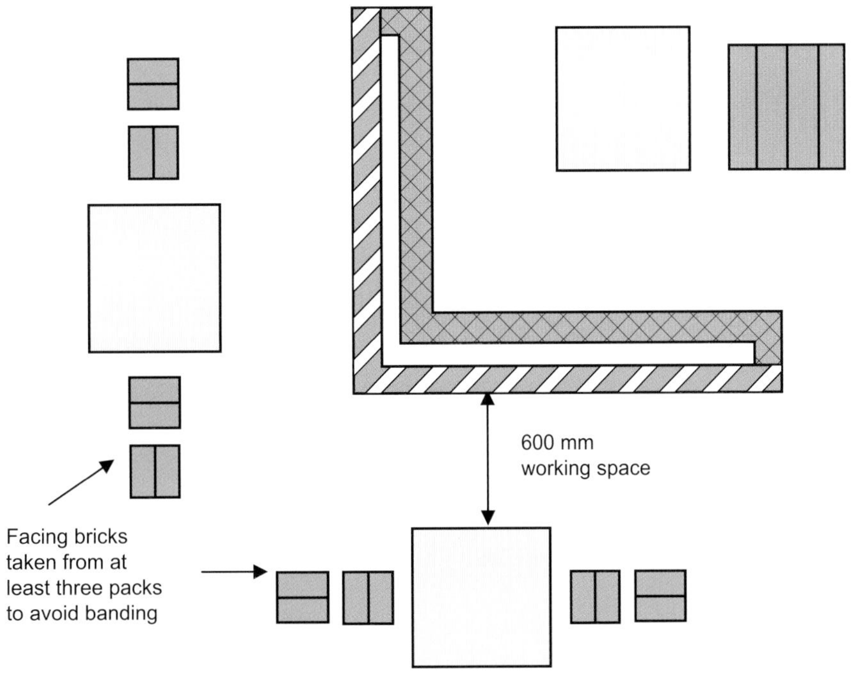
600 mm
working space
Facing bricks
taken from at
least three packs
to avoid banding

FIGURE 7.20
Suggested layout for a cavity quoin

at this stage that the proposed elevation of the building must be carefully studied.

All buildings will have doors and windows built into them and this will lead to a long wall being split up into smaller walls.

It is essential that the bonding is set out at ground level to avoid the problems of broken or reverse bond when the windows are bedded into position later.

Figure 7.21 shows a typical elevation of a building with openings.

DRY BONDING

The normal practice is to dry bond the bricks from each end of the wall, placing the 'broken bond' under the door or window opening, where it will be lost for a while.

Setting out facework is usually the responsibility of the supervisor who will, after consulting the architect, determine the detailed bond pattern and the location, if any, of broken or reverse bonds.

Setting out the brickwork is completely different from setting out the building which is done before excavation begins.

One of the main purposes of setting out facework is to create a matching and balanced appearance of bricks, particularly at the reveals on either side of the door and window openings and the ends of the wall.

It is essential that no straight joints occur.

REVEALS

The position of all window openings and 'reveal' bricks should be identified when setting out the first two courses (Figure 7.22).

This ensures that perpends can continue unbroken for the full height of the wall.

BROKEN BOND

Reveal bricks provide fixed points between which the bonding is set out. The usually short lengths of brickwork between windows offer little scope for adjusting the widths of cross-joints in order to avoid broken bond.

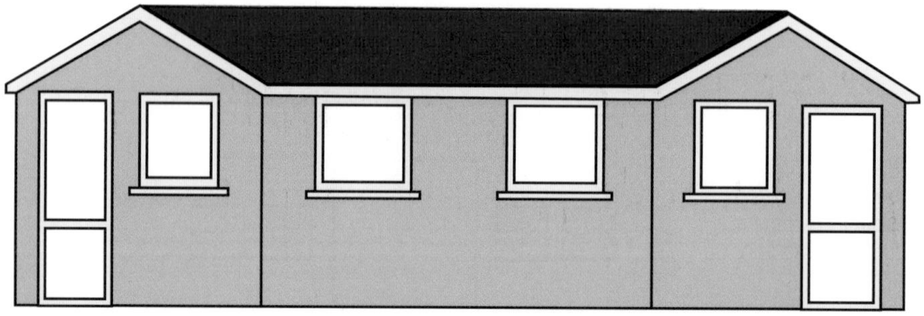

FIGURE 7.21
Semi-detached bungalow

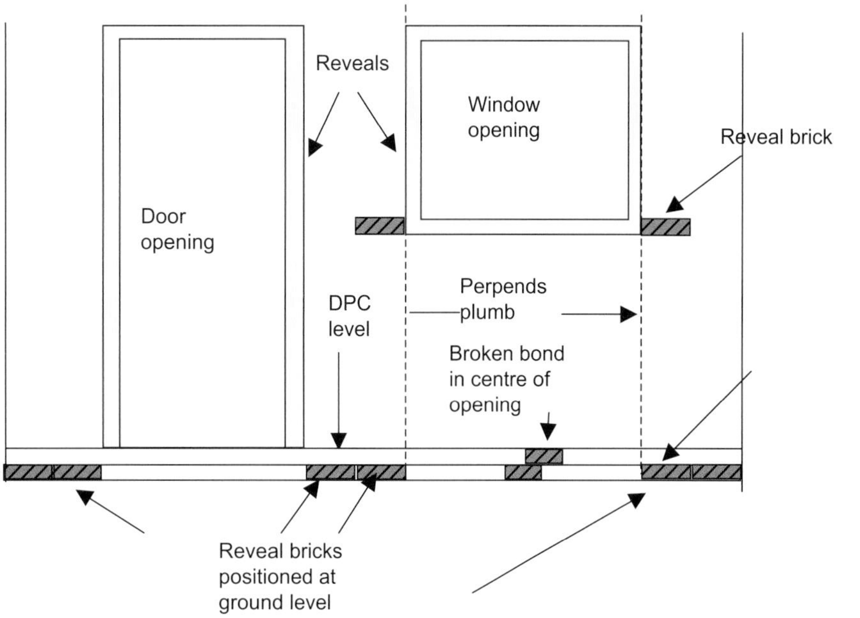
Reveals
Window opening
Reveal brick
Door opening
DPC level
Perpends plumb
Broken bond in centre of opening
Reveal bricks positioned at ground level

FIGURE 7.22
Bond set out on first two courses

Broken bond can sometimes be avoided by 'tightening' or 'opening' the joints. In doing so, bricklayers should work to the brick size plus a 10 mm joint = 225 mm.

It is far better if broken bonds can be avoided altogether, but where openings and piers do not allow for brick sizes, the broken bonds should be carefully set out. Once their position has been decided, this should be maintained throughout the height of the wall.

As shown in Figure 7.22, the broken bond under the window will disappear once the window is built in.

On no account should a closer be built in the middle of the wall; it should only be placed next to the quoin header.

Broken bond and possibly wasteful cutting can be avoided if the overall length of walls and the widths of doors, windows openings and brickwork between the openings are all multiples of a brick stretcher.

The bonding either side of reveals will also match symmetrically at each course.

In practice, the ideal situation seldom occurs and a satisfactory solution is dependent on bricklaying skills.

REVERSE BOND

Occasionally it is possible, when setting out facework, to avoid broken bond at the centre of a wall or window pier by 'reversing' the bond.

This means breaking Rule 3 (see Figure Table 6.2 in Chapter 6). The end bricks do not correspond (Figure 7.23).

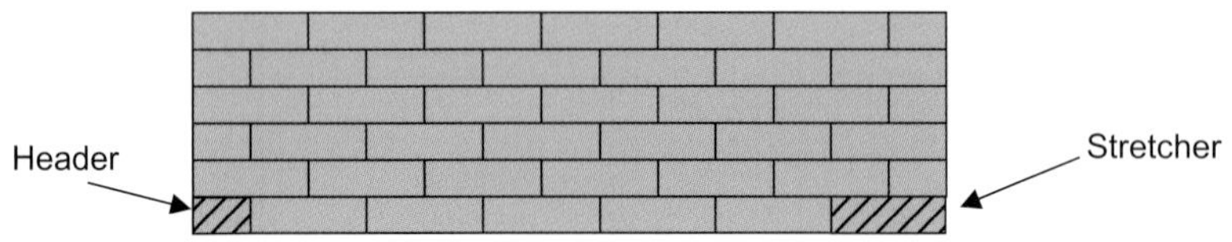
Header
Stretcher

FIGURE 7.23
Reverse bond

Erecting quoins in cavity walls

Cavity walls consist of two leaves which in modern construction are usually facework on the outer leaf and blockwork on the inner leaf (Figure 7.24).

Once the corners have been erected and checked for accuracy the remainder of the wall can be run in to a line.

Thermal insulation

When cavity walling was first used, it was common practice to install air bricks at the top and bottom, at intervals around the whole perimeter of the building, to ventilate this 50–75 mm wide space to remove damp air. Since the 1950s, however, cavity walls have become sealed, with insulating material built in as work proceeds, to improve the thermal insulation value of the cavity space.

Two ways that a bricklayer may be told to install thermal insulation in cavity walling are shown in Figures 7.25 and 7.26.

In a fully filled system, flexible fibre batts of insulation completely occupy the cavity space. In a partially filled system, stiffer boards of insulating material half-fill the cavity, but a 25–35 mm air space is retained as well. Both batts and boards are supplied in purpose-made sizes to fit neatly between layers of wall ties and are approximately 900–1200 mm in length.

All insulation batts should be fixed staggering the vertical joints, and butting the vertical and horizontal joints as tightly as possible. Insulation batts can be cut with a sharp knife, but always ensure that the cut is square and forms a perfectly tight joint.

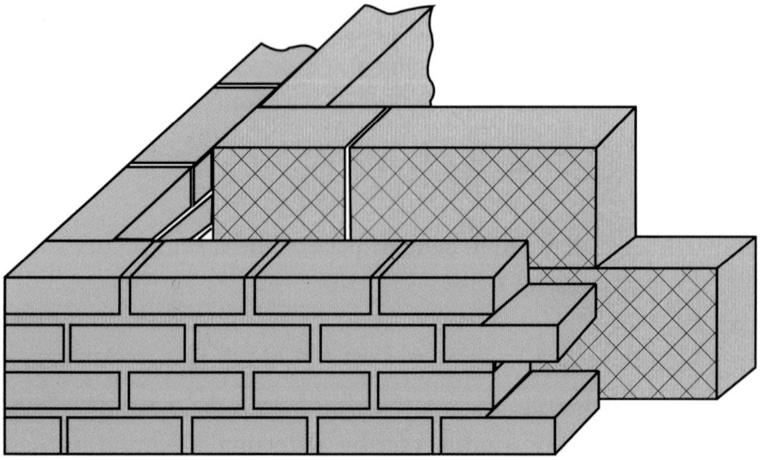

FIGURE 7.24
Cavity quoin

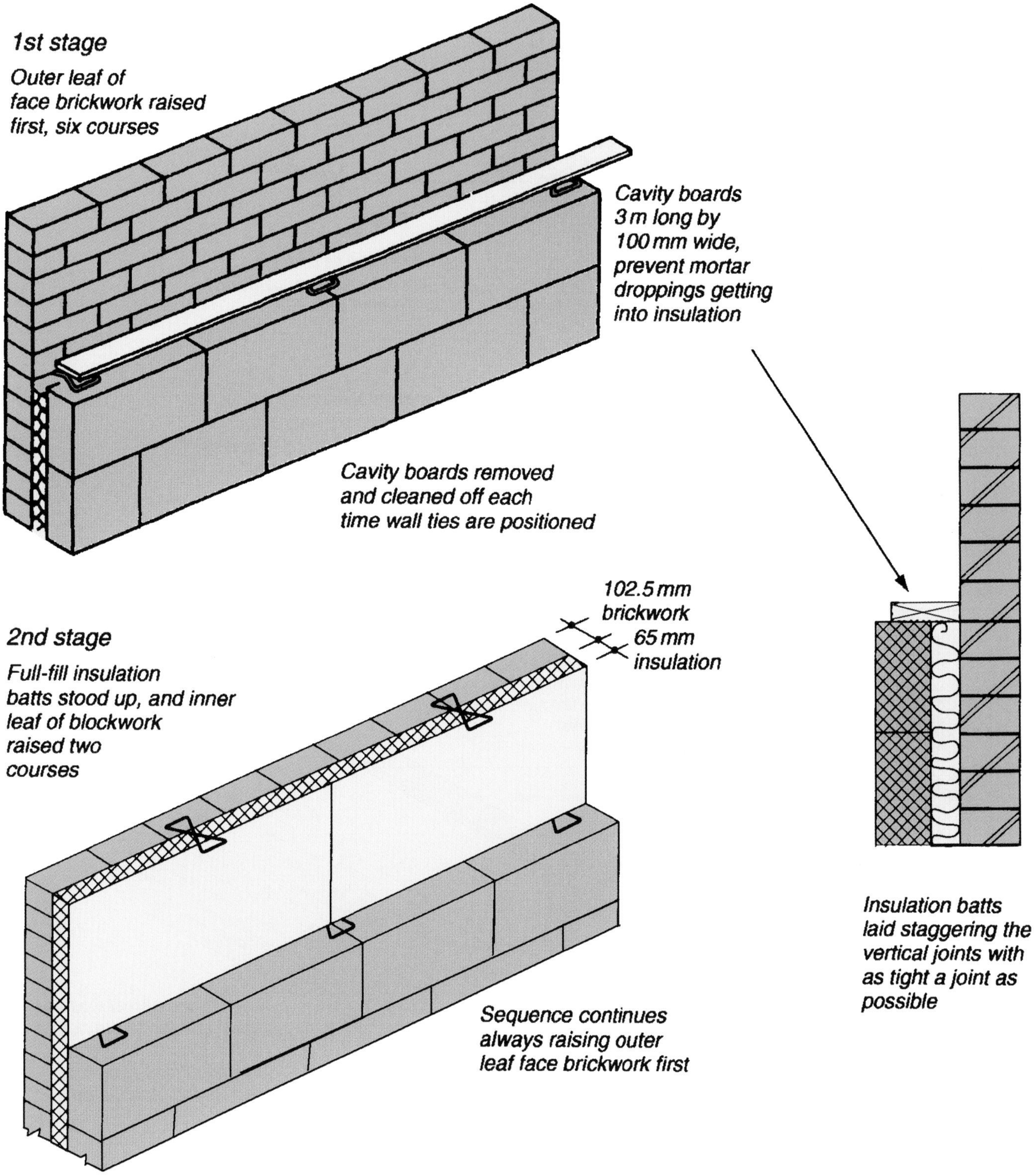
1st stage
Outer leaf of face brickwork raised first, six courses
Cavity boards 3 m long by 100 mm wide, prevent mortar droppings getting into insulation
Cavity boards removed and cleaned off each time wall ties are positioned
102.5 mm brickwork
65 mm insulation
2nd stage
Full-fill insulation batts stood up, and inner leaf of blockwork raised two courses
Sequence continues always raising outer leaf face brickwork first
Insulation batts laid staggering the vertical joints with as tight a joint as possible

FIGURE 7.25
Fully filled cavity insulation system

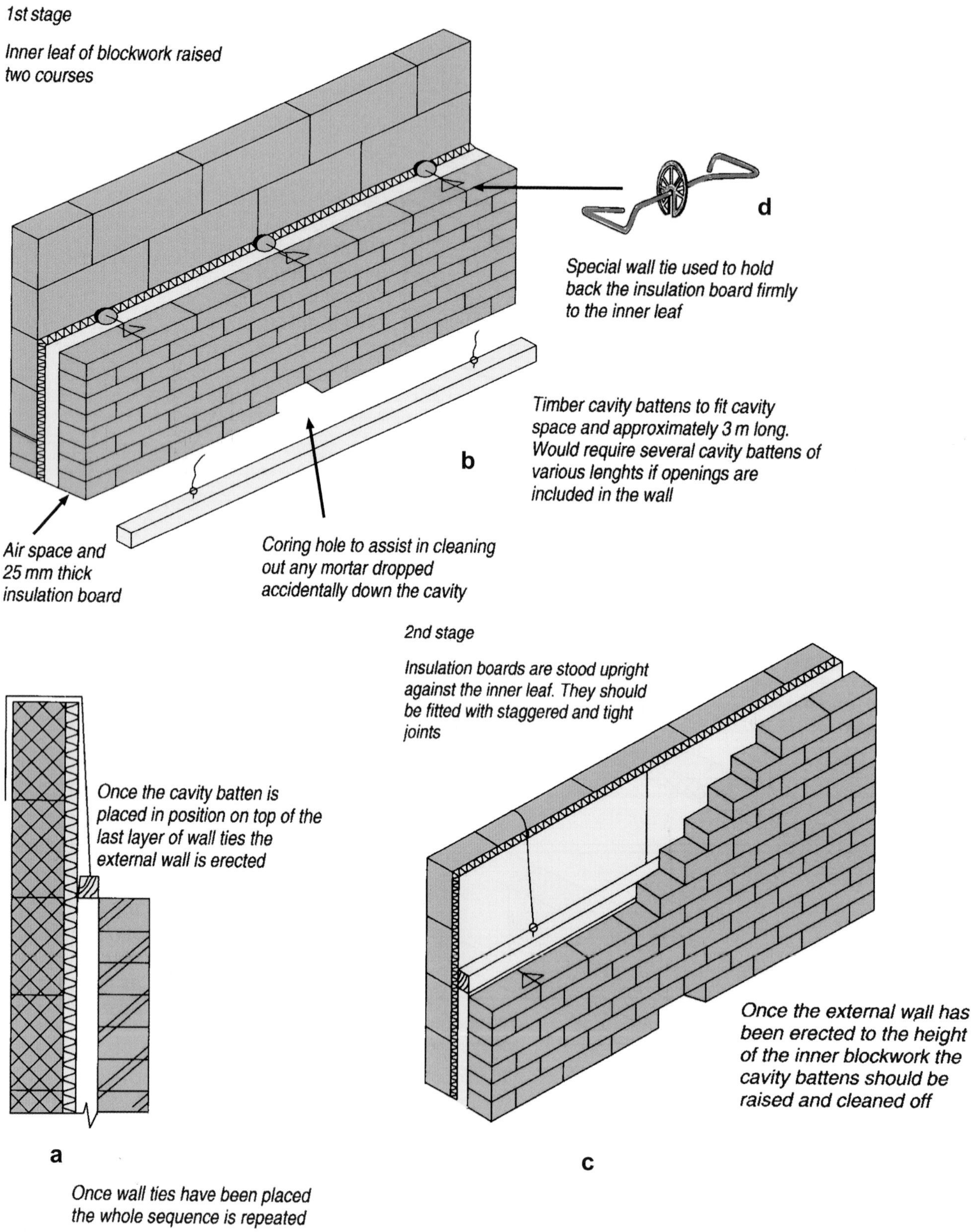
1st stage
Inner leaf of blockwork raised two courses
d
Special wall tie used to hold back the insulation board firmly to the inner leaf
b
Timber cavity battens to fit cavity space and approximately 3 m long. Would require several cavity battens of various lenghts if openings are included in the wall
Air space and 25 mm thick insulation board
Coring hole to assist in cleaning out any mortar dropped accidentally down the cavity
2nd stage
Insulation boards are stood upright against the inner leaf. They should be fitted with staggered and tight joints
Once the cavity batten is placed in position on top of the last layer of wall ties the external wall is erected
Once the external wall has been erected to the height of the inner blockwork the cavity battens should be raised and cleaned off
a
c
Once wall ties have been placed the whole sequence is repeated

FIGURE 7.26
Partially filled cavity insulation system

Insulation batts for full-fill insulation are available in various thicknesses and can be made from mineral fibre, which is soft and flexible. Insulation for partial cavity fill is made from more rigid insulation boards such as expanded polystyrene bead board. The boards in partial cavity fill should only be fixed with special wall ties with plastic clips to hold the insulation back against the internal blockwork.

For cavity walling to be effective, wall ties, insulation and cavity gutters must be kept free of mortar droppings as work proceeds. If, owing to carelessness and poor supervision, cavities are not kept clean, then dampness will be able to cross the cavity through porous mortar droppings.

Cavity battens or boards, approximately 3 m long, raised and cleaned off every six courses as work proceeds, are the best way of preventing mortar droppings falling into cavity walling (Figure 7.25). Alternatively, where fully filled cavity insulation is specified, plain battens without lifting wires are used (Figure 7.26).

Coring holes

It is usual to leave temporary openings, called coring holes, in the outer leaf, over all cavity trays. These holes, of one-brick size, are for the removal of any mortar droppings, tools, etc., that have gone past the cavity battens (see point A in Figure 7.27). Cavity ties and trays should be inspected and cleaned in this way at the end of every day's work.

Where possible, however, it is better if one-block-sized coring holes are left out of the inner leaf (see point B in Figure 7.27). This is recommended as it

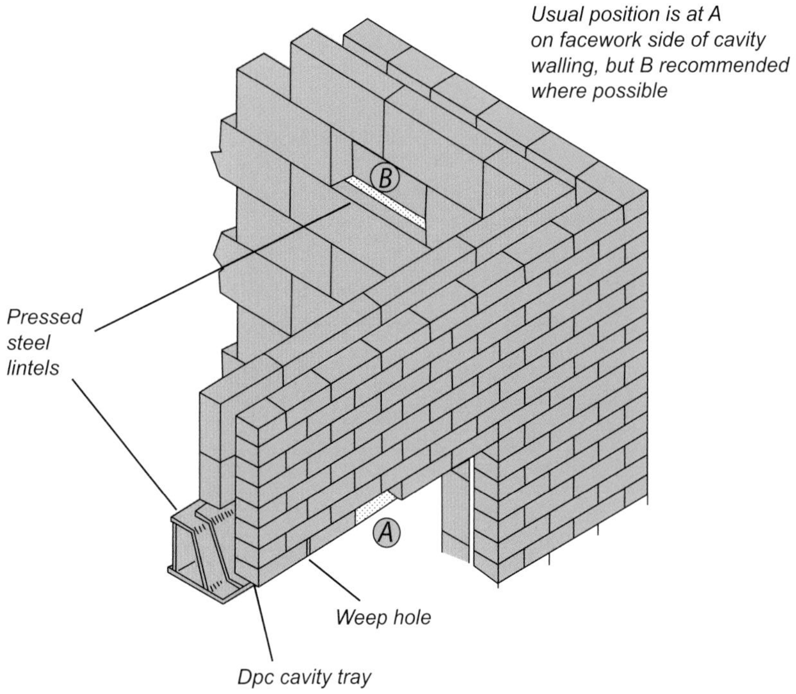
Usual position is at A on facework side of cavity walling, but B recommended where possible
B
Pressed steel lintels
A
Weep hole
Dpc cavity tray

FIGURE 7.27
Location of coring holes

provides a bigger temporary opening. It also avoids the risk of the slight difference in mortar colour that would highlight the coring holes in the finished facework. The coring holes are sealed up when the external scaffolding is removed.

Temporary bedding bricks or blocks in sand at point A or B provide support for those above, and make for easy removal to form the coring holes.

Coring holes should only be regarded as a back-up procedure, and not a substitute for cavity battens or boards.

See also Figure 7.26, which shows a coring hole at ground level

Note

Always try to keep the cavity clean.

Openings

Leaving a door or window opening in a solid brick wall is a relatively simple job, as it merely requires making allowance for the correct space for a frame, plus a lintel or arch across at the right height.

Leaving the same opening in a cavity wall, however, is a much more complicated process.

Jambs

The sides of openings are termed jambs, with the face known as the reveal. In cavity walls these could be constructed either square or rebated.

The cavity can be closed at the opening by using a suitable frame or by turning the inner leaf towards the outer leaf.

When the inner leaf abuts the outer leaf a vertical DPC must be inserted to prevent moisture passing through. The vertical DPC should extend into the cavity; 150 mm DPC is advisable when 100 mm blocks are being used.

Detailed provision must be made for dealing with:

- dampness, which may soak through the sides or reveals of any opening (Figures 7.28 and 7.29)
- rain, which may soak through any sill or threshold of an opening (Figures 7.30–7.34)
- rain water, which may run down inside the cavity from above the opening (Figures 7.35–7.37).

Reveals

Reveals require vertical DPC (Figure 7.29).

Sills

The function of the sill is to shed the rain water from the window frame. The sill is very often built into the window frame, but occasionally the frame will not have a sill and the bricklayer will have to construct one.

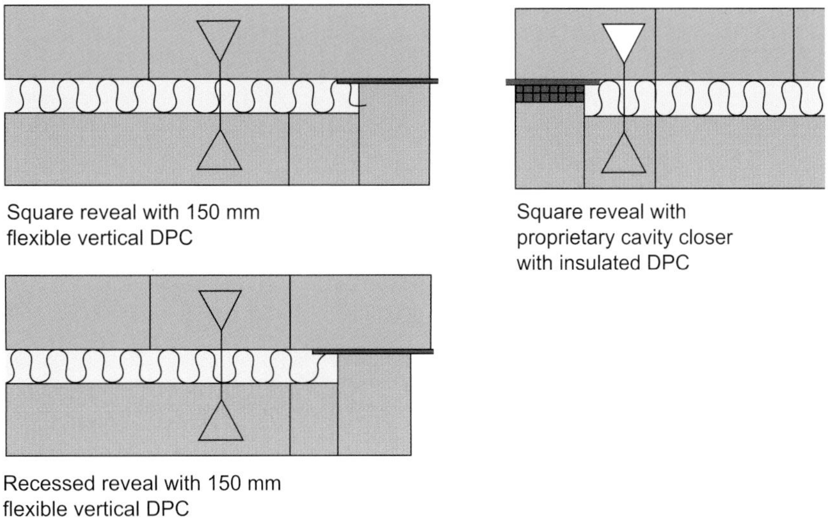
Square reveal with 150 mm
flexible vertical DPC
Square reveal with
proprietary cavity closer
with insulated DPC
Recessed reveal with 150 mm
flexible vertical DPC

FIGURE 7.28
Plans showing location of vertical DPC at reveals

A DPC tray should be built under sills (Figures 7.30 and 7.31).

Sills can also be constructed from creasing tiles (Figure 7.32).

Thresholds

As with window frames, door frames can have purpose-made bottoms fitted during manufacture which are known as thresholds. An example is shown in Figure 7.33.

Some frames are also available without a bottom and will require the bricklayer to construct a threshold.

Bricks and tiles used for either sills or thresholds should be made from well-burnt bricks able to withstand severe weather conditions.

Engineering quality bricks are most suitable as they are water resistant and have very good wearing properties.

The mortar for this type of work should be equal to the strength of the bricks: one part of cement to three parts of sand.

Bricks on edge are laid sloping to provide a weathered finish. A galvanized steel water bar should be inserted to prevent water blowing under the door. See Figure 7.34 for details.

Damp-proof course cavity trays

Bearing in mind that rain water will penetrate the outer skin of brickwork, trays or gutters of flexible bituminous felt must be built into cavity walls to collect the water and prevent dampness from reaching the inner skin of blockwork (see Figure 7.35a, showing a patent steel lintel bridging an opening).

These cavity gutters of DPC material are needed above every opening in cavity wall construction and at every floor level of a multistorey building where cavity walling is used as the external cladding.

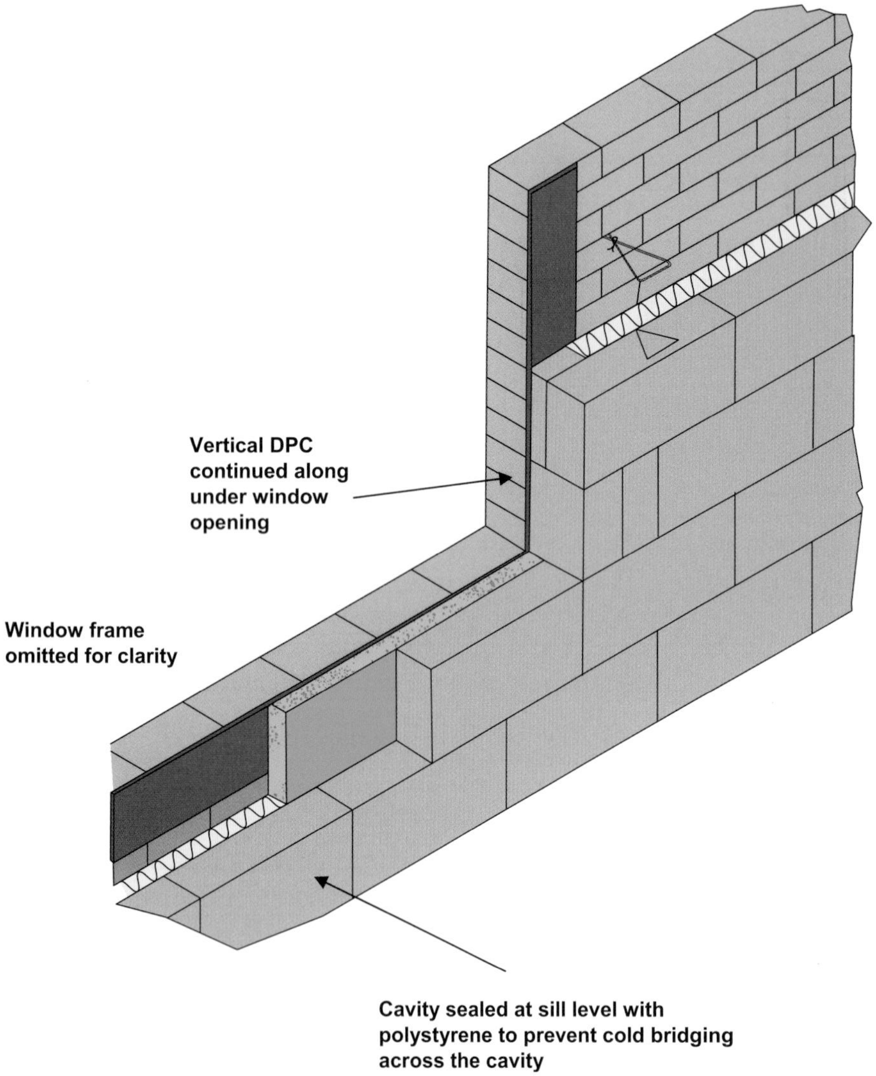
Vertical DPC continued along under window opening
Window frame omitted for clarity
Cavity sealed at sill level with polystyrene to prevent cold bridging across the cavity

FIGURE 7.29
Details showing location of vertical DPC at reveals

Purpose-made stop ends of DPC material (Figure 7.35b) are glued at either end, to prevent run-off into wall insulation and to encourage drainage through weep holes.

Running laps at the end of one roll of DPC and the next, including internal and external angles, must be effectively sealed and jointed, so as to remain watertight. Details are shown in Figure 7.36.

Purpose-made polyethylene units may be specified to provide permanent support to running laps in DPC cavity gutters (Figure 7.37).

Rolls of DPC should always be stored on end to avoid squashing and distortion, which would make the material more difficult to flatten out when used. In cold weather, the rolls should be stored in a warmer place to make bedding the material easier.

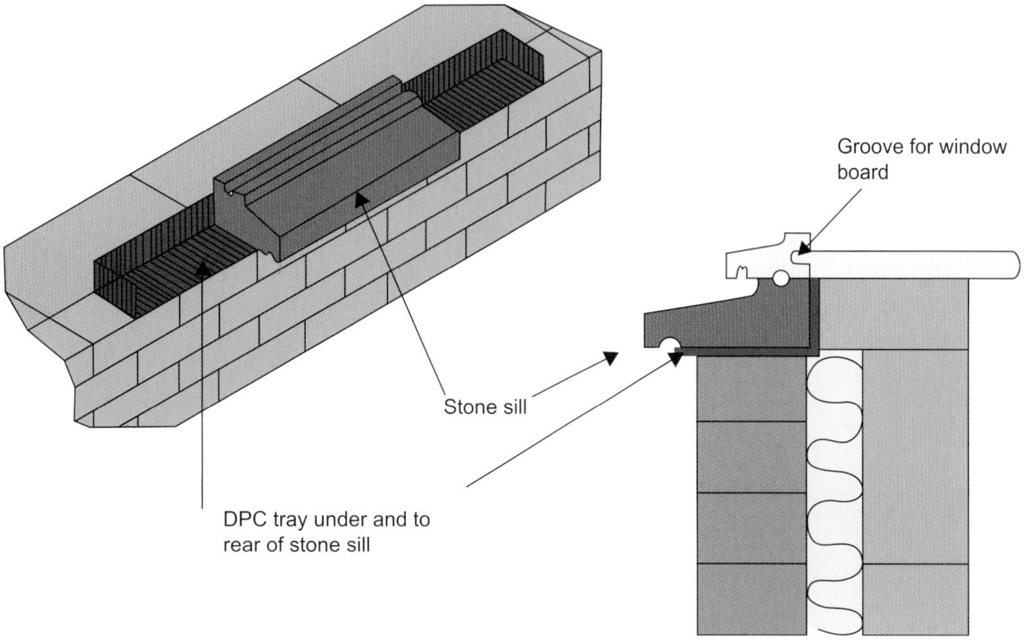
Groove for window board
Stone sill
DPC tray under and to rear of stone sill

FIGURE 7.30
Construction at timber frame on stone sill showing cavity tray

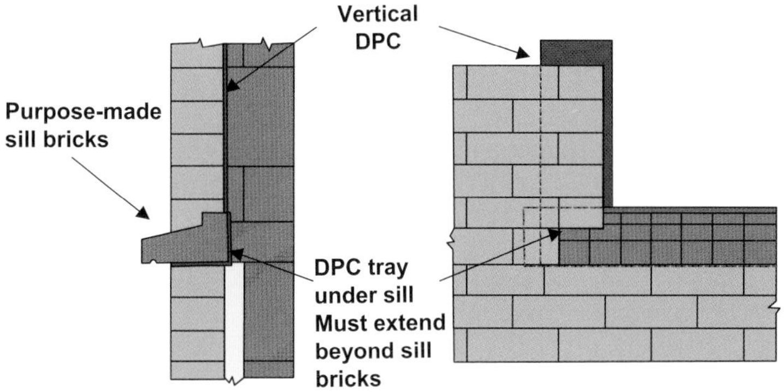
Vertical DPC
Purpose-made sill bricks
DPC tray under sill Must extend beyond sill bricks

FIGURE 7.31
Purpose-made sill bricks at window opening

Weep holes

Empty or 'open' cross-joints are left at intervals of 900 mm in the outer leaf of cavity walling, at the level of cavity gutters, to act as permanent drainage points. These weep holes may simply be left as a cross-joint free of mortar, or formed with plastic or nylon fibre inserts.

Both types of inserts are available in a limited number of colours to match mortar joint colour. Where a light colour mortar is specified, open weep holes appear black and will be conspicuous if not in matching locations above and below every window opening. On the external elevations of a building, it is important that the bricklayer takes care to keep the pattern of weep holes under and over openings symmetrical (Figure 7.38).

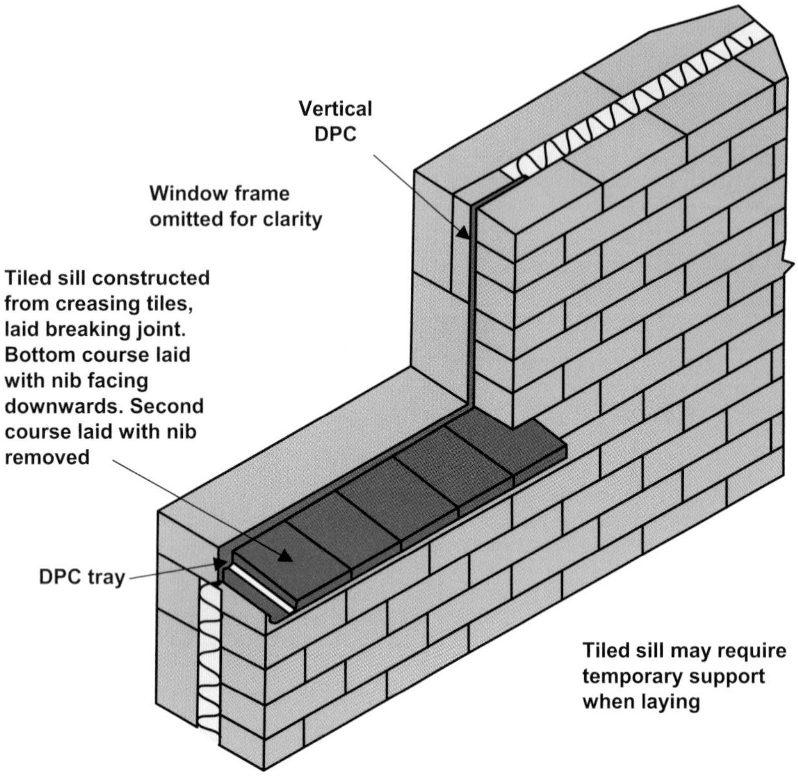
Vertical
DPC
Window frame
omitted for clarity
Tiled sill constructed
from creasing tiles,
laid breaking joint.
Bottom course laid
with nib facing
downwards. Second
course laid with nib
removed
DPC tray
Tiled sill may require
temporary support
when laying

FIGURE 7.32
Construction of tiled sill

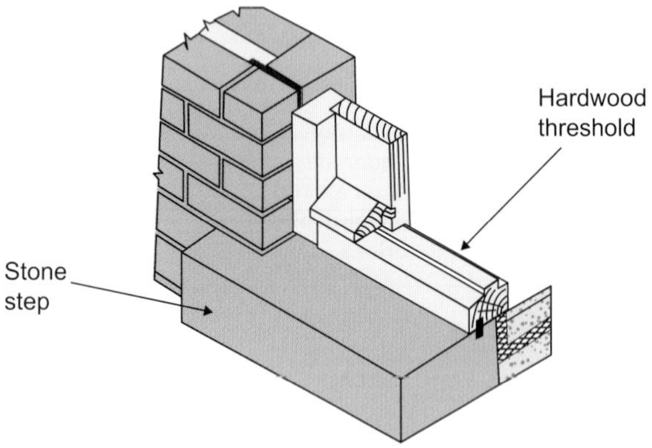
Hardwood
threshold
Stone
step

FIGURE 7.33
Door frame with bottom threshold on stone step

Cavity details at eaves level

To prevent birds that have entered the roof spaces from nesting in wall insulation, it is usual to close or seal off cavity walls at the top.

Blocks laid flat are a convenient way of doing this. This also spreads the load of a pitched roof between both inner and outer skins of brick masonry (Figure 7.39).

The roof rests on a timber wall plate which has galvanized restraint straps to secure it to the inner leaf.

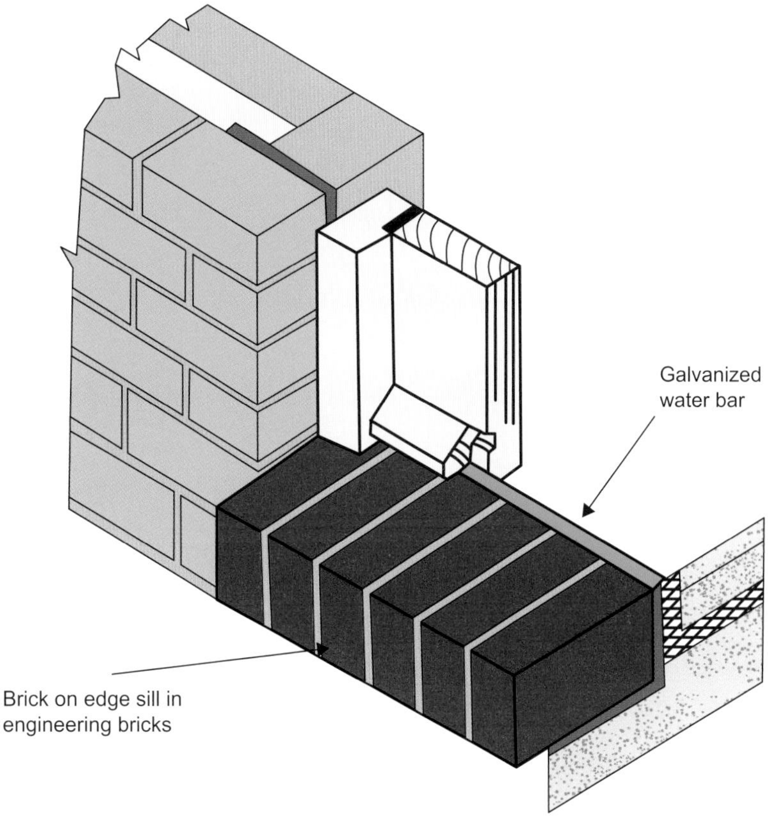
Galvanized steel
water bar
Brick on edge sill in
engineering bricks

FIGURE 7.34
Brick sill details

Openings

PURPOSE

Openings are formed in cavity walls to allow door and windows to be fitted. Openings could also be produced as access or decorative detail without a frame being fitted.

Frames

Frames may be fitted as the work proceeds or after the wall is completed.

FRAMES BUILT IN AS THE WORK PROCEEDS

The bricklayer has to build frames in as the work proceeds and this is often the preferred method for the bricklayer.

The frames can act as a profile, which will reduce the amount of plumbing.

The frame should be placed in position and held in place with timber battens (Figure 7.40). Window frames can be supported in position with a batten (Figure 7.41).

The following factors need to be considered when building in frames:

- Ensure that the horns are cut and splayed back before positioning the frame.

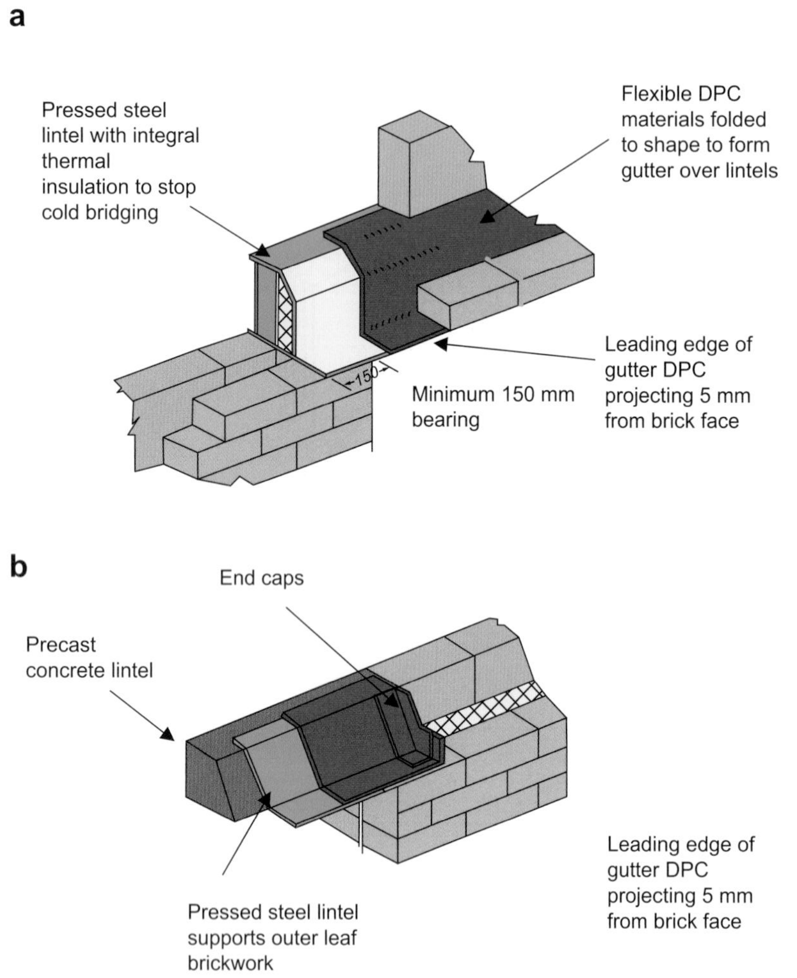
a
Pressed steel lintel with integral thermal insulation to stop cold bridging
Flexible DPC materials folded to shape to form gutter over lintels
150
Minimum 150 mm bearing
Leading edge of gutter DPC projecting 5 mm from brick face
b
End caps
Precast concrete lintel
Pressed steel lintel supports outer leaf brickwork
Leading edge of gutter DPC projecting 5 mm from brick face

FIGURE 7.35
(a) Typical cavity tray or gutter; (b) purpose-made end caps glued to cavity tray

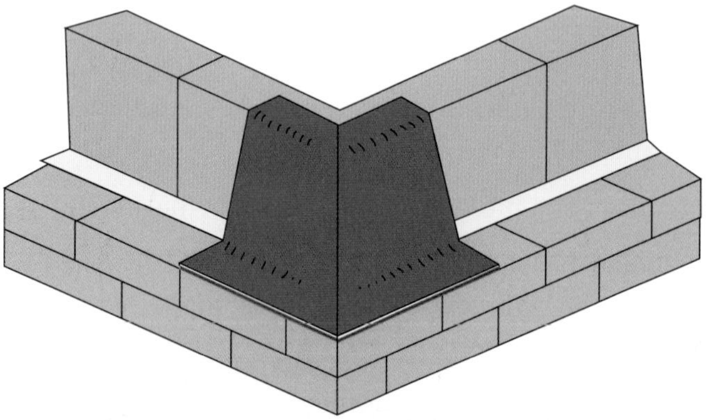

FIGURE 7.36
Purpose-made DPC tray corner unit

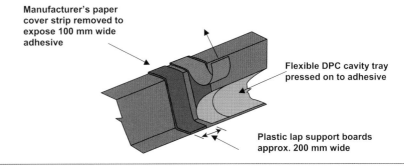

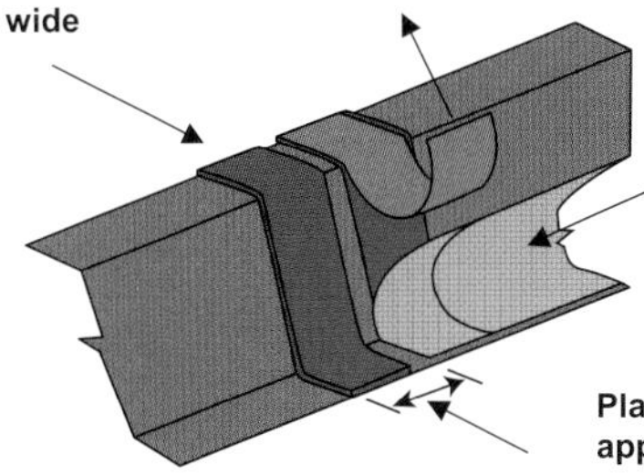
Manufacturer's paper cover strip removed to expose 100 mm wide adhesive
Flexible DPC cavity tray pressed on to adhesive
Plastic lap support boards approx. 200 mm wide

FIGURE 7.37
Patent rigid plastic support board for running laps in DPC trays

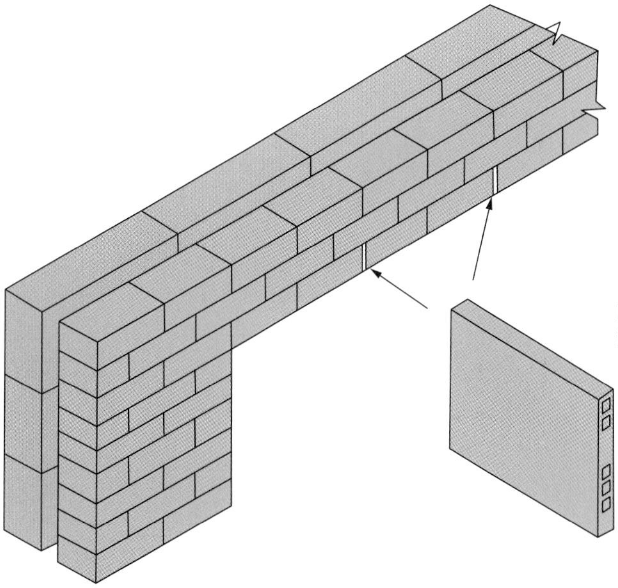
Plastic or nylon inserts

FIGURE 7.38
Position of weep holes and plastic inserts

- Before setting the frames in their correct position it is good practice to check the diagonal dimensions to prove the squareness of the frame.
- Check the drawing for the position of the frame and ensure the rebates are facing the right way.
- Set up frames on an even bed, and plumb and level and brace securely.

See Figure 7.42 for details

On small frames one batten would be sufficient, but on larger frames two or more may be required.

Fixing frames

The frames should be fastened in position as the wall is being built. This can be achieved using frame cramps (Figure 7.43). If this method is used care should be taken to ensure that the frames are plumb and level, or the joiner would have problems hanging the doors correctly.

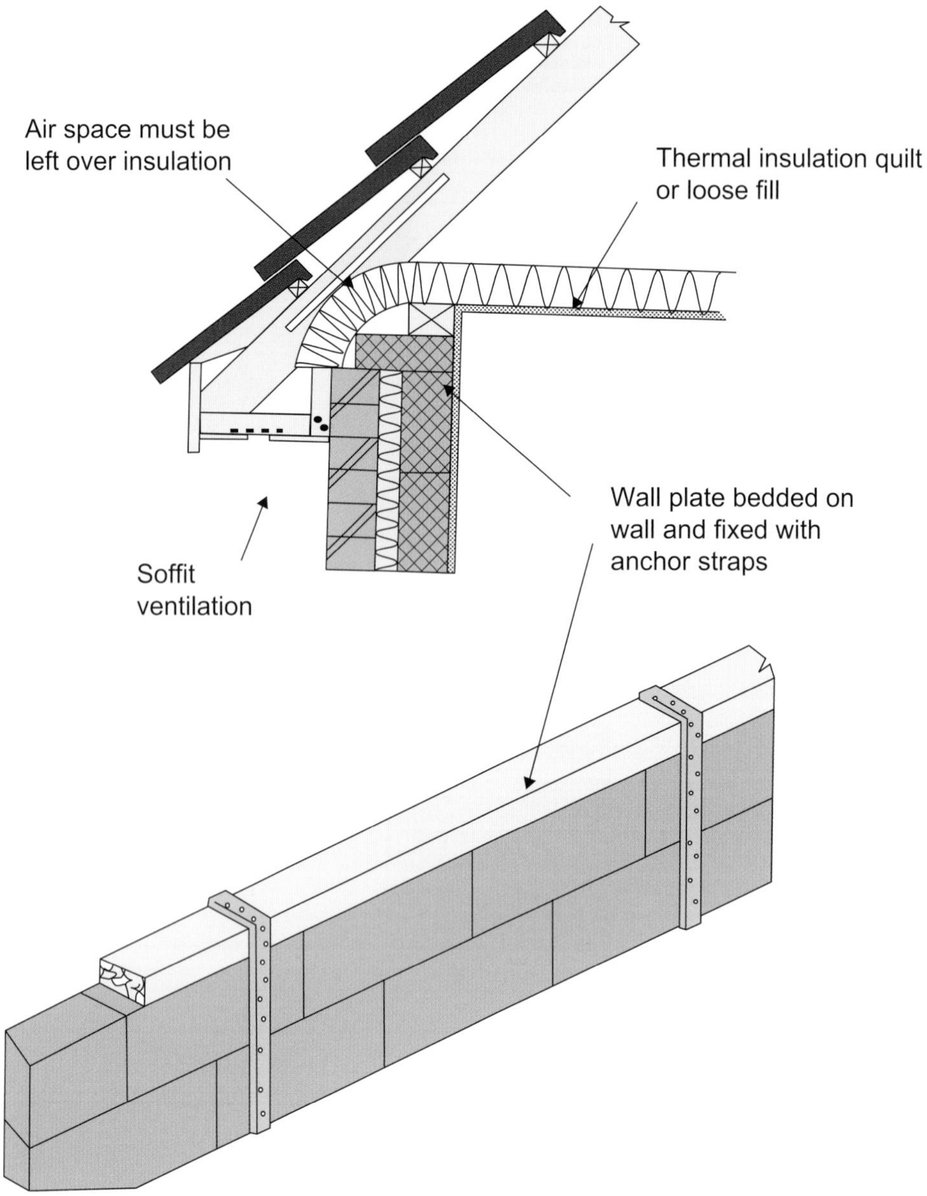
Air space must be left over insulation
Thermal insulation quilt or loose fill
Wall plate bedded on wall and fixed with anchor straps
Soffit ventilation

FIGURE 7.39
Sealing cavities at eaves level

FIXING FRAMES AFTER THE WALL IS COMPLETED

With the increase in hardwood and plastic window and door frames it has become common practice to leave openings in cavity walls for the frames to be fitted at a later date.

If mistakes are made it can be a very expensive exercise to cut away brickwork to fit new frames.

Temporary frames

If frames are fitted at a later stage then wooden jigs are built in to provide the correct opening for the frames.

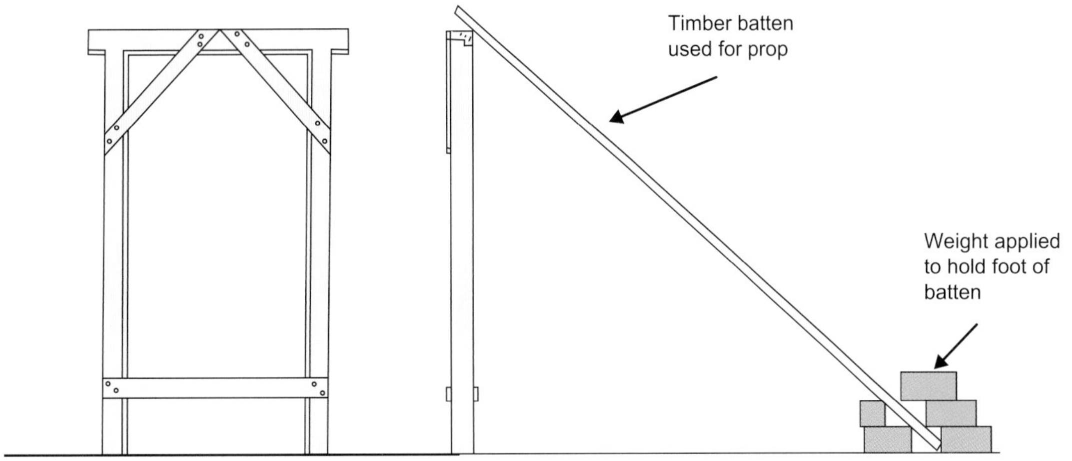
Timber batten
used for prop
Weight applied
to hold foot of
batten

FIGURE 7.40
Door frame support

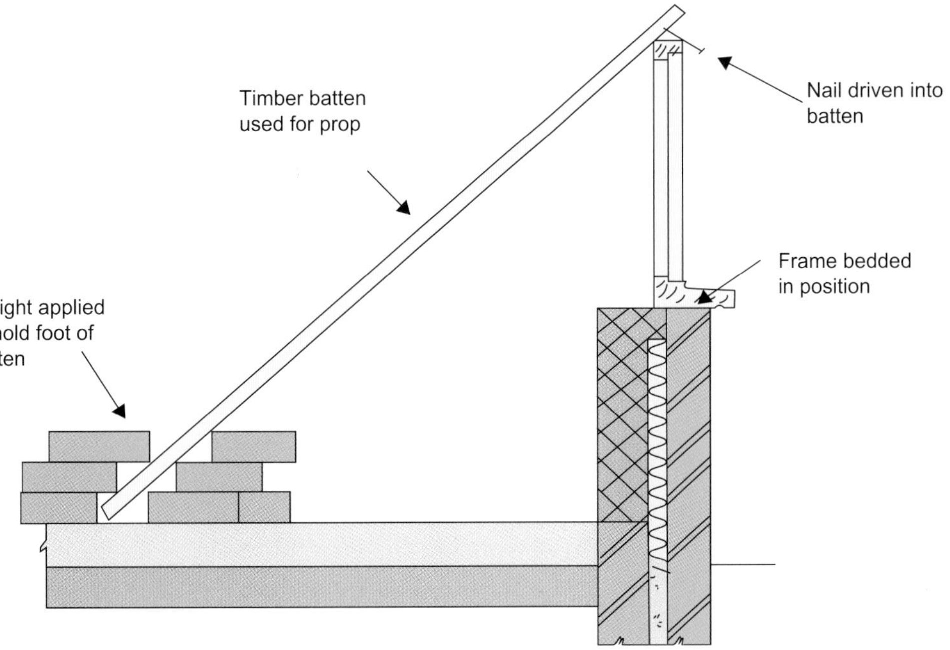
Timber batten
used for prop
Nail driven into
batten
Frame bedded
in position
Weight applied
to hold foot of
batten

FIGURE 7.41
Window frame support

Timber frames are produced exactly to the size of the proposed frame. They should be braced at the corners to make sure that they are square (Figure 7.44).

PINCH ROD

Another method of forming an opening is by using a pinch rod (Figure 7.45). This is a piece of timber cut to the exact size of the opening required and is used to check each course.

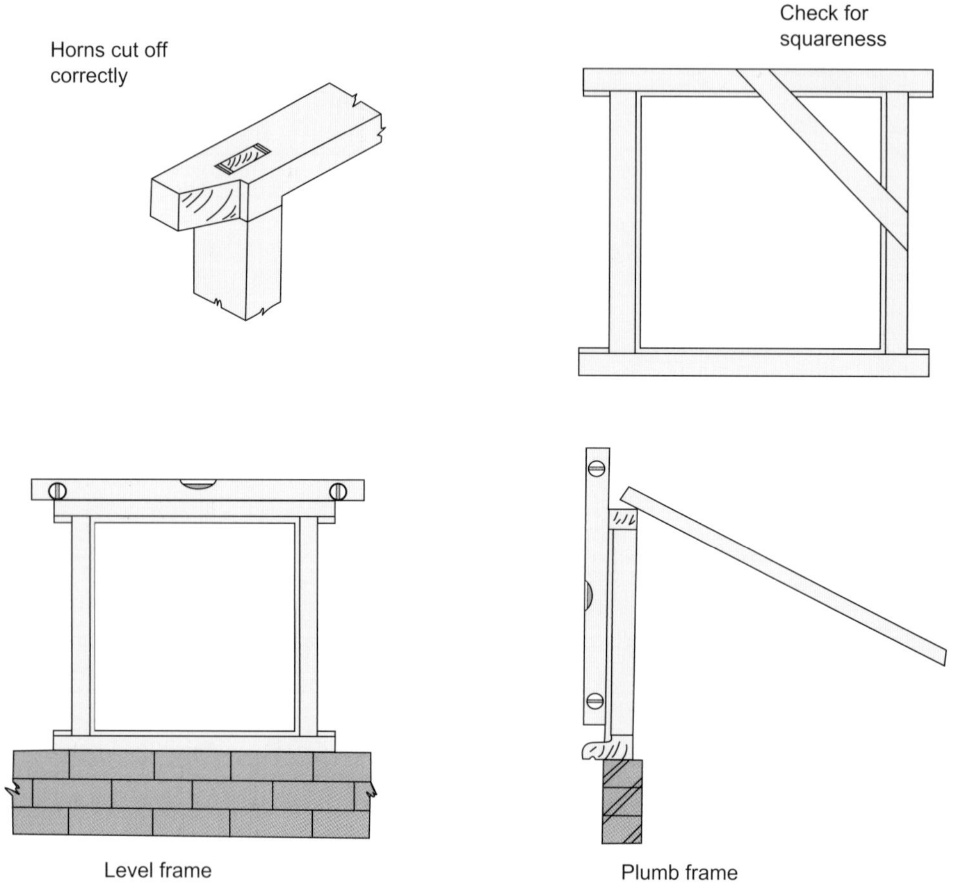
Horns cut off correctly
Check for squareness
Level frame
Plumb frame

FIGURE 7.42
Setting frames in position

As the brickwork is plumbed on each course the pinch rod is held in place to ensure that the opening is not reduced or enlarged.

PROTECTION OF FRAMES

Where site traffic has to pass through buildings after the door frame has been fixed it is essential to protect the jambs and where necessary the threshold of the frame (Figure 7.46).

The internal face of the door frame should be protected with timber to prevent knocks which would be expensive to remedy.

If the door has a precast or cast in situ sill then this also needs special protection.

Concrete, stone or terrazzo steps can be framed in timber during the remainder of the building work and only removed when cleaning prior to handover.

Where brick steps have been constructed these should also be protected from damage by falling waste materials by covering with timber or similar material.

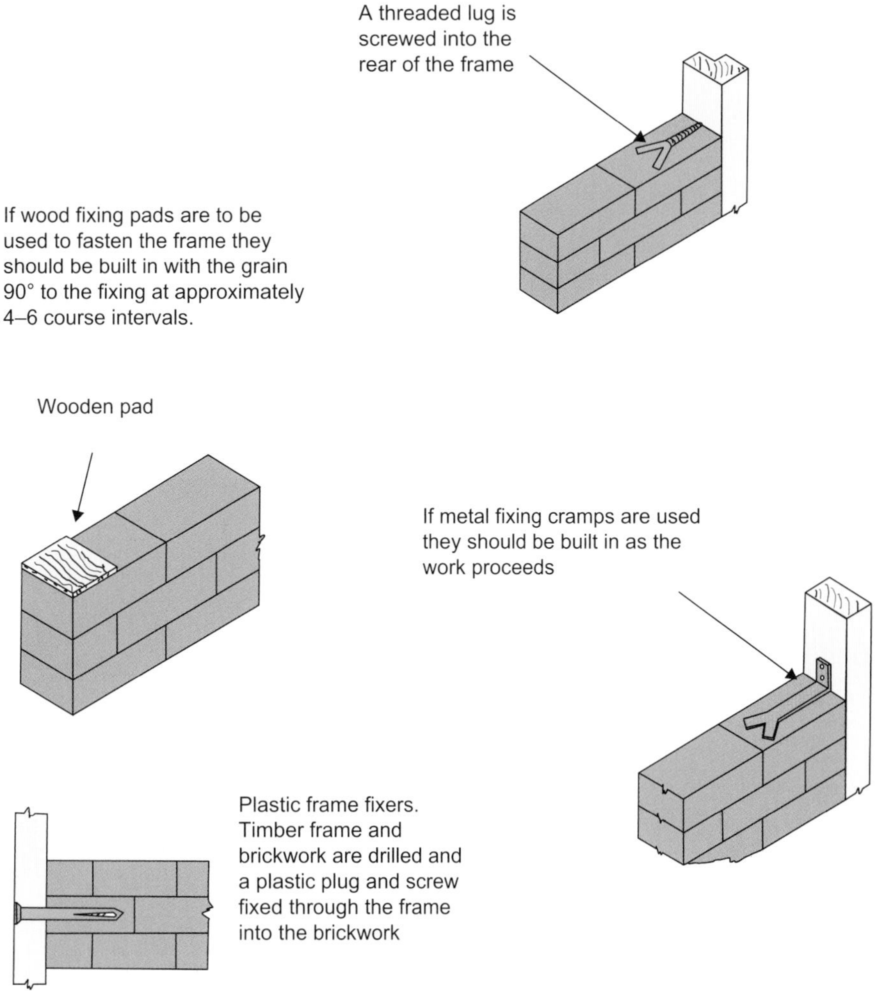
A threaded lug is screwed into the rear of the frame
If wood fixing pads are to be used to fasten the frame they should be built in with the grain 90° to the fixing at approximately 4–6 course intervals.
Wooden pad
If metal fixing cramps are used they should be built in as the work proceeds
Plastic frame fixers. Timber frame and brickwork are drilled and a plastic plug and screw fixed through the frame into the brickwork

FIGURE 7.43
Methods of fixing frames

Junction walls

It is often inconvenient to build junction walls at the same time as the main wall, so there will be a need during construction to make provision for work at a later date.

Indents can be left in the main wall to receive the junction wall later. These indents should be marked out correctly and kept plumb throughout the building of the block wall. The minimum lap at T junctions should be quarter lap. An allowance of 20 mm is usually made over the width of the block to allow the blocks on the junction wall easy access into the indent (Figure 7.47).

Another acceptable method of providing a tie to junction walls is metal reinforcement (Figure 7.48). This is built into every alternate bed joint. This method could help to avoid lots of cutting at a junction of two block walls. This method is recommended for bonding block walls together when both walls are constructed of different block types, especially when the blocks have different shrinkage properties.

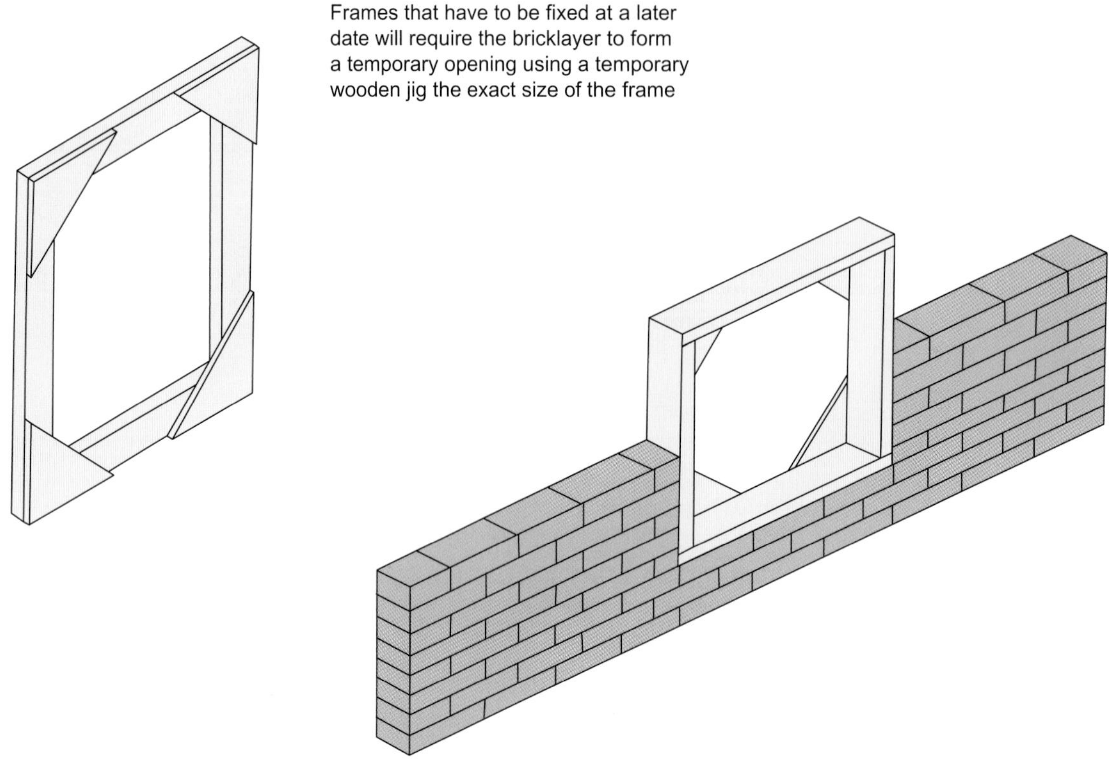
Frames that have to be fixed at a later date will require the bricklayer to form a temporary opening using a temporary wooden jig the exact size of the frame

FIGURE 7.44
Building in temporary frames

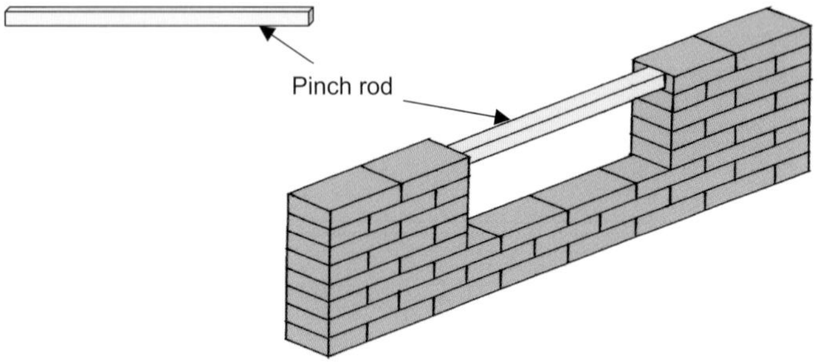
Pinch rod

FIGURE 7.45
Use of pinch rod

Bonding to brick walls

There will be a need to bond block walls to brick walls. This can be done by leaving indents in the brickwork or by fixing a proprietary wall connector (Figure 7.49).

Extensions to cavity walling

It is occasionally necessary to leave part of a building down and, in preparation for its future extension, toothings are formed as the work proceeds. Toothings are shown in Figure 7.50.

The recommended method of erecting a cavity quoin is shown in Figure 7.51.

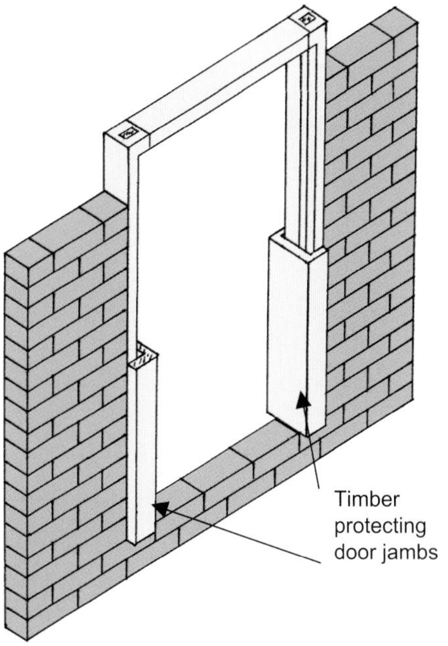
Timber
protecting
door jambs

FIGURE 7.46
Protection of frames

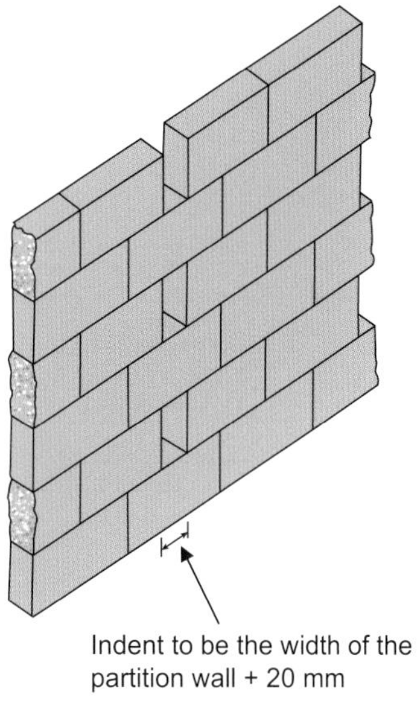
Indent to be the width of the
partition wall + 20 mm

FIGURE 7.47
Indents

Vertical movement joints

The bricklayer must make allowances, when constructing walls, for the expansion and contraction of the brickwork due to changes in air temperature and/or moisture content of the bricks.

A completely straight joint should be formed throughout the full height and thickness of the wall.

These vertical movement joints should be 10–15 mm wide on the face, and are needed at approximately 9 m intervals to prevent stress from developing

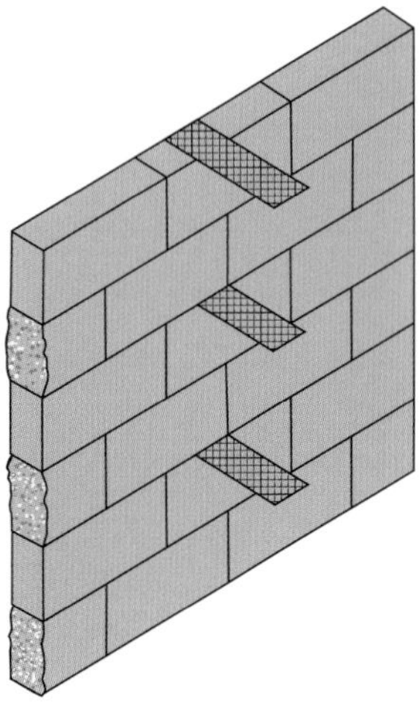

FIGURE 7.48
Reinforcing mesh

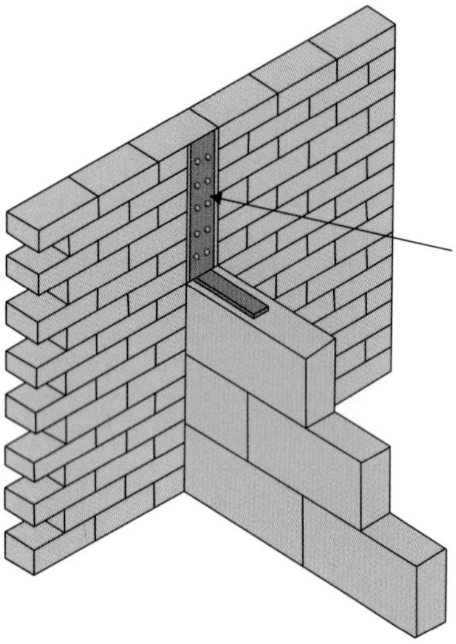
Proprietary wall connector bolted to brick wall and anchor ties are built into the new block wall

FIGURE 7.49
Proprietary wall connector

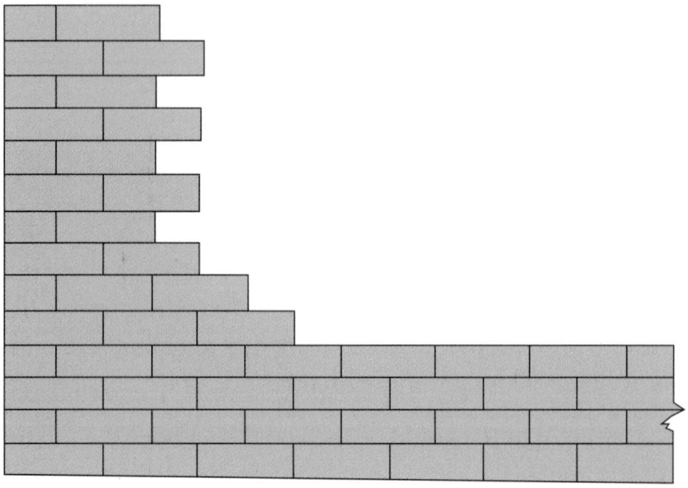

FIGURE 7.50
Toothing out at stopped end

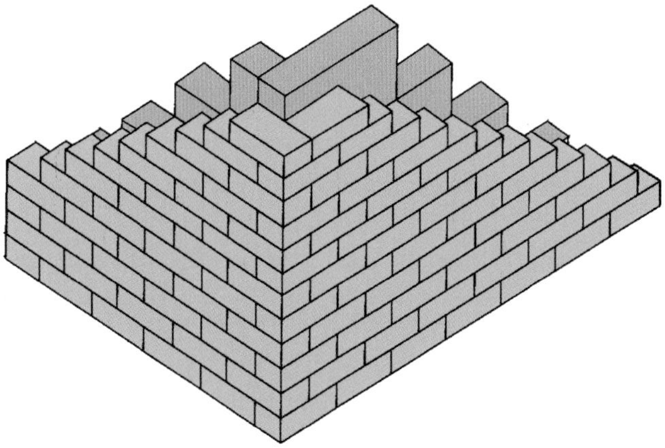

FIGURE 7.51
Cavity quoin

in the wall due to expansion. A soft filler of expanded plastic foam strip is built in, to stop mortar bridging the 10–15 mm space.

To reinforce the obvious weakness caused by these unbonded, vertical joints, special stainless steel slip-ties are built in every fourth course as the work proceeds (Figure 7.52) to keep the ends of the walling permanently in line. Standard cross-cavity wall ties must not be used at these points.

The expanded plastic foam filler is kept back (or cut back), 12 mm from both faces of the brickwork, so that a sealant-mastic can be applied, to seal these movement joints against rain penetration.

It is very important that vertical movement joints are taken right up through the wall. Vertical movement joints should commence at ground level, because the temperature below ground is practically constant.

Note

Compressed fibreboard is unsuitable as a vertical movement joint filler in brick walling. It is too hard, and does not compress easily when the walling expands.

Brick corbels

A corbel is the term given to a brick or series of bricks which project from a wall. Corbels are used as a decorative feature or to thicken a wall.

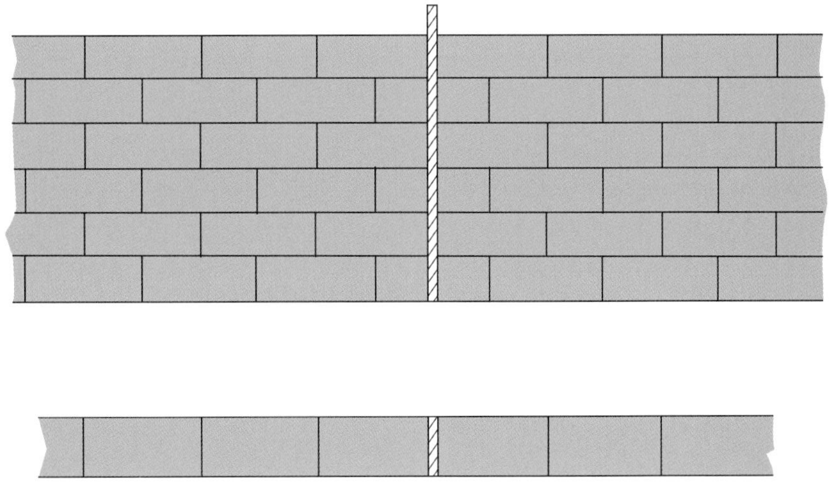

FIGURE 7.52
Formation of vertical movement joint

The building of oversailing courses forming corbels – in 56 mm oversailers for the laying and in 28 mm for bond arrangement – is not an easy operation to perform.

In the case of 56 mm corbels, headers should be used wherever possible; if the use of stretchers cannot be avoided, they should be the last bricks laid on that course and should be well bedded, a cross-joint being placed throughout the length of the brick.

In the case of 28 mm corbels, the greatest possible lap must be maintained; if special attention is given to this point, and common sense applied, no difficulty should be experienced in carrying out this work.

Both internal and external straight joints are sometimes unavoidable, but they should be reduced to a minimum and the specified bond should be adhered to, if possible. The final shape of the brickwork must be considered; this will continue to a greater height than the corbel and must be practical and not involve unnecessary cutting.

When a corbel is being built it must be kept well tailed down to prevent overturning; that is, the back bricks must be laid first to bond the previous corbel before attempting to lay the next oversailing course.

The maximum distance permitted by the current Building Regulations is equal to the thickness of the wall immediately below the corbel. See Figure 7.53 for details.

Cutting to rake

A raking cut is required on walls that have to be finished with a sloping end, generally on gable ends.

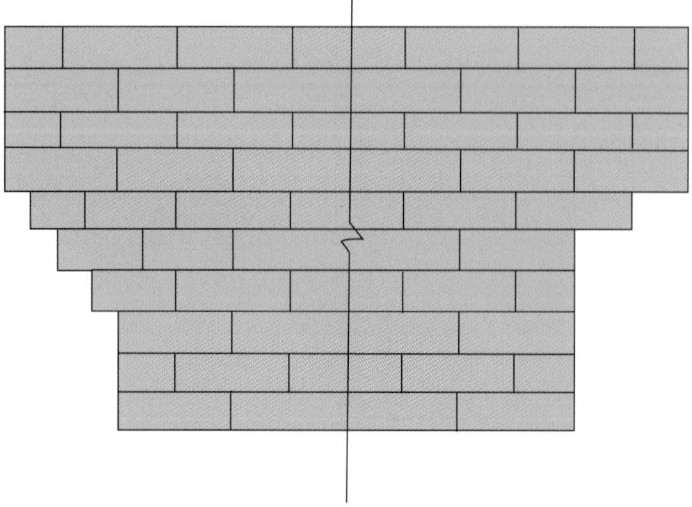

FIGURE 7.53
Two methods of corbelling out brickwork

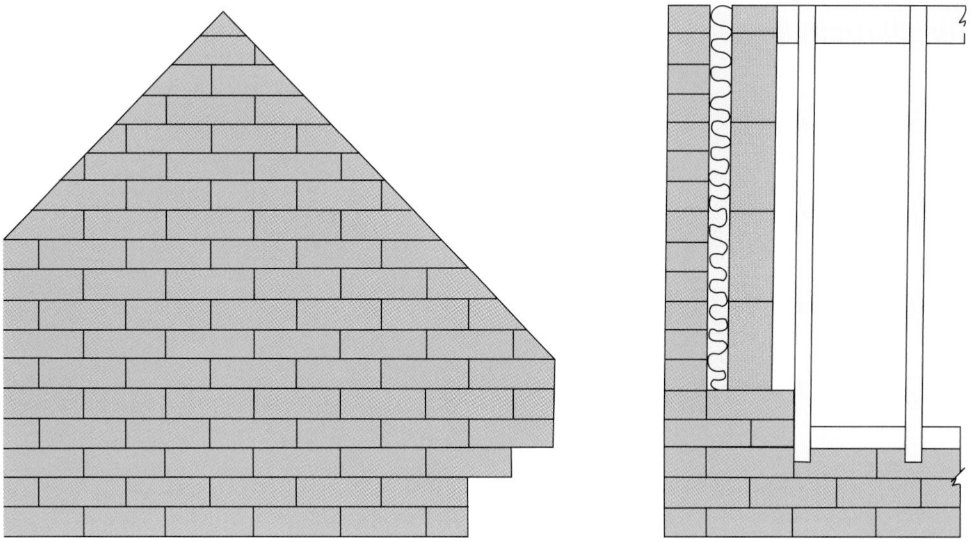

FIGURE 7.54
Typical gable end construction

There are three important points to remember when working on gable ends:

- the kind of finish that is required at the eaves
- how to maintain plumb and level
- cutting the bricks to the required rake.

A typical gable end is shown in Figure 7.54. This shows corbels at the start of the gable.

Gable ends usually start with projected corbels (Figure 7.55). These can be built in brick, tiles or precast concrete.

Corbels require setting out to ensure correct bond.

The joiner will have erected part of the roof on the wall plate and timber laths should be nailed to the rafters at the top and bottom to allow a line

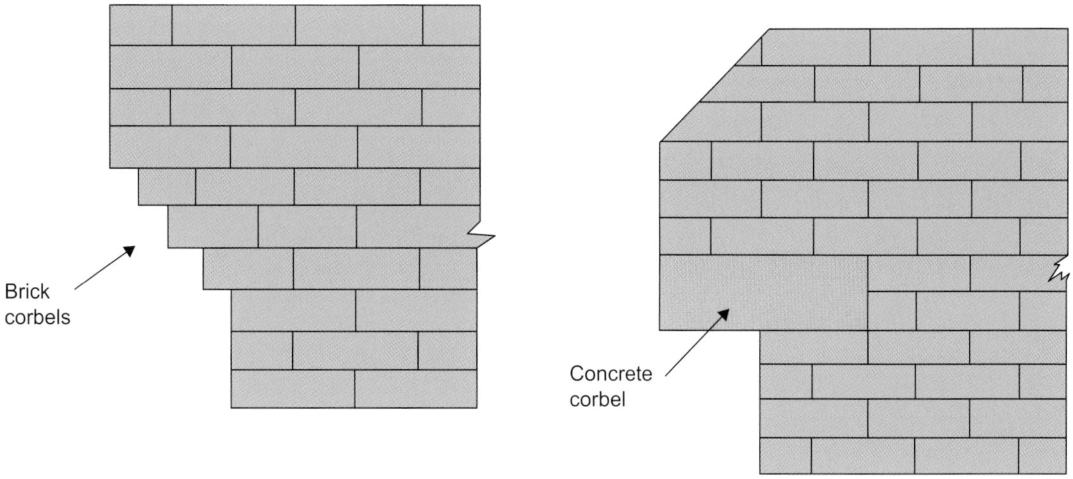
Brick corbels
Concrete corbel

FIGURE 7.55
Types of corbel

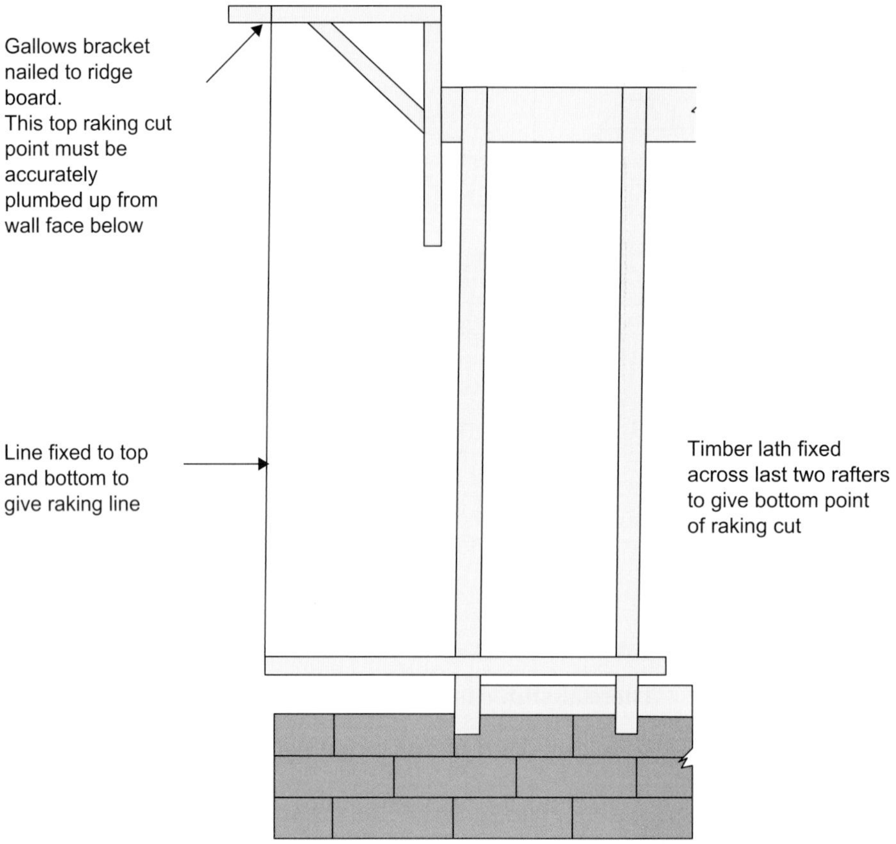
Gallows bracket nailed to ridge board.
This top raking cut point must be accurately plumbed up from wall face below
Line fixed to top and bottom to give raking line
Timber lath fixed across last two rafters to give bottom point of raking cut

FIGURE 7.56
Setting out for raking cut

to be attached. A gallows bracket could be used at the top, nailed to the ridge board (Figure 7.56).

The correct position of the timber laths must be ascertained, both for angle of line and for plumb above the main brickwork.

If this procedure is carried out correctly, the remaining brickwork need not be plumbed.

Deadman

Deadman is the term given to a temporary brick pier bedded to gauge and plumbed to the face of the wall. This acts as a corner and allows a horizontal line to be extended beyond the line of the raking cut (Figure 7.57).

A profile could also be used (Figure 7.58).

The brickwork must be raked back to obtain the correct cut and a fixing for the line and pins.

Procedure for raking cutting

1. Rake back the brickwork below the sloping line and pins (Figure 7.59).

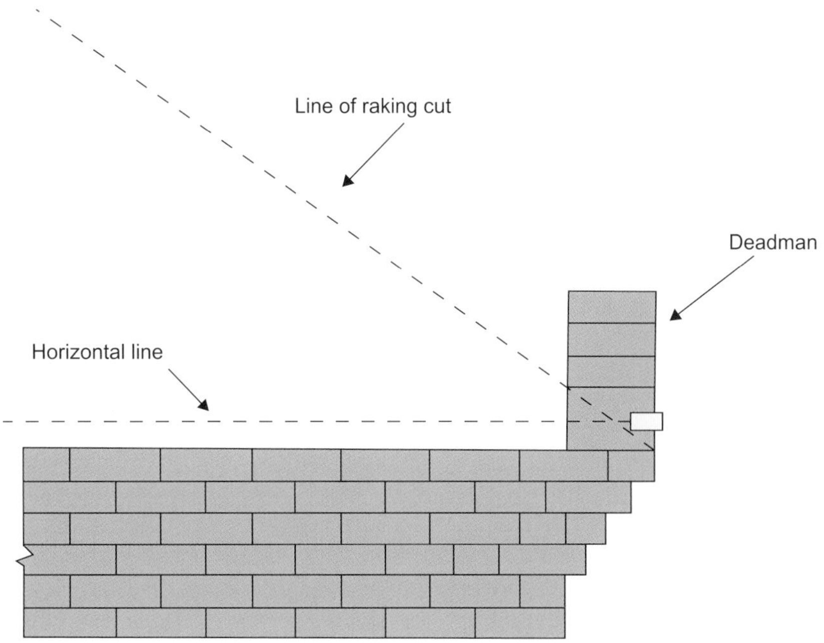
Line of raking cut
Deadman
Horizontal line

FIGURE 7.57
Use of dead man to support lines

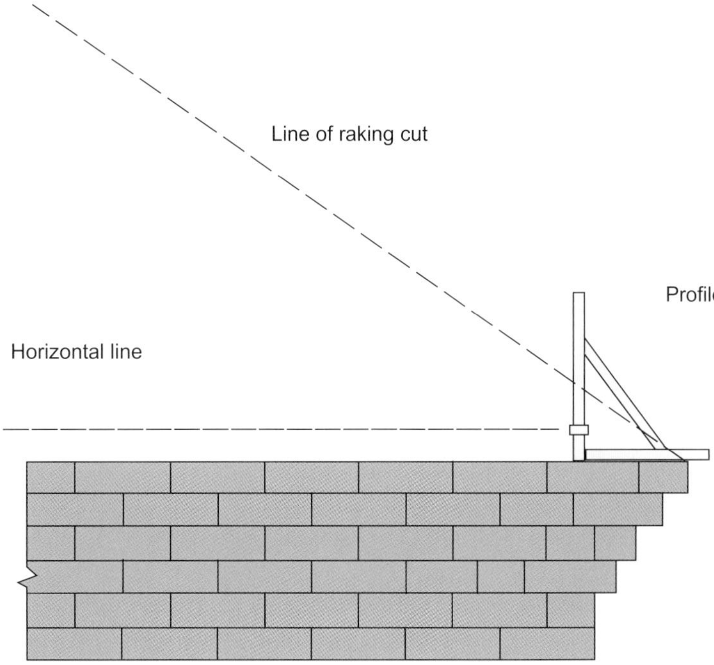
Line of raking cut
Profile
Horizontal line

FIGURE 7.58
Use of profile to support lines

2. Set the sliding bevel at the necessary angle of the raking cutting.
3. Measure the top edge of the cut brick required.
4. Deduct 10 mm from one cross-joint.
5. Mark in pencil the remaining measurement on the brick to be cut, back and front.

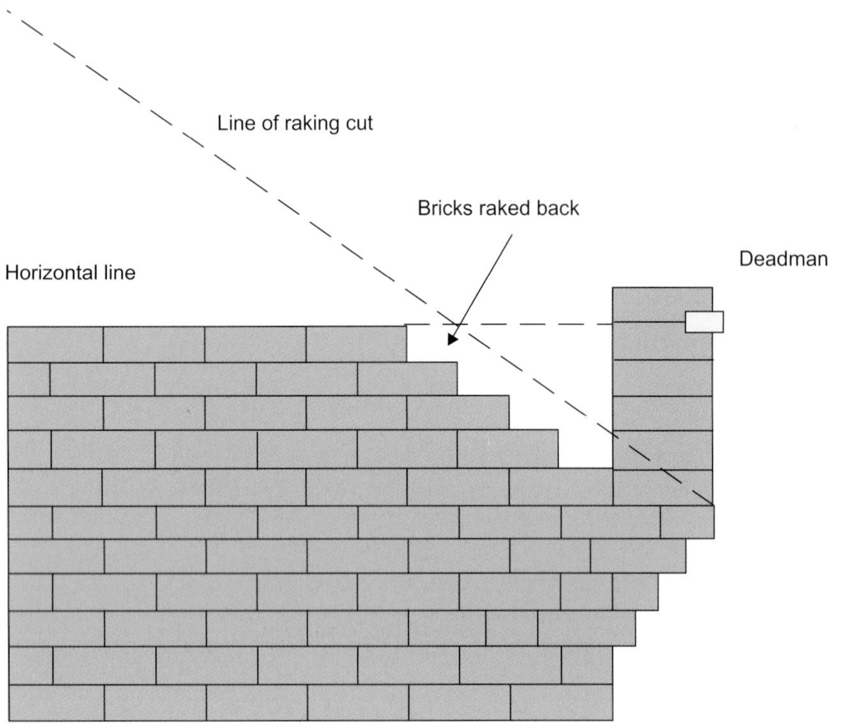
Line of raking cut
Bricks raked back
Deadman
Horizontal line

FIGURE 7.59
Brickwork raked back before cutting to rake

6. Draw in pencil the angle of sloping (raking) cut, using the sliding bevel, also back and front, taking care to match the slope on the face.
7. Cut the brick with hammer and bolster.
8. Trim the cut face with a scutch or comb hammer, slightly undercutting if possible to avoid any projections that would prevent the coping or capping from being bedded properly.
9. Carefully bed the cut brick to line and level.

ALTERNATIVE METHOD, IF SLIDING BEVEL IS NOT AVAILABLE

1. Rake back and tooth out as before.
2. Temporarily bed a brick, carefully, to line and level, as shown in Figure 7.60 (bottom drawing).
3. Apply two pencil marks on the brick face to indicate the angle of cut.
4. Remove the brick from its temporary bed, and cut it with hammer and bolster; trim with scutch; and permanently bed the cut brick.

If perforated wire-cut facing bricks or engineering bricks are being used, these will have to be cut on a masonry bench saw. Mark five or six at a time, indicating clearly the waste part of each brick to be cut. Number each one for easy identification, and send them for cutting on the site masonry bench saw.

Each brick to be cut should be bedded onto the wall in its correct position and then marked in pencil ready to be cut.

Use softing to prevent damage and cut with hammer and bolster or mechanical saw.

Procedure for marking and cutting bricks to rake

1. Set the sliding bevel at the necessary angle of the raking cutting.
2. Measure the top edge of the cut brick required.
3. Deduct 10 mm for one cross-joint.
4. Mark in pencil the remaining measurement on the brick to be cut, back and front.

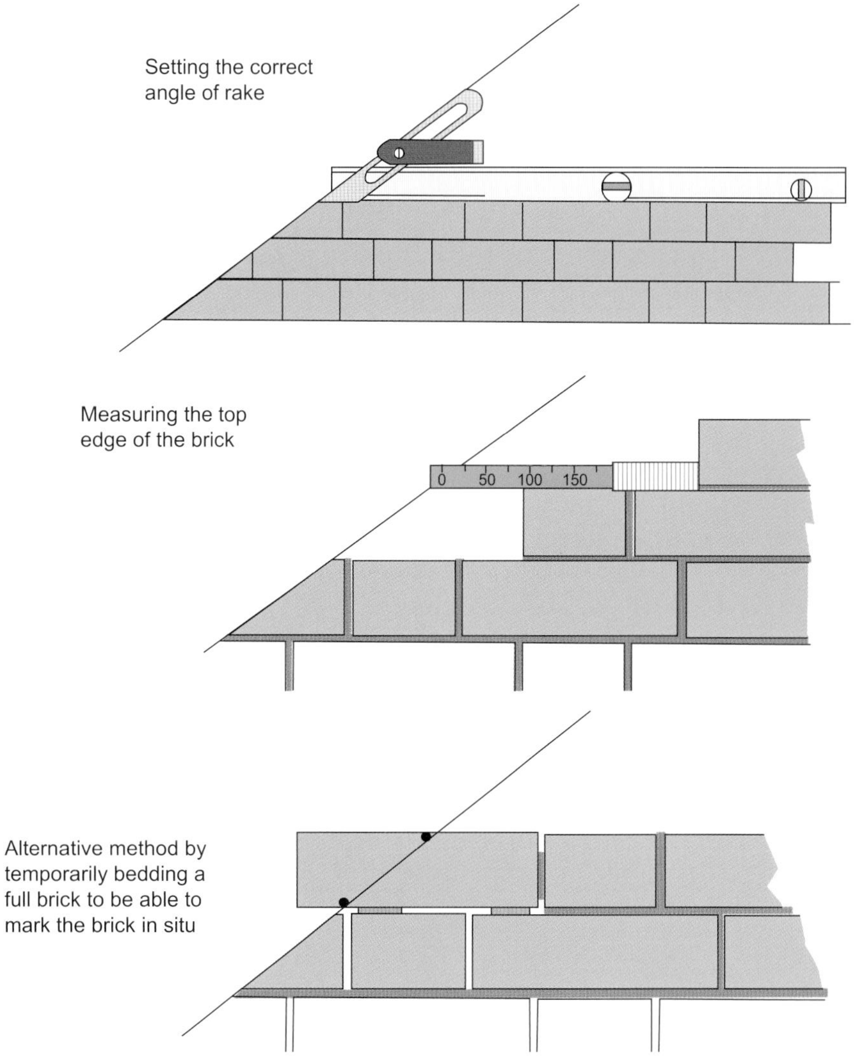
Setting the correct angle of rake
Measuring the top edge of the brick
0 50 100 150
Alternative method by temporarily bedding a full brick to be able to mark the brick in situ

FIGURE 7.60
Marking the cut bricks

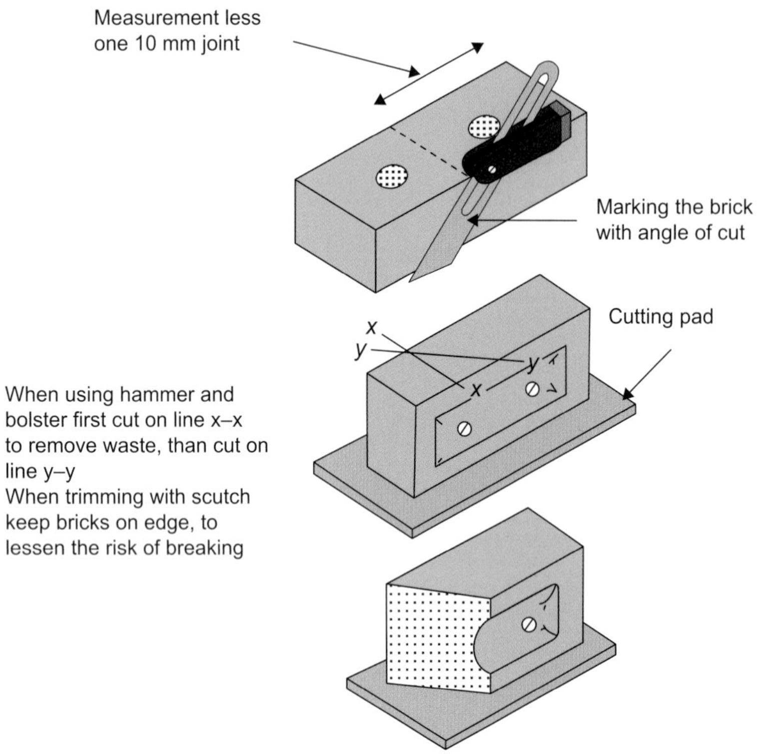
Measurement less one 10 mm joint
Marking the brick with angle of cut
Cutting pad
x
y
y
x
When using hammer and bolster first cut on line x–x to remove waste, than cut on line y–y
When trimming with scutch keep bricks on edge, to lessen the risk of breaking

FIGURE 7.61
Marking and cutting the bricks

5. Draw in pencil the angle of the sloping (raking) cut, using the sliding bevel, also back and front, taking care to match the slope on the face (Figure 7.60).
6. Cut the brick with hammer and bolster or mechanical masonry saw if available.
7. Trim the cut face with a scutch or comb hammer if necessary to avoid any projections.
8. Carefully bed the cut brick to line and level. See examples of marking in Figure 7.61.

Multiple-choice questions

Self-assessment

This section of the book is designed to allow you to check your level of knowledge. The section consists of revision questions for this chapter. The questions are all multiple choice and have four possible answers. The answers are to be found at the end of the book.

The main type of multiple-choice question will be the four-option multiple-choice question. This will consist of a question or statement, known as the stem, followed by a choice of four different answers, called the responses. Only one of these responses is the correct answer; the others are incorrect and are known as distracters.

You should attempt to answer the questions by choosing either (a), (b), (c) or (d).

Example

The person employed by the local authority to ensure that the Building Regulations are observed is called the:

(a) clerk of works
(b) building control officer
(c) council inspector
(d) safety officer

The correct answer is the building control officer, and therefore (b) would be the correct response.

Cavity walls

Question 1 What is the main benefit of cavity wall construction?

(a) decoration
(b) insulation
(c) ventilation
(d) condensation

Question 2 Which of the following is the most common bond used for the external leaf of a cavity wall?

(a) English bond
(b) Flemish bond
(c) stretcher bond
(d) Dutch bond

Question 3 Identify the type of wall tie shown:

(a) double triangle

(b) butterfly

(c) fishtail

(d) stainless steel

Question 4 Which is the correct horizontal distance for wall ties in cavity wall construction?

(a) 300 mm

(b) 600 mm

(c) 900 mm

(d) 1200 mm

Question 5 Name the item of equipment used to prevent mortar dropping down the cavity:

(a) cavity strap

(b) cavity trap

(c) cavity rod

(d) cavity batten

Question 6 Identify the item of equipment shown:

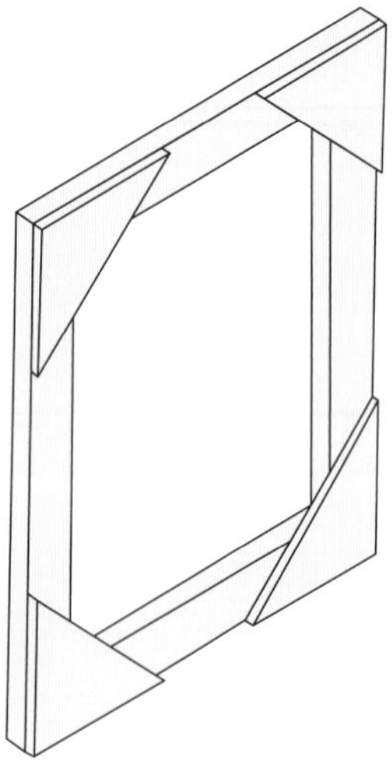

(a) window

(b) door

(c) jig

(d) frame

Question 7 Which of the following is the correct minimum width of a cavity?

(a) 50 mm
(b) 55 mm
(c) 60 mm
(d) 75 mm

Question 8 When would a sliding bevel be used?

(a) when building in frames
(b) when cutting bricks to a rake
(c) when setting out corbels
(d) when building a quoin

CHAPTER 8

Cladding

This chapter will cover the following NVQ and Diploma units:

- NVQ VR42
- CC 2046K

This chapter is about:

- Interpreting building information
- Adopting safe and healthy working practices
- Selecting materials, components and equipment
- Preparing and erecting brickwork and blockwork to pre-erected structures

The following NVQ performance criteria will be covered:

- Performance criterion 1: Interpretation of information
- Performance criterion 2: Safe work practices
- Performance criterion 3: Selection of resources
- Performance criterion 4: Minimize the risk of damage
- Performance criterion 5: Meet the contract specification
- Performance criterion 6: Allocated time

The following Diploma outcomes will be covered:

- Know how to construct masonry to timber-framed structures
- Know how to construct masonry to concrete-framed structures
- Know how to construct masonry to steel-framed structures

Building information

Definition of cladded walls

The term cladding is used when materials are used to face a pre-erected building.

The standard form of construction for the external walls of brick buildings is cavity walling, which was dealt with in Chapter 7.

This chapter deals with brick cladding to frames of timber, concrete and steel.

Cladding to timber frames is relatively straightforward and is usually restricted to two- or three-storey buildings.

With multistorey buildings, the total loading of all the floors would be too much for normal 275 mm wide cavity walling to support safely. In these situations, a permanent frame of reinforced concrete or structural steel is erected first, before external cavity walling is built around it (Figure 8.1).

Functions of cladded structures

Cladding has several advantages over traditional brick walling:

- The framework can be erected very quickly and possibly waterproofed before the external cladding is required.
- The external brick cladding can be erected without causing any delay to the inside work.
- The whole building will have to be made accessible with scaffolding, so this can be waterproofed, if necessary, to prevent any delay with the brick cladding.

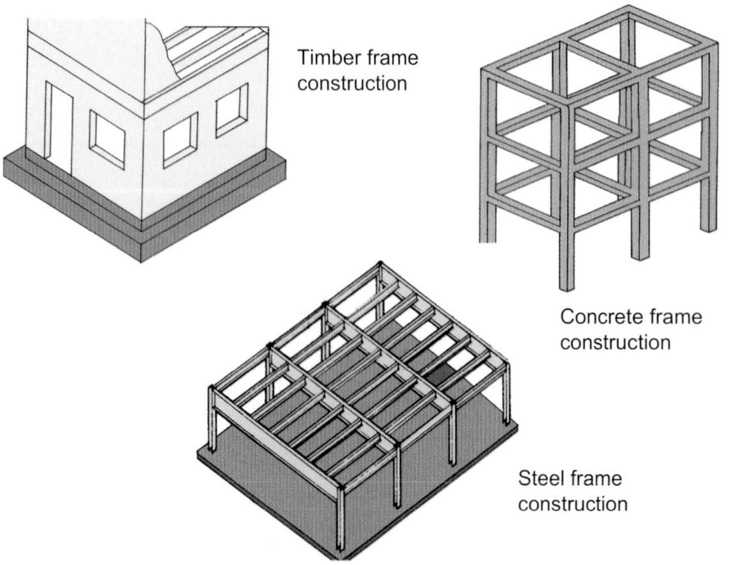
Timber frame construction
Concrete frame construction
Steel frame construction

FIGURE 8.1
Types of frame construction

Design qualities

It is important that the designers of cladded buildings keep the following points in mind:

- Allowances must be made for the thermal movement of the structure and for any movement of the structure due to loading.
- Any movement could cause cracks on the facework and allow dampness in.

Constructional requirements

The brickwork should be securely fixed and supported by the frame.

Each floor level will require a horizontal damp-proof course (DPC) tray.

Adopting safe and healthy working practices

Safety legislation has already been covered in Chapter 2. Please refer back to this chapter if necessary.

Particular areas to consider when building cladding to pre-erected frames are:

- personal protective equipment
- manual handling materials
- mechanical tools and equipment
- working at heights.

Resources

This chapter deals with various walling components and materials, including:

- standard bricks and specials
- blocks
- lintels
- damp-proof courses
- door and window frames
- insulation
- expansion and movement jointing material
- wall fixings
- air bricks and ducts
- types of mortar.

Most of these materials have been dealt with in previous chapters. Please refer back to these if necessary.

Bricks

Two main types of brick are required: normal facings and brick slips (Figure 8.2). Brick slips are required to fill in the wall in front of concrete beams.

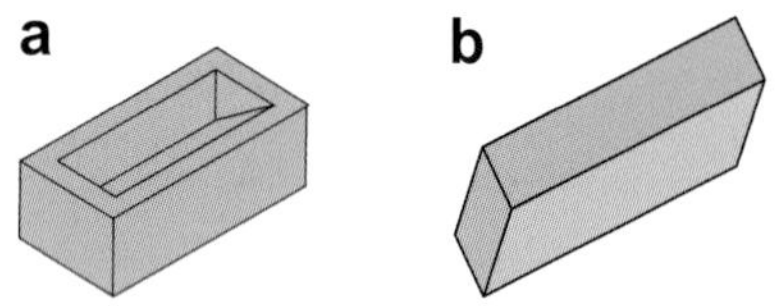
a
b

FIGURE 8.2
Types of brick required for cladding: (a) frogged brick; (b) brick slip

Blocks

Some buildings may be designed for brick facings, while others may require faced blockwork. These can also be divided into two categories, load-bearing and lightweight (Figure 8.3).

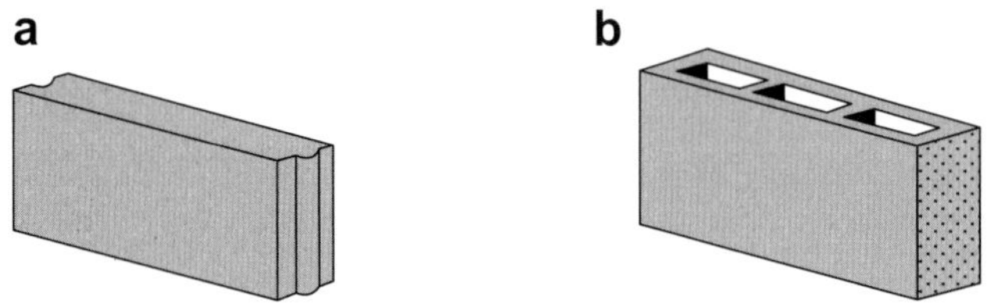
a
b

FIGURE 8.3
Types of block: (a) load-bearing; (b) lightweight

Lintels

Lintels are required to span across window and door openings.

Concrete or steel lintels can be used inside a building, where they are covered with plaster. Steel lintels are mainly used outside (Figure 8.4).

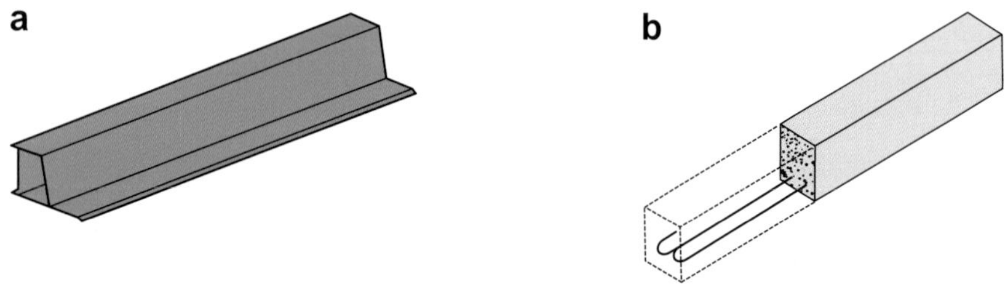
a
b

FIGURE 8.4
Types of lintel: (a) steel lintel; (b) concrete precast lintel

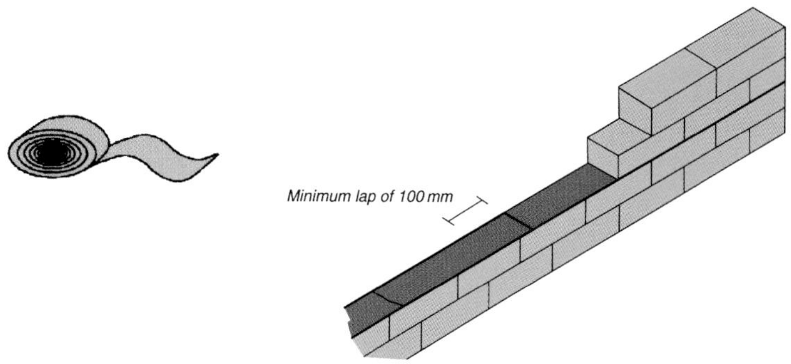
Minimum lap of 100 mm

FIGURE 8.5
Damp-proof course

Barriers to damp

The DPC is used as a barrier to rising damp.

The most common type of DPC is the flexible type (Figure 8.5), available in widths of 100, 150, 300 and 450 mm.

The damp-proof materials used vary from flexible sheet material made from bitumen with a hessian or fibre base to rigid material such as slate.

The current Building Regulations specify that the external wall must have a DPC at a height of not less than 150 mm above the finished surface of the adjoining ground.

Cavity trays

The DPC is also used as a cavity tray in cavity walls.

One purpose of providing a cavity between the two leaves of an external wall is to prevent damp penetration.

It is, however, necessary to close the cavity in some situations, particularly around openings for windows and doors (Figure 8.6). In this case, special precautions must be taken to resist damp penetration.

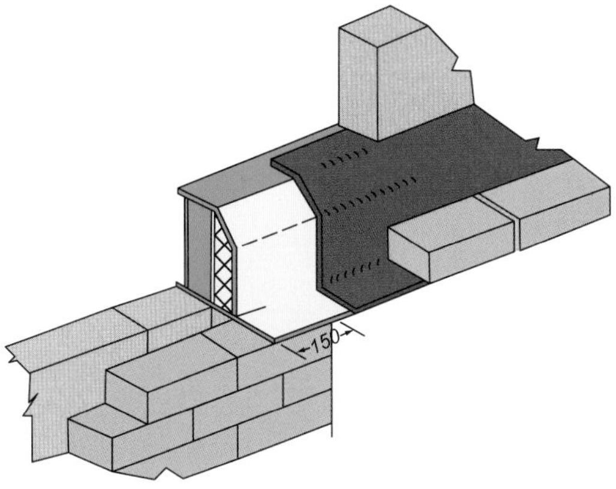
150

FIGURE 8.6
Cavity tray

It is also necessary to prevent water in the ground from rising up the two leaves of the wall.

Weep holes must be placed at 900 mm centres to allow any water to escape from the cavity.

Components

There are several components that a bricklayer will have to fix as the work proceeds.

WINDOW AND DOOR FRAMES

These are built into the walls by the bricklayer as the wall proceeds.

The frames usually have horns on the sill and head. These should be partly cut off, which helps to secure the frame.

Special cramps should also be used to secure the frame every fourth course (Figure 8.7). Door and window frames are also shown in Figure 8.7.

CAVITY WALL INSULATION SLABS

Slabs are used in cavity walls to provide insulation. They are made of fibreglass or polystyrene, and vary in thickness according to the requirements of the construction (Figure 8.8).

Special wall ties have to be used to secure the insulation slabs.

EXPANSION JOINTS

When dealing with continuous lengths of cladding it is necessary to allow for expansion of both the bricks and the frame.

Open vertical joints should be incorporated every 6–9 m and filled with bituminized foamed polyurethane compound or similar.

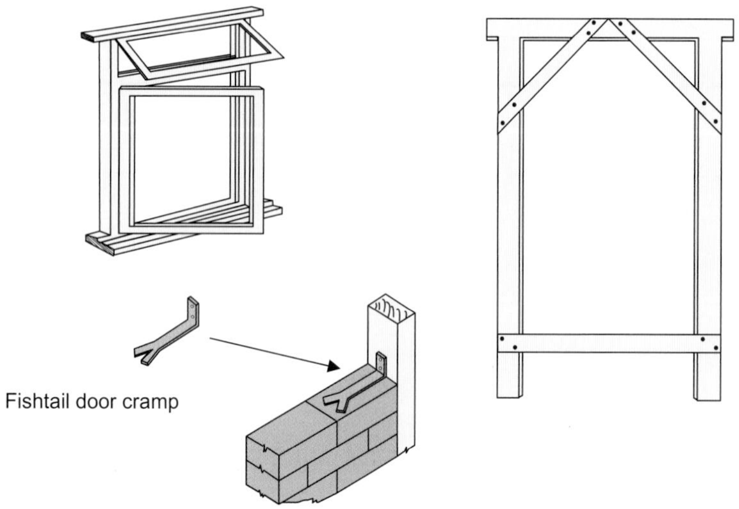
Fishtail door cramp

FIGURE 8.7
Door and window frames

Glass fibre in rolls
Polystyrene in slabs

FIGURE 8.8
Types of cavity insulation

Compression joints should also be incorporated; usually at each floor level. These joints should be below the support fixing and are formed by inserting a soft elastic material such as butyl rubber.

FIXINGS

Brickwork cladding to timber, steel and concrete frame needs to be secured to the frame.

The fixings change according to the material used to construct the frame. A selection of fixings is shown in Figure 8.9.

REINFORCEMENT

Several types of reinforcement may be laid into the bed joints of the brickwork to increase the strength (Figure 8.10).

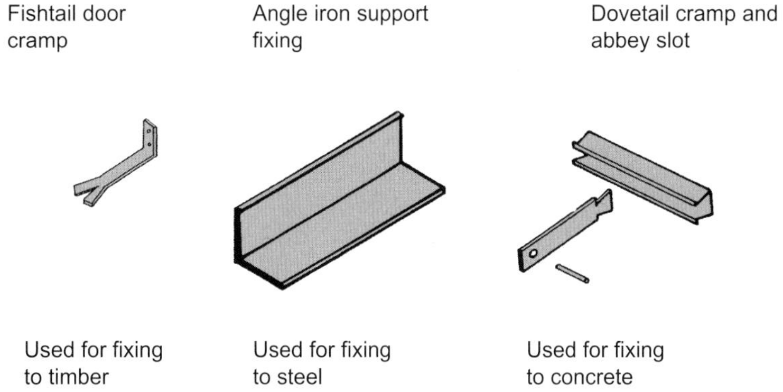
Fishtail door cramp
Angle iron support fixing
Dovetail cramp and abbey slot
Used for fixing to timber
Used for fixing to steel
Used for fixing to concrete

FIGURE 8.9
Fixings

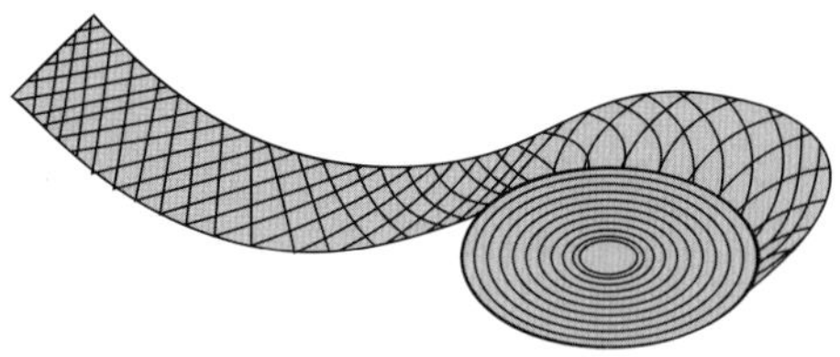

FIGURE 8.10
Flexible mesh

AIR BRICKS AND CAVITY LINERS

Air bricks are special bricks built into the outer skin of a cavity wall to allow air to pass through to ventilate the cavity. A cavity liner may be inserted, allowing the air to pass through the cavity to ventilate a hollow floor. See Figure 8.11 for details.

MORTAR

The mortar used for cladding should not be too hard or it may cause spalling of the bricks if any movement takes place.

The colour of the mortar should be constant and blend in with the cladding.

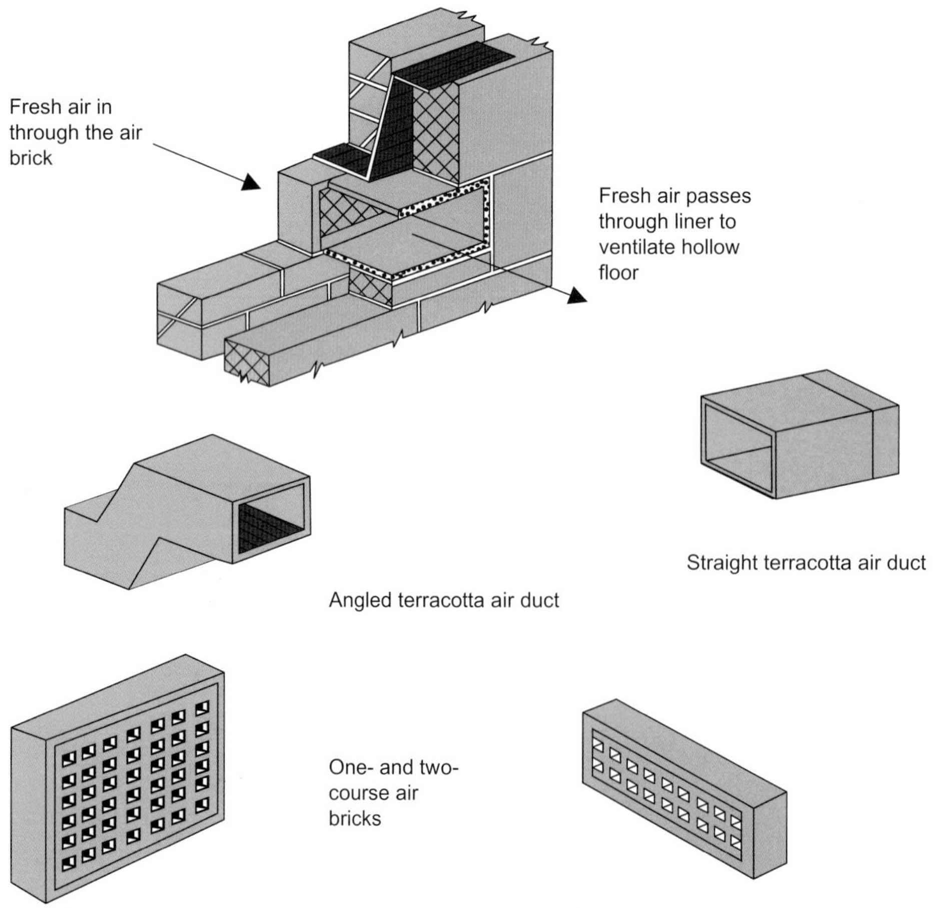
Fresh air in through the air brick
Fresh air passes through liner to ventilate hollow floor
Straight terracotta air duct
Angled terracotta air duct
One- and two-course air bricks

FIGURE 8.11
Air-brick details

Cladding brickwork

The standard form of construction for the external walls to frame buildings is cavity walling.

This means that the bricklayer builds only the external skin of masonry, whether it is brick, block or any other local material. The outer skin is usually 102.5 mm thick face brickwork or facing quality blockwork.

The requirements for cavity wall brickwork and cladding framed buildings are very similar.

Cavity wall brickwork may be load bearing, that is, used to support the load from floors and the roof.

A single skin of 102.5 mm face brickwork is used as brick cladding to framed buildings. The inner skin is formed from timber, steel or concrete storey-height panels, which support floor and roof loads.

All the craft requirements of the bricklayer contained in this chapter apply equally to all applications of cavity walling. As the most common bond is stretcher bond the basic skills gained from Chapter 7 will help with cladding to framed buildings.

The main difference lies in the method of fixing the brickwork to the frame. This will be explained later.

Framed structures

Brick cladding can be used to cover timber, concrete and steel frames.

Timber frames are usually restricted to two or three storeys, while concrete and steel can be used for multistorey buildings.

TIMBER FRAMES

Brick cladding to low-rise structures is carried out in exactly the same way as for traditional cavity walling. The only difference is the way in which the brick cladding is fixed to the timber frame. The foundation, ground works and work up to DPC level are similar to traditional construction.

The timber frame is erected on the concrete slab, protected from rising damp by the horizontal DPC. The whole frame can be erected, including the roof. This can then be waterproofed and work can carry on inside the building while the brick cladding is completed.

The prefabricated timber frame is produced off site in a factory under ideal weather conditions. It normally consists of 100 × 50 mm timber with an outer sheathing of plywood or similar material. Insulation is enclosed with an internal lining of plasterboard.

Skeleton frames constructed in timber may be fabricated from solid timber sections, built-up sections or glued and laminated sections (Figure 8.12).

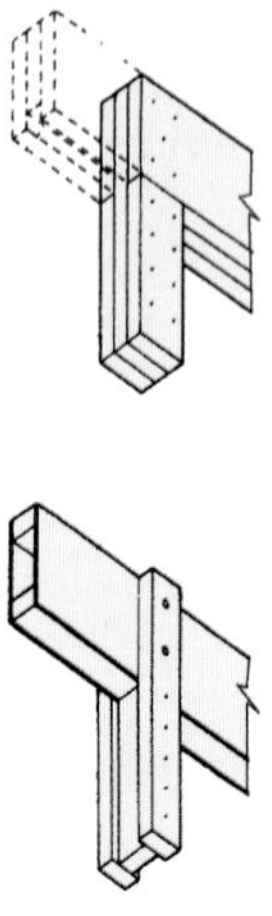
Solid, square or rectangular sections are generally the most economical to adopt
With the larger members it is necessary to form them by combining a number of smaller sections of timber
This can be done by bolting or nailing several pieces together to form built up sections
Built-up solid beam
Plywood beam
Built-up I-section beam

FIGURE 8.12
Timber sections

There are three main types of system for assembly on site:

- balloon frame system (Figure 8.13).
- platform frame (Figure 8.14).
- stick system (Figure 8.15).

The balloon system consists of prefabricated frames two storeys high. They arrive on site and are erected off the concrete slab in one stage. The timber

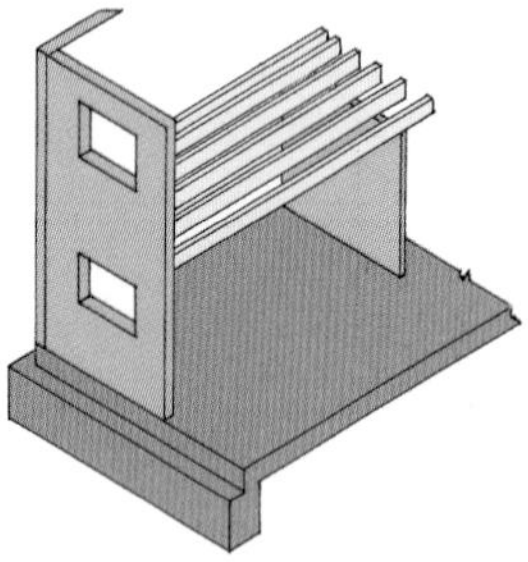
Two-storey height panel
Joists supported on inner studded wall and front panel

FIGURE 8.13
Balloon frame

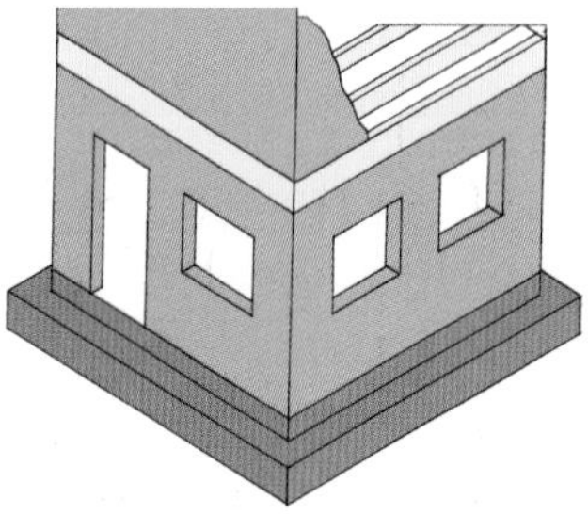
Single-storey height panel
Joists supported on panels

FIGURE 8.14
Platform frame

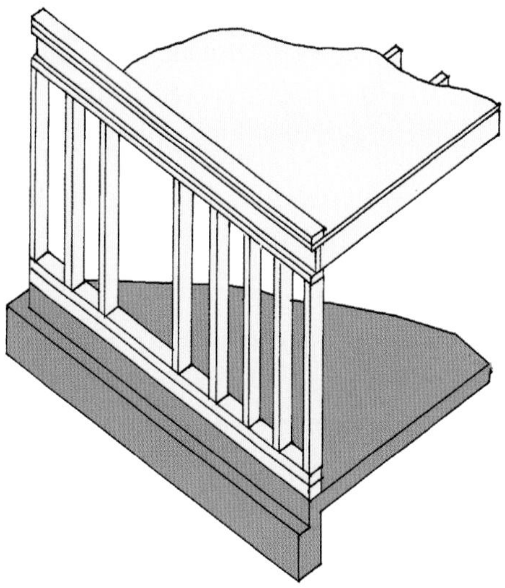

FIGURE 8.15
Stick system: frame constructed on site with individual members

joists are suspended from the framework. Gable ends are then erected on top of the two-storey frame if required.

The platform system consists of prefabricated panels of single storey height only. These are easier to handle on site. The first floor framework supports the floor joists and the second storey is erected on top.

The stick system is erected on site from individual members and not prefabricated off site. This system takes longer to erect but can overcome small site problems.

Most small-scale timber-framed structures take the form of post and beam construction in which resistance to racking and distortion of the frame under working load is provided by the infill panels.

The roof of a timber frame construction will almost certainly be a roof truss (Figure 8.16). Trussed rafters are fabricated from members of uniform thickness jointed by gluing or nailing, using plywood or, in the case of nailing, punched metal plate gussets.

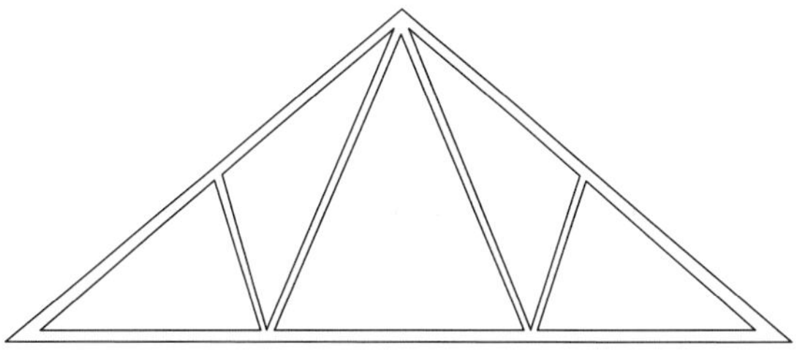

FIGURE 8.16
Roof truss

Brick cladding

Timber wall frames are erected on a foundation of conventional walls or a concrete base or raft to which a sole plate is bolted (Figure 8.17).

The frame is usually assembled as a storey-height frame, in sections around the building. Openings for windows and doors are included in the frame and the frame is covered with a sheathing of timber boards or plywood and a breather barrier.

A sheathing of 9.5 mm plywood sheeting or 25 mm tongue-and-groove boarding is applied to the outside of the frame, which gives it rigidity, provides a face area for fixing back the brick cladding and improves the thermal insulation by sealing spaces between the timber studs.

A vapour barrier is placed on the inner, warm side of the wall frame to exclude any water vapour passing through the lining which may condense within the cavities. Suitable materials are non-ferrous metal foils, polythene sheeting and especially prepared papers, which are fixed to the framing under the lining; alternatively, aluminium foil-backed plasterboard may be used for the lining. For the barrier to be effective, all joints must be sealed as it must be continuous and imperforate.

A moisture barrier is provided by a building paper stapled to the sheathing or outer faces of the studs with good overlaps at joints. It functions as a second line of defence against penetration of wind-driven rain or snow, or moisture from an external brick veneer.

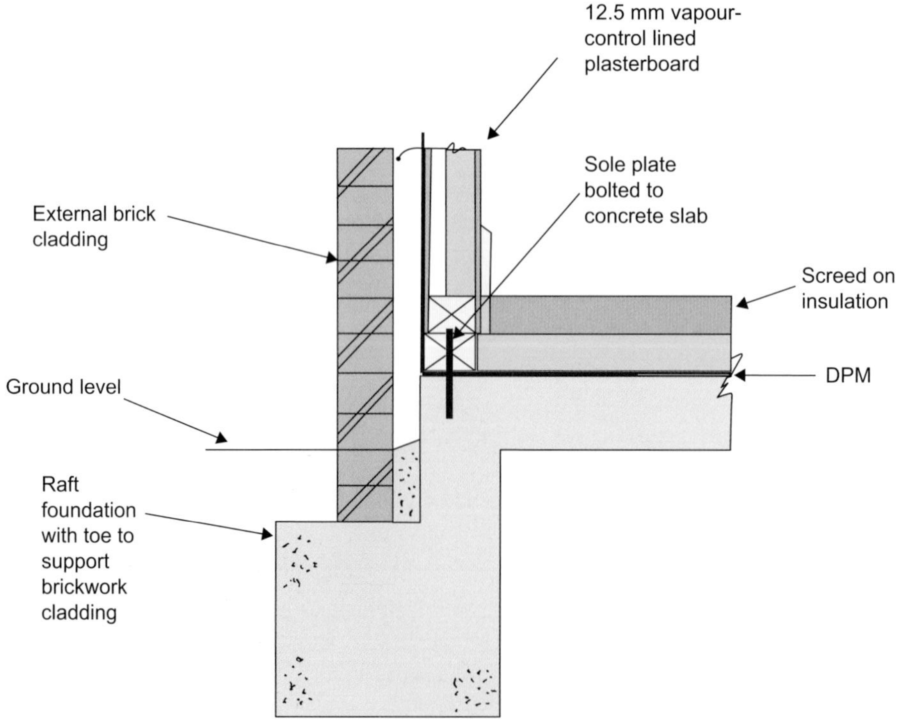
12.5 mm vapour-control lined plasterboard
Sole plate bolted to concrete slab
External brick cladding
Screed on insulation
DPM
Ground level
Raft foundation with toe to support brickwork cladding

FIGURE 8.17
Detail at ground level

The brickwork is fastened to the timber frame with galvanized steel cramps fixed every fifth course (Figure 8.18).

Figure 8.19 shows how the cladding is terminated at the eaves and window head, and at sill level.

Figure 8.20 shows a typical section through a timber frame construction including part of the brick cladding to the frame.

CONCRETE FRAMES

In multistorey buildings, the total loading of all the floors would be too much for normal 275 mm wide cavity walling to support safely. In these situations, a permanent frame of reinforced concrete or structural steel is erected first, then external cavity walling is built around it.

One of the main differences between steel and concrete frames and timber frames is that timber frames provide a total structure, whereas steel and concrete provide only a frame. It is therefore necessary to provide internal walls to both steel and concrete frames. These are known as non-load-bearing brick panel walls.

Support for brick cladding

Although the bricklayer builds this cavity walling in exactly the same way as for low-rise structures, when used on multistorey buildings, both leaves of brick masonry must be supported at every floor level. This ensures that permanent pressure does not build up on the brick masonry at ground level. Cavity wall cladding to multistorey buildings is described as non-load-bearing, because it does not carry the weight of floor slabs resting upon it.

Support for the outer skin of brick cladding using reinforced concrete edge beams (Figures 8.21 and 8.22) has generally been replaced by stainless steel angles or brackets (Figures 8.23 and 8.24)

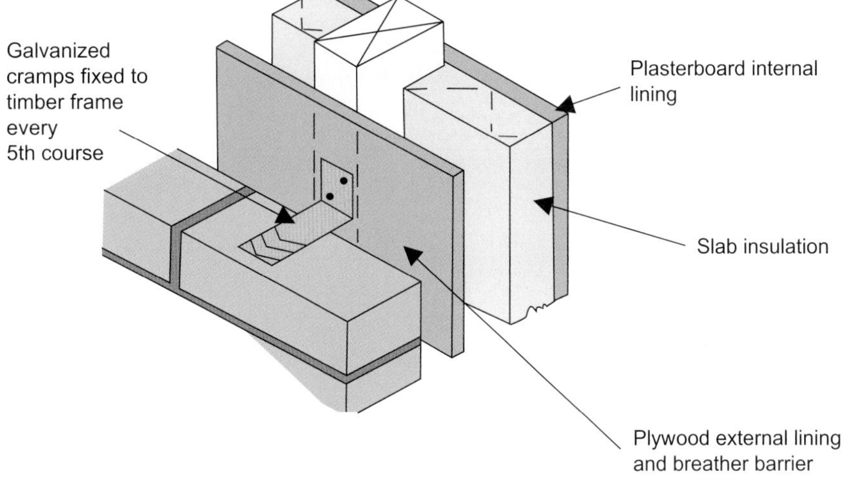
Galvanized cramps fixed to timber frame every 5th course
Plasterboard internal lining
Slab insulation
Plywood external lining and breather barrier

FIGURE 8.18
Fixing brickwork to timber frame

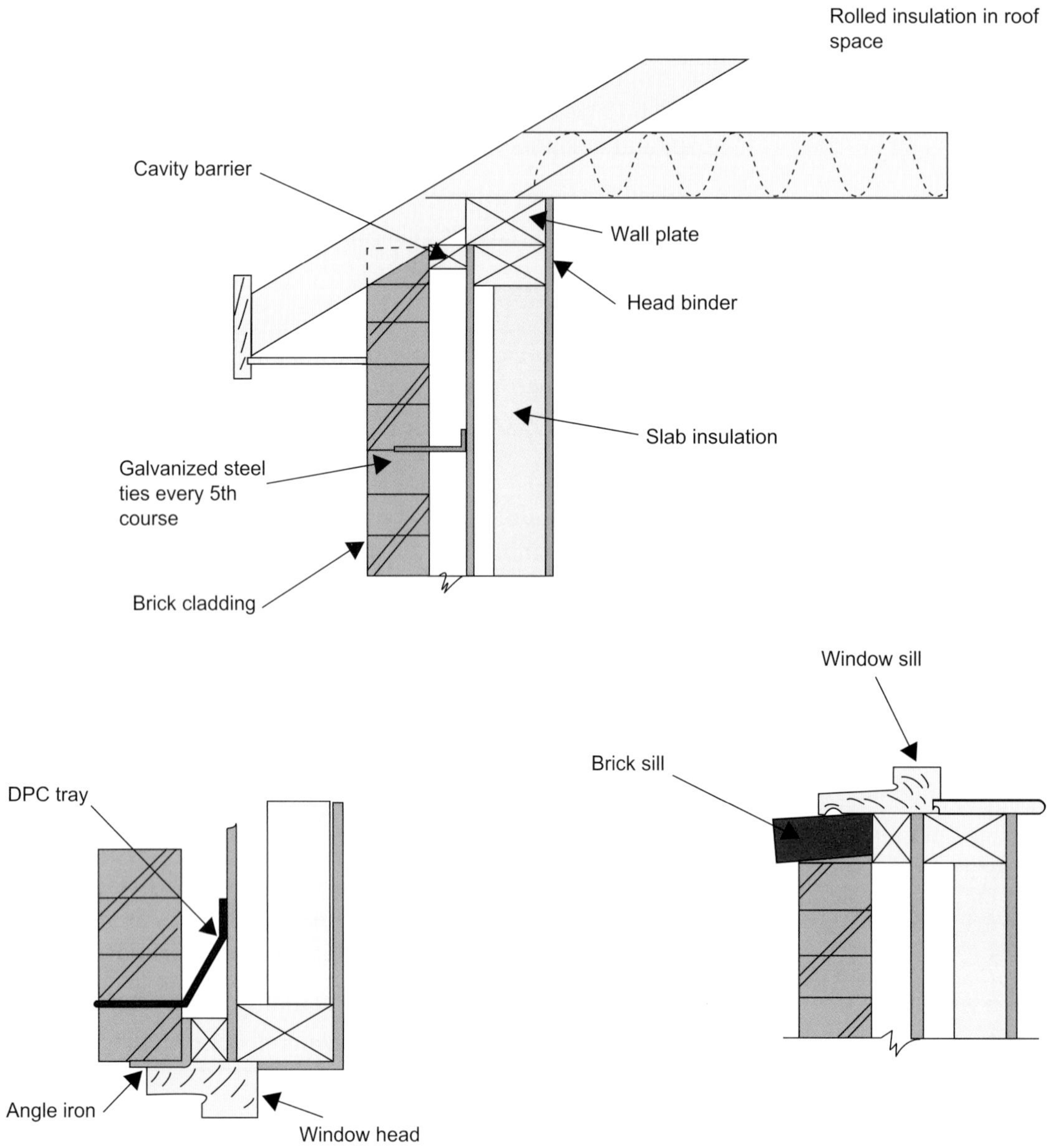
Rolled insulation in roof space
Cavity barrier
Wall plate
Head binder
Slab insulation
Galvanized steel ties every 5th course
Brick cladding
Window sill
Brick sill
DPC tray
Angle iron
Window head

FIGURE 8.19
Details at eaves, window head and sill level

You will notice that Figures 8.21–8.26 show a soft-joint filler of polyethylene foam strip under the supporting steel angle or concrete toe beams. This very important compression joint must be provided under each supporting angle of stainless steel, or reinforced concrete toe beam, around the external perimeter of a building at every floor level.

The horizontal compression joint, completely clear of mortar, permits different rates of expansion and contraction (up or down) to take place without restriction, between the building and the brick masonry cladding (Figure 8.23).

Unless these horizontal compression joints are provided at every floor level, serious cracking will result when the outer brickwork tries to expand and/or

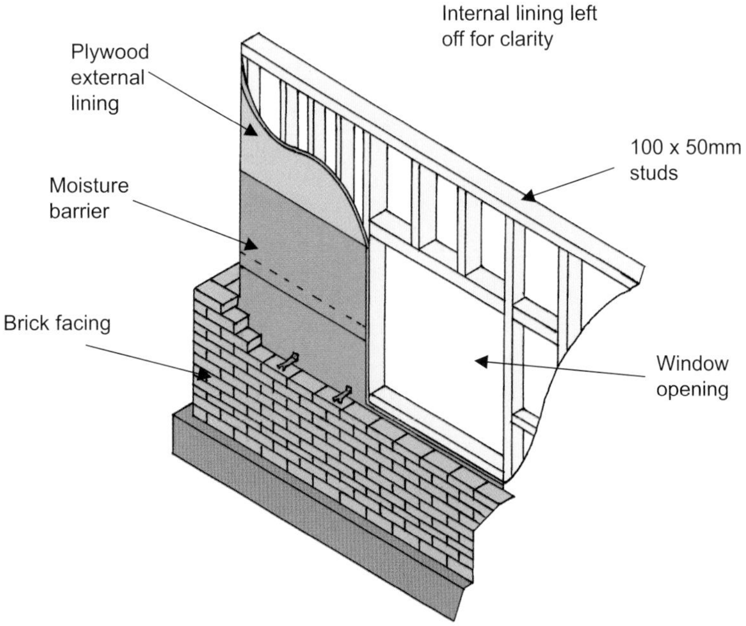
Internal lining left off for clarity
Plywood external lining
100 x 50mm studs
Moisture barrier
Brick facing
Window opening

FIGURE 8.20
Details of cladding to timber framework

the load-bearing structural frame shrinks slightly over the years (Figure 8.25).

When brick slips are used to cover the concrete frame they need to be fixed to the concrete. Several methods can be adopted; one method, using galvanized steel angles, is shown in Figure 8.26. The galvanized angle is bolted to

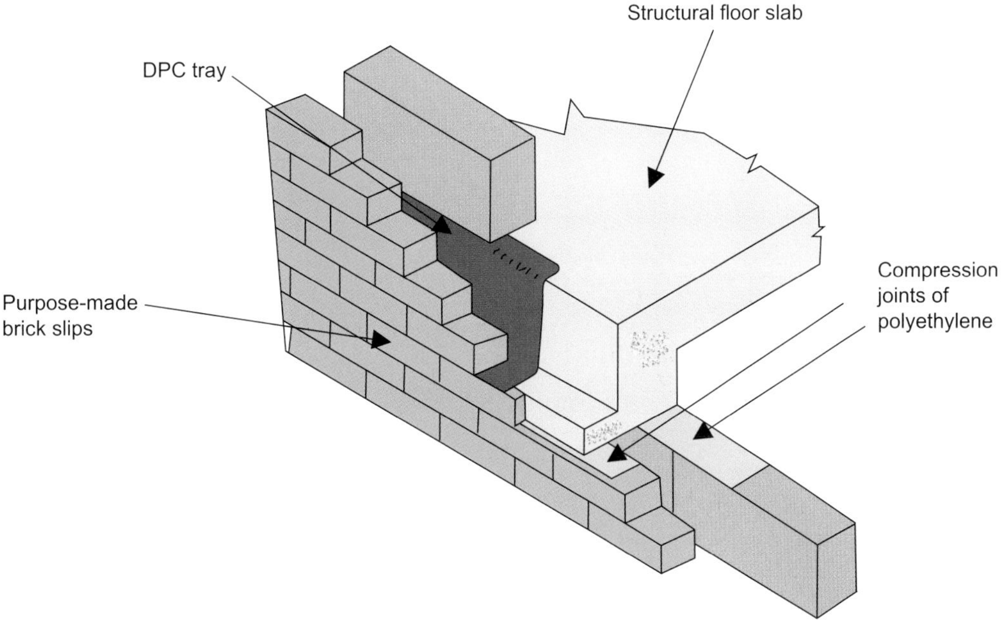
Structural floor slab
DPC tray
Compression joints of polyethylene
Purpose-made brick slips

FIGURE 8.21
Support of cavity wall brick cladding on reinforced concrete toe beam

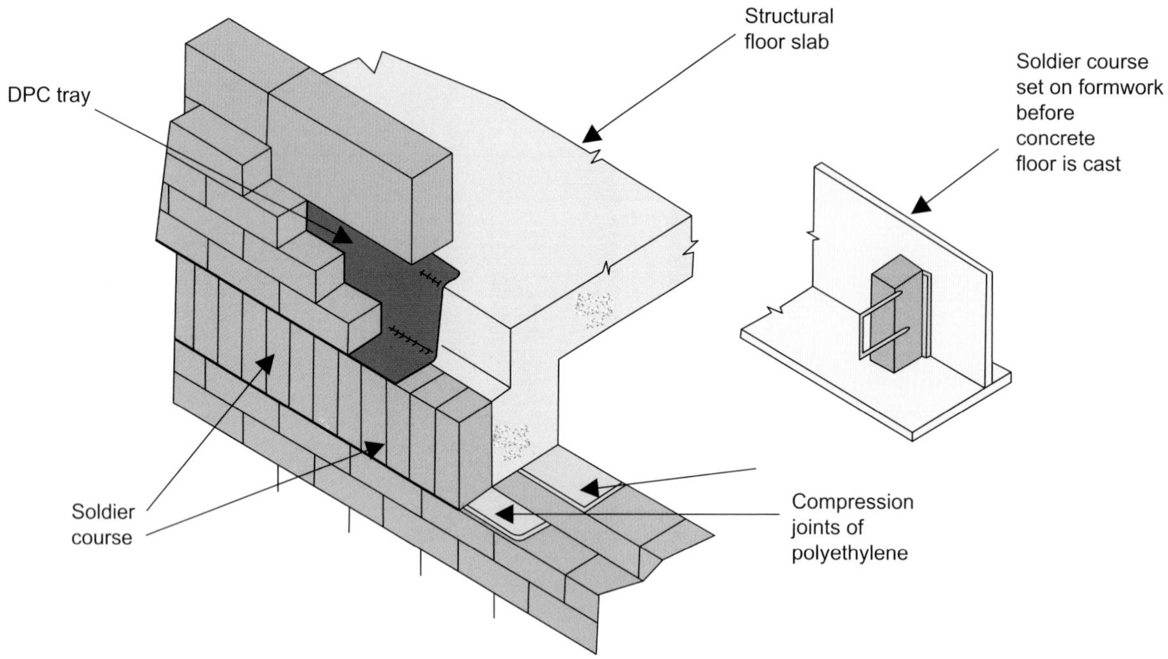
Structural floor slab
Soldier course set on formwork before concrete floor is cast
DPC tray
Soldier course
Compression joints of polyethylene

FIGURE 8.22
Outer leaf of cavity wall facework supported on soldier course of bricks, cast in situ with reinforced concrete structural floor slab and edge beam

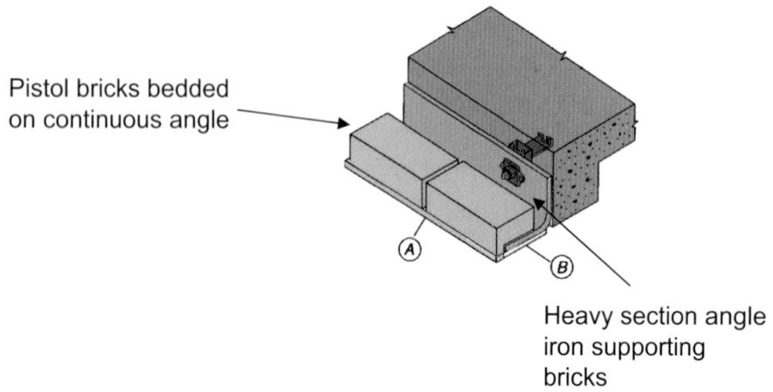
Pistol bricks bedded on continuous angle
A
B
Heavy section angle iron supporting bricks

FIGURE 8.23
Heavy section stainless steel angle

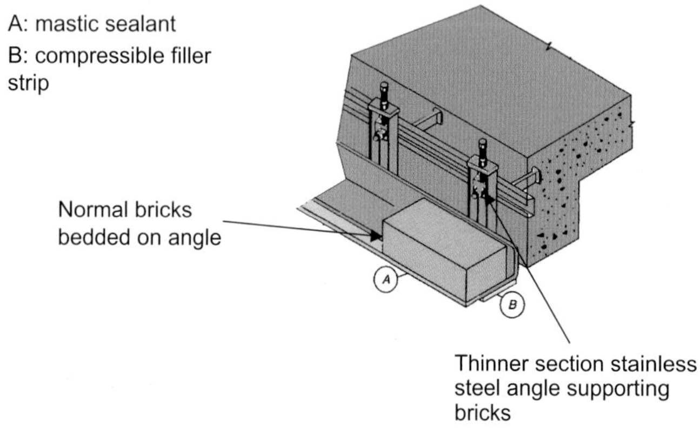
A: mastic sealant
B: compressible filler strip
Normal bricks bedded on angle
A
B
Thinner section stainless steel angle supporting bricks

FIGURE 8.24
Thinner section stainless steel angle with adjustable support

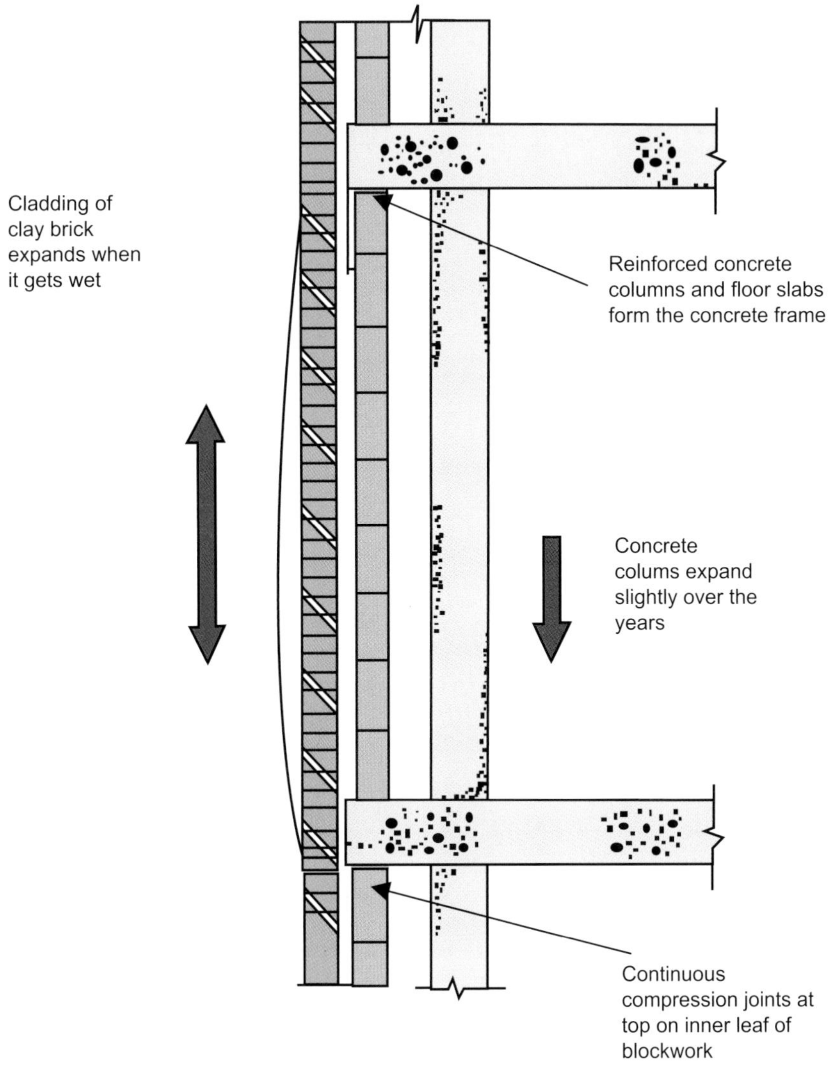
Cladding of
clay brick
expands when
it gets wet
Reinforced concrete
columns and floor slabs
form the concrete frame
Concrete
colums expand
slightly over the
years
Continuous
compression joints at
top on inner leaf of
blockwork

FIGURE 8.25
The importance of compression joints

the concrete frame and the brick slips are bedded into position when the whole building has had time to settle. Fixing the brick slips is one of the last jobs for the bricklayer when working on cladding to concrete frames.

Pistol bricks

To disguise the presence of the supporting steel angle within the thickness of a mortar bed joint, the first course of bricks may be rebated (Figure 8.27). These 'pistol bricks' can be produced as bricks of special shape if the manufacturer is given sufficient notice. Alternatively, they can be cut on a masonry bench saw.

Before the importance of horizontal compression joints at every floor level of multistorey buildings was fully realized, great problems developed from the use of brick 'slips', which were used to disguise concrete toe beams.

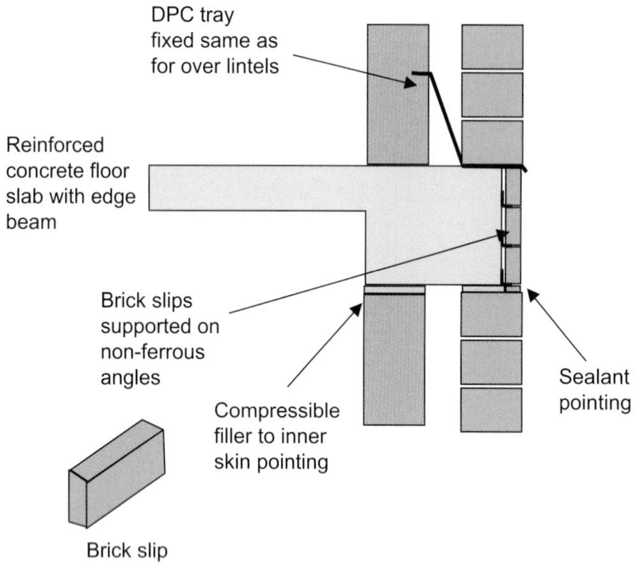
DPC tray
fixed same as
for over lintels
Reinforced
concrete floor
slab with edge
beam
Brick slips
supported on
non-ferrous
angles
Compressible
filler to inner
skin pointing
Sealant
pointing
Brick slip

FIGURE 8.26
Fixing brick slips

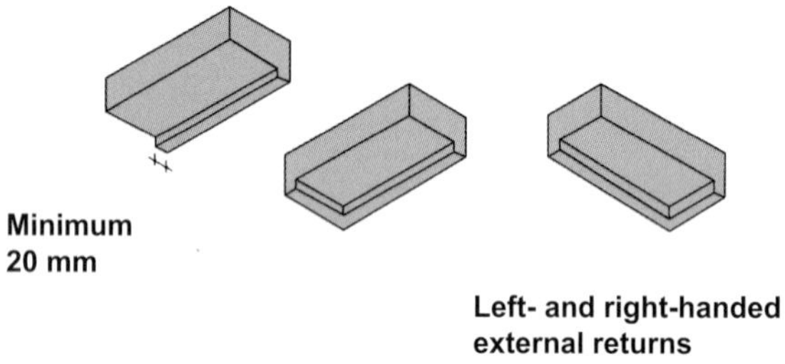
Minimum
20 mm
Left- and right-handed
external returns

FIGURE 8.27
Pistol bricks used to hide thickness of heavy section support angles

For these reasons, stainless steel angles or brackets are currently preferred for the support of brick cladding.

Brick panels

Occasionally the architect will design the building to have the concrete beams and columns exposed.

The brickwork is used simply as an infilling panel between the beams and columns. These panels are constructed in the same way as ordinary walls. They need to be adequately supported at floor levels, support their own weight and be weather resistant. Cavity trays and flashings are used to this end.

Brick panel walls are generally constructed with cavities and an insulation block inner leaf to comply with thermal insulation regulations. As in traditional cavity walls, the external leaves are linked together with wall ties.

The wall is supported on the lower concrete floor slab and a DPC incorporated in a similar manner to lintel construction, with the external face of the

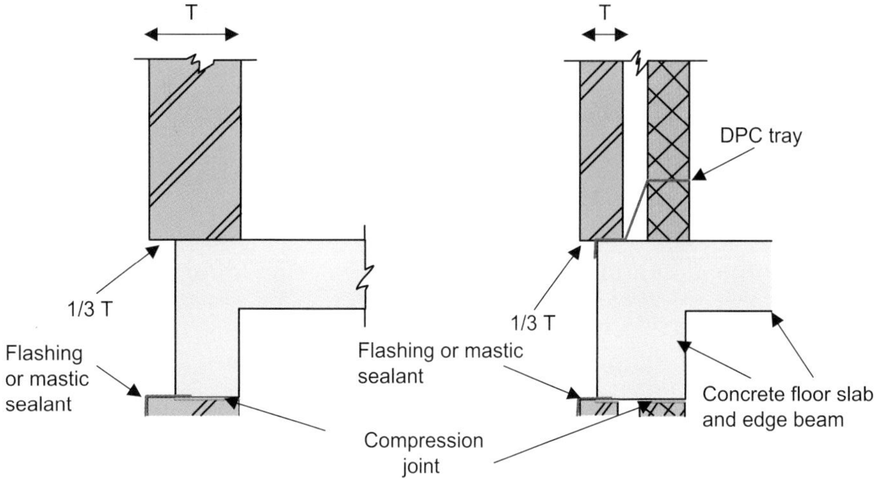
T
T
DPC tray
1/3 T
1/3 T
Flashing
or mastic
sealant
Flashing or mastic
sealant
Concrete floor slab
and edge beam
Compression
joint

FIGURE 8.28
Maximum permissible overhang

wall being set forward a small amount beyond the edge of the slab (Figure 8.28). The head of the panel must also be provided with suitable damp-proofing.

Figures 8.21–8.26 showed examples where the concrete frame is covered. Figure 8.28 shows the concrete beam exposed. Where the column is exposed the panel wall is fixed to the side of the concrete column (Figure 8.29).

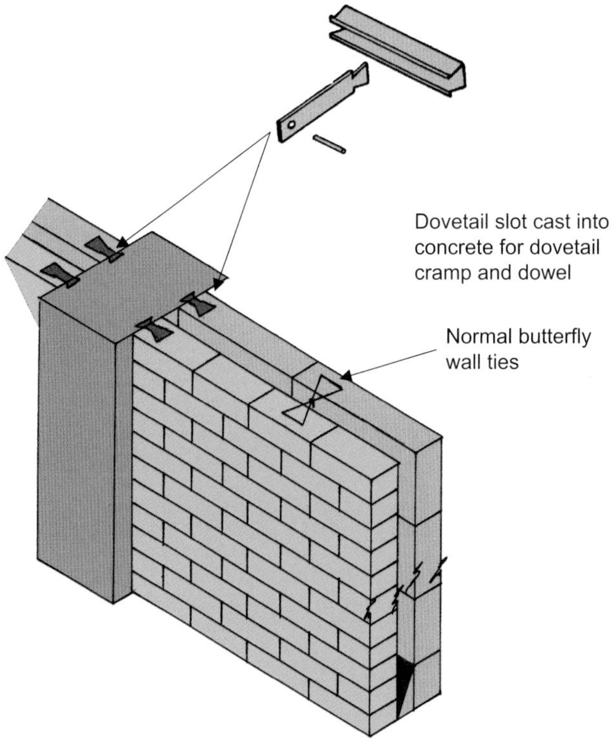
Dovetail slot cast into
concrete for dovetail
cramp and dowel
Normal butterfly
wall ties

FIGURE 8.29
Fixing to exposed columns

Special dovetail abbey slots are cast into the columns into which dovetail cramps are inserted to tie the walls to the concrete column. A cheaper method involves casting butterfly wall ties into the concrete column and bending them out when required.

Vertical movement joints

The outer leaf of face brickwork in cavity wall cladding to a framed building will expand and contract owing to changes in temperature and moisture content.

The horizontal compression joints shown in Figure 8.21–8.26 allow up-and-down movements in the brickwork to take place freely.

Allowance must also be made for sideways expansion and contraction, however, so that the outer leaf of brickwork does not cause damage to itself or the structure.

A typical vertical movement joint, with expanded plastic foam joint filler projecting, can be seen in Figure 8.30. This shows 102 mm thick brick cladding to a concrete-framed building.

Vertical movement joints should be provided at intervals between 9 and 12 m for clay brickwork, and not exceeding 7–9 m for sand lime (calcium). Since vertical movement joints form a 'straight joint' weakness in a wall, they must be strengthened with stainless steel slip-ties every fourth course.

Typical details of brick cladding to concrete frames are shown in Figure 8.31.

STEEL FRAMES

Small building frames are fabricated from sections, either rolled to shape from ingots of hot steel or cold formed to shape from steel strip (Figure 8.32).

Hot-rolled steel sections are standardized in shape and size. 'I'-section girders used as beams are known as universal beams.

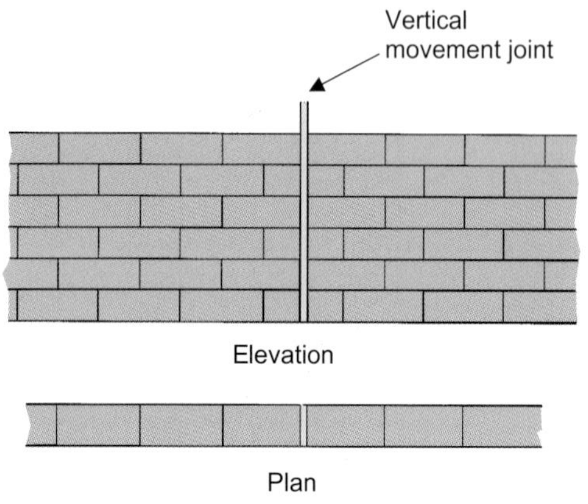
Vertical movement joint
Elevation
Plan

FIGURE 8.30
Brick cladding showing the position of a vertical movement joint

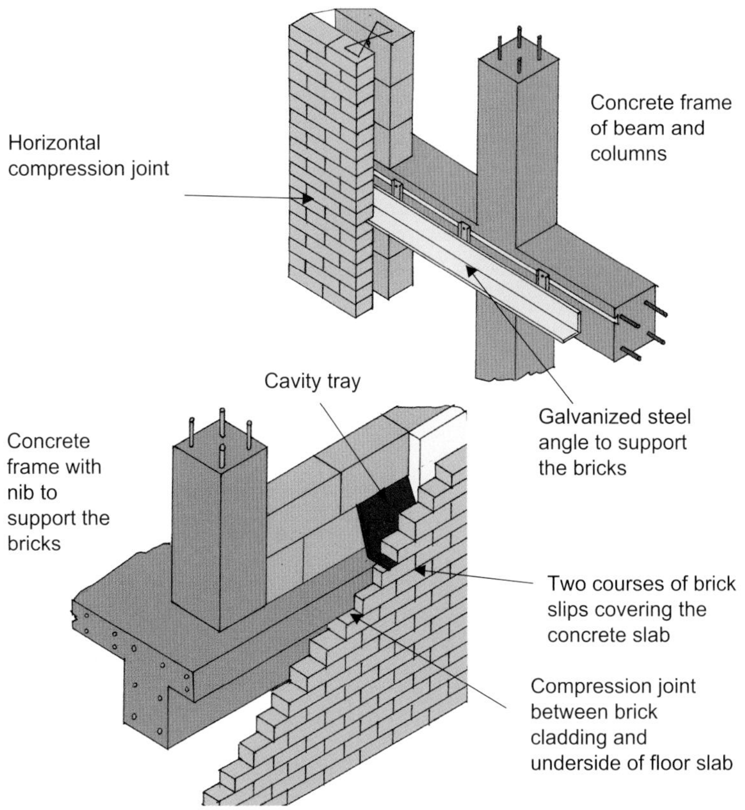
Horizontal compression joint
Concrete frame of beam and columns
Cavity tray
Galvanized steel angle to support the bricks
Concrete frame with nib to support the bricks
Two courses of brick slips covering the concrete slab
Compression joint between brick cladding and underside of floor slab

FIGURE 8.31
Various details of brick cladding to a concrete frame

Those suitable for use as columns have non-tapered flanges that are much wider relative to the depth of the web than the universal beam, and are known as universal columns.

Hot-rolled members of a steel frame are connected by welded joints or by steel bolts and rivets.

Panel walls to steel frames

Panel walls provide the envelope in skeleton-frame construction.

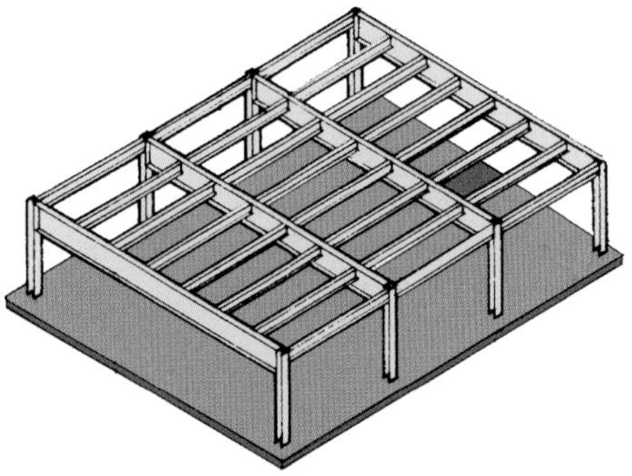

FIGURE 8.32
Typical steel frame

They are designed to carry only their own weight together with wind loading on their face, and are supported at each floor. Further restraint is provided by the columns and occasionally by the floor above.

Cavity wall construction is used, as for normal cavity walls (Figure 8.33).

BRICK PIERS AND COLUMNS

Occasionally it will be necessary to clad columns and piers, and separate skills for attached and detached piers will be required.

A free-standing brick pier or isolated pier is a pillar of brickwork that stands alone. It can be used to carry the ends of beams over a large span, to support garden gates, etc., or to clad steel or concrete columns.

Piers can vary considerably in size and bond and can be very difficult to build because of the plumbing points.

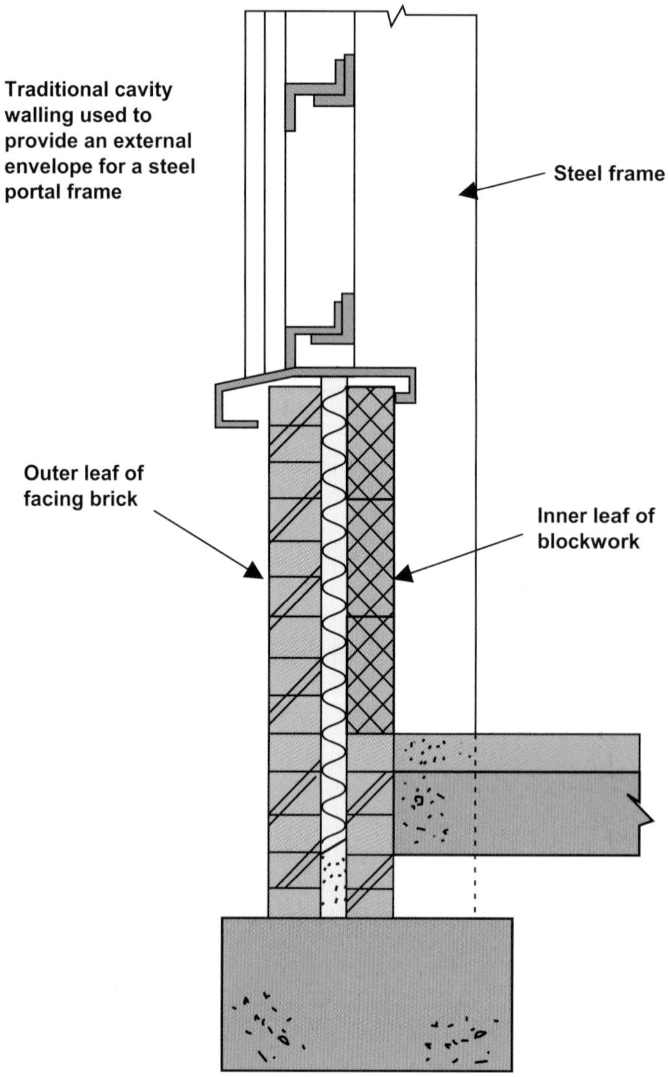
Traditional cavity walling used to provide an external envelope for a steel portal frame
Steel frame
Outer leaf of facing brick
Inner leaf of blockwork

FIGURE 8.33
Brick cladding to a steel frame

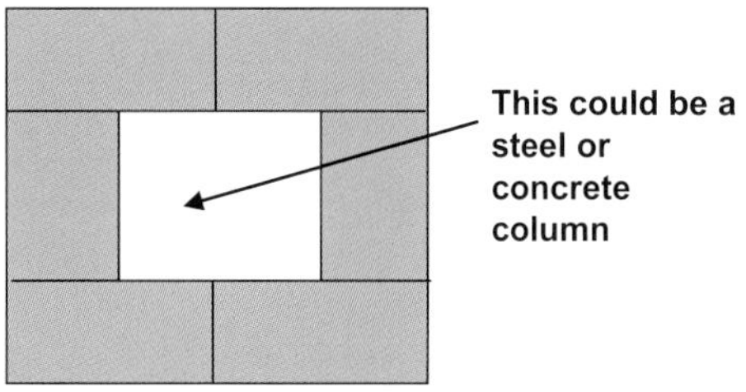
This could be a steel or concrete column

FIGURE 8.34
Two-brick pier in stretcher bond

The bond specified for a particular pier may have to be amended slightly to avoid the excessive cutting required by some bonds. Excessive cutting is expensive in terms of not only materials but also time. When there are many cuts the bricklayer has to take time to ensure that all the cuts are the same, so as to avoid any irregularity in the perpends or the plumbing.

Small section piers are often the most difficult to build because of the limited number of perpends.

If several piers are in line, the piers at either end should be built first and then a line attached to either end pier so that the remainder of piers can all be built in line, both front and back.

Always set the piers out square and to brick sizes.

Hollow piers

Piers are often constructed in stretcher bond and are constructed around the various columns (Figure 8.34). There is no need to fasten the brickwork to the columns as the brickwork will secure itself as it wraps around the column.

Multiple-choice questions

Self-assessment

This section of the book is designed to allow you to check your level of knowledge. The section consists of revision questions for this chapter. The questions are all multiple choice and have four possible answers. The answers are to be found at the end of the book.

The main type of multiple-choice question will be the four-option multiple-choice question. This will consist of a question or statement, known as the stem, followed by a choice of four different answers, called the responses. Only one of these responses is the correct answer; the others are incorrect and are known as distracters.

You should attempt to answer the questions by choosing either (a), (b), (c) or (d).

Example

The person employed by the local authority to ensure that the Building Regulations are observed is called the:

(a) clerk of works
(b) building control officer
(c) council inspector
(d) safety officer

The correct answer is the building control officer, and therefore (b) would be the correct response.

Cladding

Question 1 Identify the following fixing used to tie back brick cladding to timber frames:

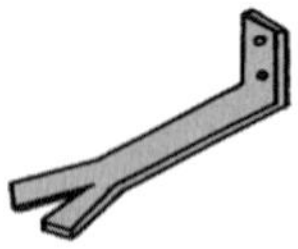

(a) tie wire
(b) fishtail cramp
(c) fishtail
(d) butterfly tie

Question 2 What is the maximum overhang for the bricks when the concrete beams are exposed?

(a) 1/2 thickness of the wall
(b) 1/3 thickness of the wall
(c) 1/4 thickness of the wall
(d) 1/5 thickness of the wall

Question 3 Which of the following is the correct position for a compression joint?

(a) between the top of the wall and the underside of a concrete beam

(b) on top of the concrete beam and under the brick wall

(c) between the top of the brick wall and the bottom of the next brick wall

(d) anywhere in the brick wall

Question 4 Identify the type of fixing for brick to concrete:

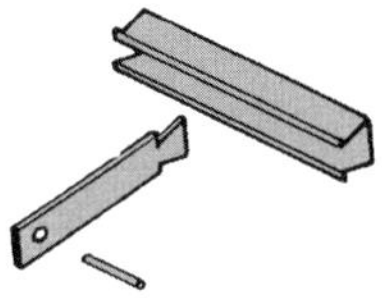

(a) fishtail cramp and slot

(b) butterfly cramp and slot

(c) abbey cramp and dovetail slot

(d) abbey slot and dovetail cramp

Question 5 An expansion joint should be produced on long lengths of cladding at intervals not exceeding:

(a) 3 m

(b) 6 m

(c) 9 m

(d) 12 m

Question 6 Identify the type of brick used when cladding concrete frames:

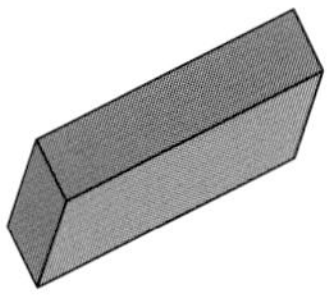

(a) pistol brick

(b) brick header

(c) brick slip

(d) brick stretcher

Question 7 Fixing cramps for cladding should be made of:

(a) non-ferrous metals

(b) steel

(c) galvanized iron

(d) plastic

Question 8 Cladding units are design to carry:

(a) all loads applied on the structure

(b) no loads applied on the structure

(c) some loads applied on the structure

(d) only live loads applied on the structure

CHAPTER 9

Thin Joint Masonry

This chapter will cover the following NVQ and Diploma units:

- NVQ VR44
- CC 2046K

This chapter is about:

- Interpreting building information
- Adopting safe and healthy working practices
- Selecting materials, components and equipment
- Preparing and erecting thin joint block masonry structures

The following NVQ performance criteria will be covered:

- Performance criterion 1: Interpretation of information
- Performance criterion 2: Safe work practices
- Performance criterion 3: Selection of resources
- Performance criterion 4: Minimize the risk of damage
- Performance criterion 5: Meet the contract specification
- Performance criterion 6: Allocated time

The following Diploma outcome will be covered:

- Know how to construct buildings using thin joint blockwork

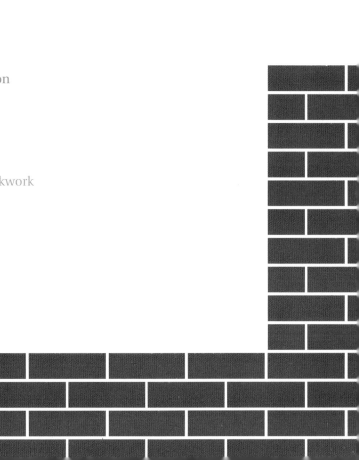

Thin joint masonry information

Continuing research has enabled leading block manufacturers to develop a thin joint mortar which in conjunction with a larger block size enables internal walls to be constructed more quickly.

This system is an alternative to using 10 mm sand and cement mortar bed and cross-joints.

It is essential that all parties have access to various sources of up-to-date information. It is also very important that the student understands where information can be found.

Information is available from various manufacturers, many of whom will provide on-site demonstrations.

Manufacturers' information

Any efficient office should have up-to-date information regarding new and existing products from the various manufacturers who produce materials of interest to them.

Many manufacturers produce technical information which is free for the asking. These papers, brochures, leaflets, etc. (Figure 9.1), should be stored in the office in an easily retrievable system.

Technical information can be produced in several formats. When items of equipment are purchased they will come with a manufacturer's information sheet.

The information may be:

- operating instructions – how to use the item
- safety guidelines – power supply, personal protective equipment (PPE) to be worn and recommended checks
- technical information – mechanical details and possible outputs.

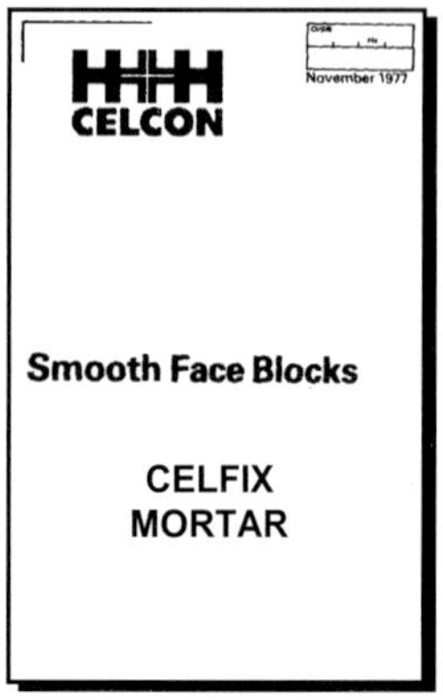
CELCON
November 1977
Smooth Face Blocks
CELFIX
MORTAR

FIGURE 9.1
Manufacturer's literature

Normal hand tools are not usually provided with manufacturer's instructions. Most textbooks explain hand tools in detail and the safety procedures to follow while using them.

Information on setting-out equipment is provided by the manufacturer to ensure that the item of equipment is used correctly.

Adopting safe and healthy working conditions

It is essential to wear the appropriate PPE when laying thin joint masonry, in the same way as for normal masonry.

When the thin joint system is used it is necessary to cut the blocks accurately to maintain the thin joint. The traditional method of bolster and club hammer will not give an accurate cut, so a saw or mechanical method should be used.

When components are cut by machine the equipment must be set up correctly and safely and the operator must have had sufficient training and acquired the required certificate. The appropriate PPE must be worn.

The main type of hand-held equipment for cutting building components is the angle grinder or disc cutter. A table version of a disc cutter is used in workshops and on building sites when there is a great deal of cutting to be carried out.

Please refer back to Figure 6.1 in Chapter 6 for further information.

Remember

It is the employer's responsibility to ensure that you are correctly trained to use power tools. The item of equipment provided must be maintained and safe.

Employees, in turn, have a duty to inform their employer of any work situation that presents a risk to themselves or their workmates.

Resources

As with any new system there will be several new materials to become familiar with. It is essential to practise using these new materials and methods before starting a new project.

Blocks

These are available in various shapes and sizes to cover all types of situation.

Blocks for thin joint masonry have sharp and clear arrises and excellent dimensional tolerances and are produced with a plain face.

Standard blocks are available from 200 mm upwards with a face size of 610 × 140 mm and are designed for solid wall and foundation construction.

Blocks are also produced in various thicknesses. They are suitable for the construction of solid and cavity walls and are available with a face size of 610 × 215 mm.

The blocks are produced to a tolerance to allow them to be used in thin joint systems and all cut blocks need to be cut to the exact size, allowing only 3 mm for the joints.

A masonry saw or mechanical saw may be used. Do not use a club hammer and bolster chisel as this would lead to joint widths greater than 3 mm.

Jumbo blocks are available in various thicknesses with a face size of 610 × 270 mm. They are suitable for the construction of cavity walls. These larger format blocks can increase the speed of construction while improving the thermal performance of the building by reducing cold bridging through the mortar. See Figure 9.2 for details.

Mortar

The key to all thin joint systems is the mortar. The mortar is designed specifically for use with the special blocks.

The dry mix should always be added to the water (approximately 5–6 litres of clean water per bag) and mixed until a thick, smooth consistency is achieved.

Note

Mortar can be remixed, but *do not add more water.*

THIN JOINT MORTAR

A thin joint mortar has been developed and is supplied as a dry premixed powder in 25 kg bags (Figure 9.3). This replaces the traditional 10 mm sand and cement mortar.

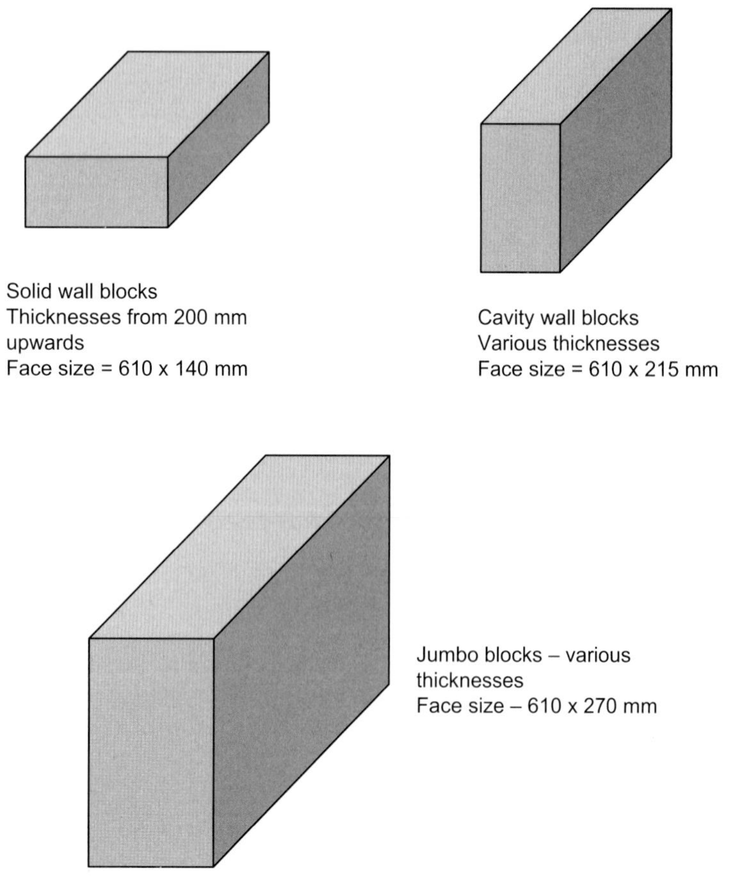
Solid wall blocks
Thicknesses from 200 mm upwards
Face size = 610 x 140 mm
Cavity wall blocks
Various thicknesses
Face size = 610 x 215 mm
Jumbo blocks – various thicknesses
Face size – 610 x 270 mm

FIGURE 9.2
Types of block

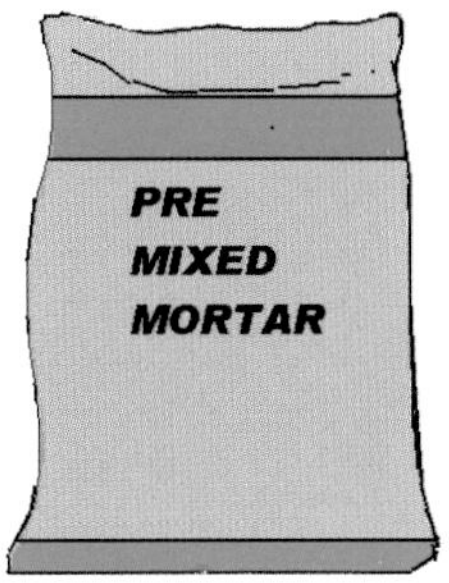
PRE
MIXED
MORTAR

FIGURE 9.3
Bagged dry mix

A special scoop is used, which creates a joint thickness of 2 mm.

The mortar will remain workable in the bucket or tub for a number of hours, but reaches initial bond strength on the block within 10–20 minutes and full design bond strength within 1–2 hours.

Tools and equipment

Before setting out any work the equipment should be carefully checked for accuracy. New tools and equipment may need to be purchased and practised with. A selection is shown in Figure 9.4. Other items that may be required can be found in Figure 9.5.

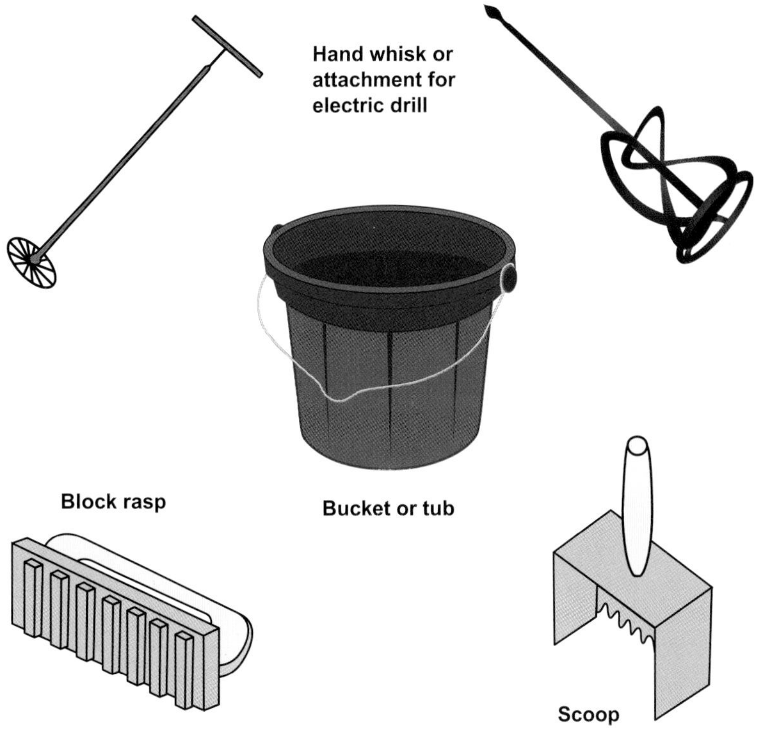
Hand whisk or
attachment for
electric drill
Block rasp
Bucket or tub
Scoop

FIGURE 9.4
Main tools and equipment required

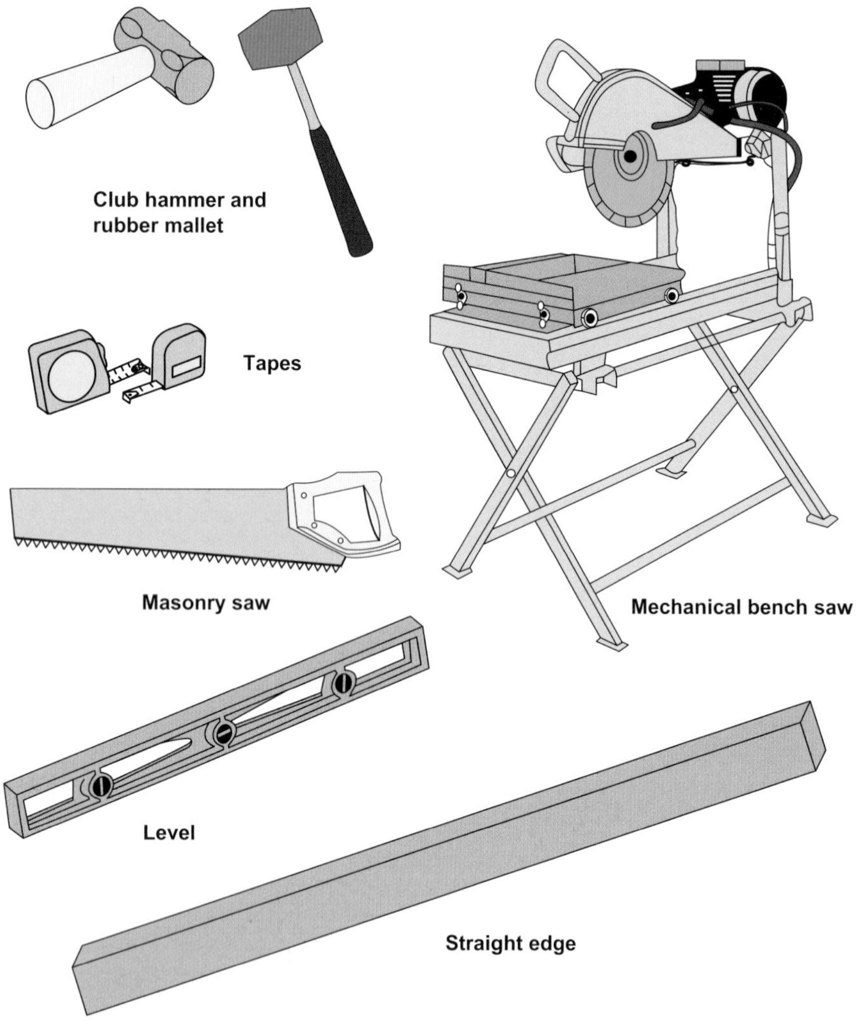
Club hammer and rubber mallet
Tapes
Masonry saw
Mechanical bench saw
Level
Straight edge

FIGURE 9.5
Other tools and equipment required

Thin joint blockwork

A revolutionary system of blockwork, the 'thin joint system', introduces a new material called aircrete. Aircrete blocks combine thermal and sound insulation qualities with load-bearing capabilities.

Aircrete blocks are also:

- highly resistant to the passage of water
- fire resistant, including surface spread
- frost resistant, with no reduction in freeze–thaw conditions
- resistant to sulphate attack up to class 4 soil conditions.

The thin coat system is in use throughout Europe.

Benefits of the thin joint system

- Speed:
  - reduced construction time
  - follow-on trades can start sooner
  - speed and ease of fixing secondary insulation
  - internal work can proceed while external work is being constructed.
- Quality:
  - improved thermal insulation
  - airtight construction
  - reduction in site waste
  - improved stability
  - accuracy of walls allows for thin coat plaster finishes
  - increased bond strength.
- Flexibility.

SPEED

The system enables the structure of a building to be constructed more quickly and to a better quality than traditional construction, allowing follow-on trades to start work sooner in a weatherproof environment, while retaining the flexibility of on-site construction.

The thin joint system allows construction times equivalent to off-site system build solutions.

The system gives increased bond strength together with ease of mixing and laying the mortar with precision-made blocks.

QUALITY

The thin joint system leads to improved quality of thermal insulation, stability and accuracy, and a reduction in waste.

FLEXIBILITY

As with traditional building methods, the construction is carried out on site. The system is easy to extend or adapt as the work proceeds.

Applications of thin joint systems

Thin joint systems can be used in the following constructions:

- cavity walls (inner and outer skins)
- solid external walls
- semi-exposed conditions

- partitions
- separating walls
- foundations
- industrial and commercial buildings.

CAVITY WALLS

Cavity walls in thin joint systems are built very differently from traditional block and brick cavities.

The inner skin can be completed without the insertion of traditional wall ties – these being too thick. This leaves the inner block wall clean and free from dangerous protruding ties.

When the external skin is built the insulation can be inserted and special helical wall ties can be driven into the thin joint system at the appropriate level.

The inner leaf can be built first, taking the outer leaf off the critical path of the building programme.

SOLID EXTERNAL WALLS

The superstructure of a dwelling can often be completed in 2–3 days.

The use of thin joint systems increases the U values by up to 20 per cent compared with conventional joints in solid walls, so the wall can meet thermal regulations without the need for additional insulation.

PARTITIONS

Excellent sound and fire insulation is provided by the thin joint system. With the speed of the build and mortar setting more quickly then traditional mortar, it is possible for the walls to be finished, i.e. plastered, without delay.

SEPARATING WALLS

Separating and flanking walls constructed with the thin joint system satisfy the sound insulation requirements of the current Building Regulations.

Building thin joint systems

It is assumed that the candidate has previous knowledge of traditional blockwork.

The approach to using the thin joint system is different from that for the sand and cement system. As mentioned previously, the blocks are specially made to a tolerance to allow them to be used in thin joint systems with a 2 mm joint. The stability of the walls is greater than with sand and cement; therefore greater height can be reached with very little or no support required.

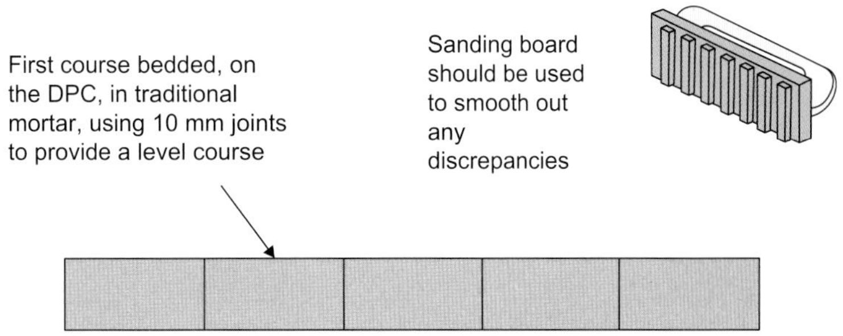
First course bedded, on the DPC, in traditional mortar, using 10 mm joints to provide a level course
Sanding board should be used to smooth out any discrepancies

FIGURE 9.6
First course should be perfectly level

Whichever method of foundation has been used under the thin joint system the first course of thin joint blockwork is very important. With traditional foundations the first course is usually laid onto the damp-proof course (DPC) using traditional sand and cement mortar to achieve a level datum to work to.

The first course is the most important, as height cannot be gained or reduced with thin joint mortar as easily as with traditional mortar. This course should be allowed to set overnight. If it is not perfectly level; any irregularities can be sanded down using a sanding board or rasp (Figure 9.6).

If the first course is laid on a level concrete raft, the thin joint system can be used, ensuring that it is true to level and plumb. It is essential that the blocks are laid accurately, true to level (± 1 mm) and vertical.

Once the first course has been laid, allowed to set and made perfectly level, the quoins can be erected as for normal blockwork, keeping to half-bond where possible. The recommended method is to lay three courses at a time (Figure 9.7).

Building with the thin joint system is a very fast and easy process. The skills required to build with the thin joint system are very similar to those for traditional blockwork, but regular checks must be made for level, plumb and line. It is difficult to make any adjustments to height as the work proceeds.

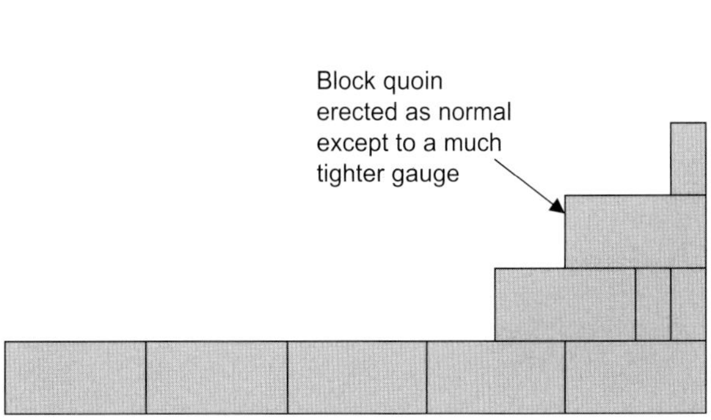
Block quoin erected as normal except to a much tighter gauge

FIGURE 9.7
Erecting a thin joint quoin

Before applying any mortar ensure that all surfaces are clear of loose materials. Avoid using wet blocks as this slows down the setting time.

PLUMBING BLOCKWORK

Take care when plumbing thin joint blockwork as the wall rises more quickly and the mortar sets more rapidly than when using traditional blockwork.

On no account tap the blocks sideways – this will only result in the bed joint opening up on one side. Always plumb the wall by tapping down on the high side of the block (Figure 9.8).

LAYING TO LINE

Again this procedure is the same as for normal blockwork but the gauge is much less: 2 mm joints rather than 10 mm (Figure 9.9).

BONDING

Half-bond should be used as much as possible. When wall lengths do not work exactly to block lengths then broken or reverse bond should be used.

Broken bond

Broken bond may have to be adopted when wall lengths are not equal to block sizes. Most block walls will require some use of broken bond.

It is important never to use cuts less than half a block as this can create a weak plane in the wall and if drying shrinkage occurs it may result in cracks appearing in the wall (Figure 9.10).

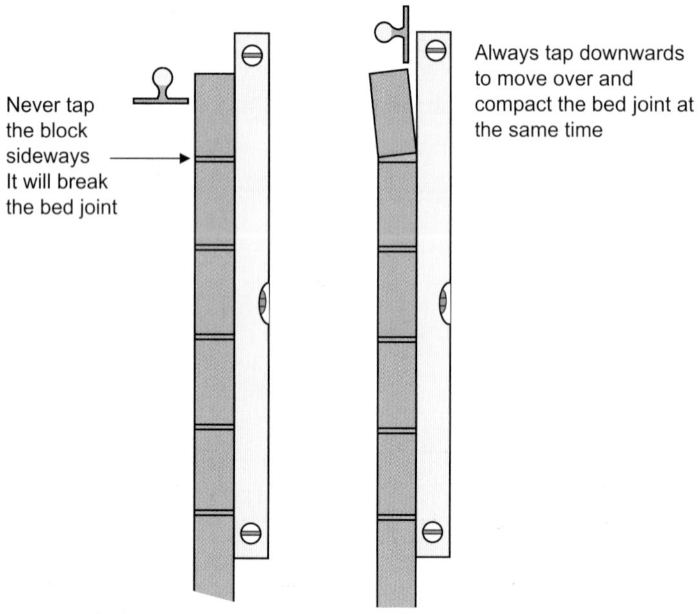
Never tap
the block
sideways
It will break
the bed joint
Always tap downwards
to move over and
compact the bed joint at
the same time

FIGURE 9.8
Plumbing thin joint blockwork

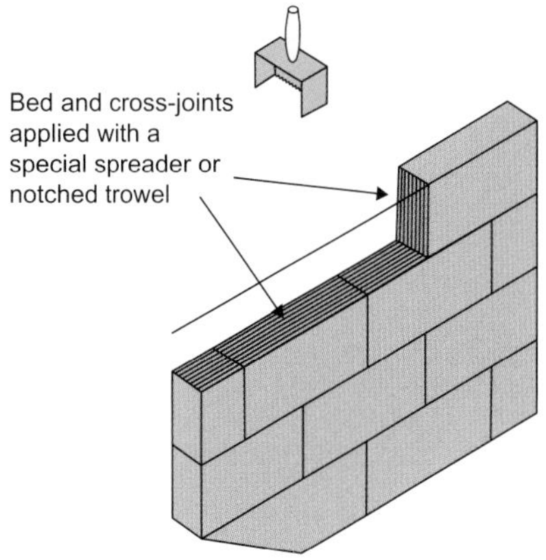
Bed and cross-joints applied with a special spreader or notched trowel

FIGURE 9.9
Laying to line

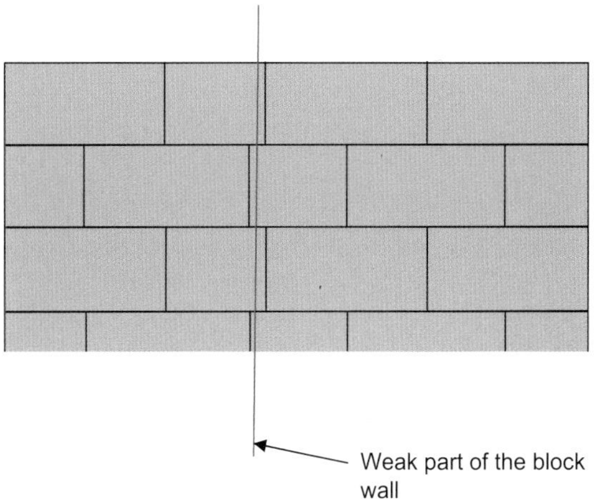
Weak part of the block wall

FIGURE 9.10
Broken bond

Blocks can be cut with a club hammer and bolster chisel, a masonry saw or a mechanical bench saw (Figure 9.5). Try to make use of any available special blocks before starting to cut. Always try to achieve a neat cut that retains the 10 mm cross-joint. Joints that are too large can again result in cracks appearing in the wall due to drying shrinkage.

Broken bonds of less than half-lap can be avoided using blocks (Figure 9.11).

Reverse bond

In reverse bond the blocks at each end of the wall are different. The front elevation in Figure 9.12 shows reverse bond being used.

When cut blocks are being used it is more economical to use three-quarter, half- and quarter blocks provided by the manufacturer. Cut blocks should always be of the same material and must be accurately cut and smoothed down.

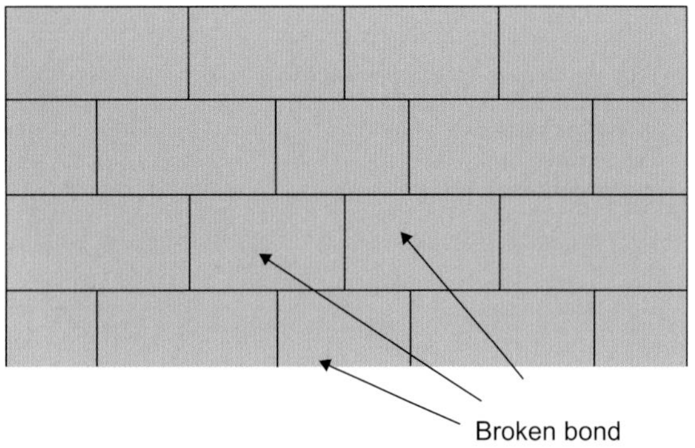
Broken bond

FIGURE 9.11
Broken bond

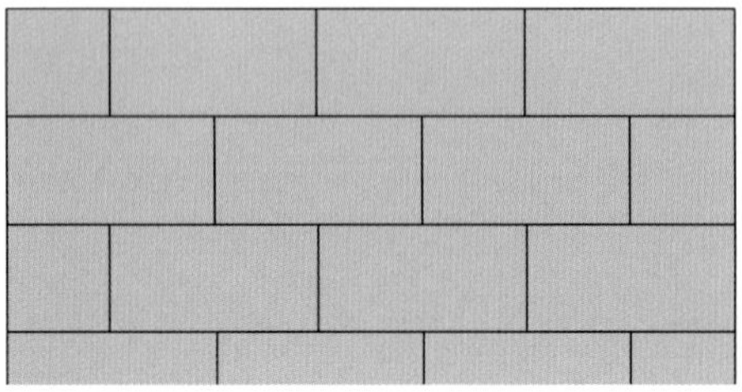
End block different to opposite end block to avoid broken bond

FIGURE 9.12
Reverse bond

JUNCTION WALLS

It is often inconvenient to build junction walls at the same time as the main wall, so there will be a need during construction to make provision for work at a later date.

Indents can be left in the main wall to receive the junction wall later. These indents should be marked out correctly and kept plumb throughout the building of the block wall. The minimum lap at T junctions should be quarter lap. An allowance of 20 mm is usually made over the width of the block to allow the blocks on the junction wall easy access into the indent (Figure 9.13).

Another acceptable method of providing a tie to junction walls uses flat strip shear ties (Figure 9.13). These are built into the main wall on every alternate bed joint. If the inner leaf has been completed before the outer leaf then helical ties can be driven through the inner leaf into the junction wall by at least 50 mm (Figure 9.14).

These methods can help to avoid a lot of cutting at a junction of two block walls.

CAVITY WALLS

The main benefit of constructing the inner blockwork leaf first is that a weathertight environment is created in which follow-on trades can work.

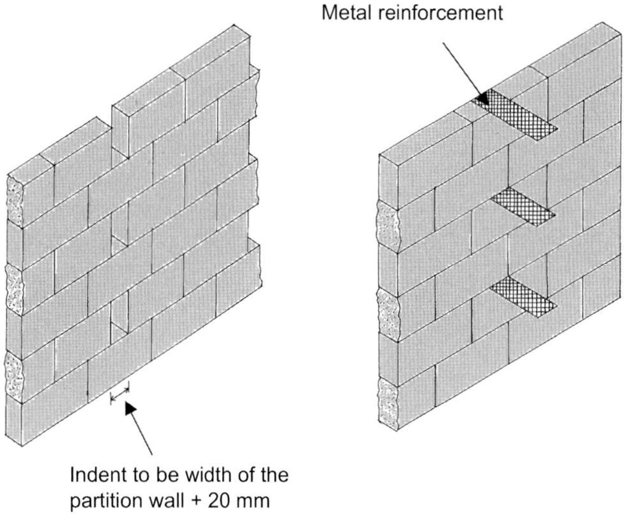
Metal reinforcement
Indent to be width of the partition wall + 20 mm

FIGURE 9.13
Junction details

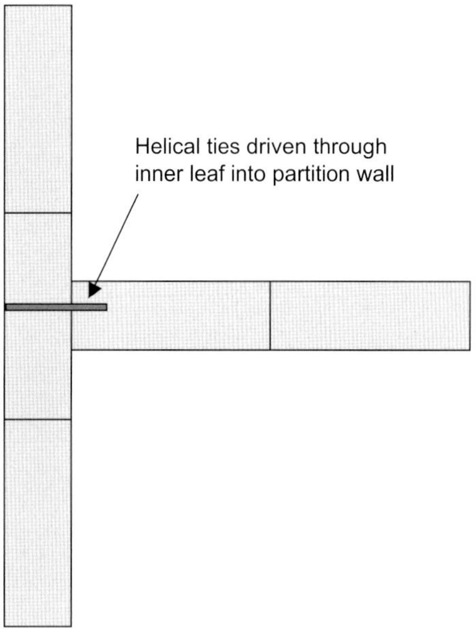
Helical ties driven through inner leaf into partition wall

FIGURE 9.14
Pinning junction wall

The inner leaf is erected by building corners and running in as for normal blockwork. The mortar bed should remain workable for 6–9 minutes and sets within 10 minutes. Any minor adjustments should be made as soon as possible.

Cross-joints can be applied with the same equipment or the blocks could be dipped into the mortar to apply the cross-joint.

As laying proceeds, blocks should be pressed firmly against the mortared face of the preceding block while being lowered onto the mortared bed below (Figure 9.15).

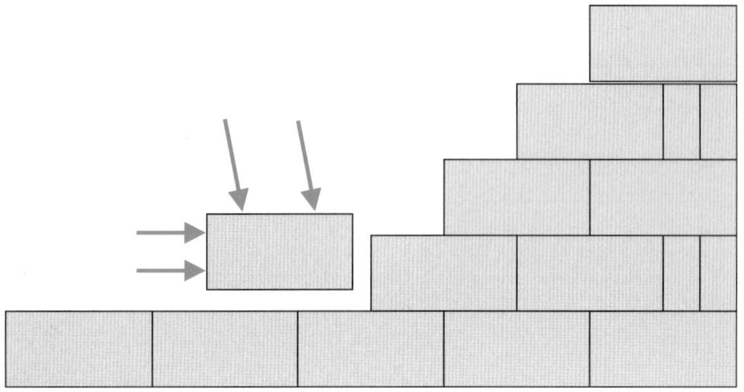

FIGURE 9.15
Laying thin joint blocks

To assist in levelling, a rubber hammer or mallet can be used to tap the block into position, especially to close the perpend joints, and reduce mortar thickness. This will also help to ensure that all the joints are fully filled.

Any excess mortar should be removed from the face of the blocks.

Always stack blocks adjacent to the wall being built for increased productivity. Building to profiles is recommended to improve the accuracy of level, line and speed (Figure 9.16).

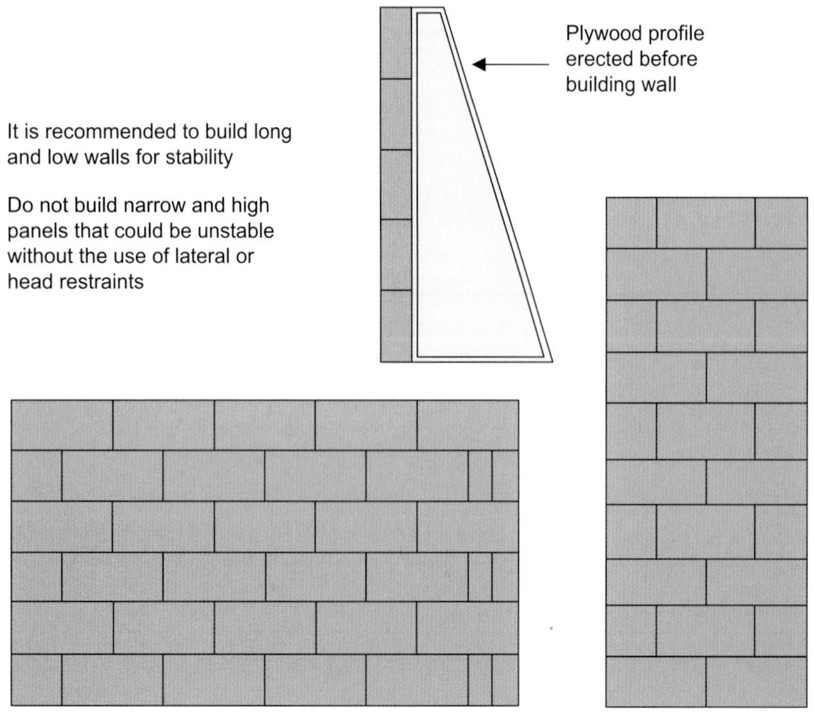
Plywood profile erected before building wall
It is recommended to build long and low walls for stability
Do not build narrow and high panels that could be unstable without the use of lateral or head restraints

FIGURE 9.16
Maintaining stability

WALL TIES

Because of the difference in gauge traditional wall ties cannot be used.

If the inner wall is erected before the outer leaf then wall ties are not required. This is an added safety feature as no wall ties will be projecting during the erection of the outer leaf. The insulation and cavity wall ties can be fitted when the external leaf is erected.

One of the main differences from conventional building methods is the height to which the wall can be built. The inner leaf can be built to wall plate level and the inner work can continue while the outer leaf of the cavity wall is being built.

Wall ties are placed in the normal positions, but owing to the bed joint not coinciding, special helical wall ties (Figure 9.17) are driven into the face of the blocks.

INSULATION

Insulation can be placed against the inner blockwork and the wall ties driven through the insulation. If partial insulation is used, special clips are inserted as normal. It is important to ensure the block wall has set and is stable before driving in wall ties.

The benefits of thin wall system can be greatly enhanced if jumbo block units are used. These are available as blocks 440 mm long and 430 mm high, although sizes vary between manufacturers.

The current Building Regulations require that heat loss through mortar joints must be taken into consideration when calculating the U value for walls. The use of thin joint masonry minimizes the effect of heat loss as a result of the 2 mm bed and cross-joints.

If the inner leaf has been completed before the outer leaf the cavity is kept cleaner and also the insulation is cleaner.

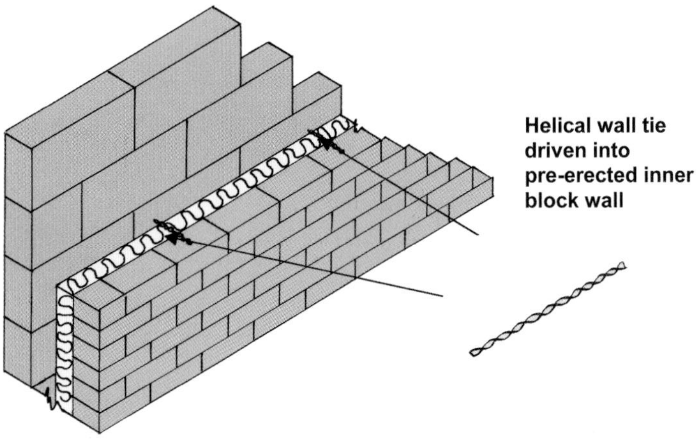
Helical wall tie
driven into
pre-erected inner
block wall

FIGURE 9.17
Helical wall tie

MOVEMENT JOINTS

Provision for movement should be considered at the design stage.

Walls in excess of 6 m long should be designed as a series of panels separated by movement joints at not more than 6 m centres. The position of the first movement joint should be within 3 m of a fixed end or corner.

Wall panels are built butting against the next panel, leaving a dry joint or a 10 mm gap that can be filled with fibreboard.

Where lateral stability demands continuity across the joint, proprietary flat strip shear ties should be set in at 450 mm maximum vertical centres (Figure 9.18). These should be set parallel to the plane of the wall, allowing the plain end to become debonded if movement occurs within the blockwork panel, thus creating a movement joint.

REINFORCEMENT

For areas where high stress is expected, such as openings or where blocks are under concentrated imposed loads, horizontal bed-joint reinforcement is recommended. It should be of adequate length to distribute stresses to nearby movement joints or into adjacent blockwork on each side of the opening.

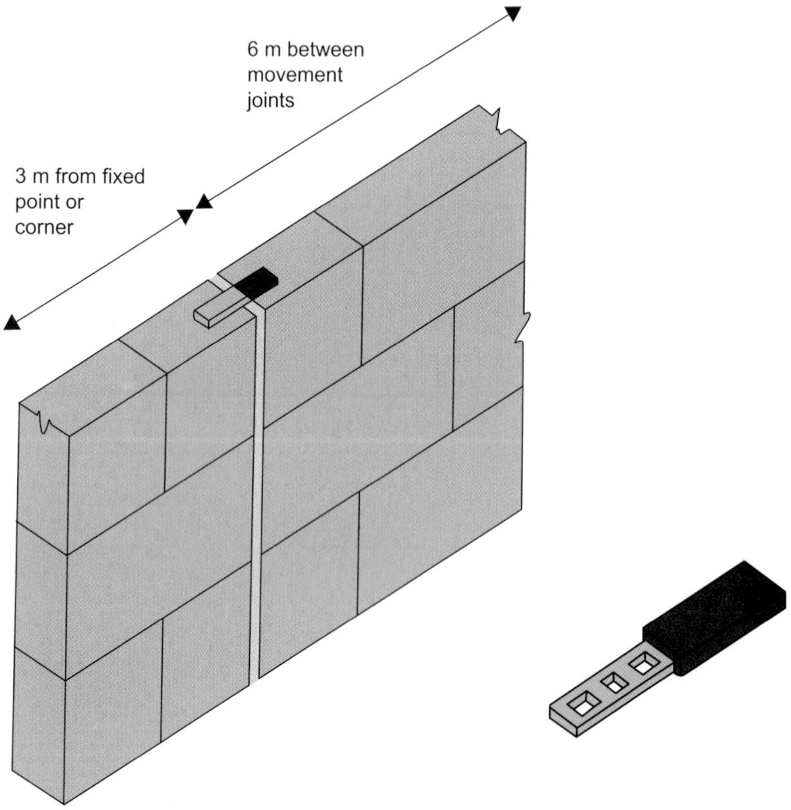
6 m between
movement
joints
3 m from fixed
point or
corner

FIGURE 9.18
Movement joints and shear tie

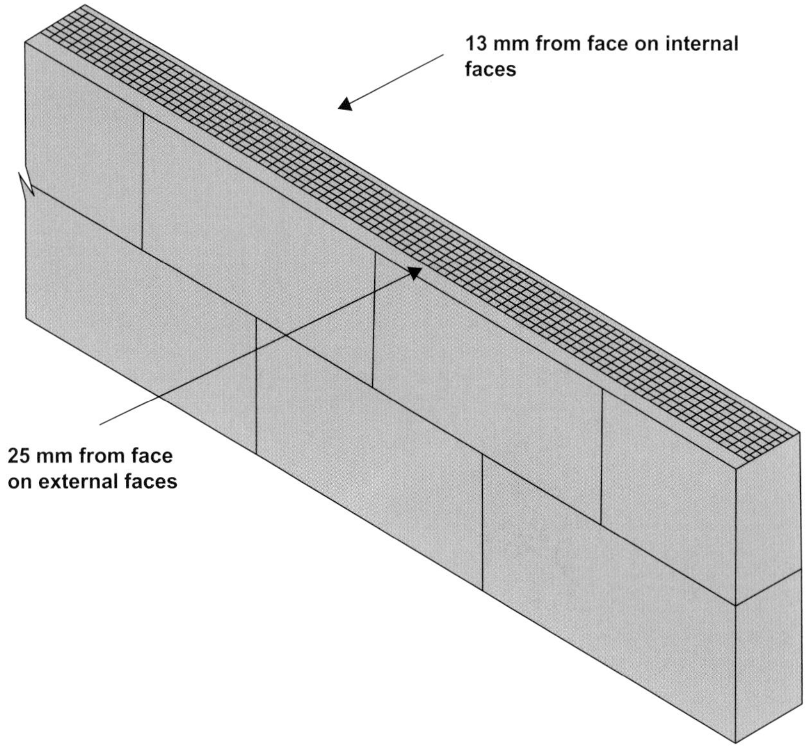
13 mm from face on internal faces
25 mm from face on external faces

FIGURE 9.19
Reinforcement

Remember

Only bed-joint reinforcement that has been specifically manufactured for use with thin layer mortar masonry should be used with the Celcon thin joint system.

Reinforcement should normally have a mortar cover of a minimum 13 mm for internal faces and 25 mm for external faces (Figure 9.19).

The reinforcement should be laid on the blocks and then covered with a scoop of mortar. The next course of blocks should be laid while the mortar is still wet.

Reinforcement should be overlapped by a minimum of 100 mm where required.

OPENINGS

When forming window openings, the blockwork can be taken up in full-block heights and then cut blocks used to build up to sill level, lintel level and wall plate level.

LINTELS

Several types of lintel can be used with thin joint systems, including steel box and combined lintels (Figure 9.20).

For cavity wall construction the box lintel is recommended. If a combined lintel is planned then the lintel will require propping until the outer leaf has been built to support the lintel and prevent it rotating.

For openings in solid walls the combined lintel is recommended.

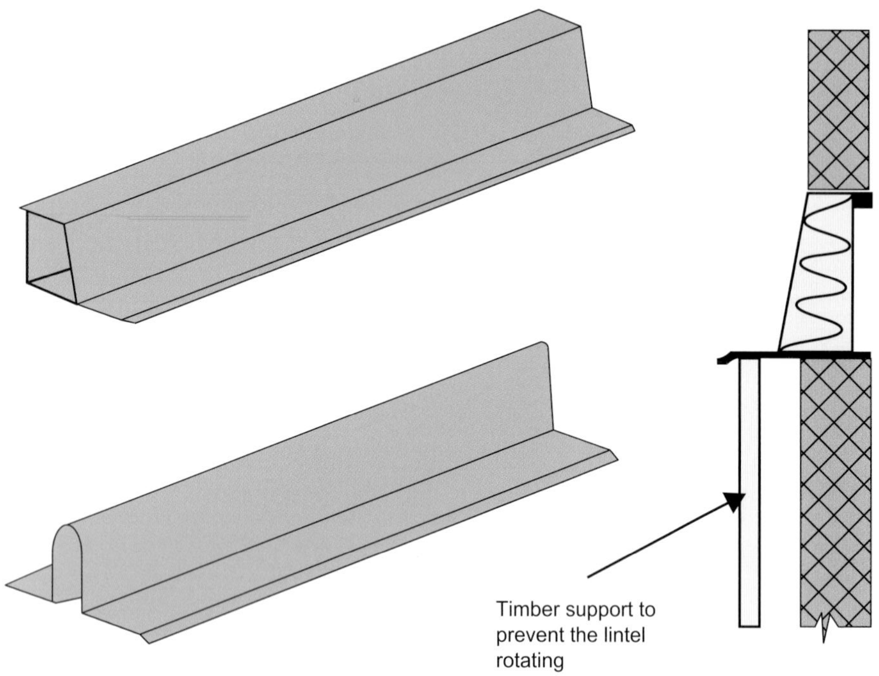
Timber support to
prevent the lintel
rotating

FIGURE 9.20
Lintels for cavity walls

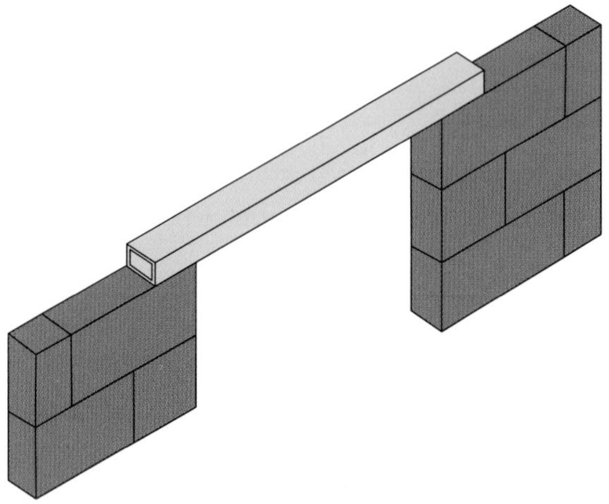

FIGURE 9.21
Lintels for internal walls

Openings in internal walls can be bridged with various types of lintel (Figure 9.21).

DAMP-PROOF COURSES

These are positioned as for traditional blockwork.

Multiple-choice questions

Self-assessment

This section of the book is designed to allow you to check your level of knowledge. The section consists of revision questions for this chapter. The questions are all multiple choice and have four possible answers. The answers are to be found at the end of the book.

The main type of multiple-choice question will be the four-option multiple-choice question. This will consist of a question or statement, known as the stem, followed by a choice of four different answers, called the responses. Only one of these responses is the correct answer; the others are incorrect and are known as distracters.

You should attempt to answer the questions by choosing either (a), (b), (c) or (d).

Example

The person employed by the local authority to ensure that the Building Regulations are observed is called the:

(a) clerk of works
(b) building control officer
(c) council inspector
(d) safety officer

The correct answer is the building control officer, and therefore (b) would be the correct response.

Thin joint masonry

Question 1 Identify the following item used with thin joint masonry:

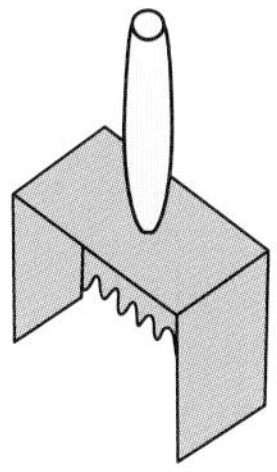

(a) trowel
(b) scoop
(c) template
(d) box square

Question 2 How is the thin joint mortar mixed?

(a) by machine in a cement mixer

(b) by hand with a shovel

(c) by hand in a bucket

(d) delivered to site ready mixed

Question 3 Which of the following bonds is used for thin joint masonry?

(a) English bond

(b) Flemish bond

(c) Dutch bond

(d) stretcher bond

Question 4 Which of the following is the recommended bed joint thickness?

(a) 2 mm

(b) 4 mm

(c) 6 mm

(d) 10 mm

Question 5 Identify the following wall tie:

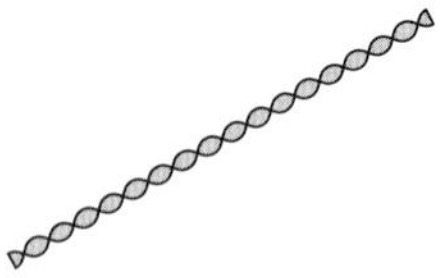

(a) helical tie

(b) spiral tie

(c) twisted tie

(d) steel tie

Question 6 Compared with traditional blockwork, what is the main benefit of thin joint masonry?

(a) cheaper to build

(b) easier to build

(c) slower to build

(d) faster to build

Question 7 Where should a shear tie be used?

(a) at a junction with another wall

(b) at a corner

(c) across a movement joint

(d) across the cavity

Question 8 How are wall ties fixed to tie brick outer leaves to inner leaves in thin joint masonry?

(a) placed on a corresponding course

(b) driven into the thin block wall

(c) screwed to the thin block wall

(d) stuck to the thin block wall

CHAPTER 10

Bridging Openings

This chapter will cover the following NVQ and Diploma units:

- NVQ VR40
- CC 2048K

This chapter is about:

- Interpreting building information
- Adopting safe and healthy working practices
- Selecting materials, components and equipment
- Preparing and erecting brickwork and blockwork structures

The following NVQ performance criteria will be covered:

- Performance criterion 1: Interpretation of information
- Performance criterion 2: Safe work practices
- Performance criterion 3: Selection of resources
- Performance criterion 4: Minimize the risk of damage
- Performance criterion 5: Meet the contract specification
- Performance criterion 6: Allocated time

The following Diploma outcomes will be covered:

- Know how to plan and select resources for practical tasks
- Know how to form openings in cavity walls

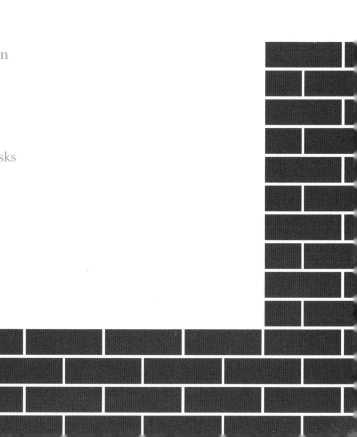

Building information

Whenever openings are formed some form of structure is needed to bridge the opening. This chapter deals with bridging openings with lintels and arches.

It is important that the student can understand drawings and schedules to be able to understand which method is used and where.

A typical lintel schedule is shown in Table 10.1. This explains to the builder what type of lintel is required and where it is fitted, and allows the builder to preorder the correct lintels for the works.

If the openings are to be bridged with arches the builder will require special arch centres for the bricklayers to construct the arches as the work proceeds.

Description and brief history

All buildings have openings of some description to allow access and egress and admit light and ventilation. Whenever an opening is made in brick or blockwork walls a weakness could occur in the walls if the work is not carried out correctly.

Today's modern buildings include larger openings than before so it is essential to take extra care when forming the opening. The most appropriate method of bridging the opening must be selected to maintain the maximum amount of stability and strength in the wall.

These openings have to be bridged to support the walling above in a way that will not cause any cracking to the structure either above or below the opening. There are several methods available, the simplest of which is to use a lintel. The lintel carries the weight of the load above the opening and distributes this load to the abutments either side.

Early civilizations tried various ways of supporting brick and stone masonry across openings in their buildings (Figures 10.1 and 10.2). The

Table 10.1 Lintel schedule

Job Number		Site Address		
Location	Position and size	Type		
		Steel boxed lintel	Precast concrete lintel	Corrugated steel
Dining	External 1.500	✓		
	Internal 1.100			✓
Lounge	External 2.000	✓		
	Internal 1.100			✓
Kitchen	External 1.500	✓		
	Internal 1.100			✓
Study	External 2.000	✓		
	Internal 1.100			✓
Garage	External 1.500	✓		
	Internal 1.100		✓	

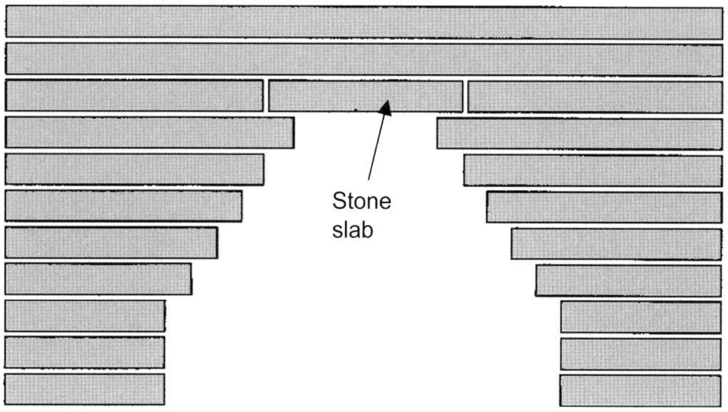
Stone
slab

FIGURE 10.1
Spanning an opening by corbelling

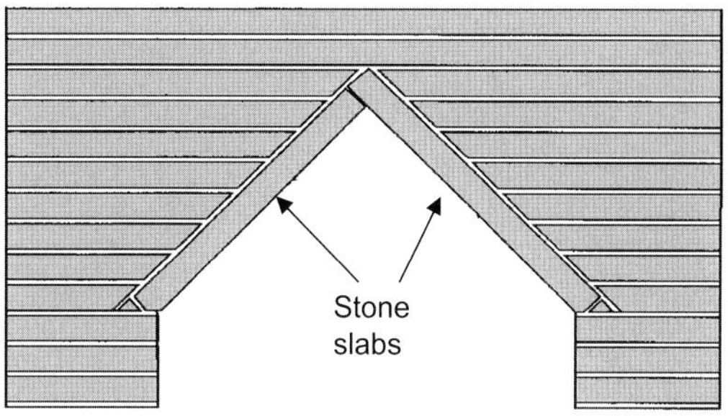
Stone
slabs

FIGURE 10.2
Alternative early method of spanning an opening

pyramid-building Egyptians and ancient Greeks used lots of columns and stone lintels, but it was the later Roman builders who developed the idea of forming arches from separate blocks of stone, bricks or tiles mortared together.

Properly bonded brickwork can bridge across openings in a series of offsets (Figure 10.3). This is called natural bracketing and means that in theory, only the area of brickwork within the 60 degree triangular shape in

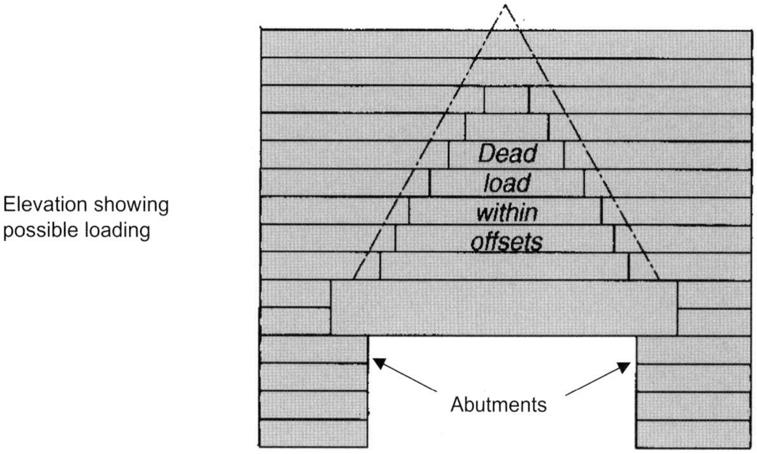
Elevation showing
possible loading
Dead
load
within
offsets
Abutments

FIGURE 10.3
Natural bracketing of bonded brickwork above a lintel

Figure 10.3 needs supporting by a lintel. The greater the span of an opening, the larger this triangular piece will be.

Lintels

Definition

Lintels are straight beams of concrete or steel that are made strong enough to support the dead load of bricks or blocks above. Because lintels are straight, they have a slight tendency to bend or deflect when bricks or blocks are bedded on top. This loading results in tensile stress in the lower part of a lintel, plus compressive stress in the upper part (Figure 10.4).

Steel rods are cast into lower portion of concrete lintels to absorb this tension, so that the lintel does not crack on the underside when loaded. Concrete in the upper part of the lintel is very good at resisting compressive stress.

Reinforced concrete lintels

Concrete lintels can be made in any one of five ways, depending on the following site considerations:

- How large do the lintels need to be?
- How much will each weigh (mass in kg)?
- How are they to be moved around site and to upper floors?
- Are they to be lifted into place by hand or mechanically?
- Will they be seen in the finished work, or plastered?

PRECAST LINTELS

These lintels are made in a mould at ground level. When the concrete has hardened for at least a week, they are lifted and bedded in place.

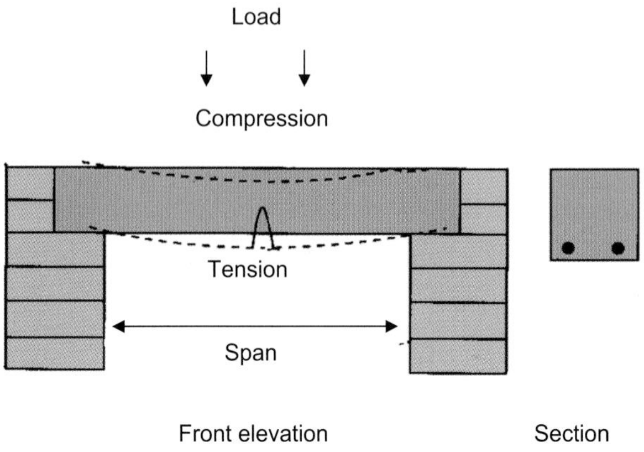
Load
Compression
Tension
Span
Front elevation
Section

FIGURE 10.4
Typical stresses in a lintel

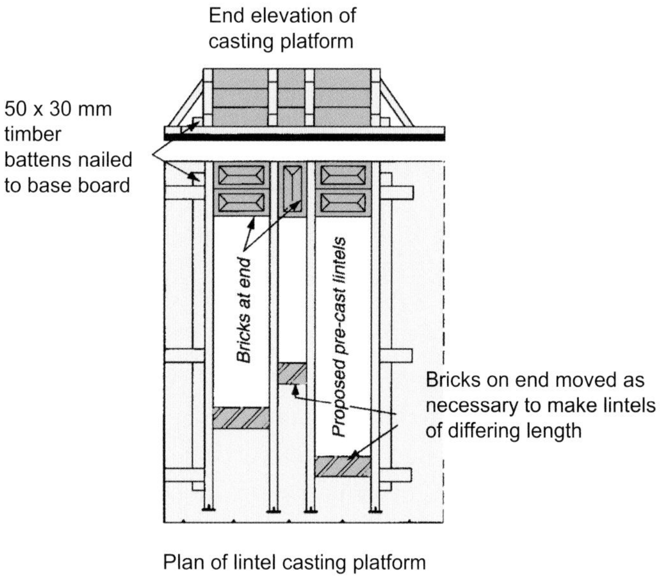
End elevation of casting platform
50 x 30 mm timber battens nailed to base board
Bricks at end
Proposed pre-cast lintels
Bricks on end moved as necessary to make lintels of differing length
Plan of lintel casting platform

FIGURE 10.5
Method of casting lintels on site

Figure 10.5 shows a typical site casting platform, set perfectly level both ways. Bricks or blocks are used as spacers, to obtain lintels of exactly the right width.

The inside surfaces of these lintel boxes should be lightly coated with mould oil to make dismantling or 'striking' easier. When filling the mould a 25–30 mm layer of wet concrete should first be spread over the bottom before placing reinforcing steel bars.

The remainder of the concrete should be added and thoroughly compacted by manual or mechanical vibration.

The upper surfaces should be trowelled smooth and marked clearly as the 'Top' surface, so that the steel bars will be towards the soffit when permanently bedded (Figure 10.6).

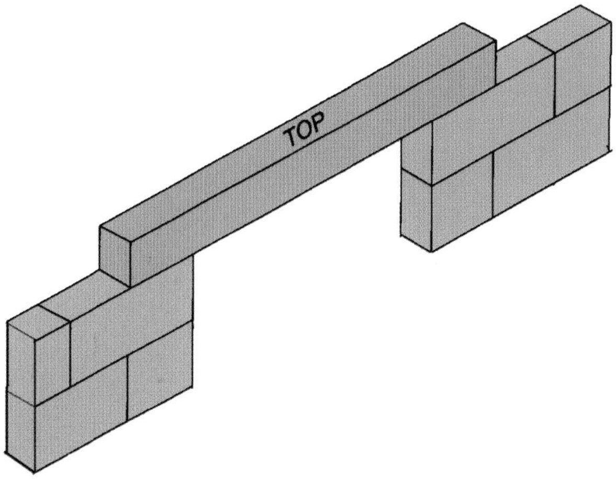
TOP

FIGURE 10.6
Precast reinforced concrete lintel in position

FACTORY-MADE REINFORCED CONCRETE LINTELS

For larger requirements precast reinforced concrete lintels can be ordered from a specialist supplier to suit the range of openings in a building. Care should be taken that a fair face is specified where lintels will be permanently exposed, and that the top surface of all lintels delivered to site is clearly marked (Figure 10.6).

PRESTRESSED CONCRETE PLANK LINTELS

These special reinforced concrete lintels can only be made in a factory.

The high-tensile steel strands are stretched by hydraulic jacks before the concrete is compacted in the lintel mould. When the concrete has thoroughly hardened, the stretched steel reinforcement is carefully released and keeps the whole lintel section in a permanent state of compression. These lintels are known as plank lintels because they are only 65 mm deep for all spans up to 1800 mm (Figure 10.7).

CAST IN SITU LINTELS

These are used in cases where a lintel would be much too heavy to lift by hand if cast at ground level. Instead, a timber box or mould is constructed exactly where the lintel is required. This formwork (Figure 10.8) is made strong enough to withstand the pressures of filling with wet concrete and thoroughly compacting it.

The concrete is poured around steel reinforcing bars placed approximately 25 mm up from the soffit (Figure 10.9).

BEDDING LINTELS

Lintels must be set upon a wet mortar bed joint at each side, so as to spread the load evenly over the whole bearing surface. When tapping down into position the spirit level should be held against the underside of the lintel in case the top edge is not parallel with the soffit.

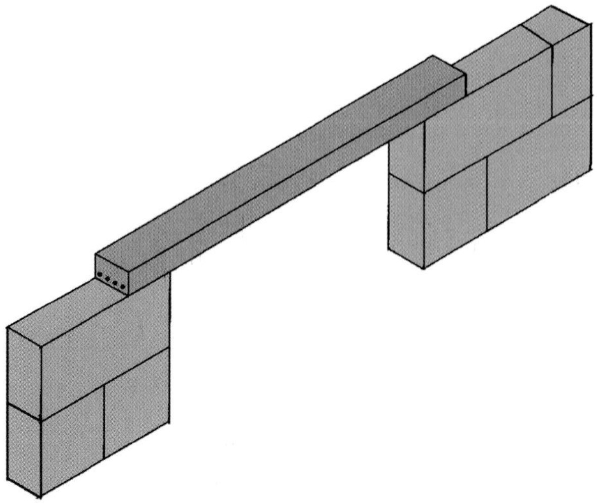

FIGURE 10.7
Prestressed concrete plank lintel in position

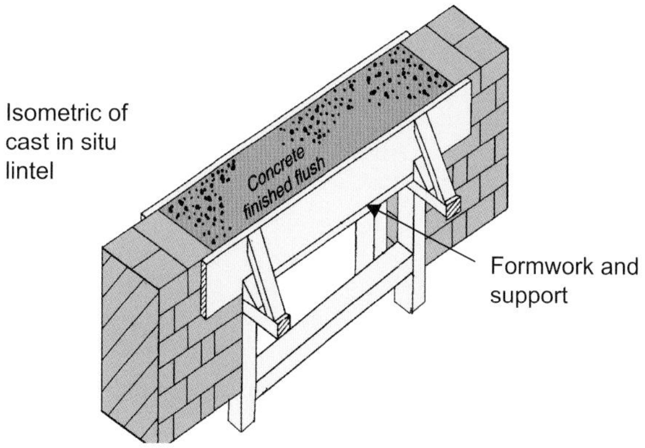
Isometric of
cast in situ
lintel
Concrete
finished flush
Formwork and
support

FIGURE 10.8
Formwork for a cast in situ lintel

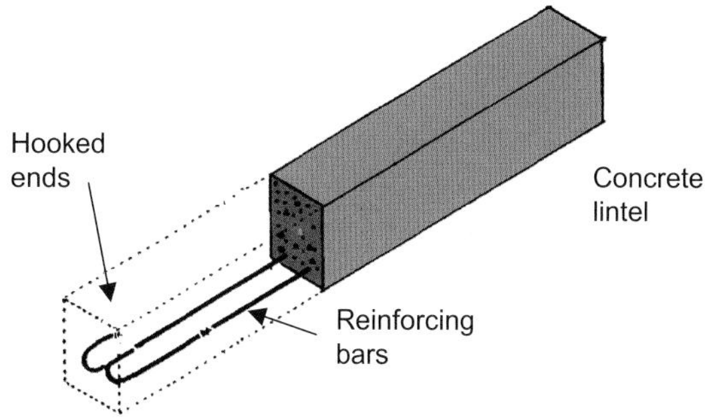
Hooked
ends
Concrete
lintel
Reinforcing
bars

FIGURE 10.9
Reinforcement in lower part of lintel

Steel lintels

Sheet steel can be pressed into shape in a factory to make lintels that are much lighter to lift than concrete, and which can withstand both the tensile and compressive stresses.

These are available in several shapes and sizes.

INTERNAL STEEL LINTELS

Internal steel lintels are designed mainly for 100 mm walls but other sizes are available. They can be boxed, corrugated or channelled (Figures 10.10–10.12).

EXTERNAL STEEL LINTELS

External steel lintels are designed mainly for cavity walls and are available in various designs (Figures 10.13–10.15).

LINTEL BEARINGS

A solid seating or bearing is required at each end of a lintel, to support the concentrated pressure of one half of the total load resting on the lintel.

This bearing at each side should be a minimum of 150 mm for lintel spans up to 1800 mm, with 225 mm minimum up to a 3 m span. A structural

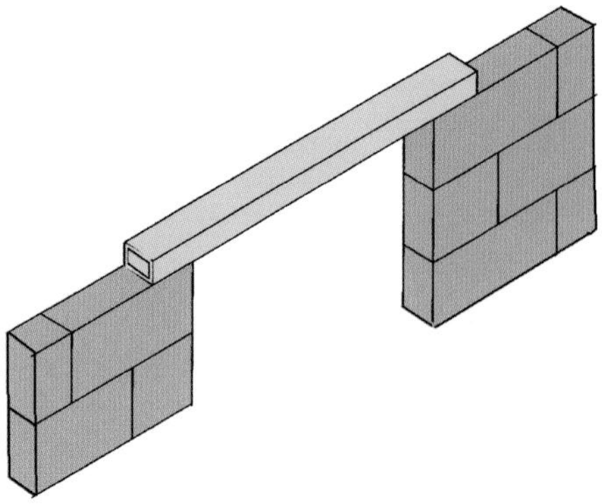

FIGURE 10.10
Boxed steel internal lintel

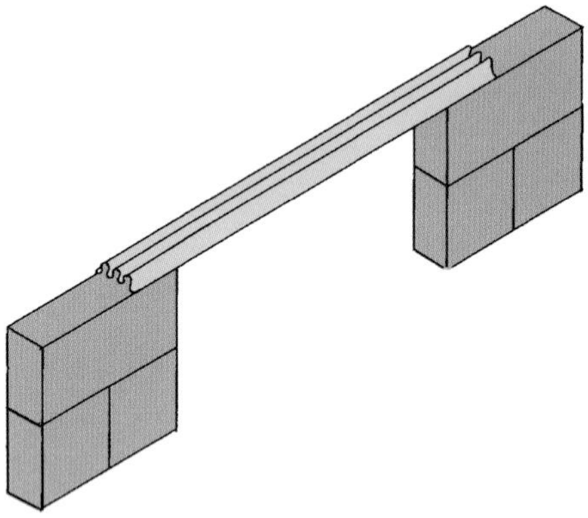

FIGURE 10.11
Corrugated steel internal lintel

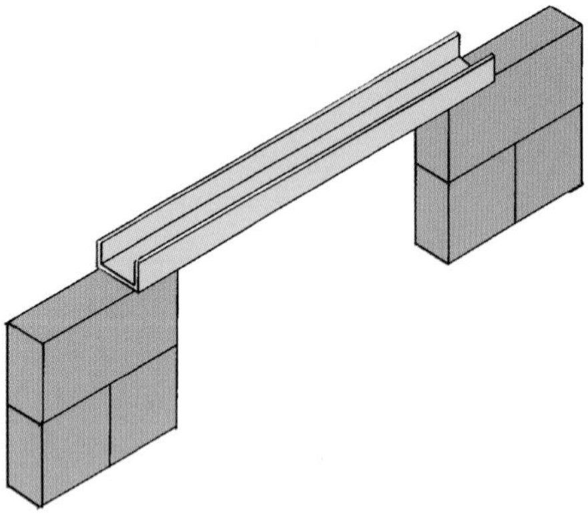

FIGURE 10.12
Channelled steel internal lintel

engineer will be required to calculate the safe bearing area needed for larger span lintels. BS5628:1985 Part 3, 'Workmanship', recommends that this bearing level should coincide with a whole block rather than a half or cut block, for greater stability where lintels are set in block walling. See Figure 10.16 and all previous drawings.

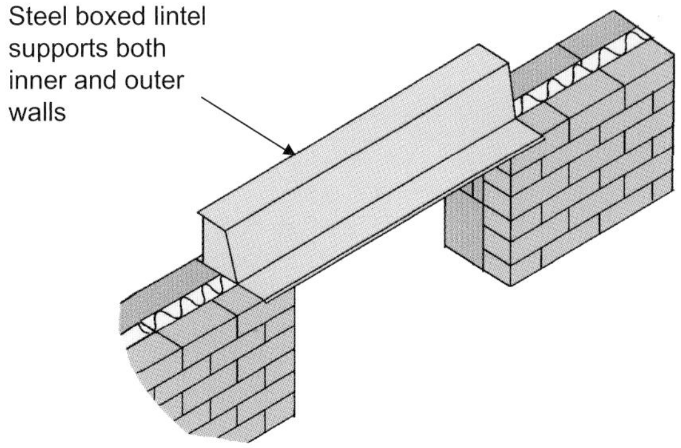
Steel boxed lintel
supports both
inner and outer
walls

FIGURE 10.13
Steel section lintel for cavity walls

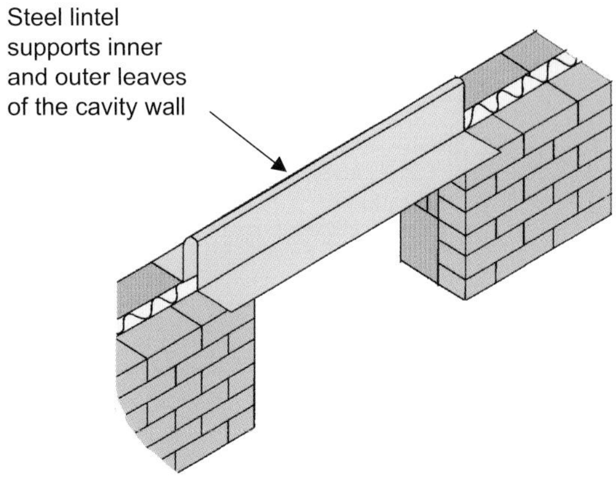
Steel lintel
supports inner
and outer leaves
of the cavity wall

FIGURE 10.14
Bent steel lintel for cavity walls

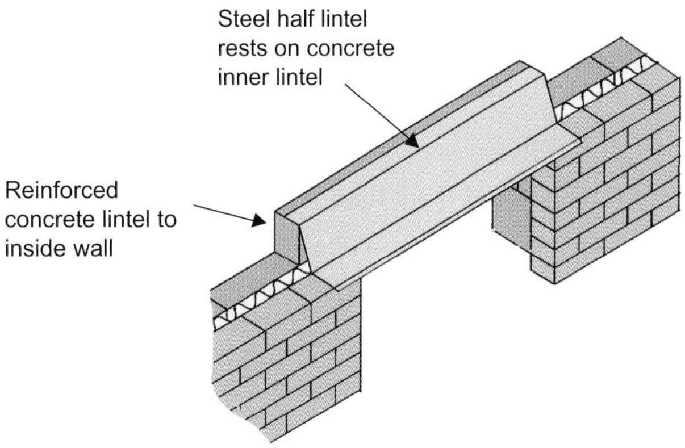
Steel half lintel
rests on concrete
inner lintel
Reinforced
concrete lintel to
inside wall

FIGURE 10.15
Half steel lintel for cavity walls

Reinforced brick lintel

The bricklayer calls this a soldier arch, as the bricks are placed on end, resembling a file of soldiers.

Care should be taken to avoid laying the bricks at an angle. A small boat level should be used to check the work, as explained in Chapter 6 for soldier courses (Figure 10.17).

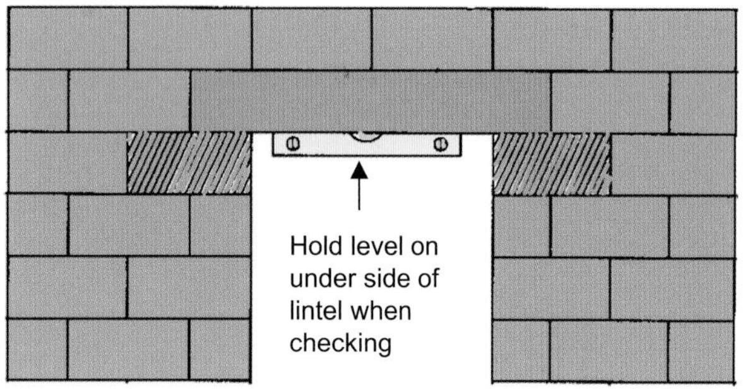
Hold level on
under side of
lintel when
checking

FIGURE 10.16
Whole block under lintel bearings to spread the load

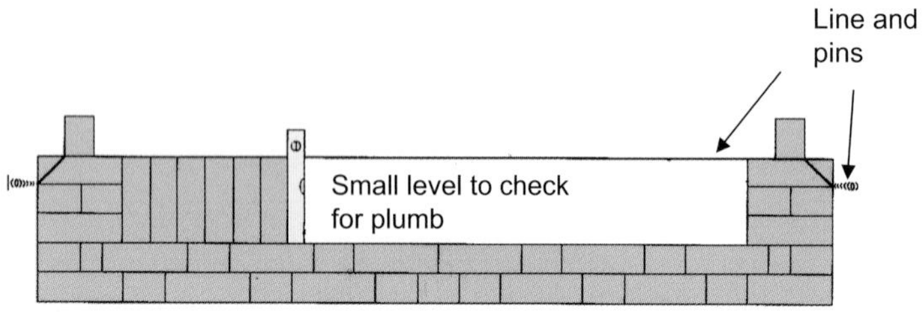
Line and
pins
Small level to check
for plumb

FIGURE 10.17
Checking soldier course bricks

Before setting the lintel, temporary supports are placed (Figure 10.18). Several methods of reinforcing brick lintels are shown in Figure 10.18: at 'A', where a 225 mm soffit occurs, steel bar reinforcement is placed throughout the length of the lintel, with stainless steel ties positioned at intervals of approximately 225 mm; at 'B' the brick lintel is erected with wire reinforcement built in and left projecting, to allow the concrete lintel to be cast in situ around it, thus giving adequate stability.

This type of lintel, if erected truly horizontal and plumb, causes an optical illusion to occur at a distance of 3 m or more from it; it appears to sag. This appearance can be overcome by giving the lintel a slight camber or by giving a curvature to the soffit along its length, together with a slight skewback. The adjustment must be made carefully and with the consent of the architect.

Brick arches

Arch construction is a more decorative means of spanning openings. True arches are made to curve upwards when looked at in elevation, so that they are always in a state of compression, wedged between abutments. As bending under load cannot take place, there will be no tensile stress in the arch.

All arches are formed on a temporary support called an arch centre, which must not be removed until the jointing mortar has been allowed to harden

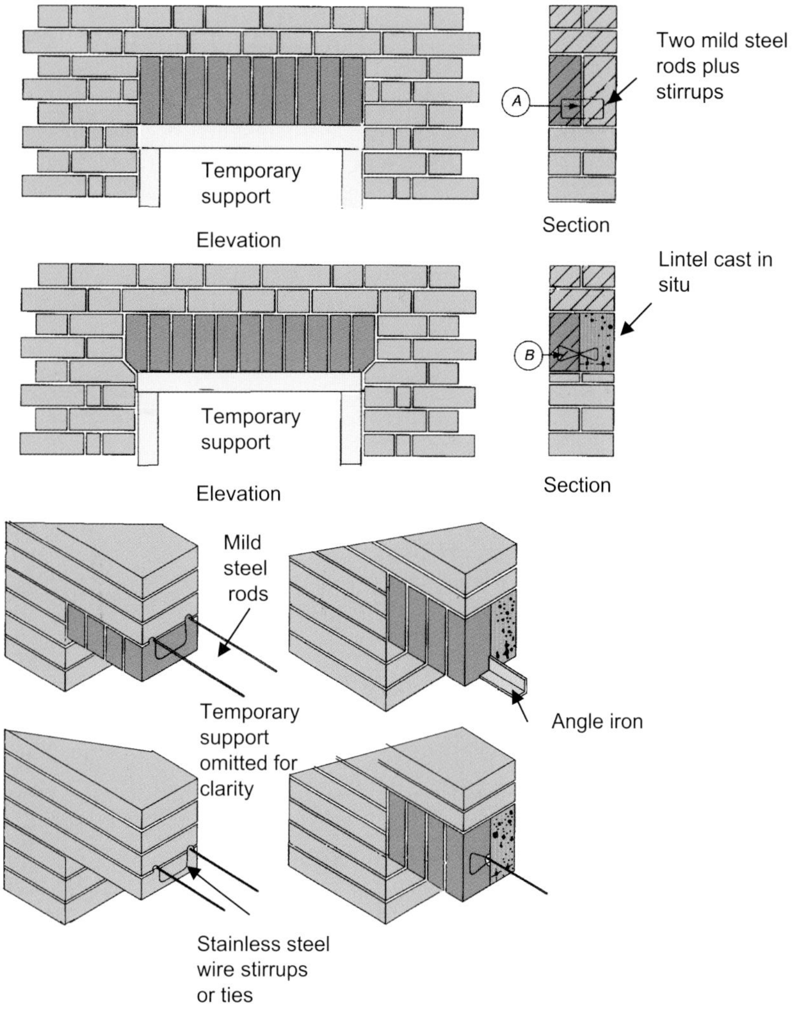
Two mild steel rods plus stirrups
A
Temporary support
Elevation
Section
Lintel cast in situ
B
Temporary support
Elevation
Section
Mild steel rods
Temporary support omitted for clarity
Angle iron
Stainless steel wire stirrups or ties

FIGURE 10.18
Soldier arches and reinforced lintels

for at least a week. Only when the arch centre has been removed does the arch become self-supporting.

Terminology

Arches comprise small units bonded together around a curve.

If an arch is constructed correctly it will not require any additional reinforcement, as brick lintels do, because the units are wedge shaped and any load applied on to the arch will tighten the units together.

The curved shape of an arch distributes the load of the walling above an opening down through the abutments on either side of the opening.

Arches are classified generally into three groups according to their shape, the number of centres and the method of cutting and preparing the arch.

The names of the parts of an arch are shown in Figure 10.19.

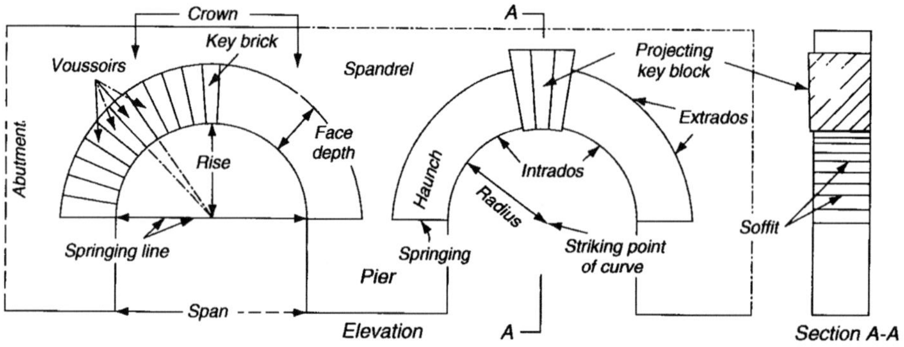
Crown
Key brick
Voussoirs
Spandrel
Abutment
Face depth
Rise
Springing line
Span
Projecting key block
Extrados
Haunch
Intrados
Radius
Springing
Pier
Striking point of curve
Soffit
Elevation
A
A
Section A-A

FIGURE 10.19
Craft terms in arch construction

Arch types

The three types of arch are:

- rough arches
- axed arches
- gauged arches.

ROUGH RINGED ARCH

This is the type of arch that will be dealt with in Level 2.

This type of arch is the simplest to construct. It is formed of uncut bricks and its shape is controlled by the type of turning piece or centre adopted.

The joints are wedge shaped and thus play an important part in the stability of the arch.

Their use is usually restricted to semi-circular and segmental arches (Figure 10.20).

The arch is constructed of a number of half-brick rings, hence the name ringed arch, the number of half-bricks varying according to the size of the opening.

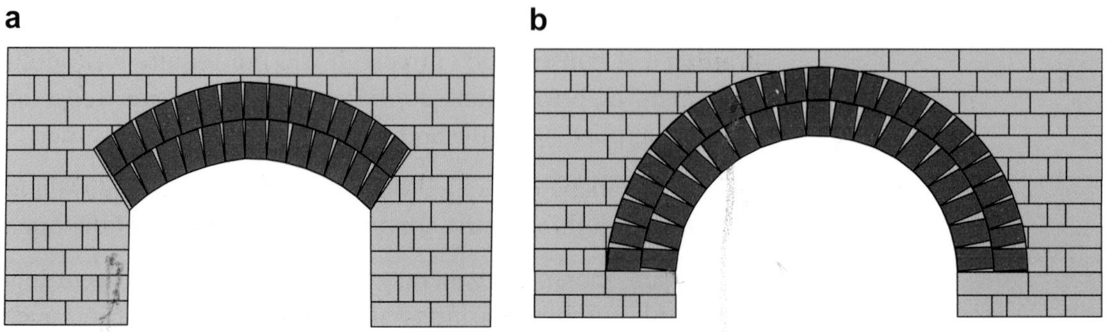
a
b

FIGURE 10.20
Rough ring arches: (a) segmental arch; (b) semi-circular arch

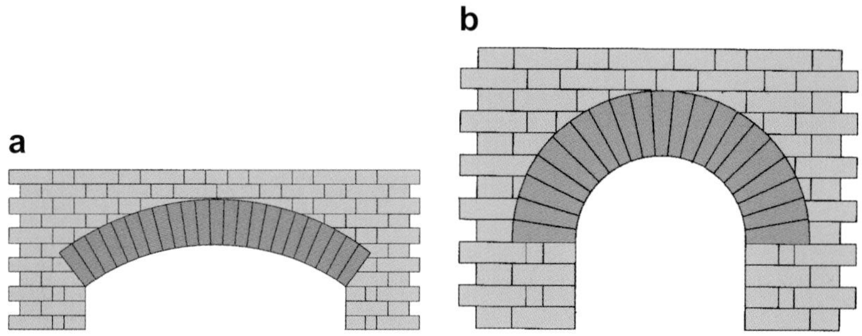
a
b

FIGURE 10.21
Axed arches: (a) segmental arch; (b) semi-circular arch

This method is adopted to reduce the size of the mortar joint at the extrados of each separate ring.

AXED ARCHES

This type of arch is formed by bricks cut to a wedge shape using an ordinary facing brick (Figure 10.21). It can be of the same colour and texture as the wall facings, or of some contrasting colour and probably of finer texture.

The tools used for cutting and trimming are the hammer, bolster and scutch, with a carborundum block.

The thickness of the mortar joint varies from 4 to 10 mm.

The operations of cutting and setting will be explained later in this chapter. It is also possible to bond the face of an axed arch.

GAUGED ARCHES

These arches are sometimes known as rubbed arches and have a much finer joint between the bricks. It is usually 2 mm and made from lime putty (Figure 10.22).

This work requires specially made clay bricks called red rubbers. These bricks contain approximately 30 per cent fine sand mixed with the clay,

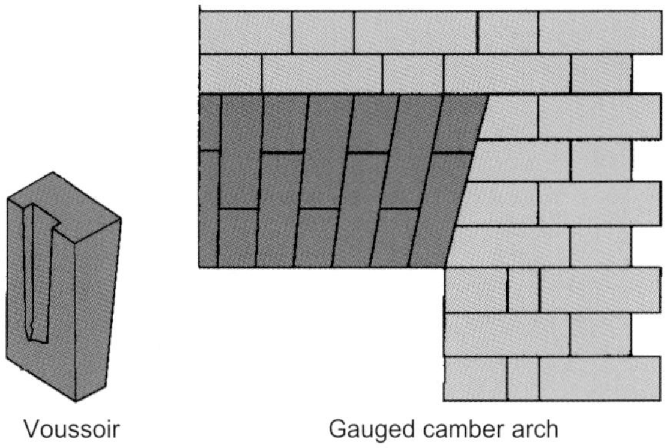
Voussoir
Gauged camber arch

FIGURE 10.22
Gauged arch

and are carefully fired so that the same even orange–red colour and texture is present throughout the body of each brick.

The high content of sand allows these bricks to be hand sawn and rubbed, on a block of York stone, to permit joints of 2 mm thickness.

Gauged work demands the very highest level of skills from the bricklayer in setting out, cutting, rubbing and bedding these bricks, which are 'white-line' jointed, using a putty made only from freshly slaked lime and water.

SEMI-CIRCULAR ARCHES

As the name implies the arch forms a semi-circle. Any load placed on the walling above the arch is transferred vertically down through the arch onto the abutments.

Figure 10.23 shows the various parts of a semi-circular arch.

Geometry of semi-circular arches

Information required:

- the span – assume 150 mm (using a scale of 1:10)
- depth of face – assume one brick in two rings.

Construct a 150 mm span and bisect to find the centre (Figure 10.24).

Place a compass point on the centre point and set the radius to half span. Scribe the arch. This curve is the intrados of the arch.

The depth can be set out, say 22.5 mm, and the extrados drawn.

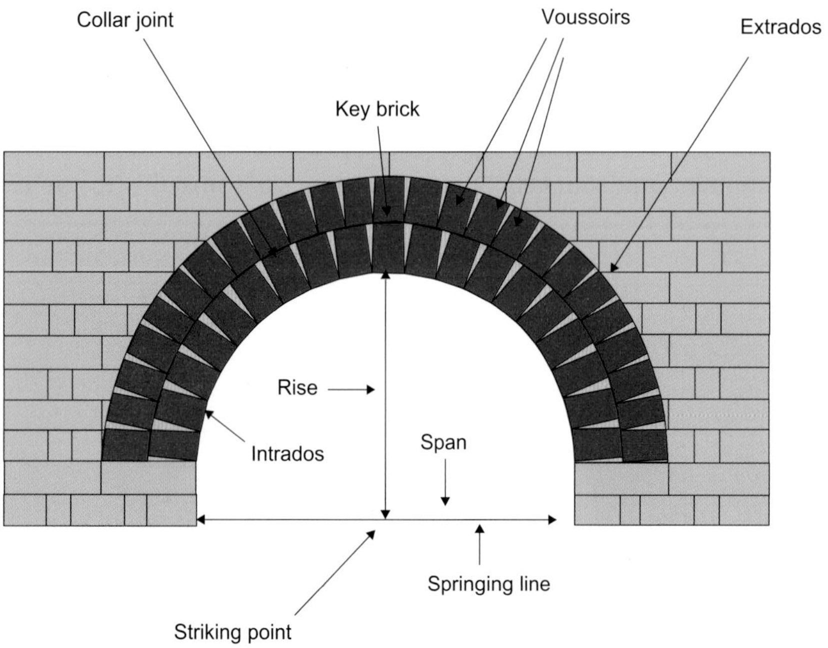
Collar joint
Voussoirs
Extrados
Key brick
Rise
Intrados
Span
Springing line
Striking point

FIGURE 10.23
Semi-circular arch details

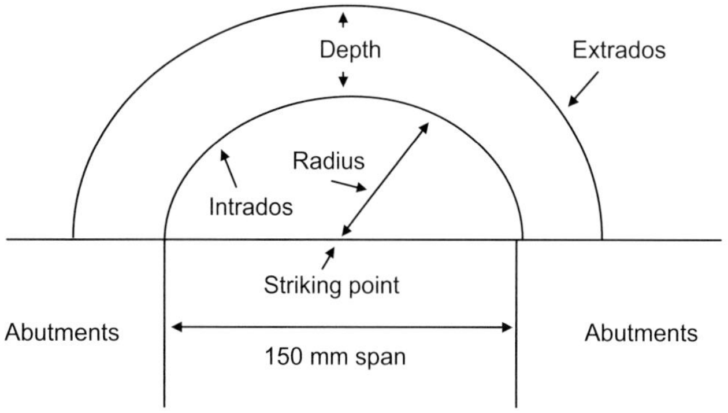
Depth
Extrados
Radius
Intrados
Striking point
Abutments
150 mm span
Abutments

FIGURE 10.24
Geometry of a semi-circular arch

The centre brick, known as the key brick, is marked out first. When marking out in situ, a brick can be placed in the required position at the top of the arch, allowing space for a joint either side (Figure 10.25).

Use dividers to set out the voussoirs of the arch from the key bricks back down to the springing line either side.

Note

The bricks are set out on the intrados for a rough ring arch.

The dividers should be set to the width of a brick plus a 10 mm joint. This should always give an odd number of voussoirs.

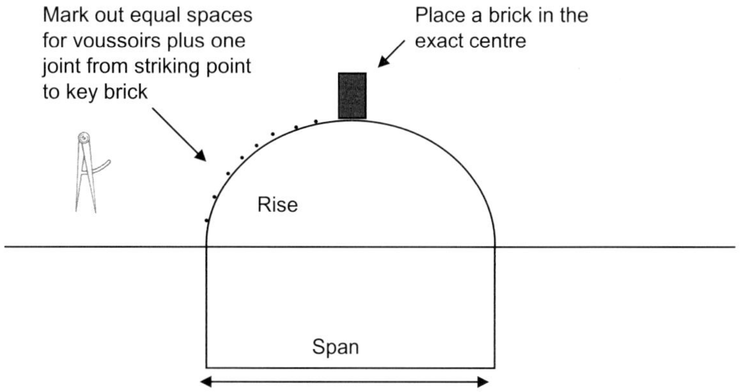
Mark out equal spaces for voussoirs plus one joint from striking point to key brick
Place a brick in the exact centre
Rise
Span

FIGURE 10.25
Setting out the key brick and the number of voussoirs

When drawing the voussoirs the key brick should be marked out first in each ring.

Draw a circle equal to the width of the brick around the striking point.

A line is then drawn from either side of the circle through the marked out voussoirs to give the wedge-shaped joint (Figure 10.26).

Note

If the spacing does not work exactly first time, reduce the dividers to produce a tighter joint.

NEVER widen the dividers.

Constructing a semi-circular arch

The method of constructing a semi-circular arch is similar for all types (Figure 10.27).

Set out the opening with a tolerance of 5 mm.

Use the arch centre as a guide or produce a pinch rod, to check the abutments for plumb as the work proceeds.

Remember

The voussoirs should be marked out on the intrados of the arch when a rough arch is required.

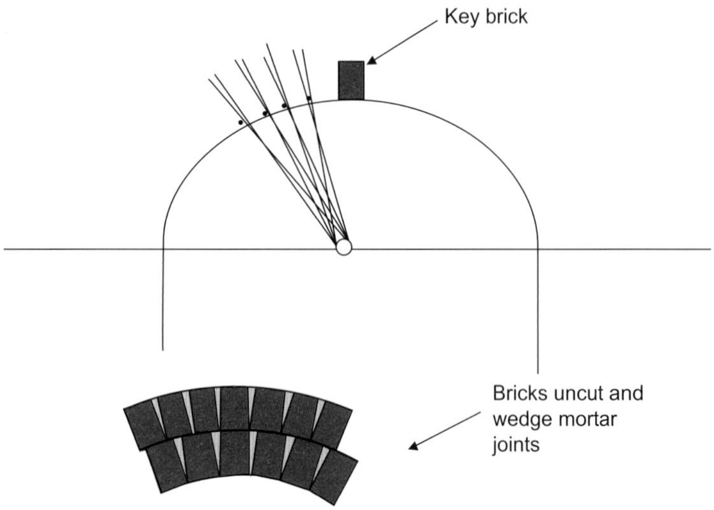
Key brick
Bricks uncut and wedge mortar joints

FIGURE 10.26
Marking out the wedge-shaped joints

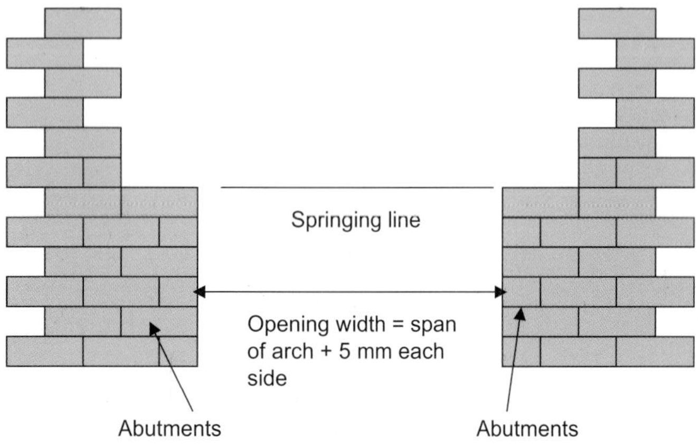
Springing line
Opening width = span of arch + 5 mm each side
Abutments
Abutments

FIGURE 10.27
Building up the abutments to receive the arch centre

Build the abutments up to the springing line. Position the centre supported on timber props and folding wedges. These will be dealt with later in this chapter.

It is advisable to build the abutments to the height of the crown of the arch as support for the lines.

Position and fix the arch centre using folding wedges for easy removal after the arch has been built.

Set dividers to the thickness of a brick plus a normal joint and step out around the arch centre, not forgetting to have a key brick in the centre (Figure 10.28).

Pull a line through from the corners and check that the centre is correctly placed and not interfering with the line.

Lay a few voussoirs on either side of the arch; never build one side completely.

It may be necessary to raise the line while the voussoirs are being laid.

Lay a second layer, if required, as before, keeping the key brick in the centre.

Raise the line course by course to complete the cutting around the extrados of the arch.

Always ensure that the joints are full to provide maximum strength to the arch. Strike the centre as soon as the arch is complete to allow the joints to contract uniformly when setting (Figure 10.29).

Always remember to joint the soffit of the arch immediately the centre is dropped.

SEGMENTAL ARCHES

As the name implies, a segmental arch forms part of a circle.

Any load placed on the walling above the arch is transferred vertically down through the arch onto the abutments.

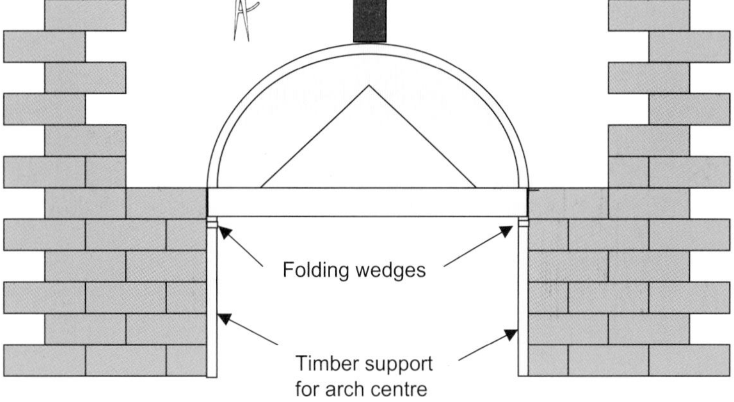
Folding wedges
Timber support
for arch centre

FIGURE 10.28
Setting the arch centre and marking out the voussoirs

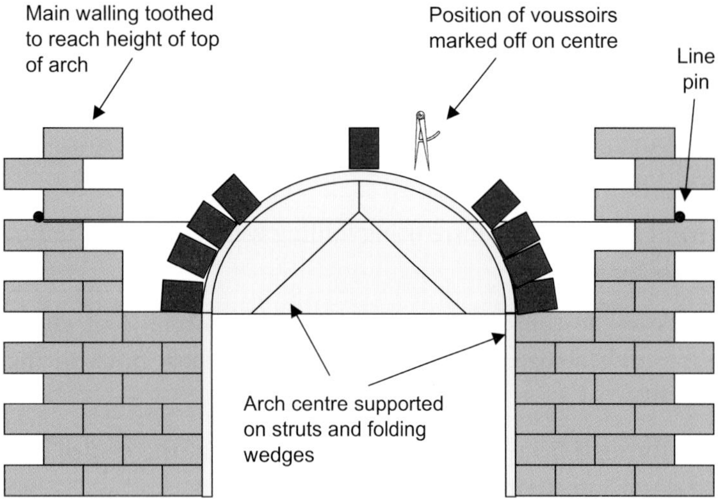
Main walling toothed to reach height of top of arch
Position of voussoirs marked off on centre
Line pin
Arch centre supported on struts and folding wedges

FIGURE 10.29
Completing the arch

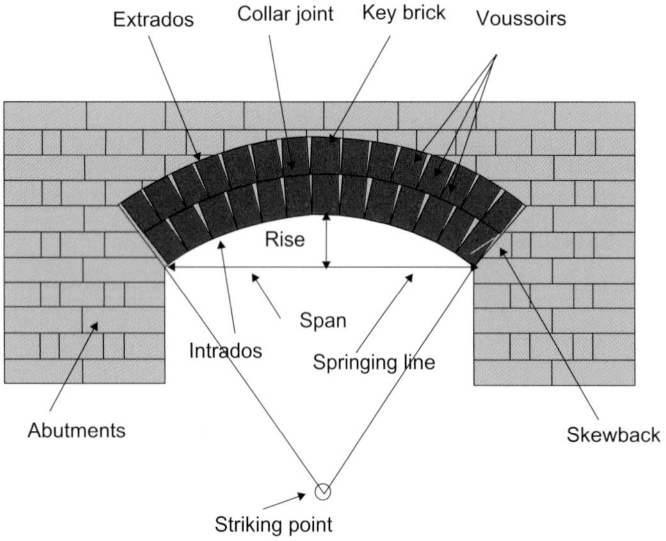
Extrados
Collar joint
Key brick
Voussoirs
Rise
Span
Intrados
Springing line
Abutments
Skewback
Striking point

FIGURE 10.30
A two-ring, rough, segmental arch

Figure 10.30 shows the various parts of a segmental arch. Recap on previous information if necessary.

Geometry of segmental arches

Again, the span must be known, but in the case of the segmental arch we also need to know the rise.

The rise to any segmental arch is normally one-sixth of the span.

First draw the span – assume it to be 900 mm – and set up a perpendicular bisector.

Mark off the rise: 150 mm. Join AB and bisect.

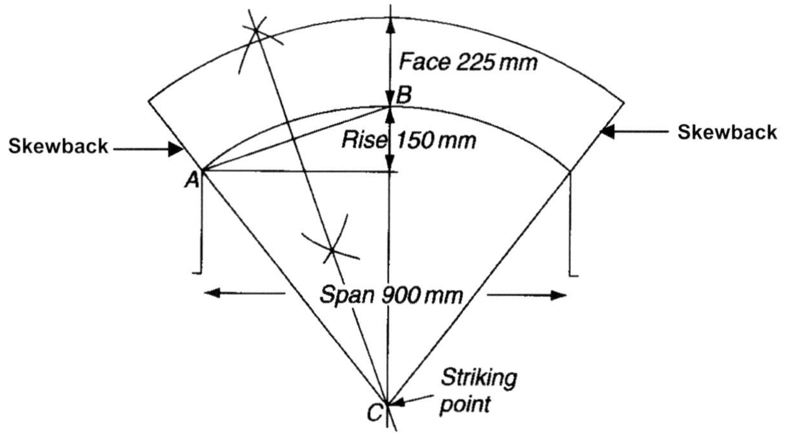
Face 225 mm
B
Skewback
Rise 150 mm
Skewback
A
Span 900 mm
Striking point
C

FIGURE 10.31
Geometric setting out of segmental arch

The point where this bisector intersects the centre line, C, is the striking point of the required arch.

A face depth of 225 mm is shown in Figure 10.31.

The angle produced by the line from the striking point, passing through the springing point to the extrados, is known as the skewback.

To ascertain the number of voussoirs and the position of the joint lines in the axed arch, set the dividers at the width of a brick plus 10 mm, and place the points equidistant on each side of the centre line, thus forming the key brick and step round the intrados.

If the last step fails to connect with the springing point at the first attempt, make a further attempt by slightly closing the dividers.

This procedure is exactly the same as for the semi-circular arch.

The angle of skewback can now be set using one of two methods.

Place a brick on the centre adjacent to the springing point and set a sliding bevel to the angle produced.

Pull a line from the striking point through the springing line at the point of support and set the bevel to this angle (Figure 10.32).

Cutting skewbacks

For cutting the skewbacks in preparation for the building of a segmental arch, always use a gun template.

Its function is to keep the angle of the skewback constant, especially when there are several arches to be constructed (Figure 10.33).

Constructing a segmental arch

The method of constructing a segmental arch is similar for all types.

Set out the opening with a tolerance of 5 mm.

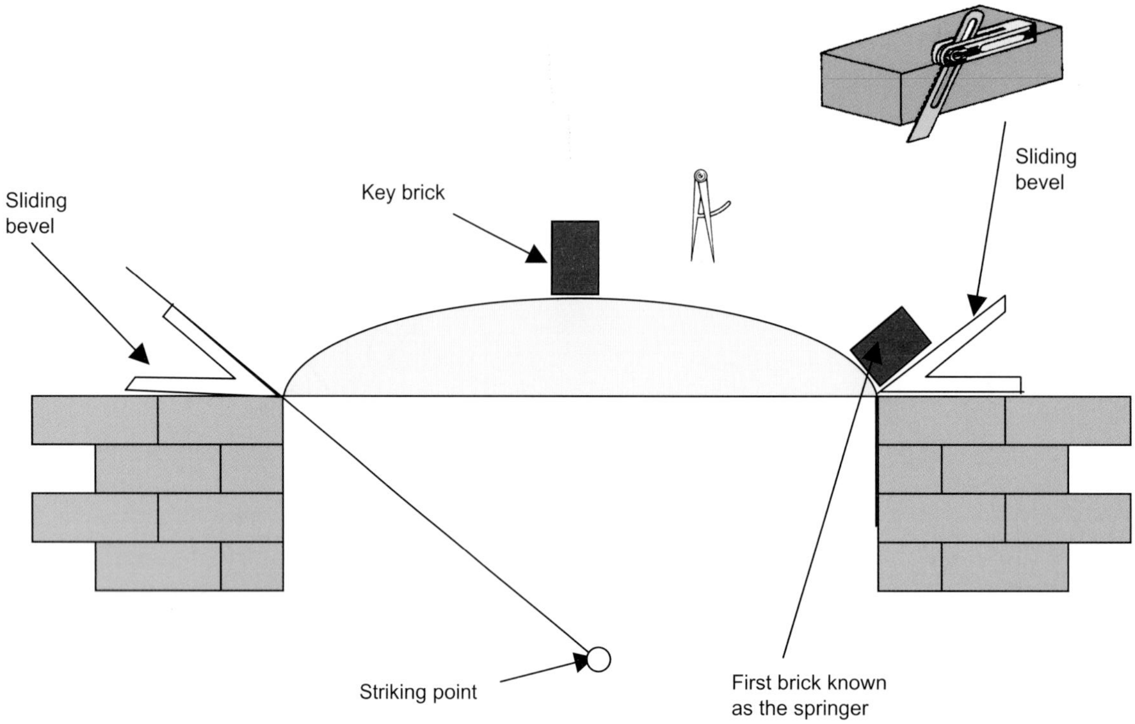
Sliding
bevel
Key brick
Sliding
bevel
Striking point
First brick known
as the springer

FIGURE 10.32
Setting out skewback

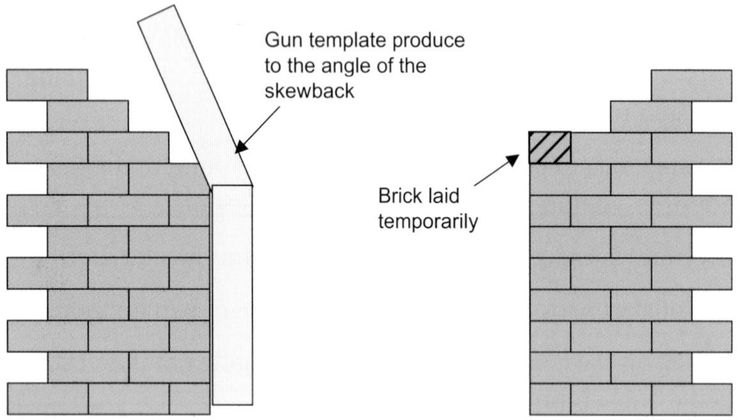
Gun template produce
to the angle of the
skewback
Brick laid
temporarily

FIGURE 10.33
Use of the gun template

Use the arch centre as a guide or produce a pinch rod, to check the abutments for plumb as the work proceeds.

Build the wall up to the springing line, and position the centre supported on timber props and folding wedges. Extend the abutments to a height above the top of the arch to support the lines (Figure 10.34).

Skewbacks must be marked and cut (Figure 10.35).

Pull a line from the striking point through the springing line at the point of support and set the bevel to this angle (Figure 10.32) or use a gun template.

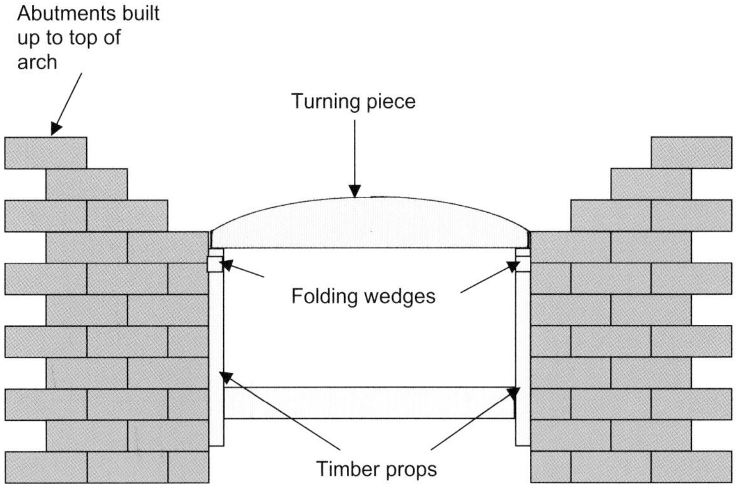
Abutments built up to top of arch
Turning piece
Folding wedges
Timber props

FIGURE 10.34
Setting the turning piece for a segmental arch

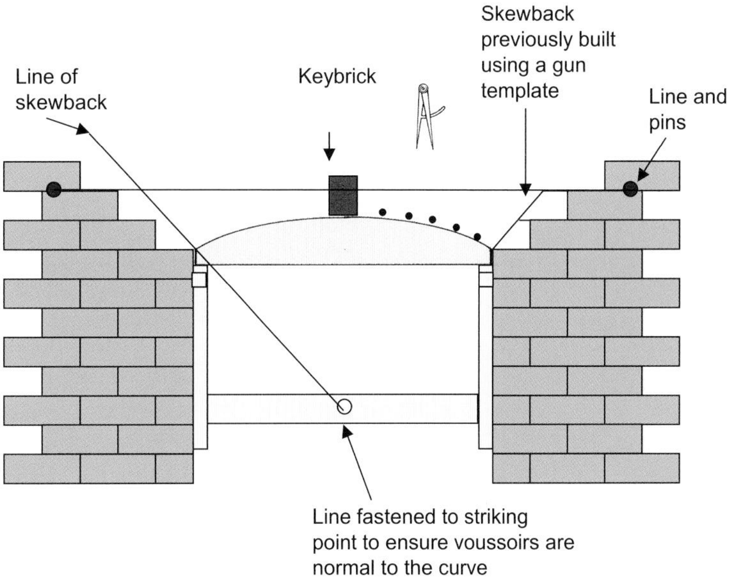
Skewback previously built using a gun template
Line of skewback
Keybrick
Line and pins
Line fastened to striking point to ensure voussoirs are normal to the curve

FIGURE 10.35
Setting the skewbacks and voussoirs for a segmental arch

Set dividers to the thickness of a brick plus half a normal joint and step out around the arch centre, not forgetting to have a key brick in the centre.

Pull a line through from the corners and lay the voussoirs to the line.

Lay second layer if required as before, keeping the key brick in the centre. Raise the line course by course to complete the cutting around the extrados of the arch. Strike the centre as soon as the arch is complete to allow the joints to contract uniformly when setting.

Always remember to joint the soffit of the arch.

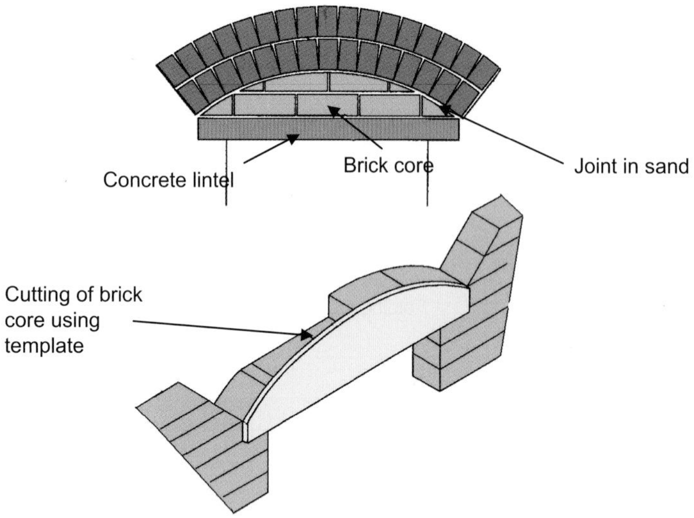
Concrete lintel
Brick core
Joint in sand
Cutting of brick core using template

FIGURE 10.36
Formation of relieving or discharging arch

RELIEVING OR DISCHARGING ARCH

This is a form of ringed arch, built over a lintel, and turned on a shaped brick core (Figure 10.36). It relieves the lintel of any point load and discharges the weight to the abutments.

The joint between the brick core and the soffit of the arch must be of sand, as this ensures the proper discharging of loads. This sand joint is pointed on the face only.

The shape of the brick core is obtained using a template.

Types of support

No attempt will be made here to describe temporary arch supports over large spans. Only those encountered by the bricklayer in the course of everyday work will be dealt with.

Turning piece

A turning piece is made from solid timber for arches of limited span.

For an arch with a 225 mm soffit, two 75 mm timbers can be placed side by side. Alternatively, a turning piece could be constructed from plywood and packed out with timbers. See examples in Figure 10.37.

Arch centres

Arch centres are usually produced for larger spanning arches. They can either be open or closed lagged:

- An open-lagged centre is framed from light timber and used for the turning of ringed arches (Figure 10.38).

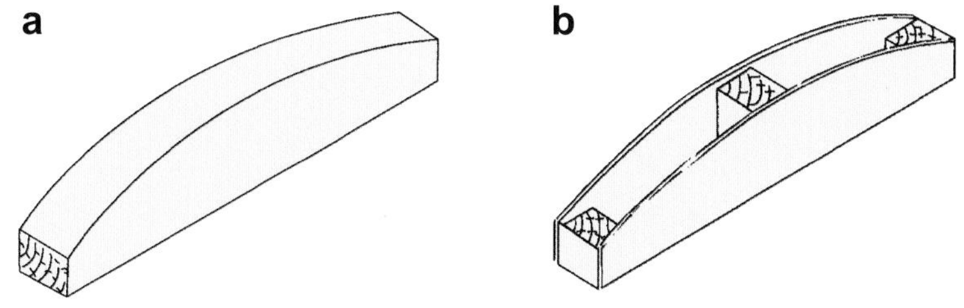
a
b

FIGURE 10.37
Arch turning pieces: (a) solid turning piece; (b) built-up turning piece

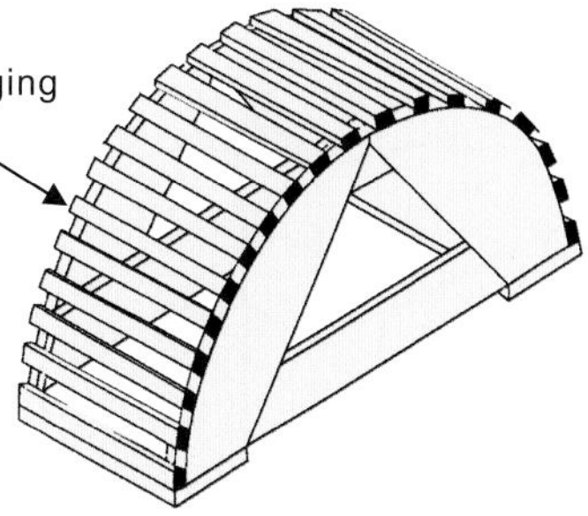
Timber lagging

FIGURE 10.38
Open lagged centre for semi-circular arch

- A close-lagged centre is suitable for the turning of an axed or gauged arch. Its use facilitates the marking of the voussoir positions (Figure 10.39).

When using temporary arch supports, folding wedges are placed directly under the centre (Figure 10.40). This facilitates the removal of the centre or turning piece and, in particular, avoids the chipping of the arch face on the intrados arris, which invariably occurs if wedges are not used.

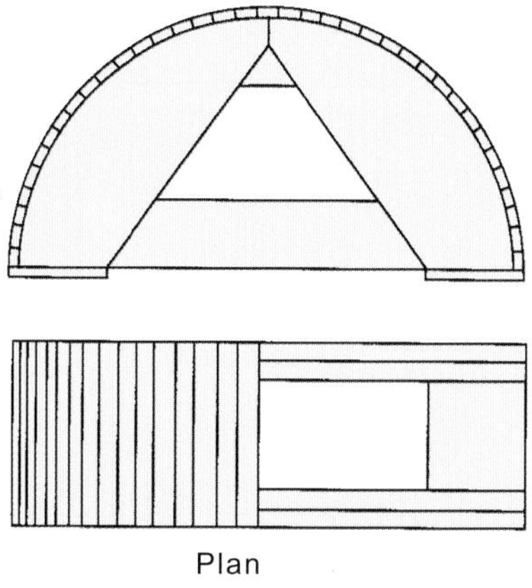
Closed lagging
Plan

FIGURE 10.39
Closed lagged centre for semi-circular arch

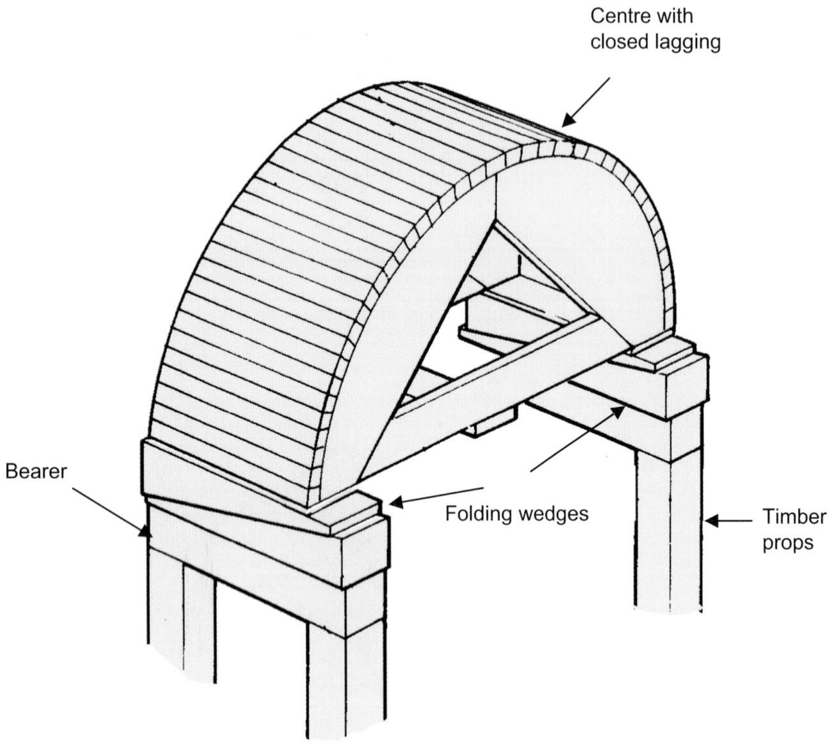
Centre with
closed lagging
Bearer
Folding wedges
Timber
props

FIGURE 10.40
Centre supported on timber props and folding wedges

Modern methods

There are also new methods of supporting arches in both solid and cavity walls. These arch formers are available in plastic and steel.

The centres are built in and remain as part of the structure.

A semi-circular centre is shown in Figure 10.41, but segmental centres are also available. A version that sits on a steel lintel is shown in Figure 10.42.

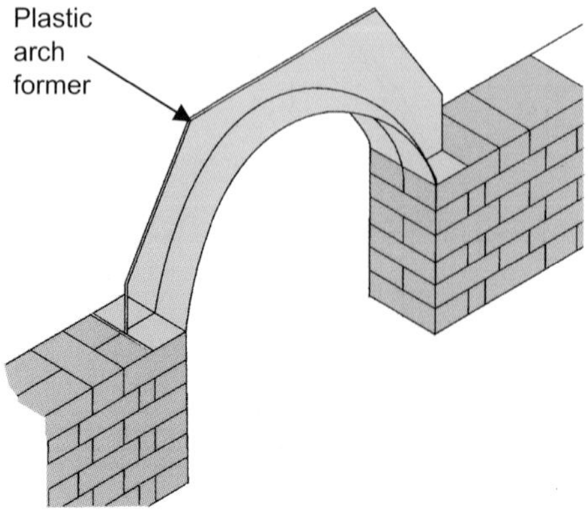
Plastic
arch
former

FIGURE 10.41
Plastic arch former for solid walls

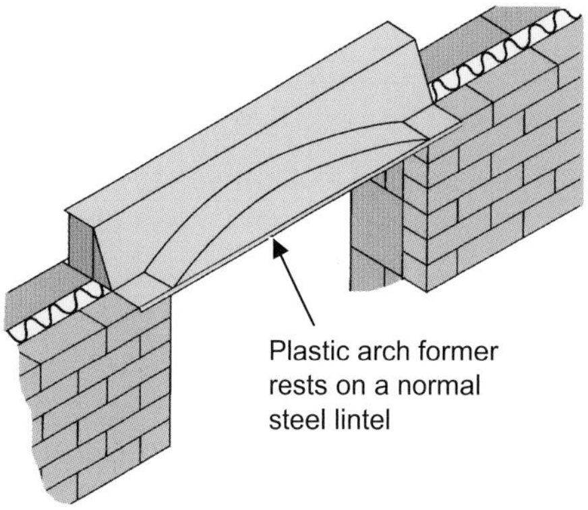
Plastic arch former
rests on a normal
steel lintel

FIGURE 10.42
Plastic arch former sitting on a boxed steel lintel on cavity walls

Multiple-choice questions

Self-assessment

This section of the book is designed to allow you to check your level of knowledge. The section consists of revision questions for this chapter. The questions are all multiple choice and have four possible answers. The answers are to be found at the end of the book.

The main type of multiple-choice question will be the four-option multiple-choice question. This will consist of a question or statement, known as the stem, followed by a choice of four different answers, called the responses. Only one of these responses is the correct answer; the others are incorrect and are known as distracters.

You should attempt to answer the questions by choosing either (a), (b), (c) or (d).

Example

The person employed by the local authority to ensure that the Building Regulations are observed is called the:

(a) clerk of works
(b) building control officer
(c) council inspector
(d) safety officer

The correct answer is the building control officer, and therefore (b) would be the correct response.

Bridging openings

Question 1 What are the bricks laid vertical over openings known as?

(a) soldier arch
(b) vertical arch
(c) flat arch
(d) brick arch

Question 2 Which of the following is the correct name for the bricks in an arch?

(a) key bricks
(b) voussoirs
(c) soldiers
(d) soffit bricks

Question 3 What is the timber support to a semi-circular arch known as?

(a) formwork
(b) springer
(c) centre
(d) turning piece

Question 4 What is the timber support to a segmental arch known as?

(a) formwork

(b) springer

(c) centre

(d) turning piece

Question 5 Which of the following items of equipment can be used for setting out arch curves?

(a) gauge

(b) turning piece

(c) trammel

(d) datum

Question 6 What is the name given to the slope of brickwork cut to receive an arch?

(a) skewback

(b) springer

(c) closer

(d) voussoir

Question 7 An arch built with wedge-shaped joints is known as a:

(a) cut arch

(b) rough arch

(c) axed arch

(d) soldier arch

Question 8 Where should reinforcement be placed in a concrete lintel?

(a) at the top

(b) at the bottom

(c) just up from the bottom

(d) just below the surface

CHAPTER 11

Domestic Drainage

This chapter will cover the following NVQ and Diploma units:

- NVQ VR43 – Optional Unit

This chapter is about:

- Interpreting building information
- Adopting safe and healthy working practices
- Selecting materials, components and equipment
- Preparing, laying and testing foul and surface drainage

The following NVQ performance criteria will be covered:

- Performance criterion 1: Interpretation of information
- Performance criterion 2: Safe work practices
- Performance criterion 3: Selection of resources
- Performance criterion 4: Minimize the risk of damage
- Performance criterion 5: Meet the contract specification
- Performance criterion 6: Allocated time

The following Diploma outcomes will be covered:

There is no comparable Diploma Level 2 unit but the chapter will give the student an insight into domestic drainage.

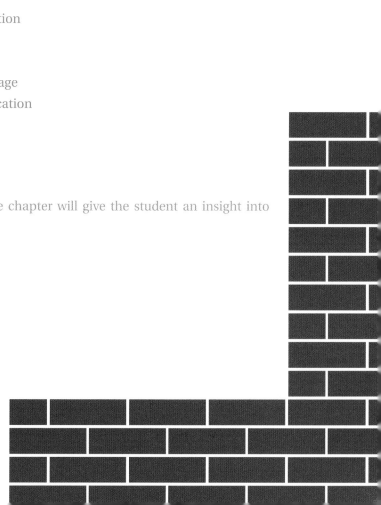

Interpreting technical information

There are numerous sources of information available and they include the following: drawings, programmes of work, schedules, specifications, policies, mission statements, manufacturers' technical information, organizational documentation, and training and development records and documents.

It is important that the student understands where information can be found.

Information sources

Information can be gathered from numerous sources and can be either written, oral or in drawing format.

WRITTEN

Manufacturers' technical information

Any efficient office should have up-to-date information regarding new and existing products from the various manufacturers who produce materials of interest to them.

Many manufacturers produce technical information which is free for the asking. These papers, brochures, leaflets, etc., should be stored in the office and sent to site as required.

ORAL

Verbal communication can include talking on the telephone, mobile or walkie-talkie. This form of communication is instant and feedback can be obtained immediately.

DRAWINGS

Drawings are one of the simplest forms of communication available. They are used extensively in the construction industry for transmitting ideas and information.

The types of drawing used range from simple sketches, to scale plans, graphs, product range drawings and charts. The type of drawing used will for the most part depend on the information to be transmitted and the time it is to be used.

Simple sketches can be used on site to show how a detail is to be built at that time, but if a few days or weeks are to pass before it is to be built, then a scale drawing will be more useful.

Drawings have been discussed in detail in Chapter 3. This chapter will only deal with drawings relating to drainage.

Understanding drawings

Everyone in construction needs to be able understand and extract information from drawings.

Table 11.1 Preferred scales for building drawings

Type of drawing	Scales
Block plans	1:2500, 1:1250
Site plans	1:500, 1:200
Location drawings	1:200, 1:100, 1:50
Range drawings	1:100, 1:50, 1:20
Detail drawings	1:10, 1:5, 1:1
Assembly drawings	1:20, 1:10, 1:5

There are many types of construction drawings, which are listed in Table 11.1 along with the appropriate scales.

Production drawings

These are known as working drawings and are used to relay information to the building contractor and other members of the building team.

The working drawings can be classified as:

- location drawings
- component drawings
- assembly drawings.

LOCATION DRAWINGS

These are further classified into:

- block plans
- site plans
- location plans.

Block plans

These are used to identify the site in relation to the surrounding area (Figure 11.1).

The scale is usually 1:2500 or 1:250 and is too small to allow much more than an outline of the site and boundaries, road layouts and the other buildings in the near vicinity.

The orientation of the site is always shown with a suitable logo depicting north. The actual site should be outlined in red.

It is unlikely that dimensions would be added to these drawings.

Plans of this sort are usually based on the Ordnance Survey sheet for the area; however, if such a source is used, permission should be obtained for its reproduction.

Site plans

These are used to show the position of the proposed building on the site, together with information on proposed road, drainage and service layouts,

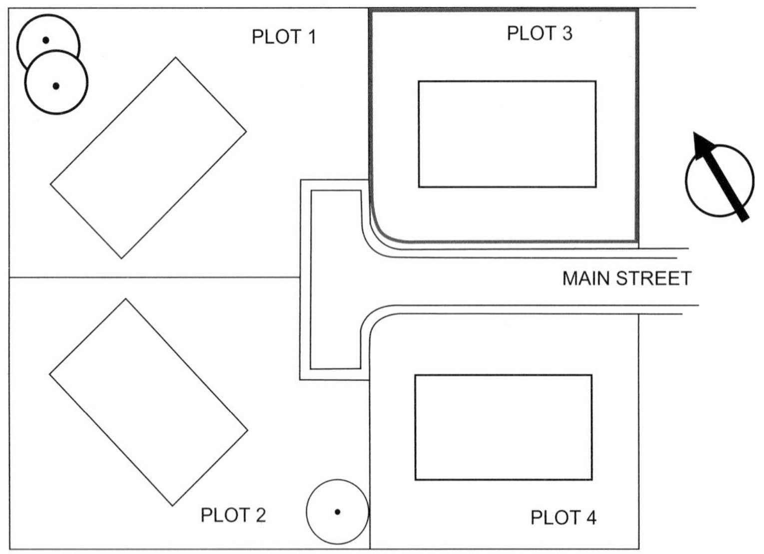
PLOT 1
PLOT 3
MAIN STREET
PLOT 2
PLOT 4

FIGURE 11.1
Block plan

and other site information such as levels. A typical site plan is shown in Figure 11.2.

Again, the orientation of the site should be shown.

Scales of 1:500 and 1:200 are often used.

The information on this drawing is used by both the design team and the contractors.

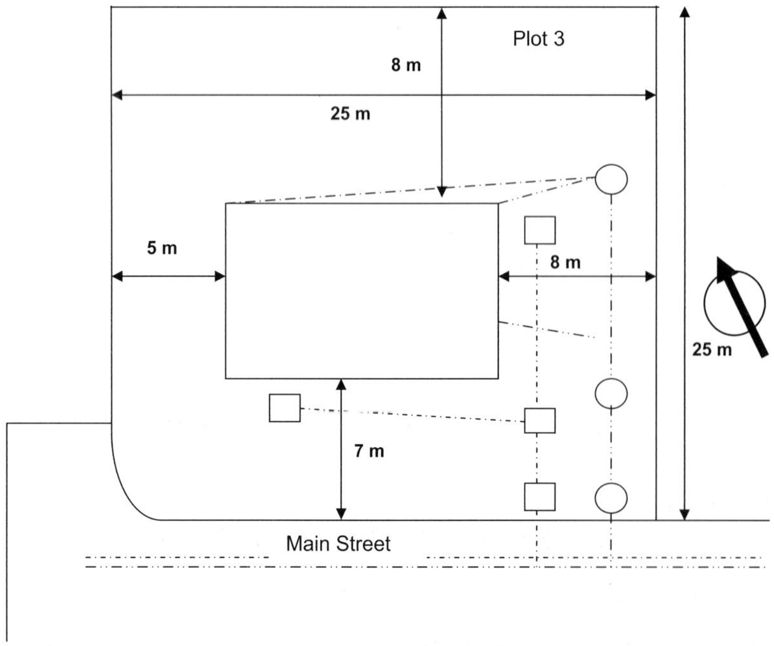
Plot 3
8 m
25 m
5 m
8 m
25 m
7 m
Main Street

FIGURE 11.2
Site plan

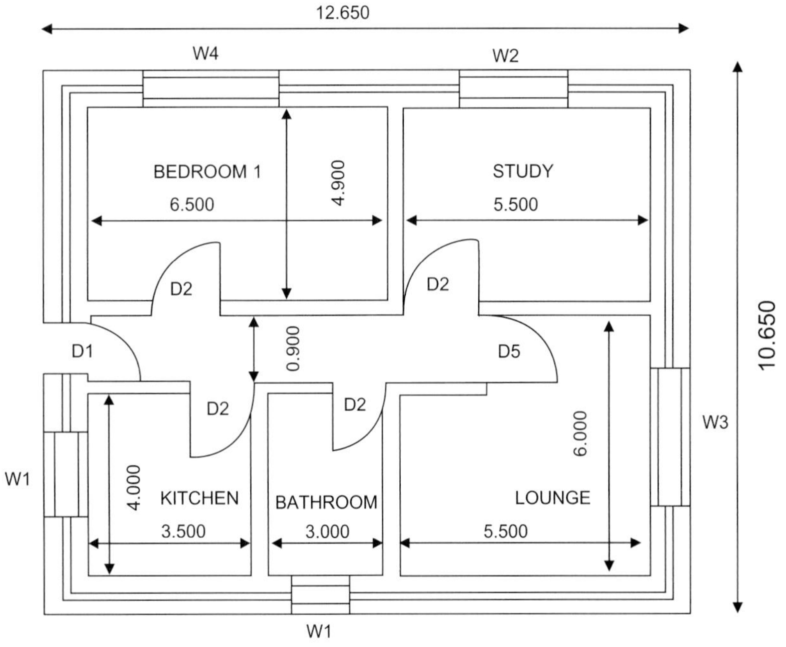
12.650
W4
W2
BEDROOM 1
6.500
4.900
STUDY
5.500
D2
D2
D1
0.900
D5
10.650
D2
D2
W3
6.000
W1
4.000
KITCHEN
3.500
BATHROOM
3.000
LOUNGE
5.500
W1

FIGURE 11.3
Location plan

The drainage layout should be used in conjunction with the drainage schedule which will give details of inspection chamber construction, cover and invert levels, and other relevant information.

Location plans

These are used to show the size and position of the various rooms within the buildings and to position the principal elements and components. A typical location plan is shown in Figure 11.3.

Plans are usually drawn to a scale of 1:200, 1:100 or 1:50.

These scales are sufficiently large to allow dimensions to be added.

Further information from the drawing includes the wall construction and position and type of doors and windows. This information is used in conjunction with door and window and drainage schedules. The way the doors are hung is also shown.

COMPONENT DRAWINGS

These are further classified as range and details drawings.

Range drawings

These show the basic sizes and reference system of standard components.

They are usually drawn to a scale of 1:100, 1:50 or 1:20.

The types of bends can be obtained from any drainage manufacturer's literature with their references, e.g. quarter bend, eighth bend or twelfth bend. This reference is all that is required by the architect when completing the

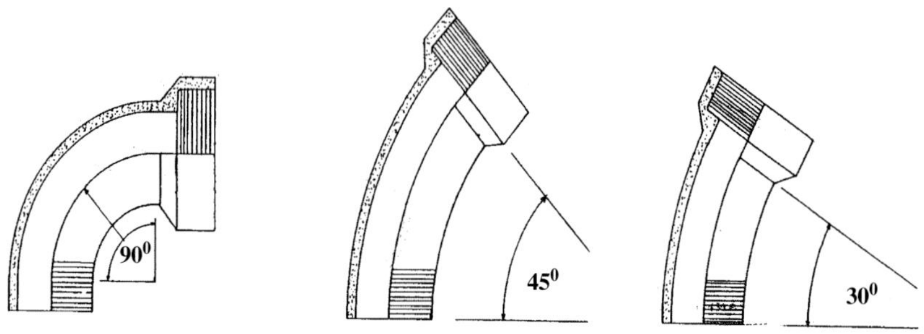
90⁰
45⁰
30⁰

FIGURE 11.4
Range of bends

working drawings, as the builder is provided with a drainage schedule for cross-referencing (Figure 11.4).

ASSEMBLY DRAWINGS

These are used by the architect to show in detail the junction between the various elements and components of the buildings (Figure 11.5).

These details are necessary for the builder to know exactly how the architect requires the construction to be completed.

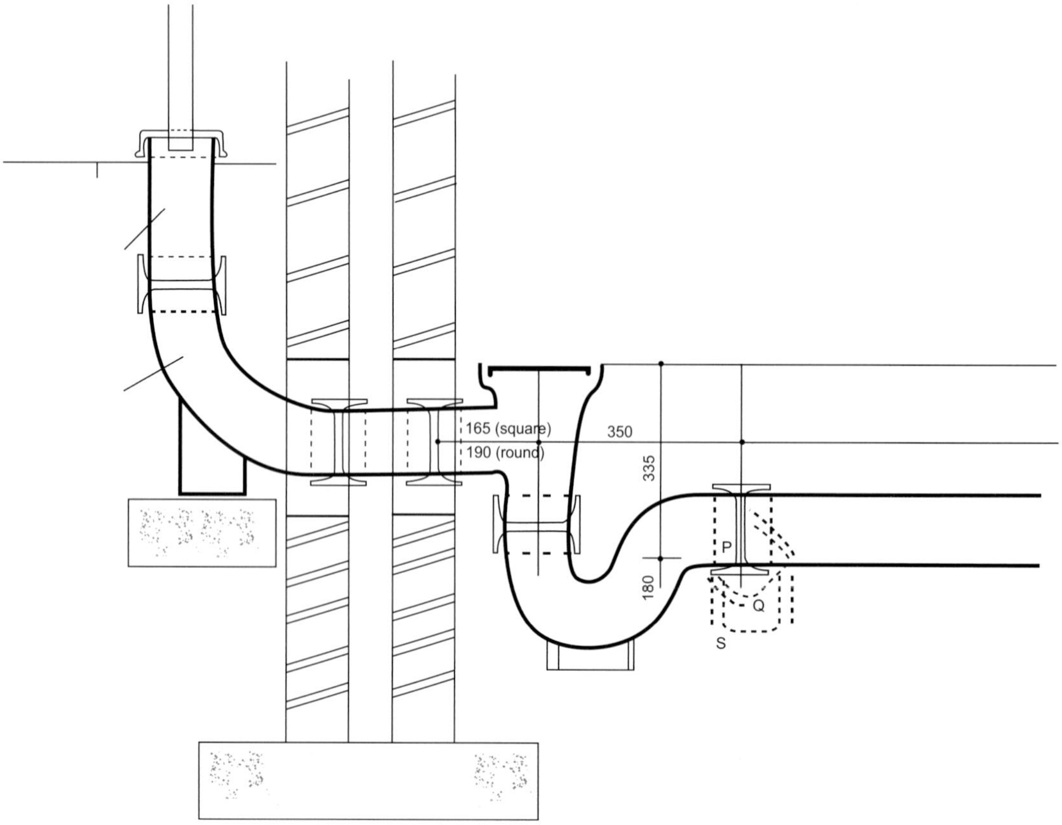
165 (square)
190 (round)
350
335
180
P
Q
S

FIGURE 11.5
Assembly drawings

The scales are usually 1:20, 1:10 and 1:5 and should be fully dimensioned and annotated.

Schedules

Schedules form an important method of communication for the design team. These are prepared for large contracts to simplify tabulated information about a range of components.

The main areas where schedules are used include:

- doors
- windows
- ironmongery
- internal finishes
- sanitary ware
- drainage
- lintels
- steel reinforcement.

Some schedules are best produced by the specialist; for example, the steel reinforcement schedule by the structural engineer and the sanitary ware schedule by the mechanical engineer.

The information received from the schedule enables the quantity surveyor to prepare the bills of quantities more easily and more accurately.

By using the schedule, the main contractor and subcontractors can carry out certain operations more quickly, without the need to communicate with the architect constantly.

The schedules are additional aids to the specifications, drawings and the bill of quantities, and are extremely useful when locating work and checking deliveries of materials and components.

Producing and reading schedules is fairly straightforward.

The manufacturer's detail of the component is used for reference and the plans of the building.

The design of a schedule can take various formats, but normally the component types are listed across the top with the various finishes, and the number of units is listed down the left-hand column.

In the inspection chamber schedule shown in Table 11.2, the inspection chambers (ICs) are listed down the left-hand column and the various designs and sizes are shown across the top.

The completed schedules will be used by the:

- quantity surveyor for cost planning and production of the bill of quantities

Table 11.2 Inspection chamber schedule

Inspection chamber	Size	Invert depth	Connections	Pipe size	Step irons	Cover	Internal finish	Comments
IC1	450 × 450	450	1	100	0	450 × 450	Brick pointed	
IC2	450 × 450	600	1	100	0	450 × 450	Brick pointed	
IC3	450 × 750	750	2	100 150	1	450 × 450	Brick pointed	
IC4	450 × 750	1000	2	150	2	600 × 600	Brick pointed	
IC5	750 × 1200	1200	3	150	2	600 × 600	Brick pointed	

- structural engineer for reinforcement bending schedules, beam, column and foundation schedules
- services engineer for builder's work details, fire stopping, sanitary ware and luminaries
- builder as part of the production information used for tendering and contract control.

Taking off quantities

At some stage in your career you will have to determine the quantity of materials required to carry out an activity. This will entail having to calculate areas and volumes.

Quantities of materials

The buyer is responsible for taking off and scheduling materials from the bills of quantities or, on smaller jobs, the working drawings.

When the site supervisor if given this role he or she must be careful to ensure that correct quantities and details of quality of materials are extracted from the drawings and specifications.

Allowances should be made for materials wastage which, from experience, is usually 5–10 per cent.

Taking off materials measurements from drawings should be carried out with care to allow for extra lengths or quantities, especially on joists, to give a suitable margin of safety.

Too many examples are noticed on building sites of long offcuts from joists, rafters and other structural members owing to generous allowances having been made at the taking-off stage. Extra members can also be seen lying around the site unused.

EXAMPLES

Bricks

The format size for a standard brick (plus joint) is 225 mm × 112.5 mm × 75 mm.

This means that a square metre of a one-brick-thick wall requires 118.5 bricks, but allowing 5 per cent for cutting and waste a more realistic figure is 125 bricks.

To calculate the total number of bricks for a one-brick wall, calculate the area in square metres and multiply by 125.

Example 1

A one-brick wall is 6.00 m long and 1.50 m high. Calculate the number of bricks required.

Answer 1

$$6.00 \times 1.50 = 9.00\ \text{m}^2$$

$$9.00 \times 125 = 1125 \text{ bricks required}$$

Concrete

To calculate the volume of concrete foundations you need three dimensions: the length of the trench, the width of the trench and the thickness of the concrete foundations.

Example 2

Calculate the concrete required for a foundation if the trench is 39.00 m long and 0.75 mm wide and the concrete is 0.15 mm thick. This would be exactly the same for pipe bedding.

Answer 2

$$39.00 \times 0.75 \times 0.15 = 4.387\ \text{m}^3$$

Drainage

The number of pipes can be calculated by dividing the length of the trench by the length of each individual pipe.

Example 3

Calculate the number of drain pipes required for a trench 36.00 m long if 0.900 mm pipes are being used.

Answer 3

$$36.00 \div 0.900 = 40 \text{ pipes}$$

Gradients

The most common gradient is stated as 1:40. If we use a drain length of 40 m it will simplify the calculations.

Example 4

Calculate the fall of a drain trench 40 m long if the gradient has to be 1:40.

Answer 4

$$\frac{40 \times 1}{40} = 1\ \text{m}$$

Example 5

Calculate the fall of a drain trench 25 metres long if the gradient has to be 1:40.

Answer 5

$$\frac{25 \times 1}{40} = 0.625\ \text{mm}$$

Safe working practices

It is essential that the correct personal protective equipment (PPE) is available and worn by all operatives working on drainage schemes.

As most domestic drainage is below ground in trenches the correct care and attention must be taken. Two types of accidents can occur with excavations: either the trench itself collapses or people fall into the trench. Always make sure that the trench is well supported, keeping heavy loads away from the edges, and erect barriers around the excavation where necessary (Figure 11.6).

Accidents should only occur in unforeseen circumstances.

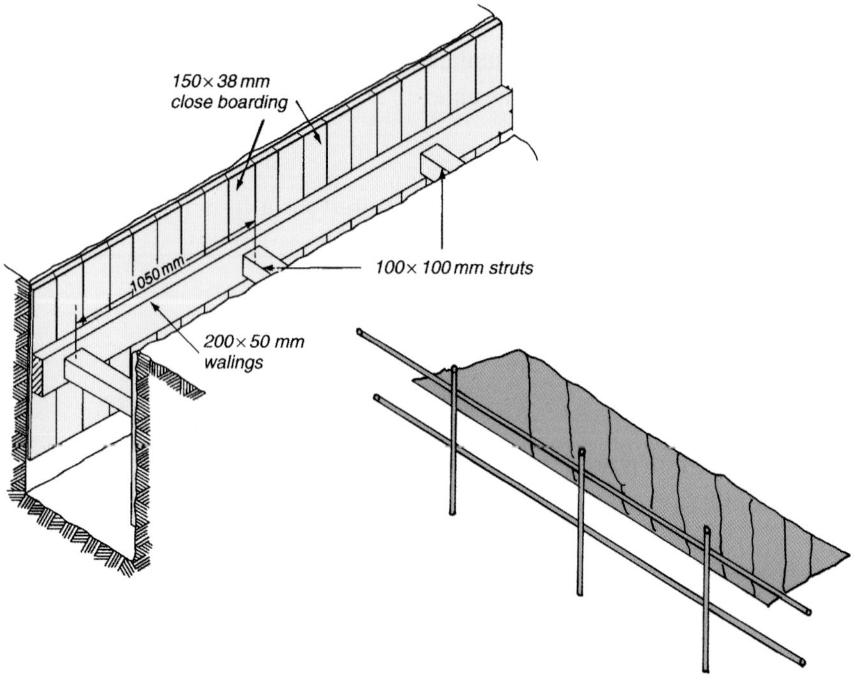
150× 38 mm
close boarding
1050 mm
100× 100 mm struts
200× 50 mm
walings

FIGURE 11.6
Timbering to a drain trench and barriers along a trench

150
SV
3
6
GPO

FIGURE 11.7
Service posts

Anyone who sees a hazard and does nothing about it is making it much more likely that an accident will occur. It is essential to make safe or report dangerous situations.

When excavating trenches the normal attention should be paid to existing underground services. These can be located by contacting the local services and with the use of a cable locater. Look out for tell-tale signs and clues in roadside service signs such as those shown in Figure 11.7.

Material and component suitability

This chapter deals with various walling and drainage components and materials, including:

- standard and engineering quality bricks
- types of cement and mortar additives
- drainage
- inspection chamber covers and frames
- step irons
- plastic and concrete inspection chambers
- bricks.

Drainage pipes

Drainage pipes can be manufactured from various materials and are classed as either rigid or flexible.

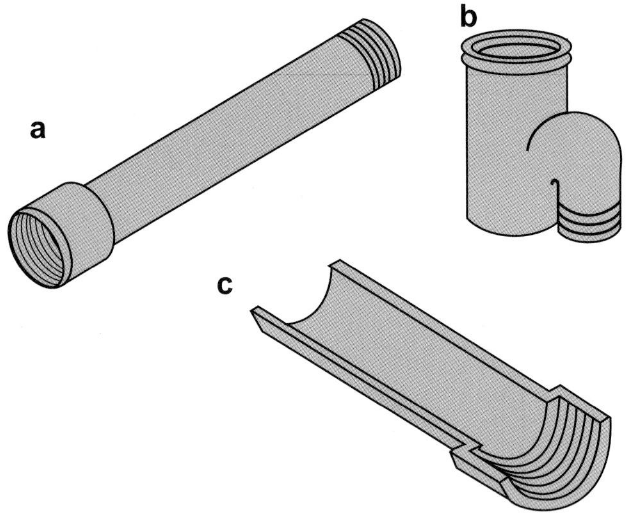
a
b
c

FIGURE 11.8
Selection of clayware: (a) pipe; (b) gully with 'S' trap; (c) channel

Clay is the main material used for rigid drain pipes in domestic work, with cast iron the main alternative. The usual materials for flexible drain pipes are pitch fibre and unplasticized PVC. A selection of clayware is shown in Figure 11.8.

Inspection chamber covers

These are produced from concrete, cast iron or steel. They come in various shapes and sizes to suit the particular situation. Figure 11.9 shows two examples.

Inspection chambers (manholes)

Inspection chambers are usually constructed in engineering bricks (Figure 11.10). Other methods on the market include plastic and precast concrete chambers.

When constructing deep inspection chambers in brick, step irons will be required (Figure 11.11).

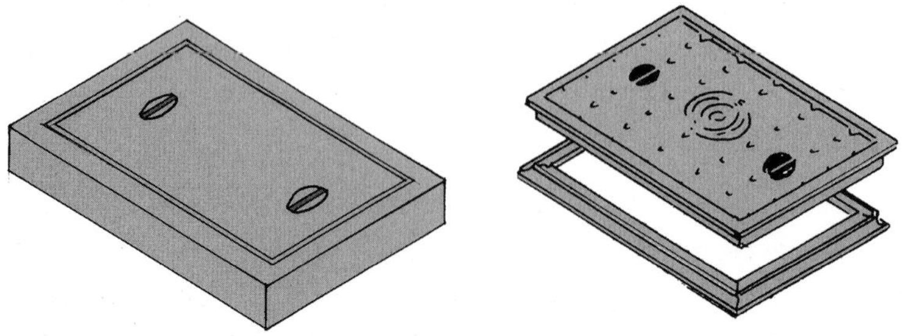

FIGURE 11.9
Concrete and steel covers and frames

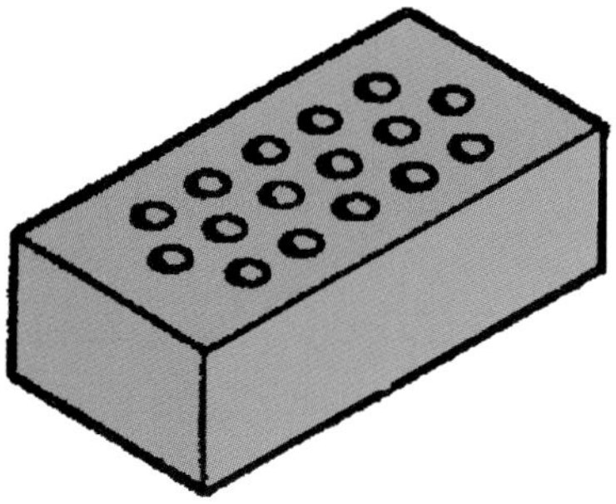

FIGURE 11.10
Engineering brick

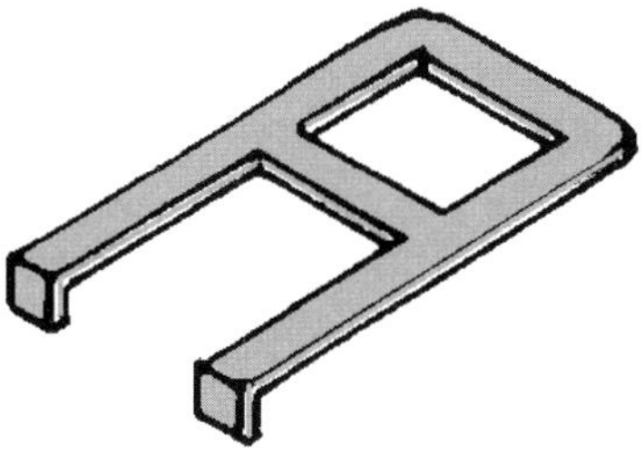

FIGURE 11.11
Galvanized steel step irons

MORTAR FOR CONSTRUCTING INSPECTION CHAMBERS

Cement and sand

These materials are used to produce mortars for construction brick inspection chambers. Cement should be ordinary Portland cement, or sulphur resisting if required, and sand should be clean building sand.

Plasticizers

Plasticizers are added to mortar mixes to increase the workability of the mix.

Accelerators

These are added to mortar or concrete mixes to speed up the initial setting time.

SELECTION OF MORTAR

As a general guide, the hardness or eventual compressive strength of mortar should be the same as the hardness of, or preferably slightly weaker than, the bricks to be laid. Therefore, if, as a result of slight foundation settlement, cracks develop, these will follow the joint lines, which can easily be cut out and repointed.

Excessively hard mortar can result in the bricks becoming fractured at settlement cracks, leading to a more extensive repair operation. Another reason for relating strength of mortar to the compressive strength of the bricks is to ensure that external walls weather evenly during the lifetime of the building, with any absorbed water evaporating at a similar rate from the surfaces of bricks and joints alike.

In addition to being hard enough to transfer loads evenly between irregular surfaces of bricks, the mortar must resist the effects of rain and frost in the long term.

Drainage systems

The purpose of a drainage system is to convey waste products from toilets, bidets, baths, hand basins and sinks from all public and private buildings.

The waste must be conveyed in a manner that will not cause any unpleasant odours or hazards to health. It is conveyed through a series of impervious drain pipes laid to falls which take the contents to a place where it can be treated and disposed of.

All drainage work must be carried out in accordance with the current Building Regulations.

Definitions

DRAIN

This includes all pipes, fittings, gullies and inspection chambers laid to remove soil water, waste water or surface water from *one* building. It is the property of one owner, who is responsible for its maintenance and repair.

PRIVATE SEWER

This includes all pipes, fittings, gullies and inspection chambers laid to remove soil water, waste water or surface water from *two or more* buildings. It is the joint property of two or more owners, who are responsible for its maintenance and repair.

PUBLIC SEWERS

This is any length of drain that is not covered by the definition of drain or private sewer and is the property of the local authority, which is responsible for its maintenance and repair.

SEWAGE

Effluent or sewage can be defined as unwanted water that has to be disposed of. It may comprise water-borne human, domestic, trade or farm waste, and it may also be surface water or subsurface water.

Types of water

There are two types of water in drainage systems:

- surface water
- foul water.

SURFACE WATER

Surface water may be collected from the roofs of buildings and the paved areas around them. This water is collected in the roof gutters and in gullies from paved areas, and discharged directly into the drainage system.

FOUL WATER

Water which is collected from sanitary appliances such as sinks, baths and toilets is said to be foul water.

Once water has been introduced into a building it will need to be discharged, whether contaminated or not. Any system that carries this effluent within the building site is called a drainage system and is the responsibility of the owner, but once this system passes to the highway it becomes a sewage system and is the responsibility of the local authority.

Systems

There are two main systems currently in use, with a possible third in some areas:

- combined system
- separate system
- partially separate system.

COMBINED SYSTEM

This system conveys waste water from both surface water and foul water in a single pipe to the sewage treatment works. This is very common in certain parts of the country and is often used when treatment is not available, e.g. when waste is fed into the sea.

It is the simplest and cheapest system, with no chance of connecting to the wrong pipe.

Drains are well flushed during heavy rainfall and there is only one pipe to maintain. Silting of the pipes may occur in dry weather unless self-cleansing velocities are reached during foul flow – not during storms. At treatment works spasmodic rushes of water during storms can cause operational difficulties. A typical combined system is shown in Figure 11.12.

SEPARATE SYSTEM

This system involves two separate pipes, one carrying the foul water to the sewage treatment works, and one carrying the surface water directly into a water course or the sea.

In house drainage it requires two sewer connections, with two pipes to maintain. Sometimes the pipe layout is difficult and costly and pipes may cross each other.

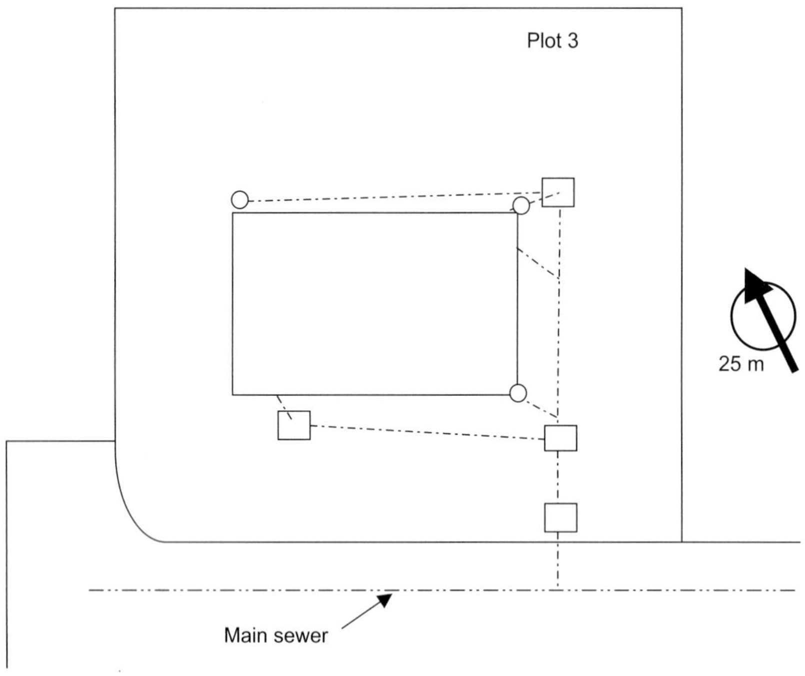
Plot 3
25 m
Main sewer

FIGURE 11.12
Combined system of drainage

The foul drain cannot be flushed by rain water, but surface water is well flushed in times of heavy rain. The treatment plant receives a flow of regular strength and sewers must be laid to correct gradients to give self-cleansing velocities. A typical separate system is shown in Figure 11.13.

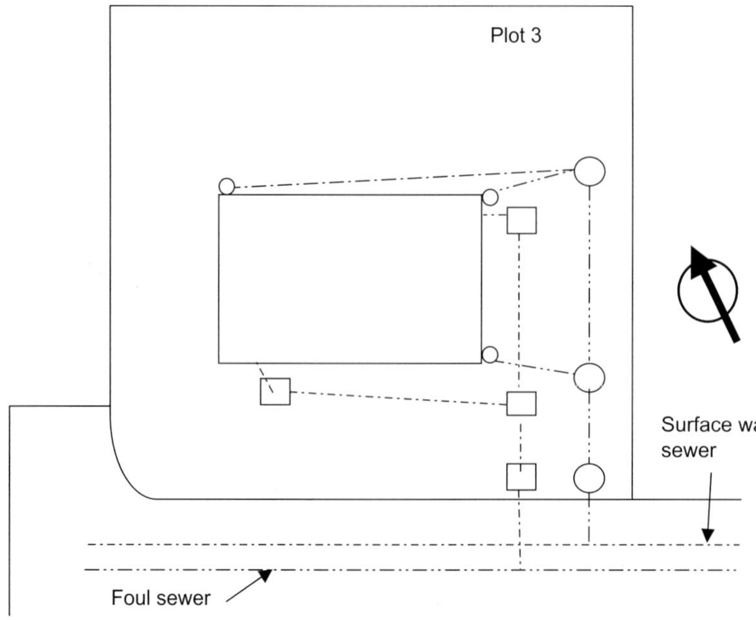
Plot 3
Surface water sewer
Foul sewer

FIGURE 11.13
Separate system of drainage

PARTIALLY SEPARATE SYSTEM

This is a combination of the two previous systems where a proportion of the surface water is conveyed to the surface water sewer, soakaway or watercourse, and the remainder is connected to the foul water system.

It is usual to connect the farthest rain water gully to the foul water system to assist in flushing out the system during heavy rain storms.

A typical partially separate system is shown in Figure 11.14.

Pipes

Pipes are manufactured from various materials, such as:

- Clay
- UPVC
- concrete
- cast iron.

Pipes and fittings must be sound and free from visible defects.

They are available in either salt glazed or ceramic glazed finishes.

All pipes manufactured in accordance with the requirements of the British Standard (BS) specification should be clearly marked, with the identifying mark of the institute stamped on them.

CLAY PIPES

Clay pipes are manufactured with or without a socket (Figure 11.15).

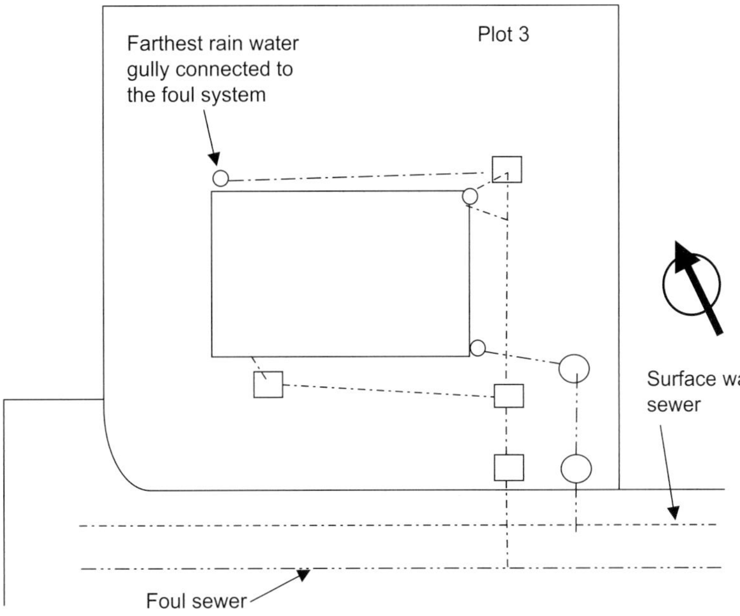
Farthest rain water gully connected to the foul system
Plot 3
1
Surface water sewer
Foul sewer

FIGURE 11.14
Partially separate system of drainage

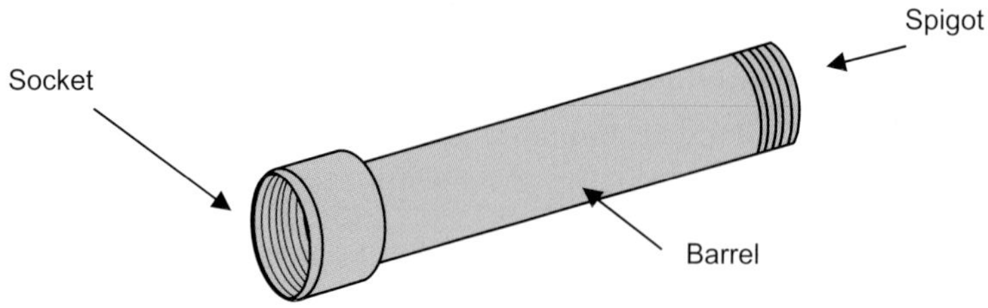
Spigot
Socket
Barrel

FIGURE 11.15
Clay pipe

Pipes must bear the manufacturer's name or trademark. They should also carry the BS kite mark.

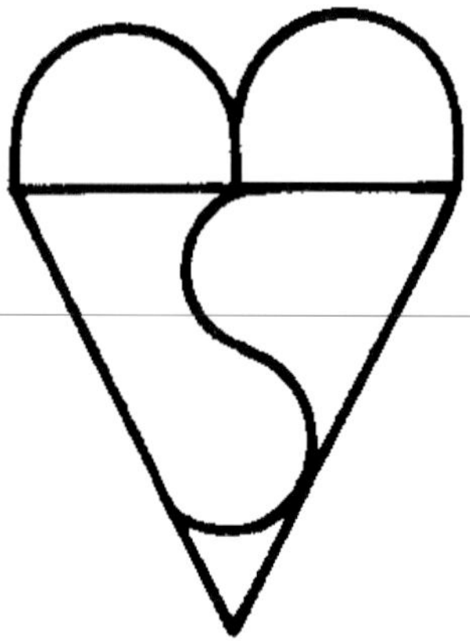

Joints

All pipes can be connected with a spigot and socket joints. Some pipes are now produced with plain ends.

> **Note**
>
> Pipes that fail the BS test are marked with a black band and are sold as 'seconds'. They can only be used for surface water drainage.

RIGID JOINTS

Each pipe is laid to line and the joint is caulked with gaskin or tarred rope. This centralizes the pipe and prevent the cement mortar from entering the barrel of the pipe (Figure 11.16).

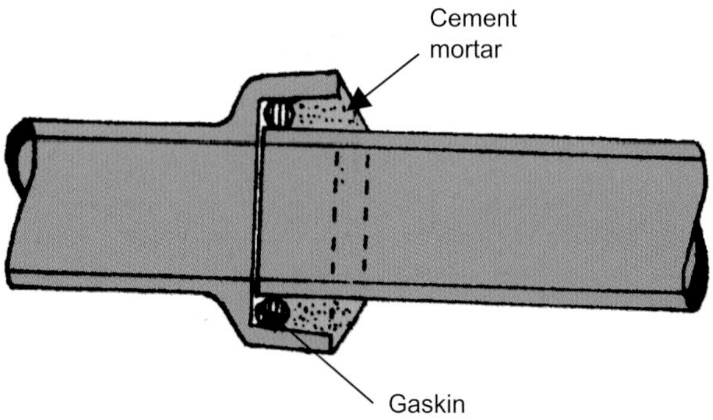
Cement
mortar
Gaskin

FIGURE 11.16
Rigid joint

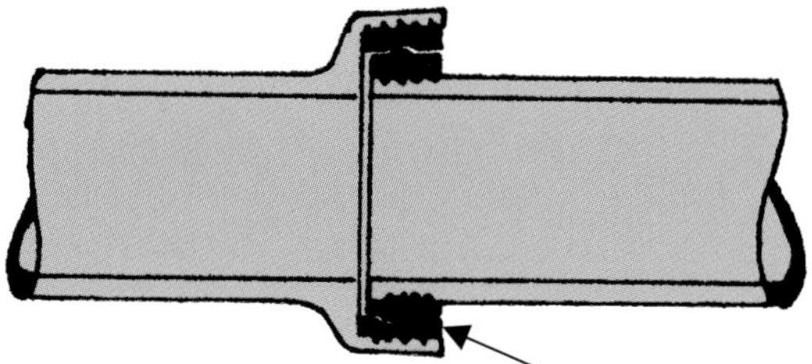
A band of resilient polyester is moulded to both pipes

FIGURE 11.17
Flexible joint

FLEXIBLE JOINTS

There are various types of flexible joint.

In some cases the jointing material is fixed to the pipes during manufacture (Figure 11.17). Others have special couplings, which when greased, slide over the ends of pipes to form a watertight and flexible joint. The pipes in this method have plain ends (Figure 11.18).

Bedding

The code of practice for building drainage recommends the following bedding for rigid and flexible pipes. Examples are shown in Figure 11.19.

FLEXIBLE BEDDING

Pea gravel is the recommended bedding for flexible pipes.

Figure 11.20 shows a typical drain trench backfilled with the correct materials

UPVC (PLASTIC) PIPES

UPVC pipes (Figure 11.21) are supplied with both ends plain for use with separate coupling. They are manufactured in longer lengths than clay pipes.

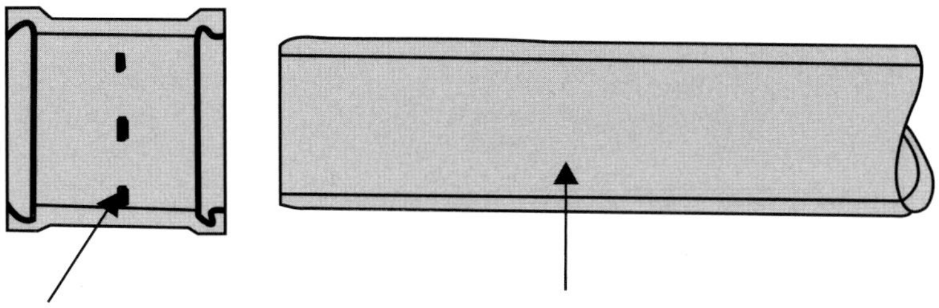
Plastic coupling with stops in the centre of each coupling
Plain-ended pipes

FIGURE 11.18
Plastic coupling

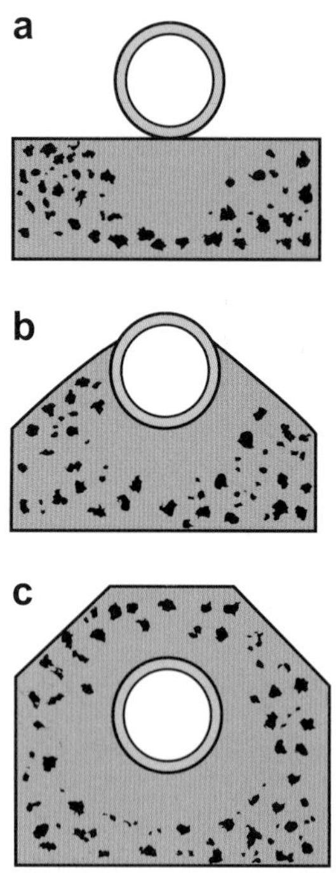
a
b
c

FIGURE 11.19
Rigid bedding: (a) bedded; (b) surrounded; (c) haunched

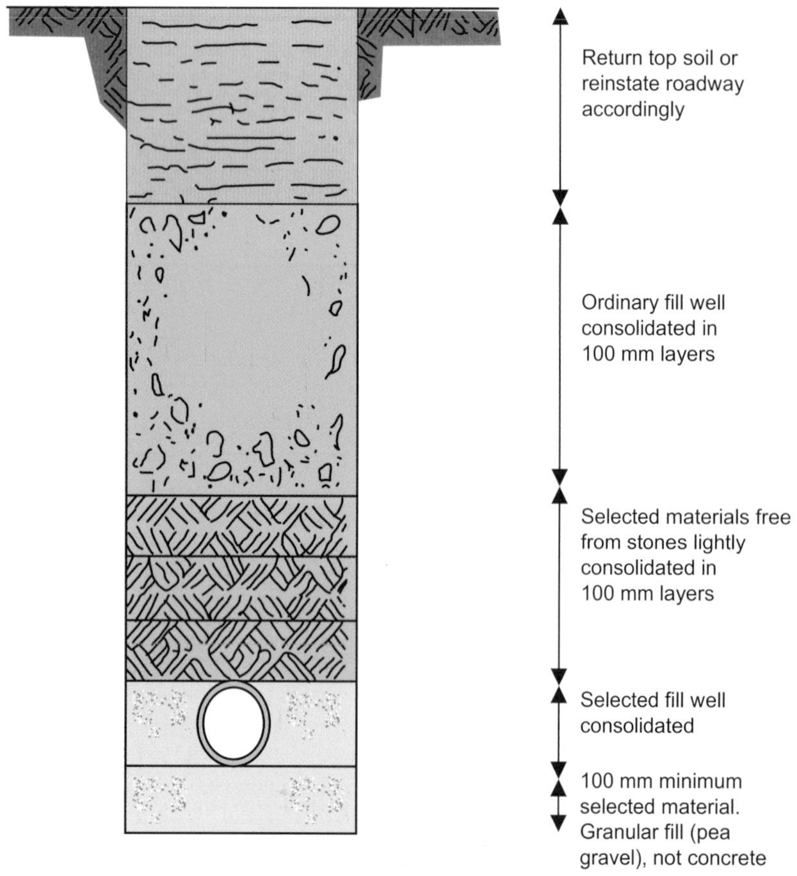
Return top soil or reinstate roadway accordingly
Ordinary fill well consolidated in 100 mm layers
Selected materials free from stones lightly consolidated in 100 mm layers
Selected fill well consolidated
100 mm minimum selected material. Granular fill (pea gravel), not concrete

FIGURE 11.20
Backfill to drain trench

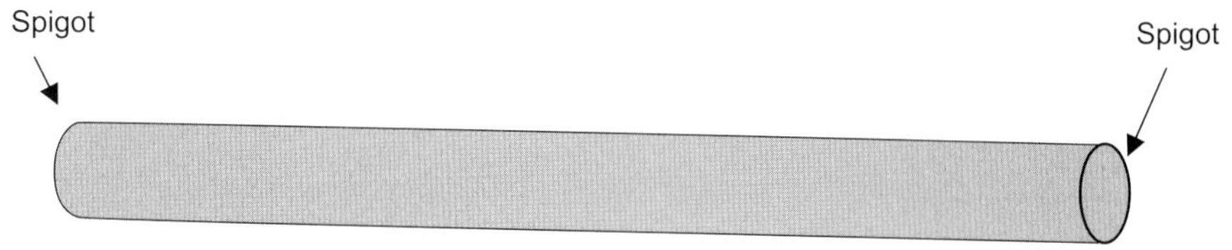
Spigot
Spigot

FIGURE 11.21
UPVC pipe

UPVC is flexible both as a pipe and at the joint. UPVC pipes must be correctly stored to prevent bending.

It does not have a crushing strength, so it must be bedded correctly to avoid continuous pressure deforming the pipe. Bedding for UPVC pipes is the same as for flexible pipes.

Pipes without sockets have to have separate couplings fitted (Figure 11.22).

Fittings

There are numerous fittings available in all shapes and sizes (Figure 11.23).

Some of the special fittings are:

- Double collars – These are used when repairing drains.
- Taper pipes – These are used when the internal diameter of the drain increases owing to the increase of flow of sewage.
- Junctions – These are used when tributary drains join the main drain without an inspection chamber being built.
- Channels – These can either be half or three-quarter section and are used in the base of inspection chambers to allow the drain to be opened for inspection and cleaning. They are also available in junctions and bends.
- Three-quarter channels – These are used to connect tributary drains to a main drain when the junction is being made within the inspection chamber. They are either right or left handed.
- Saddles – These are required when a new drain is to be connected to an existing one. A hole is carefully cut into the existing drain.

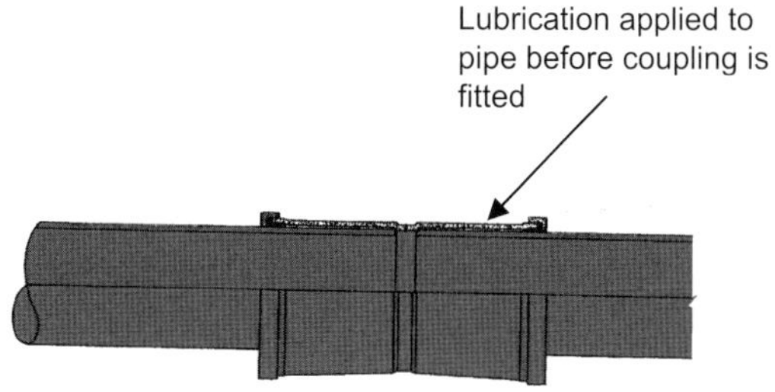
Lubrication applied to pipe before coupling is fitted

FIGURE 11.22
Coupling to pipes without collars

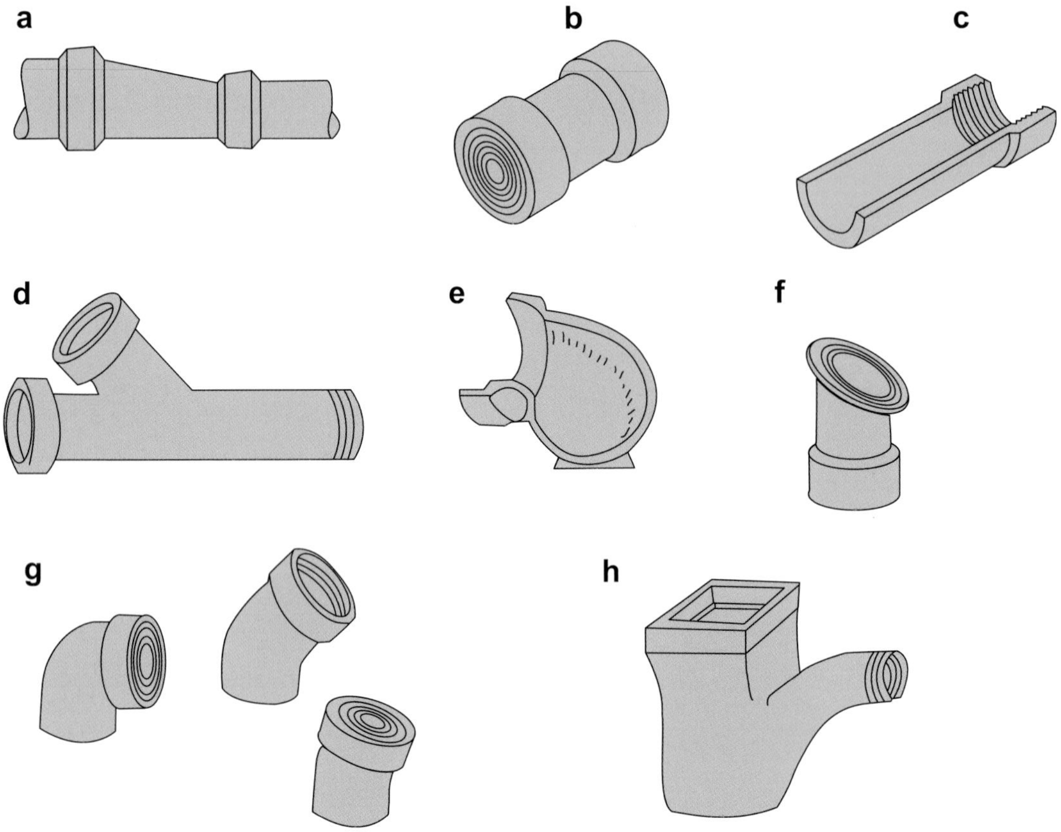
a
b
c
d
e
f
g
h

FIGURE 11.23
Various fittings: (a) taper; (b) double collar; (c) channel; (d) junction; (e) channel; (f) saddle; (g) bends; (h) gully

- When the hole is complete the saddle is bedded onto the existing pipe and allowed to set. The new drain is then connected to the saddle to allow the new drain to join the existing one.
- Bends – These are usually described by the degree of angle, such as 45 or 90 degrees. Very small quadrant bends are called knuckle bends and are used when connecting to a toilet or a rain water fall pipe.
- Gullies – Gullies are used to provide a water seal which prevents the foul smells and gases that build up in a sewer from escaping into the atmosphere.

Laying drains

As stated before, all drainage work must be carried out in accordance with the current Building Regulations.

Approved Document H2 lays down an acceptable level of performance necessary to reduce the risk of danger to the health and safety of people in the building.

1. Convey the flow of foul water to a foul outfall (public sewer).
2. Minimize the risk of blockage or leakage.
3. Prevent foul air from the drainage system entering the building.

4. Be ventilated.
5. Be accessible for cleaning blockages.

Drainage is primarily divided into two groups:

- surface water
- foul water.

Surface water is water from roofs and the surrounding ground, whereas foul water is water which is contaminated by soil, waste and trade effluent.

Surface water does not require treatment and can be discharged into a local watercourse.

Any drainage system must convey all surface water and liquid sewage away from the building in the most speedy and efficient way, possibly to the sewer, without the risk of nuisance or danger to health and safety.

When designing any system the following principles must be observed:

- Provide adequate access points.
- Keep pipework as straight as possible between access points. An access point should be provided for all bends over 45 degrees.
- Ensure that all pipework is adequately supported.
- Ensure that the pipe is laid to a self-cleansing gradient.
- The whole system must be watertight, including inspection covers.
- Drains should not run under buildings, unless this is unavoidable or in so doing they would considerably shorten the route of the pipework.

GRADIENTS

A drain should be designed with sufficient gradient to ensure that the speed of the liquid within the drain is such that it becomes self-cleansing.

A rule-of-thumb method for calculating the fall in a 100 or 150 mm diameter drain uses the following gradients:

- A 100 mm drain may have a fall of 1 in 40.
- A 150 mm drain may have a fall of 1 in 60.
- A 225 mm drain may have a fall of 1 in 90.

When levels are stated for drainage they will always refer to the invert level. It is vital that the gradient is consistent throughout the full length of a section of drain.

Changes of gradient or alignment must always occur within the confines of an inspection chamber.

A small drain trench can be set out with lines between the inspection chamber and the gully.

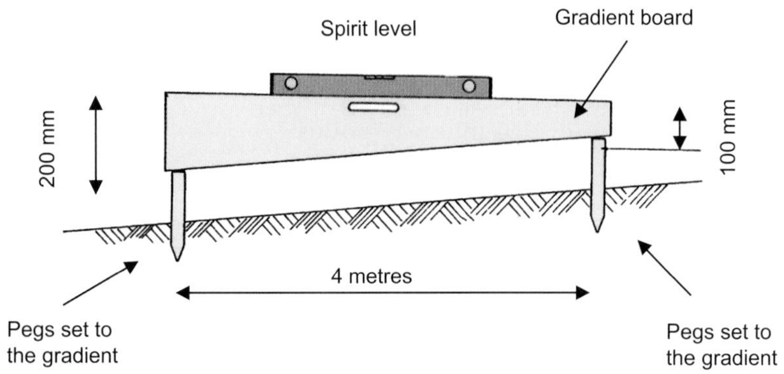
Spirit level
Gradient board
200 mm
100 mm
4 metres
Pegs set to
the gradient
Pegs set to
the gradient

FIGURE 11.24
Setting out a 1 in 4 gradient

Method 1

A tapered straight edge (gradient board) could be used to check the correct gradient (Figure 11.24). Therefore, 4000 mm/100 mm = 1 in 40.

The pegs represent the level of the granular fill or concrete bed on which the drain will be laid.

Method 2: Boning rods

Boning rods (Figure 11.25) are made of timber in sets of three, which must be identical in size.

Procedure:

1. Peg 1 is driven into the required level.
2. Peg 2 is then driven in 1 m along the trench and is given the required fall, i.e. 1:40 = 25 mm in 1 m.
3. A boning rod is placed on both pegs 1 and 2.
4. Further pegs can now be driven in at any distance apart until the height of the peg plus the third boning rod is in line with the sight line of the other two boning rods.

Method 3: Sight rails

This method is similar to the boning rod method, but this time the fixed points are at either end of the drain run (Figure 11.26).

The site rails are set out a maximum 15 m apart along the proposed drain trench.

As the excavator digs the trench the banksman checks the depth using the traveller.

The traveller acts in the same way as the third boning rod, but now extra length has been added to set out the bottom of the trench.

As excavation takes place the traveller moves down the trench establishing the correct fall of the trench bottom, which will be parallel to the line of sight.

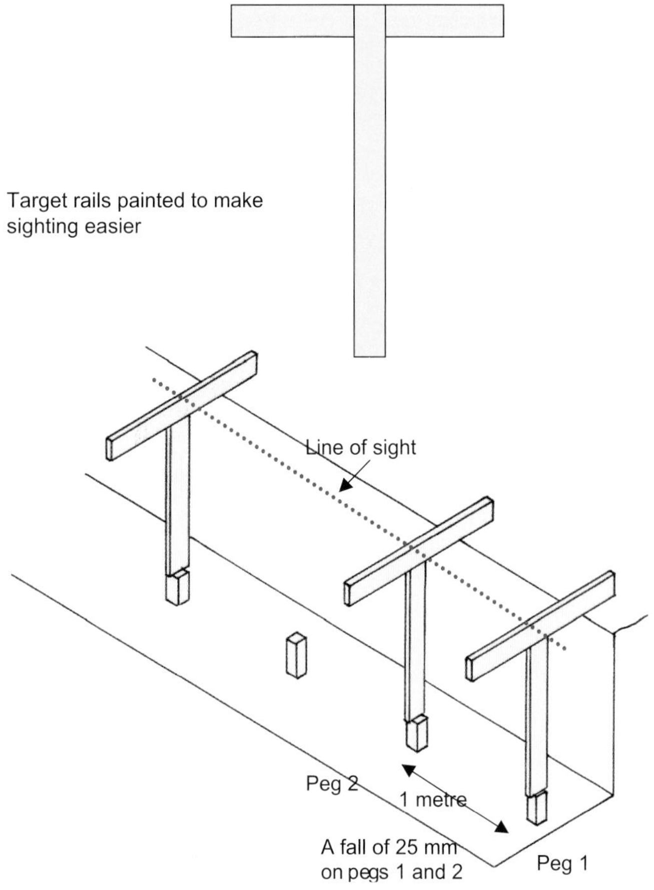
Target rails painted to make sighting easier
Line of sight
Peg 2
1 metre
A fall of 25 mm on pegs 1 and 2
Peg 1

FIGURE 11.25
Use of boning rods

When the trench has been excavated to the required gradient the bottom should be 'bottomed-up' to an even surface and uniform gradient.

The bottom of the trench is usually 100 mm deeper than required to allow the trench to be filled with 100 mm of suitable bedding material to receive the flexible drains.

Once the trench bottom is established the traveller can be shortened to invert depth and used to lay the drain pipes.

ACCESS POINTS

Access points are provided for clearing blockages from any drain runs that cannot be reached by any other means and can be either:

- inspection chambers
- rodding eyes.

All drains and sewers should have such means of access as may be necessary for inspection and cleaning. These should be placed:

- where there is a change in direction
- where there is a change in gradient

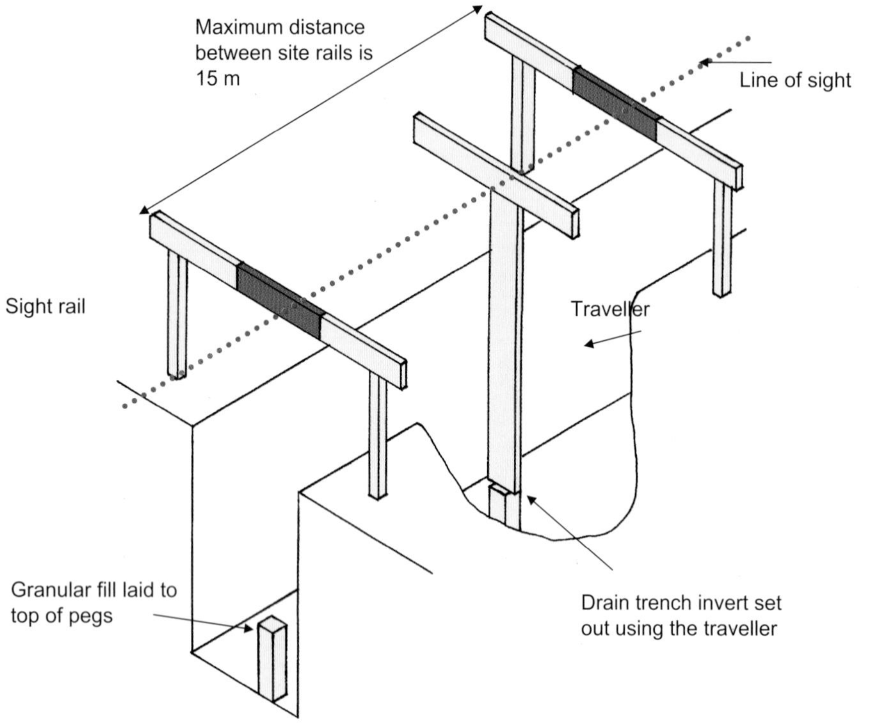
Maximum distance between site rails is 15 m
Line of sight
Sight rail
Traveller
Granular fill laid to top of pegs
Drain trench invert set out using the traveller

FIGURE 11.26
Site rails

- at the highest point of the drain
- on a drain within 12.5 m of a junction of another drain or sewer unless an inspection chamber is at the junction.

No part of a drain should be more than 45 m away from an inspection chamber

Inspection chambers can be constructed from brick, concrete and plastic.

BRICK INSPECTION CHAMBERS

Brickwork in inspection chambers is below ground and out of sight, but it must still be built correctly as the inspection chamber will have to be tested against water leaking out.

This section only provides a basic guide and operations could change around the country.

Inspection chambers are required for access to the drainage and sewerage systems. Brick chambers could also be required for access to other services.

The basic requirement of an inspection chamber is that it should be large enough to allow a person entrance and space to be able to work.

There may be branch drains entering the inspection chamber, so it should be designed according to the number of branches.

It has to be watertight and resist all possible loads from above and externally.

The recommended minimum internal sizes are:

- 450 mm long × 450 mm wide when the inspection chamber is not more than 1 m deep
- 1200 mm long × 750 mm wide if it is more than 1 m deep

When inspection chambers have branches a minimum of 300 mm should be allowed for each 100 mm and 150 mm branch on the side with the most branches.

It is also advisable to allow 600 mm at the outlet end of the inspection chamber to allow for further connections to the inspection chamber and for easy rodding. The minimum width should be 450 mm or an extra 300 mm if there is a branch.

The inspection chamber should be constructed on a concrete base, normally 150 mm thick and built with one-brick thick walls in cement mortar not weaker than 3:1. The concrete base should be at least 150 mm thick.

Some authorities may specify class B clay engineering bricks (Figure 11.10), and sulphate-resisting cement if the ground contains sulphates.

The inspection chamber is normally built in English bond, but if there is high water pressure from outside water bond may be specified.

If the inspection chamber is over 750 mm deep, step irons should be built in every fourth course beneath the opening. They should be staggered, 300 mm apart, to allow easy access.

The cover frame should be 450 × 450 mm for inspection chambers less than 1 m deep and 600 × 600 mm for those more than 1 m deep

CONSTRUCTION OF INSPECTION CHAMBERS

The main channel pipe is laid when the main run of pipes is being laid by the drain layers. Any connecting branches are also laid and completed.

The concrete base is then laid in position around the channel pipes and allowed to set.

The brickwork is set out for the inspection chamber according to the number of branches required. Approximately six courses should be built, taking care to keep the inspection chamber plumb, level and square on the inside.

Any pipes through the walls should be protected from any pressure by turning small arches over those of 150 mm and above.

The benching should be completed at this level, when it is easier. Bricks could be used to form the basic structure of the benching and then finished smoothly with a stiff concrete. The gradient should be approximately 1:12 to prevent any waste from congregating on the benching.

Once the benching has hardened the remainder of the brickwork can be laid. If the inspection chamber is over 750 mm deep, step irons should be built in to assist in access to the chamber.

The inspection chamber is usually roofed over with an in situ reinforced concrete slab with an access hole. The brick shaft can be extended until the correct height is reached and the inspection cover fitted. Smaller inspection chambers may have the brickwork corbelled inwards to receive the inspection cover, instead of a concrete slab.

Brick inspection chambers have been superseded on many construction sites by precast concrete and plastic inspection chambers, which are circular on plan. The bricklayer may still be required to construct the small chamber on top of the concrete inspection chamber to receive the inspection cover.

Brickwork inspection chambers less than 1 m deep can be built as shown in Figure 11.27. Those more than 1 m deep can be built as shown in Figure 11.28.

Figure 11.29 shows details of arches, benching and corbelling to inspection chambers.

Water bond is used to provide a watertight inspection chamber. Stretcher bond is used, but each leaf is built with bed and cross-joints staggered (Figure 11.30). Cement rendering is used between each leaf for added waterproofing.

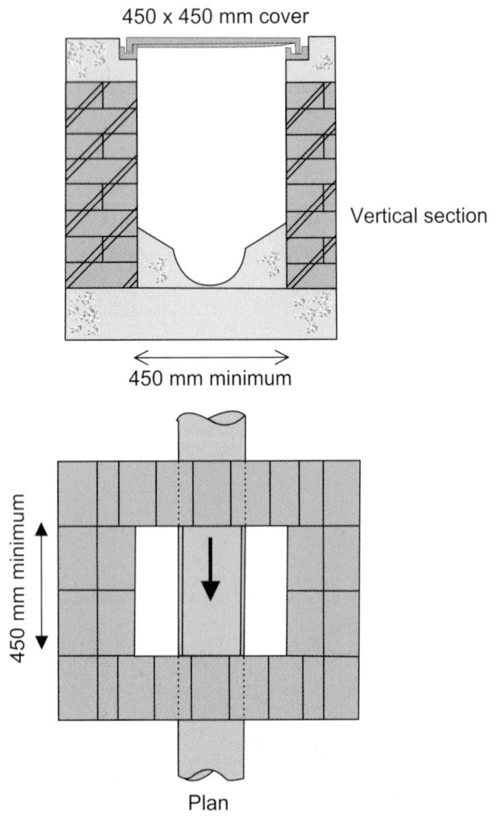
450 x 450 mm cover
Vertical section
450 mm minimum
450 mm minimum
Plan

FIGURE 11.27
Small inspection chamber

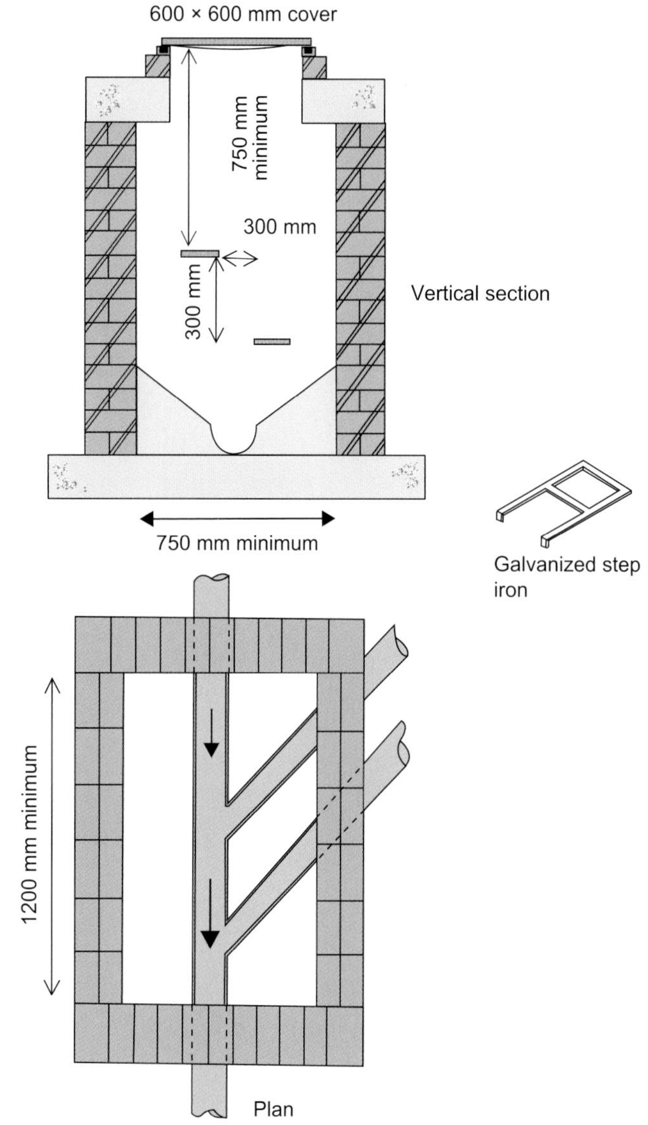
600 × 600 mm cover
750 mm minimum
300 mm
300 mm
Vertical section
750 mm minimum
Galvanized step iron
1200 mm minimum
Plan

FIGURE 11.28
Large inspection chamber

RODDING EYE

A rodding eye is used in place of an inspection chamber to enable the drain to be cleaned (Figure 11.31).

PLASTIC INSPECTION CHAMBERS

The base unit is available with several inlets, all of which are blanked off. Those required for the design will need to be knocked out.

The base is bedded on a shingle base or 150 mm wet concrete to ensure a uniform base. Raising pieces, each 650 mm deep, are added until the required height is reached. They can be cut with a saw to reach the required depth. A cover and frame are fitted to the top when the required height has been reached.

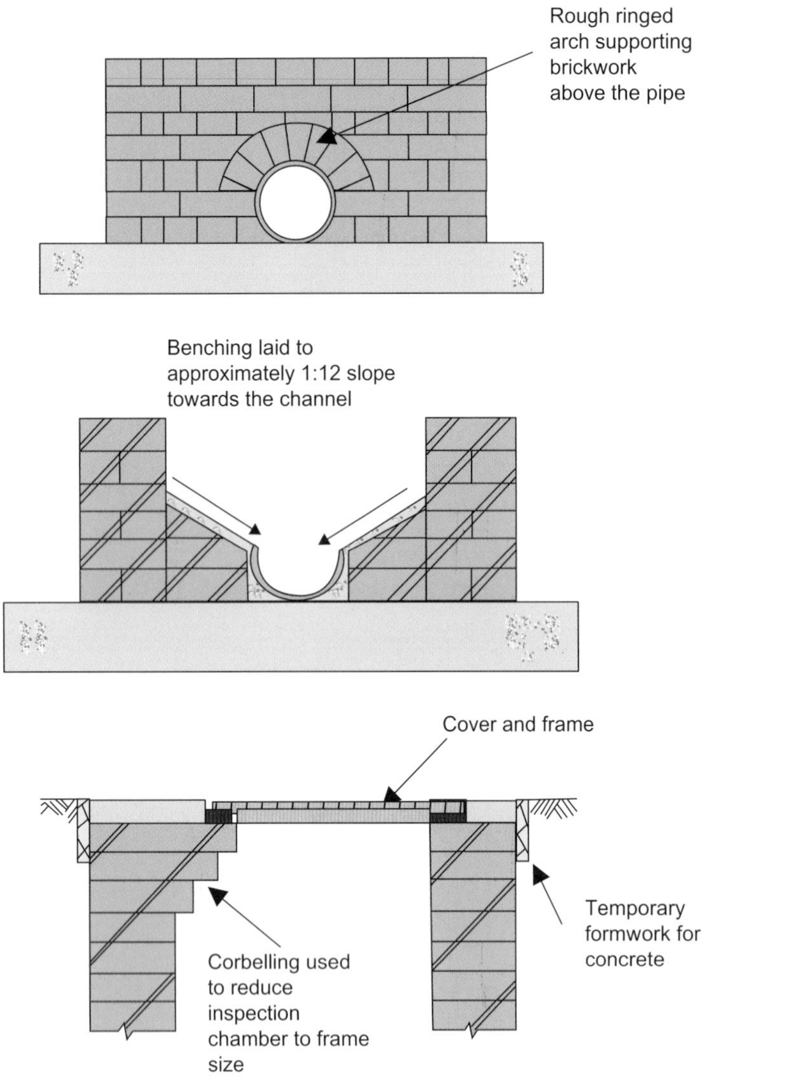
Rough ringed arch supporting brickwork above the pipe
Benching laid to approximately 1:12 slope towards the channel
Cover and frame
Temporary formwork for concrete
Corbelling used to reduce inspection chamber to frame size

FIGURE 11.29
Inspection chamber details

Precast concrete inspection chambers

Once the drain run has been completed, the channel pipes have been set and the branch connections made if required, the concrete base can be placed as for brick inspection chambers. More concrete should be placed to form the required benching to form a level base to receive the first chamber section. The chamber sections are either rectangular or circular on plan.

The first chamber section should be bedded onto the concrete base. Further chamber sections are bedded until the correct height is reached. A cover and surround are then positioned; these are available in various sizes and strengths. For deeper chambers sections are available with step irons already fitted.

Drain testing

After the pipe run has been completed it should be tested. All drains have to be tested before and after backfilling.

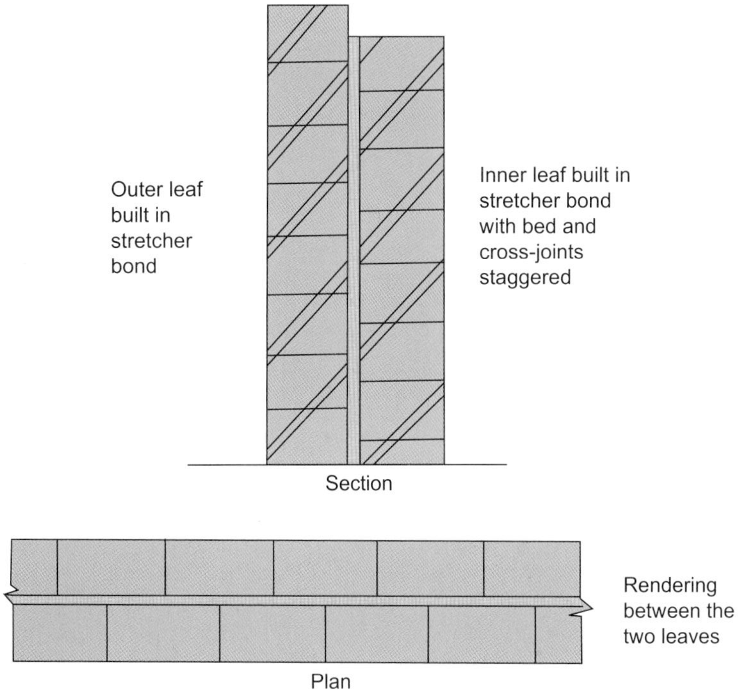
Outer leaf
built in
stretcher
bond
Inner leaf built in
stretcher bond
with bed and
cross-joints
staggered
Section
Rendering
between the
two leaves
Plan

FIGURE 11.30
Water bond

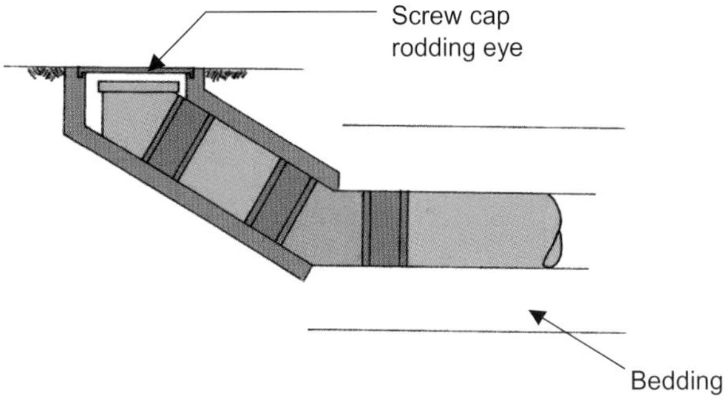
Screw cap
rodding eye
Bedding

FIGURE 11.31
Rodding eye

There are three main methods of testing for watertightness:

- water
- air
- smoke.

The most common method is the air test, as the water test could cause problems with the removal of the water after the test.

EQUIPMENT AND PROCEDURE FOR CARRYING OUT THE AIR TEST

The following equipment is required (Figure 11.32):

- Expanding drain plug – These are available in sizes from 50 to 300 mm. They are placed into the drain and tightened.

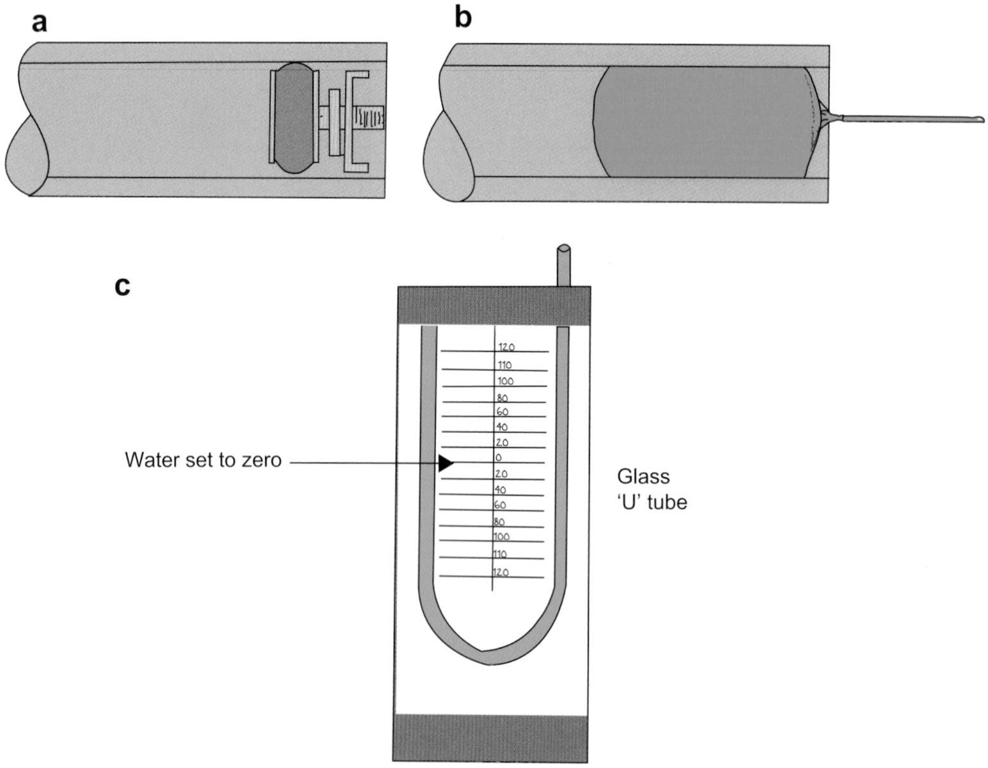
a
b
c
Water set to zero
Glass
'U' tube

FIGURE 11.32
Drain testing equipment: (a) expanding drain plug; (b) inflatable stopper; (c) U gauge

- Inflatable stopper – These are canvas-covered rubber air bags which are placed into the drain and then inflated from above.
- U gauge – The U gauge is connected to the stopper at the highest end.

Inspect the drain system for possible damage during or after laying.

Check the condition of and fit the drain plugs to all ends of the drain.

After the drain has been sealed with drain plugs, air is pumped into the drain to approximately 0.021 N/mm^2 (using a hand pump or bellows).

The U gauge is connected to the highest stopper: 100 mm of pressure should be indicated on the U gauge connected to the system. The pressure should not drop by more than 25 mm in five minutes.

PROCEDURE FOR CARRYING OUT THE SMOKE TEST

Inspect the drain system for possible damage during or after laying.

Check the condition of and fit drain plugs at the highest ends.

Before the lowest drain plug is fitted a smoke bomb is lit and placed inside the drain before fitting the last drain plug.

Any leaks will be spotted by smoke coming from the joints in the drain system.

PROCEDURE FOR CARRYING OUT THE WATER TEST

As with all methods, inspect the drain system for possible damage during or after laying.

Check the condition of and fit the drain plugs to lower ends of the drain.

If rigid joints have been used with cement mortar joints, allow 24 hours to dry before testing.

With the water test 1.5 m of head is required. This can be achieved by attaching a pipe to the highest point of the drain (Figure 11.33).

Fix the stopper to the lowest points of the drain and fill the drain with water to the top of the header pipe. Allow to settle and then check for leaks. Some water may be lost in absorption – top up the header pipe as required.

The loss of water is now checked over a period of 30 minutes and loss should not exceed 1 litre per hour, per linear metre, per metre of nominal internal diameter.

Example 6

Given a drain run, 30 m long in 300 mm diameter pipes, the permitted allowable loss per 10 minutes is:

$$\frac{1 \times 10 \times 30}{60} \times \frac{300}{1000} = 1.5 \text{ litres}$$

Testing inspection chambers

Inspection chambers are generally tested by fitting an inflatable stopper in the outlet of the inspection chamber, so the air can be removed at ground level. All the remaining inlets and outlets are also stopped up.

The chamber should then be filled with water and left to stand for eight hours. Topping up can be carried out as necessary.

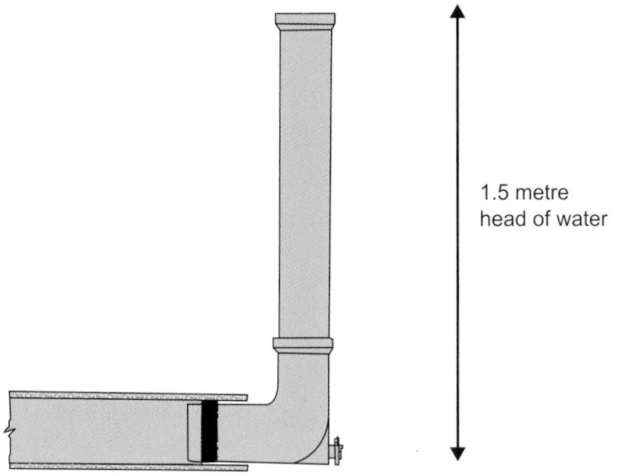
1.5 metre
head of water

FIGURE 11.33
Calculation of loss of water

The acceptable criterion is that no water should be observed issuing from the outside of the chamber.

Backfilling

The drain trench can then be backfilled in 100 mm layers for 300 mm. Subsequent backfilling can be carried out using excavated material in layers not exceeding 300 mm.

Multiple-choice questions

Self-assessment

This section of the book is designed to allow you to check your level of knowledge. The section consists of revision questions for this chapter. The questions are all multiple choice and have four possible answers. The answers are to be found at the end of the book.

The main type of multiple-choice question will be the four-option multiple-choice question. This will consist of a question or statement, known as the stem, followed by a choice of four different answers, called the responses. Only one of these responses is the correct answer; the others are incorrect and are known as distracters.

You should attempt to answer the questions by choosing either (a), (b), (c) or (d).

Example

The person employed by the local authority to ensure that the Building Regulations are observed is called the:

(a) clerk of works
(b) building control officer
(c) council inspector
(d) safety officer

The correct answer is the building control officer, and therefore (b) would be the correct response.

Domestic drainage

Question 1 Identify the following drain fitting:

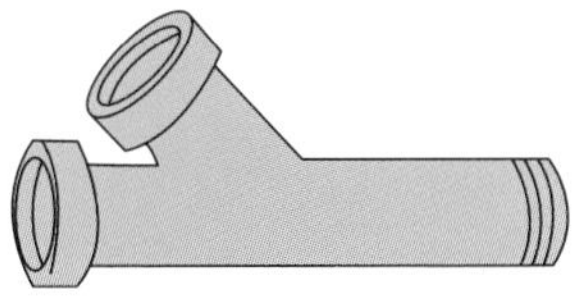

(a) junction
(b) saddle
(c) double collar
(d) bend

Question 2 The maximum distance between access points must not be greater than:

(a) 35 m
(b) 40 m
(c) 45 m
(d) 50 m

Question 3 Calculate the number of drain pipes required for a trench 13.5 m long if 0.900 mm pipes are being used:

(a) 15

(b) 13

(c) 11

(d) 9

Question 4 Which drainage system conveys waste water from both surface water and foul water in a single pipe to the sewage treatment works?

(a) separate system

(b) combined system

(c) partially separate system

(d) joint system

Question 5 What is the significance of a black band printed on a drain pipe?

(a) first class pipe

(b) can only be used in foul water systems

(c) second class pipe

(d) can only be used in surface water systems

Question 6 What is a joint caulked with tarred hemp and finished with cement mortar known as?

(a) a flexible joint

(b) a solid joint

(c) a patent joint

(d) a rigid joint

Question 7 What is the recommended gradient for a 150 mm drain?

(a) 1 in 30

(b) 1 in 40

(c) 1 in 50

(d) 1 in 60

Question 8 Identify the equipment shown, used for setting out drain trenches:

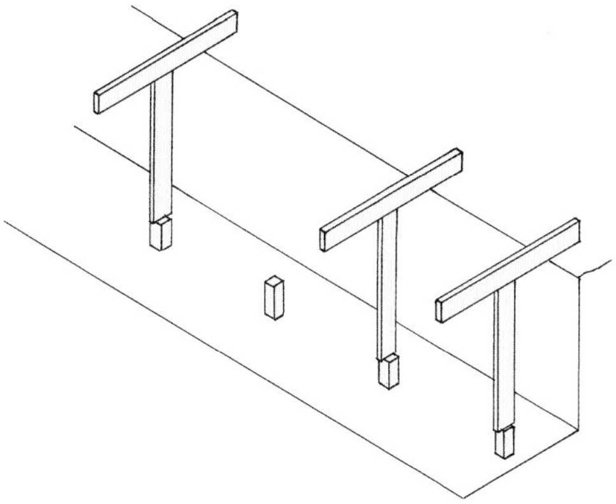

(a) gradient boards

(b) boning rods

(c) profile boards

(d) site rails

CHAPTER 12

Placing and Finishing Concrete

This chapter will cover the following NVQ and Diploma units:

- NVQ VR45

This chapter is about:

- Interpreting information
- Adopting safe and healthy working practices
- Selecting materials, components and equipment
- Preparing, placing and finishing concrete slabs and bases

The following NVQ performance criteria will be covered:

- Performance criterion 1: Interpretation of information
- Performance criterion 2: Safe work practices
- Performance criterion 3: Selection of resources
- Performance criterion 4: Minimize the risk of damage
- Performance criterion 5: Meet the contract specification
- Performance criterion 6: Allocated time

The following Diploma outcomes will be covered:

There is no comparable Level 2 Diploma unit but the chapter will give the student an insight into placing and finishing concrete.

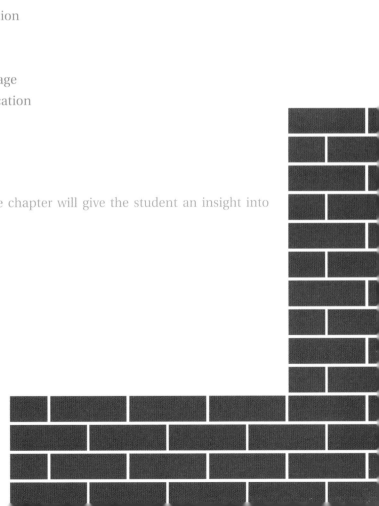

Interpreting information

This chapter deals with placing and finishing concrete. Mixing of concrete has already been dealt with in Level 1, Chapter 6. Please refer back to this if mixing details are required.

It is important to read and understand any instructions given to ensure that the concrete is mixed, transported and finished correctly.

Materials and methods are constantly being introduced into the industry and it is important that the users of these materials keep up to date with this ever-changing industry.

Types of instructions

There are many sources of information available, both oral and written. Oral information may come from your supervisor and written could be from manufacturer's information.

When mixing on site information is usually from the person in charge, who finds the information from contract documents such as the specification or bill of quantities.

When a new material or item of equipment is introduced to the industry the manufacturer will provide written instructions on its safe use.

Every designer and contractor must be able to call on a team of specialists and other back-up information.

It is essential that all parties have access to various sources of up-to-date information. No single person in isolation could satisfy the demands of all members of the design team and construction teams.

It is also very important that the student understands where information can be found. Information can be gathered from numerous sources, such as:

- manufacturers' literature
- organization handbooks and manuals
- legal documents
- general reference books.

MANUFACTURERS' TECHNICAL INFORMATION

Many manufacturers produce technical information, which is free for the asking. These papers, brochures, leaflets, etc., should be stored in the office in an easily retrievable system.

ORGANIZATION HANDBOOKS AND MANUALS

Most bodies, including professional institutions and employers' and research organizations, publish handbooks and manuals, listing details of their members and giving additional information relevant to the function of the body.

LEGAL DOCUMENTS

Constant reference must be made throughout the duration of any contract to all relevant contractual documents. It is important that these are always available. Examples are:

- the contract document
- specifications
- bills of quantities
- form of tender
- schedules.

GENERAL REFERENCE BOOKS

The industry has numerous books to meet the needs of all members of the construction industry. Publishers ensure that the industry keeps abreast of technical advances that take place in construction.

Legislation

It is important that operatives understand their responsibilities regarding current legislation while working:

- in the workplace
- below ground level
- at height
- with tools and equipment
- with materials and substances
- with movement and storage of materials
- with mechanical handling and lifting equipment.

Health and safety legislation has been covered in detail in Chapter 2.

Legislation, the main form of English law, is an Act of Parliament. Most legislation only provides a basis from which to operate.

The Health and Safety at Work Act is a typical example, where the agency carrying out the details and arrangements of the Act is the Health and Safety Commission.

Other building legislation is outlined below.

BUILDING REGULATIONS

Under the control of the Secretary of State for the Environment, the main purpose of the Building Regulations is to ensure that buildings are constructed so that they are safe and do not present danger to those who occupy them. In simple terms, they describe how a building should be constructed.

The current regulations appropriate to this chapter control the materials used in the production of concrete.

TOWN AND COUNTRY PLANNING ACTS

There are many Acts in existence that affect planning and the control of development. The main aim of them all is to safeguard the public regarding buildings and developments. They are mainly dealt with by local government offices.

OTHER LEGISLATION

The following legislation also affects the construction industry:

- Building Control Act
- Clean Air Act
- Factories Act
- Guard Dogs Act
- Historic Building Act
- Housing Act
- Road Safety Act
- Water Acts, etc.

BRITISH STANDARDS

These are issued by the British Standards Institution to lay down minimum standards for materials and components (e.g. walling blocks and doors) used in construction and other industries. They usually have the kite mark printed on them (Figure 12.1).

Standards greatly simplify the workings of construction and are quoted in:

- specifications
- Building Regulations, etc.

British Standard codes of practice

Code of good practice are issued by the British Standards Institution, to cover workmanship in specific areas, e.g. building drainage, and brick and block masonry.

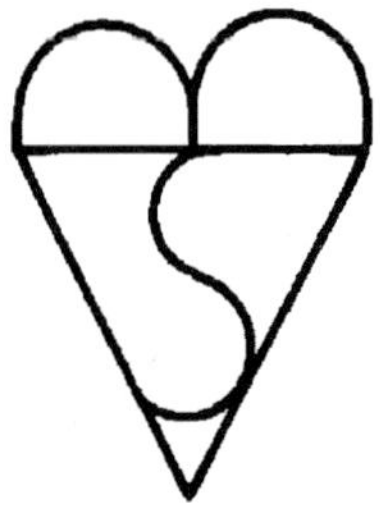

FIGURE 12.1
British Standard kite mark

AGRÉMENT CERTIFICATE

These certificates are granted by an independent testing organization, called the British Board of Agrément, stating that the manufacturer's products have satisfactorily passed agreed tests.

Subsequent to the granting of the certificate strict quality control has to be continued.

The British Standards are now being overtaken by the European Standards, with the CE logo (Figure 12.2).

Safe working practices

When dealing with concrete there could be times when operatives involved have to respond to a situation.

Your firm will have put into place a procedure to follow in such an instance. Always ensure that you follow it.

Always make sure that the appropriate personal protective equipment (PPE) is used. When working placing concrete ensure that wellington boots have steel protection.

Safety

There are three main dangers when transporting, placing and finishing concrete:

- stress injuries
- mechanical injuries
- chemical injuries.

STRESS INJURIES

Lifting and shovelling heavy materials incorrectly may lead to many injuries, in particular to the back muscles and spine.

Whenever handling materials it is essential to take care of yourself and others. Always use the correct lifting techniques.

Injuries can occur through:

- poor material handling
- poor mechanical usage.

FIGURE 12.2
European Standard mark

Material handling

Lifting or carrying heavy or awkward objects, such as bags of cement, can cause injury if performed incorrectly. Bad handling also increases fatigue (Figure 12.3).

The weight of cement bags has now been reduced to bags of 25 kg to avoid strain.

Before handling any materials always read the instructions and find out how heavy the materials are. If in doubt seek the help of other workmates.

Examples of typical materials are:

- cement = 25 kg (previously 50 kg)
- lime = 25 kg
- gypsum plaster = 40–50 kg
- wheelbarrow full of concrete = 100–150 kg
- buckets of adhesive = various depending on the size
- 1 litre of water = 1 kg.

Ensure that all formwork has nails removed before stacking. Never leave nails protruding from formwork.

MECHANICAL INJURIES

Many injuries are caused by a lack of proper maintenance and the incorrect use of tools and equipment.

Remember that there is a correct way of lifting and pushing a wheelbarrow.

FIGURE 12.3
Safe handling techniques

A few simple dos and don'ts can make using tools and equipment quite safe. The first is probably the most important!

Note

NEVER, ever put your hand inside the drum of a mixer while it is moving. This is obviously dangerous.

Catching your hand or coat sleeve in the paddle fins leads to the most spectacular injuries, from tearing off a few fingers to screwing the arm round until it is wrenched off at the shoulder.

Never put the end of the shovel or anything else into a moving drum: if it snatched, it will spin round and hit you before you know what is happening.

Remember

ALWAYS read all the manufacturer's instructions on how to operate the equipment and essential maintenance.

Always clean the equipment properly after use.

Clean equipment is a more efficient tool.

Remember

NEVER throw concrete.

ALWAYS report accidents.

Remember

Accidents do not just happen – they are caused.

CHEMICAL INJURIES

Cement, lime and plaster are caustic when wet and will cause burns to the skin, especially if it is already broken.

Eyes are particularly vulnerable to these materials, even as a dry powder because the eye is always wet.

Always read the manufacturer's instructions and warning notices printed on the bag or can, until you are completely familiar with them.

Maintaining safety

If in any doubt about safety on the building site, refer to Chapter 2 on health and safety.

Remember that any type of work undertaken by the construction industry is often difficult and hazardous.

The type of work and conditions are different on each site; consequently, the hazards are also different.

It is of utmost importance that *all* trainees are capable of using machinery and equipment efficiently and safely.

Furthermore, they should be aware of the causes of accidents and be able to take actions to deal with any accidents that may occur.

Security procedures

All sites should have arrangements in place to store items of equipment and personal belongings. These can range from a small metal or wooden cabin to the property that is being worked on.

Materials suitability

During this chapter you will be dealing with the placing and finishing of concrete – used for foundations, oversite concrete, reinforced concrete-framed building, paths and drives, etc.

Characteristics of materials

There are numerous materials with which you could come into contact while preparing the area, placing and finishing concrete, such as:

- concrete
- formwork
- reinforcement

Table 12.1 New terminology

Old standards	New standards
Mix	Concrete
Strength or grade	Strength class
Slump or workability	Consistence (target or class)
PC/OCP	CEM1
20 mm aggregate	10/20
10 mm aggregate	4/10
Sand	0/2MF

- release agents
- damp-proof membrane.

Concrete

Concrete can be either mixed on site or premixed off site.

BS 5328 has been replaced by the new European Standard BS EN 206-1:2000 Concrete Part 1, which deals with producing fresh concrete.

The main changes are in the terminology. There are some new terms and some existing terms with new meanings. Table 12.1 explains some of the changes.

Concrete is one of the few building materials that can be produced on the building site.

The main advantage of using concrete is its versatility. It can be moulded to any required shape and its load-bearing capabilities can be increased by casting in steel reinforcing bars.

There are several types of concrete.

- dense concrete
- lightweight concrete
- air-entrained concrete.

CONSTITUENTS

The three main constituents used to manufacture concrete are:

- cement
- aggregates
- water.

MANUFACTURE OF CONCRETE

The manufacture of hardened concrete involves two stages: the plastic (setting) stage and the rigid (hardening) stage. During both of these stages the chemical process of hydration occurs, where the cement reacts with the water.

The aggregate, although present, does not take part in the chemical reaction.

$$Cement + Aggregate + Water \rightarrow Concrete + Heat$$

Formwork

Apart from when concrete is used for foundations, where the ground supports the side of the concrete, formwork is required to hold the concrete in place while it sets.

The formwork and the way it is made and used play a great part in the finished appearance. Apart from its appearance, formwork usually needs to be used many times and this can only be done if it is carefully and properly handled, cleaned and stored.

The materials that can be used include timber, steel and fibreglass.

Formwork should:

- be strong enough to resist any pressure from the wet concrete
- be designed to be easily erected and dismantled without damage
- be of reasonable shape and size to be easily placed and removed
- have reasonable joints to prevent water leakage and water loss.

Formwork for small slabs, paving slabs, etc., can be constructed as shown in Figure 12.4. Formwork for the sides of large slabs may consist of timber beams supported with timber pegs (Figure 12.5a). An alternative method is used where there is a great deal of pressure from the concrete (Figure 12.5b).

Steel formwork is available but it is not as versatile as timber. It is purchased in off-the-shelf sizes. Steel formwork offers a simple means of dealing with repetitive work and can be reused quickly (Figure 12.6). All shapes and sizes are available, especially for curved work.

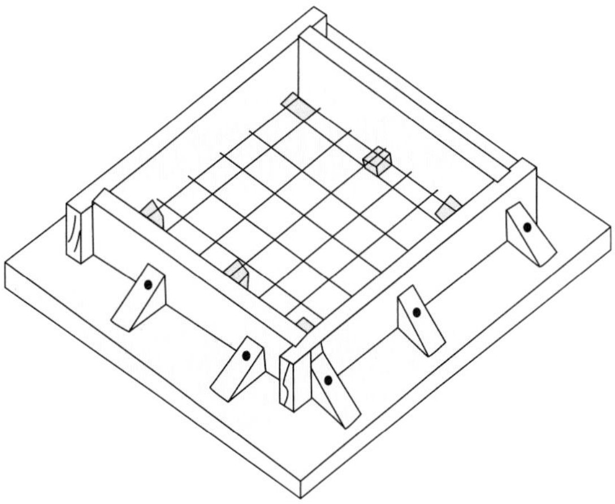

FIGURE 12.4
Formwork for small slab

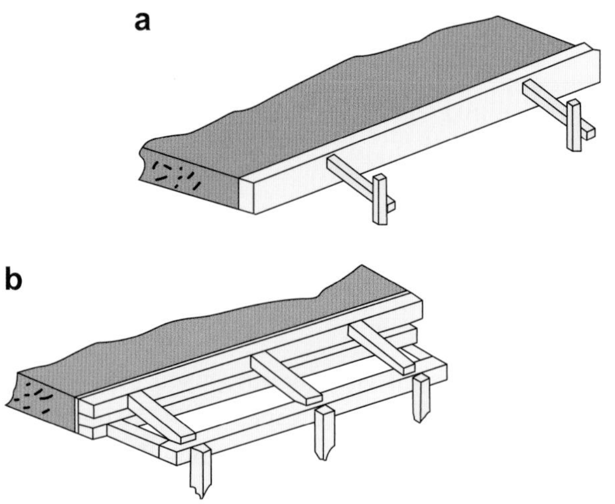
a
b

FIGURE 12.5
Timber formwork: (a) simple timber formwork; (b) timber formwork to support more pressure

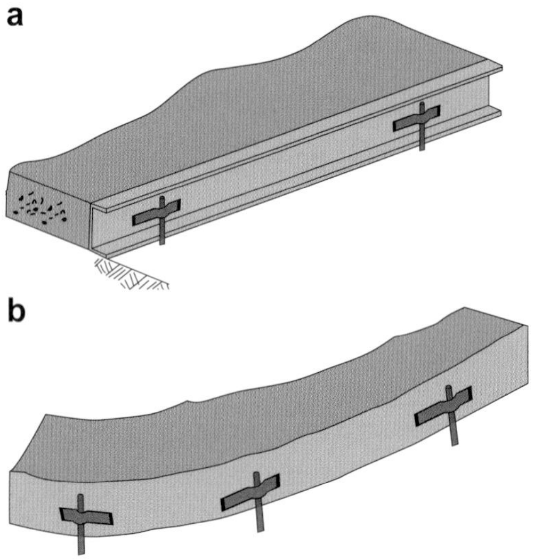
a
b

FIGURE 12.6
Steel formwork: (a) simple steel formwork; (b) curved steel formwork

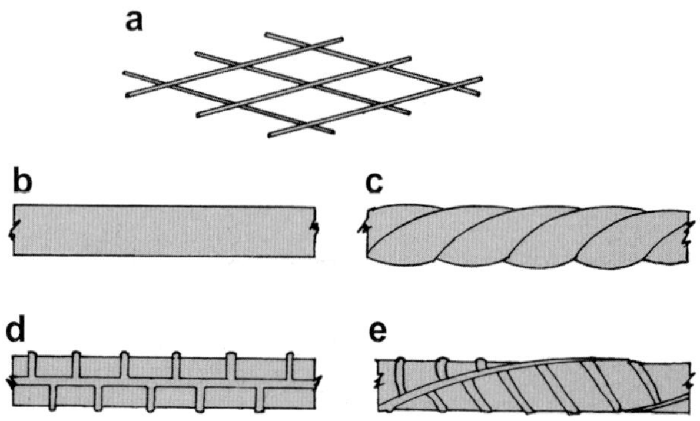
a
b
c
d
e

FIGURE 12.7
Types of steel reinforcement: (a) mesh reinforcement; (b) plain; (c) twisted; (d) ribbed; (e) ribbed and twisted

Reinforcement

The position of reinforcing within a floor slab or lintel should be specified.

All reinforcing requires cover from the concrete to protect it from rusting. Simple floor slabs are constructed with a reinforcing mat which is placed in position on spacers before the concrete is placed. It is important to ensure that the steel used for reinforcing concrete is free from mill scale, loose rust, grease or site mud, before placing it.

Various types of reinforcing bars are shown in Figure 12.7.

Release agent

This is applied to formwork to allow the removal of the formwork without damage to the face of the concrete. It is applied to the inside face of the formwork immediately before placing the concrete.

It is available as a liquid or a cream. The type used depends on the required finish. Care must be taken not to stain the finished face of the concrete.

Damp-proof membrane

When concrete is used for ground floor slabs it may be necessary to lay a horizontal membrane to protect the floor from rising damp (Figure 12.8).

Striking formwork

This is the term used when removing the formwork after the concrete has set.

Vertical sides of concrete slabs, etc., can be removed after 12 hours, but any soffits should be left for 7–14 days depending on the type of concrete.

Remember to clean and stack away all formwork.

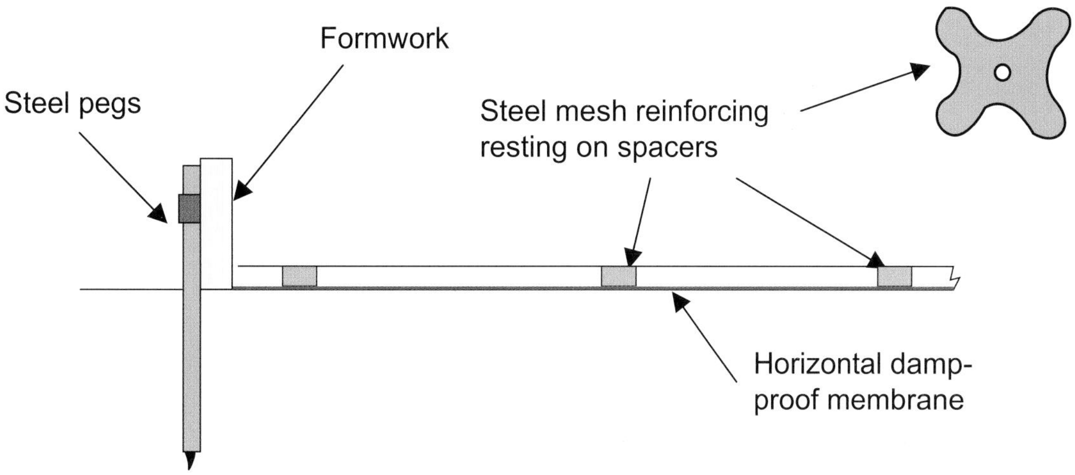
Formwork
Steel pegs
Steel mesh reinforcing resting on spacers
Horizontal damp-proof membrane

FIGURE 12.8
Simple formwork for a concrete ground floor slab

Transporting concrete

Concrete may be transported from the mixer or ready-mixed concrete lorry to the point of placement in a number of ways, some specifically developed for concrete, others being multipurpose.

The chosen method will depend on the size of the site and the position where the concrete is required. In all cases the concrete must be transported so that it does not segregate and so that it does not become contaminated with any other material after it has left the mixer.

Where concrete is to be placed below the level of supply, a chute can be used because gravity is the cheapest method of transport.

Concrete should generally be placed within half an hour of mixing.

Methods of transport range from hand wheelbarrows to concrete pumps (Figure 12.9).

- Barrow – The most common method on small building sites is the wheelbarrow.
- Dumper – There are many sizes and designs of dumper, which can be used for transporting concrete from mixers up to 600 litre capacity.

Remember

There is a danger of segregation, so concrete should not be dropped more than 1 m.

a
b
c
d
e
f

FIGURE 12.9
Transporting machinery: (a) dumper; (b) ready-mixed concrete lorry; (c) barrow; (d) lorry-mounted concrete pump; (e) crane; (f) front loader

- Ready-mixed concrete lorry – When ready-mixed concrete is prescribed, large amounts can be delivered either ready mixed or site mixed to most site positions.
- Pumps – Concrete pumps are used to transport large volumes of concrete (up to 100 m^3) in a short period, in both vertical and horizontal directions.
- Cranes – these are available in many makes and sizes according to the job in hand. Cranage can be either mobile or fixed. Cranes can span the whole building area so are excellent for the placing of concrete on large multistorey buildings. There are also several types of machine available to move concrete around construction sites, such as front loaders.

Placing concrete

The concrete mixture is usually placed into formwork to obtain the required shape.

The workability of fresh concrete is important in assessing the practicability of placing and compacting the mix, and also in maintaining consistency throughout the job.

The inside of forms should be inspected before concrete is placed to make sure that they are clean and have been treated with a release agent. Rubbish such as sawdust, shavings and wire should be blown out before the release agent is applied and the concrete is placed.

The formwork into which the concrete is to be placed should have been treated with any release agent necessary to allow its removal from the hardened concrete.

Concrete should be placed into its final position as quickly as possible.

Note

Never add too much water, or it will dry out and leave voids which will weaken the concrete.

Before concrete is placed into any mould or formwork certain preparations may have to take place.

Remember

On no account add water to a mix after it has left the mixer or ready-mixed concrete lorry.

Compacting concrete

The concrete mixture may need to be vibrated to remove any air voids formed during placing. This is known as compaction of the concrete mixture.

If concrete is to reach its maximum strength, it must be compacted so that it contains the minimum amount of unwanted air. This is easy with wet concrete but the extra water contributes to weakness. The strongest and most durable concrete is as dry as can be fully compacted by the means available. Vibration leads to better concrete.

Slabs can be compacted by tamping beams (Figure 12.10). These combine the action of a screed and a vibrator, but they are only effective for a limited depth.

In general, a slab more than 150 mm thick should be compacted with a poker vibrator and finished with a vibrating beam. Pokers vary in size,

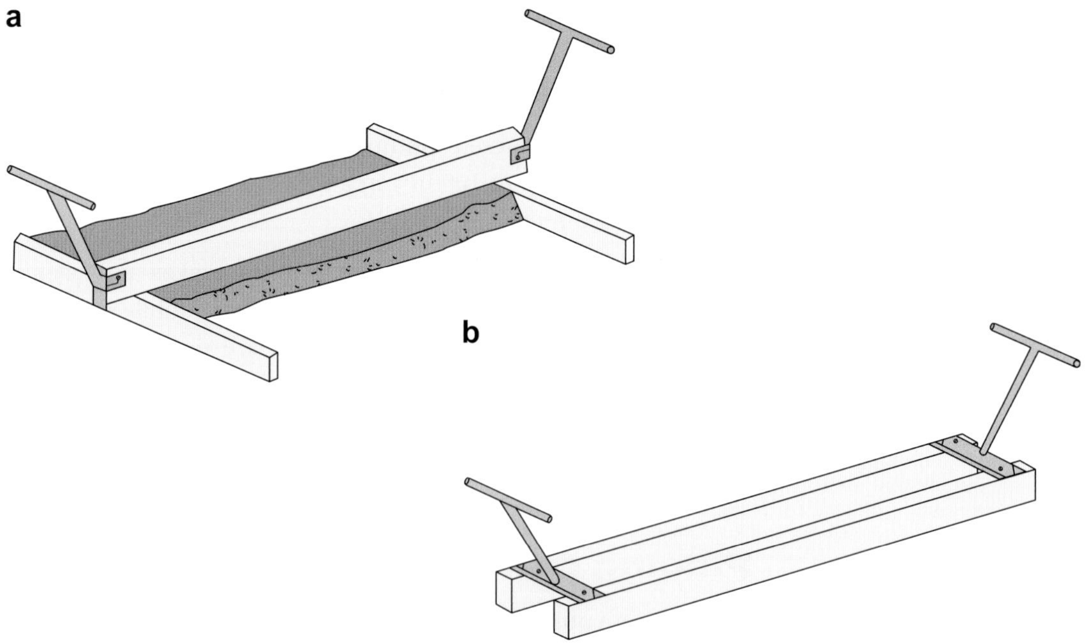
a
b

FIGURE 12.10
Tamping beams: (a) single beam; (b) double beam

usually from 25 to 75 mm in diameter. They can be either internal or external.

An external vibrator can be fixed to tamping beams to give extra vibration of the concrete. An internal poker vibrator should not be dragged through the concrete to spread it out. It should be placed vertically in the concrete, held in position until the air bubbles cease to come to the surface, then slowly withdrawn so that concrete can flow into the space it produced. This should be repeated at about 450 mm centres.

Note

Overvibration should be avoided at all costs.

Overvibration may cause segregation of the materials and form a weak layer of cement paste on the top.

Finishing concrete

In many instances concrete will not require any further treatment after the formwork has been struck. However, if the concrete will be exposed a surface finish may be required.

A good-quality concrete, well laid and finished, makes a satisfactory floor for many purposes. The concrete surface can be smooth, easily cleaned and have good resistance to abrasion at a low maintenance cost.

Most concrete finishes are applied with a trowel (Figure 12.11). A power float can be used for larger areas (Figure 12.12). The finish is usually applied after the concrete has set for at least three hours.

Many finishes can be applied with the aid of the formwork. The formwork may have a pattern on the inside that will be reproduced on the concrete. One of the most common patterns is grooves formed into the face of the concrete to relieve the monotony of plain concrete.

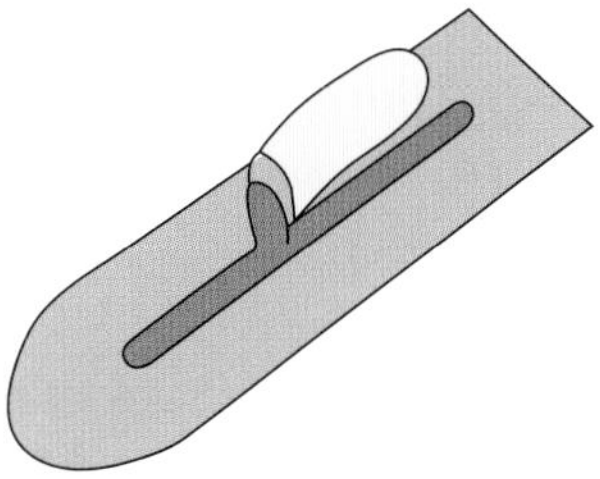

FIGURE 12.11
Concreting trowel

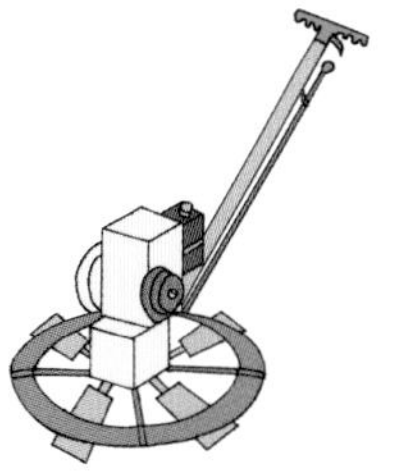

FIGURE 12.12
Power float

A retarder can be applied to the inner surface of the formwork to all the concrete finish to be brushed after the formwork has been removed.

Curing concrete

When concrete has been placed it needs protecting until it has sufficiently hardened.

Concrete requires the presence of water for the chemical reaction of hardening to take place. It is therefore essential that the water used in the mixing remains until full hydration has taken place. The temperature of the concrete is kept at about 20°C. This ensures that the cement binds the aggregate particles together and that the concrete hardens at a favourable rate.

There are various methods of curing available, depending on where the concrete has been placed:

- The concrete surface can be sprayed with water, thereby replacing any water loss.
- Hessian or straw blankets spread over the concrete surface and suitably damped provide insulation as well as a moisture-holding medium.
- Damp sand provides similar protection to hessian but does not give the same degree of insulation.
- Spray coatings form an impervious coating and may also act as a discolouring agent or a solar reflector.

If the temperature drops below 5°C, the hardening process almost stops. If the temperature is too high, the temperature difference between the concrete and the surroundings can cause cracking.

Maintenance of tools and equipment

On completion of the day's work it is essential that all items of tools and equipment are cleaned and any maintenance work is carried out. Poor tools and equipment can delay the job and cause delays.

Concrete must be removed before it hardens and forms a coating on the tools.

Washing with water is the preferred method of cleaning tools and equipment. This also applies to transporting plant and equipment such as dumpers, barrows and skips.

Maintain a clean and tidy work station:

IT SHOULD BE THE AIM OF EVERYONE TO PREVENT ACCIDENTS.

Remember, you are required by law to be aware of and fulfil your duties under the Health and Safety at Work Act.

The main contribution you as a trainee can make towards the prevention of accidents is to work in the safest possible manner at all times, thus ensuring that your actions do not put you, your workmates and the general public at risk.

Team working

The building industry relies a great deal on the co-ordination and teamwork of relatively small groups.

If any project is to be successful, not only in making a profit but also in finishing by the proposed date, all those concerned will have to work together in goodwill and harmony.

It is essential that the whole process of batching, mixing, transporting, placing and curing concrete is carried out as a team operation.

Each member of the team should be conversant with all aspects of the work and be able to fit in on any part of the construction.

Multiple-choice questions

Self-assessment

This section of the book is designed to allow you to check your level of knowledge. The section consists of revision questions for this chapter. The questions are all multiple choice and have four possible answers. The answers are to be found at the end of the book.

The main type of multiple-choice question will be the four-option multiple-choice question. This will consist of a question or statement, known as the stem, followed by a choice of four different answers, called the responses. Only one of these responses is the correct answer; the others are incorrect and are known as distracters.

You should attempt to answer the questions by choosing either (a), (b), (c) or (d).

Example

The person employed by the local authority to ensure that the Building Regulations are observed is called the:

(a) clerk of works
(b) building control officer
(c) council inspector
(d) safety officer

The correct answer is the building control officer, and therefore (b) would be the correct response.

Placing and Finishing Concrete

Question 1 What is a release agent used for?

(a) providing a finish to the concrete
(b) allowing easier striking of the formwork
(c) allowing longer time to work the concrete
(d) preventing the concrete from sticking to the shovel

Question 2 What are the spacers shown used for?

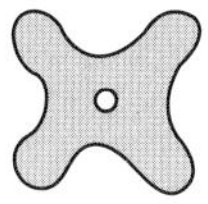

(a) reducing the amount of concrete required
(b) ensuring the correct depth of concrete
(c) preventing the reinforcement from moving
(d) preventing the reinforcement from touching the formwork or ground

Question 3 Vertical formwork provided to floor slabs can be removed after:

(a) 12 hours
(b) 24 hours
(c) 7 days
(d) 14 days

Question 4 Identify the type of reinforcement shown:

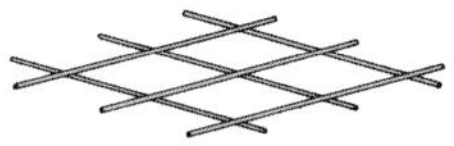

(a) plain reinforcement
(b) mesh reinforcement
(c) ribbed reinforcement
(d) crossed reinforcement

Question 5 What can be caused if concrete is dropped more than 1 m?

(a) segregation of the materials
(b) damage to the formwork
(c) damage to the surrounding concrete
(d) reduction in the setting time

Question 6 What is the maximum time for placing concrete after mixing?

(a) 15 minutes
(b) 30 minutes
(c) 45 minutes
(d) 60 minutes

Question 7 What will happen if you add water to concrete after it has been correctly mixed?

(a) it will strengthen it
(b) it will prevent it segregating
(c) it will weaken it
(d) it will make it easy to lay

Question 8 What is the purpose of tamping concrete?

(a) to remove any excess water
(b) to remove any air voids formed during placing
(c) to level the concrete
(d) to allow it to set more quickly

CHAPTER 13

Jointing and Pointing

This chapter will cover the following NVQ and Diploma units:

- NVQ VR39
- CC 2047K, 2048K

This chapter is about:

- Interpreting building information
- Adopting safe and healthy working practices
- Selecting materials, components and equipment
- Jointing and pointing brick and block structures

The following NVQ performance criteria will be covered:

- Performance criterion 1: Safe work practices
- Performance criterion 2: Selection of resources
- Performance criterion 3: Minimize the risk of damage
- Performance criterion 4: Given contract instructions
- Performance criterion 5: Allocated time

The following Diploma outcomes will be covered:

Both Diploma units require the student to know the methods available for jointing and pointing masonry.

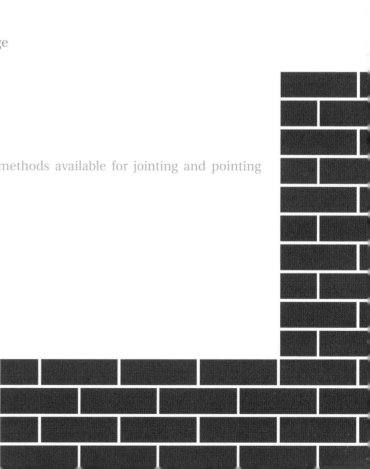

Joint finishing

The surface-finishing treatment of new facework may be a jointing or pointing operation for bricklayers, and has an important effect on the finished appearance of brickwork.

When looking at stretcher-bonded facework, 18 per cent of what you see is mortar colour.

After the bricks have been bedded it is necessary to treat the exposed joints in some way to prevent the weather getting into the structure, and to provide a decorative appearance.

Jointing

Jointing is the craft term applied when joints are finished with the same mortar as is being used for the bricklaying, while the work proceeds.

Joint finishing is usually left until a convenient moment. Bricklaying will generally stop before break times and the end of the day to leave time for jointing up.

It is important to allow sufficient time for finishing the joints correctly but the need to do so at the right times throughout the day is also extremely important.

Pointing

Pointing is the term used to describe the surface finish applied to the cross-joints and bed joints of a brick wall when raked out to a depth of approximately 12 mm, and filled with a mortar of different colour, texture and sometimes density from that used for laying the bricks.

Appearance

Good jointing can improve poor brickwork, but bad jointing can spoil good brickwork.

Careful and skilful jointing can minimize the effect of small deficiencies in bricks and bricklaying, but careless jointing can make them look worse.

Jointing up is a critical part of building facework and is not something to be rushed at the end of the day.

It considerably affects the permanent appearance of facework, as almost one-fifth of the total surface consists of mortar joints.

Mortar mixes

Where facework is to be jointed as work proceeds, the bricklaying mortar of course provides the joint finishing colour. Ironing-in bricklaying mortar made from a fine-grained building sand, for example, will leave a smoother surface than if coarser local sand is used.

If fine-grained building sand is used to produce pointing mortar, then a 'weather struck and cut' finish will polish up better and may be cut or trimmed more cleanly with the frenchman than when the sand is coarser.

Cement-rich or strong pointing mortar should be reserved for very dense class A engineering bricks only, while 1:½:4½ is suited to class B bricks.

A slightly stronger mix than 1:1:6 is in order for the majority of bricks with compressive strengths between 20 and 40 N/mm^2, e.g. 1:1:5. The reduction in sand improves the fattiness of the mortar, so that it sticks to the pointing trowel. If the fattiness of a pointing mortar needs further improvement, it is better to increase the proportion of lime rather than the cement.

Very careful and consistent batching is necessary, with strict control of mix proportions by volume using gauge boxes or buckets each time, if mortar is always to finish up the same colour and strength.

Timing

The timing is probably the most important aspect of jointing-up, particularly when making a neat flush joint without smudging the facework.

The right time to joint-up is determined by both the suction rate of the bricks and the weather conditions.

At one extreme, bricks of low water absorption that are very wet will have a low suction rate. The bricks will tend to 'float' and the mortar will dry very slowly, especially during wet or cold weather.

At the other extreme, high water absorption bricks that are very dry will have a high suction rate and the mortar will dry out very quickly. This can also affect the bond and in some cases bricklayers will dampen the bricks before they are laid.

During summer months it is necessary to joint-up every course in a length of walling. In winter months many courses can be laid before the mortar is dry enough to joint.

The mortar should be soft enough for the jointing tool to leave a smooth surface and to press the mortar into contact with the brick arrises in order to maximize rain resistance.

Trying to finish a mortar joint that is too dry and pressing too hard with the jointing tool can 'blacken' the face of the joint. Trying to joint-up too soon spreads the mortar and leaves a rough joint surface.

To achieve the correct finish the bricks should be laid to perfect gauge. The jointing tool must remain in contact with the brick arrises above and below the bed joints and each side of then cross-joints, otherwise 'tramlines' will be left.

Cross-joints should always be finished first, whatever the type of joint finish.

Types of joint finish

Weather struck joint

The weather struck joint (Figure 13.1) is a popular method of finishing joints on external facework. It gives protection against rain penetration as the slope of the joints runs water off the joint.

The upper edge of the joint is struck back with a pointing trowel. The bottom of the joint can be struck by using a straight edge and trowel. This makes irregular bricks appear straighter than they are.

Struck joint

The struck joint (Figure 13.2) is normally used on internal fair-faced work, especially where the brick or blockwork is to receive an applied decorative finish of paint, etc.

It should not be used for external work as water could collect on the upper arrises of the bricks.

The lower edge of the joint is struck with a pointing joint.

Flush joint

In a flush joint (Figure 13.3), the mortar is compressed into the joint and finished flush with the face of the brickwork. It may be used for internal or external work.

The finish is obtained by rubbing over the joints lightly with a piece of cloth.

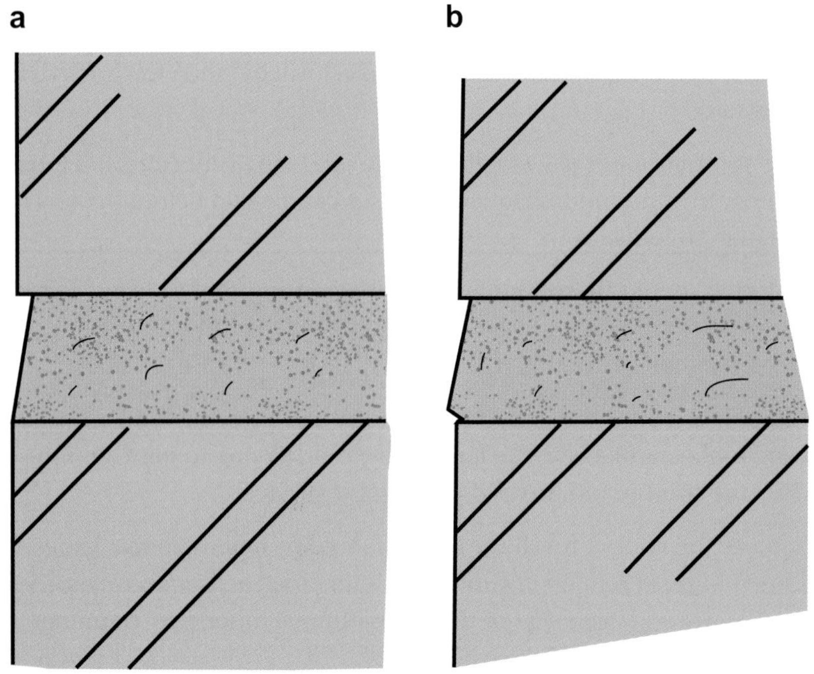
a
b

FIGURE 13.1
(a) Weather struck jointing; (b) weather cut and struck

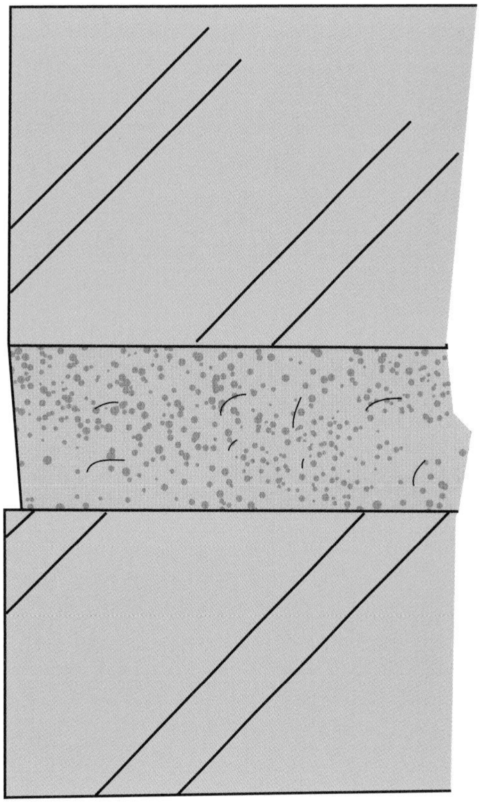

FIGURE 13.2
Struck jointing

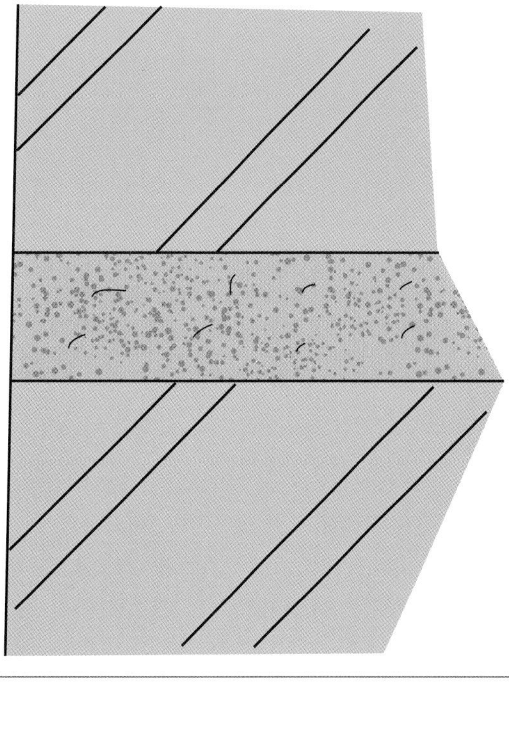

FIGURE 13.3
Flush jointing

Keyed joint – rounded or tooled

A concave finish is obtained by rubbing a suitably shaped tool over the joint (Figure 13.4). The simple tool is often a piece of bucket handle or rounded mild steel piping.

This joint is mostly used for external work.

Recessed joint

In a recessed joint (Figure 13.5), the mortar is pressed back firmly into the joint with a metal jointer or a piece of wood the exact width of the joint.

Remember

Care is needed when finishing joints at external and internal angles.

Finishing joints in internal angles must emphasize the tie bricks or bonding at these points, finishing alternately left to right, and not with a straight joint.

Take care to continue jointing under projections, copings and soffits of arches.

Tools and equipment

The equipment required to carry out the treatment of joints includes the hand hawk and pointing trowel (Figure 13.6).

A small brush will also be required to clean down the face of the components after the work has been completed and the joints have set.

Brushing brickwork before the joints have set can spoil the appearance of facework.

Pointing

Today most face brickwork is 'jointed', which means that the joints are finished as the work proceeds and should require no further attention at

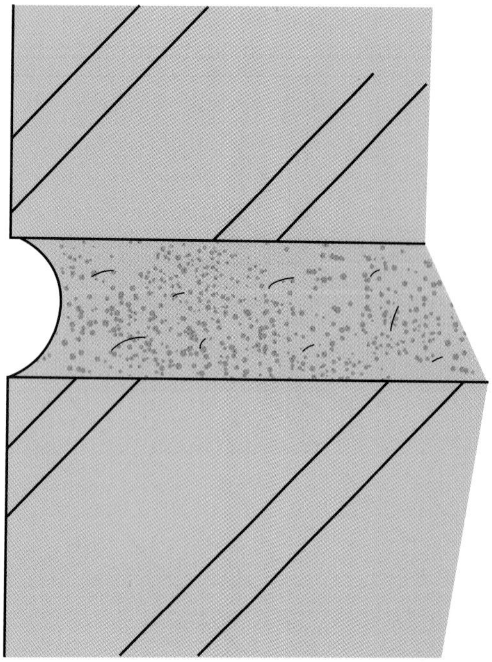

FIGURE 13.4
Keyed jointing

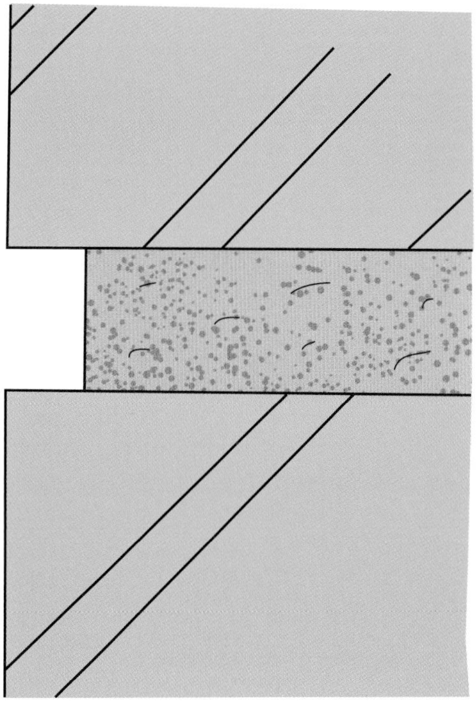

FIGURE 13.5
Recessed jointing

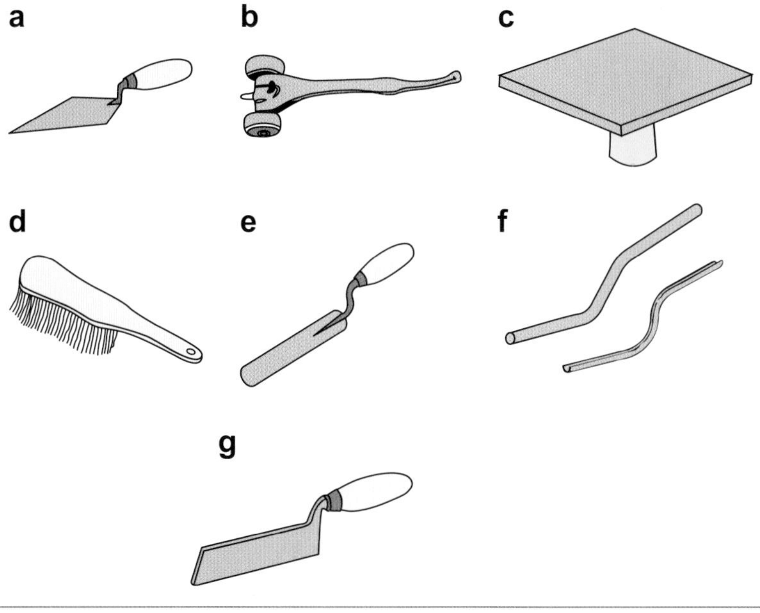
a
b
c
d
e
f
g

FIGURE 13.6
Tools and equipment used for jointing and pointing: (a) pointing trowel; (b) chariot jointer; (c) hand hawk; (d) brush; (e) finger trowel; (f) bucket handles; (g) recessed jointer

the end of the day. Occasionally, however, architects will specify that the joints shall be 'pointed' to achieve a particular effect.

When new work is to be pointed all joints are raked 12–15 mm deep on the day the wall is built, ready to receive a different mortar, or at a later date.

An architect may have requested pointing to ensure that the finish texture and colour of the joint finish are constant; or the architect may require a differently coloured mortar from the bedding mortar to form the joint finish.

Any style can be selected when repointing a wall, depending on the architect's or client's requirements.

Pointing is not very popular with bricklayers as it slows the process down and requires a great deal of patience.

Careless pointing can spoil good brickwork, but good pointing can considerably improve facework.

Before pointing begins, all loose mortar and debris should be removed from the joints with a dry brush and the work wetted down to a damp condition. Wetting down reduces the amount of water sucked into the brickwork from the mortar, which if too great would prevent complete hydration, resulting in a weak, crumbly mortar.

Repointing should start at the top and work down.

As for jointing, the cross-joints should be filled in first. The mortar is pressed firmly into the joints with the trowel. The bed joints are filled in next and also pressed firmly into the joints.

The mortar should inset approximately 1 mm at the top and 'cut projecting' the lower edge by the same amount. This will leave a sloping weathered surface which will allow rain water to fall away quickly and therefore provide better rain resistance than recessed or flush joints.

Repointing

Repointing may be required when the joints of a wall have eroded after a considerable amount of time, owing to exposed weather conditions or poor materials used in the original mortar.

If repointing is required the original mortar joints need to be removed to a depth of approximately 12 mm.

The whole area should be dusted down to remove any powder from the joints and then dampened down to provide a key for the new mortar. See Figure 13.7 for the tools and equipment required.

The existing joints will have to be removed with a plugging chisel and lump hammer. There are also many modern devices that can be set to rake out joints to various depths.

PRIOR WORK

Before old brickwork is repointed it is essential that the cause of the deterioration is established.

It is usually the result of slow erosion over many years, but if it is due to sulphate attack on the mortar then the cause should be sought out and corrected before any repointing is carried out.

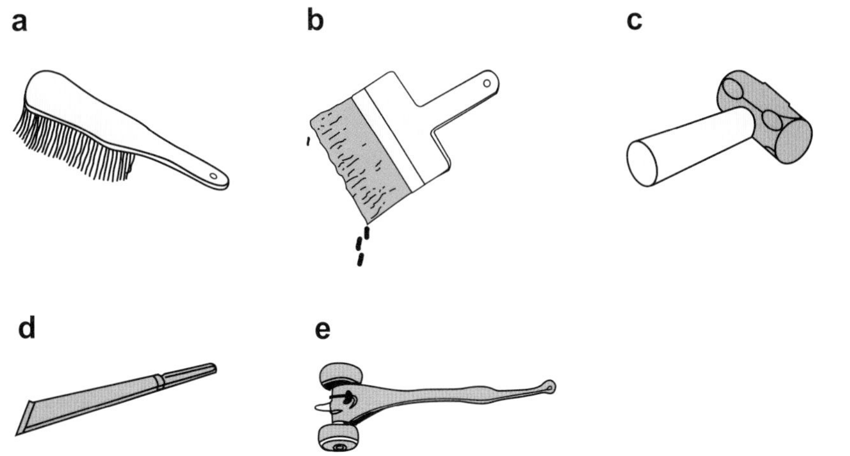
a
b
c
d
e

FIGURE 13.7
Tools and equipment for repointing: (a) dry brush; (b) wet brush; (c) lump hammer; (d) plugging chisel; (e) chariot jointer

ACCESS

The sequence of operations is virtually the same as for new brickwork, except that a working platform will have to be erected. When jointing and pointing new work the scaffold would already have been erected.

The most common method is for tower scaffold to be used (Figure 13.8), but it all depends on the height of the wall to be repointed.

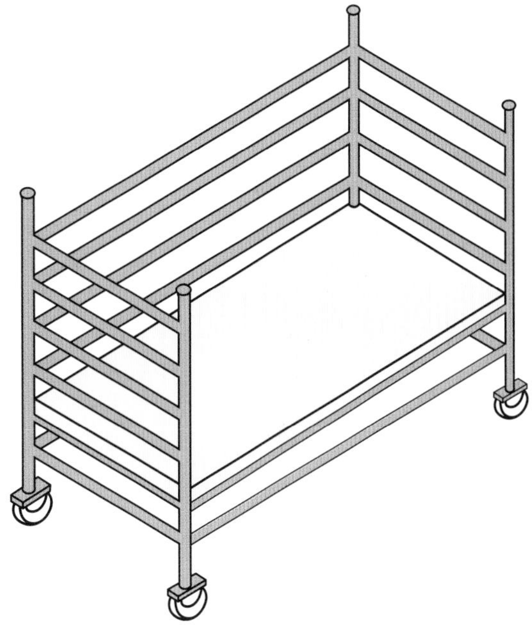

FIGURE 13.8
Small tower scaffold

Techniques

Jointing

Whichever finish is to be used, timing is very important when jointing-up as the bricks are laid throughout the working day. The mortar between bricks should be allowed to stiffen up just enough due to brick suction, so that the jointing tool can pass smoothly and cleanly. Too soon and the mortar smears and does not leave a smooth profile. Left too long before jointing, heavy pressure on the jointer leaves black metal marks on the dried mortar face.

With the single exception of tuck pointing, whether jointing or pointing, always do the cross-joints first, followed by bed joints, each time you stop bricklaying to joint-up.

Brushing-off with a soft bristle hand brush, to remove any loose crumbs of mortar, should be left until the end of the day. Brush lightly and on no account leave bristle marks in the mortar face. It is better to leave brushing until the following morning than to risk marking the joints. Take particular care when jointing face brickwork at those points shown in Figure 13.9.

Pointing

This craft operation, carried out some weeks or months after the wall has been built, requires patience and is a skill that takes time to develop. The joint finish commonly specified for pointing brickwork is weather struck and cut (see Figure 13.1b).

1. Always start at the very top of the walling to be pointed.
2. Remove any obvious hardened crumbs of mortar clinging to the wall face when the joints were raked out some weeks or months ago.

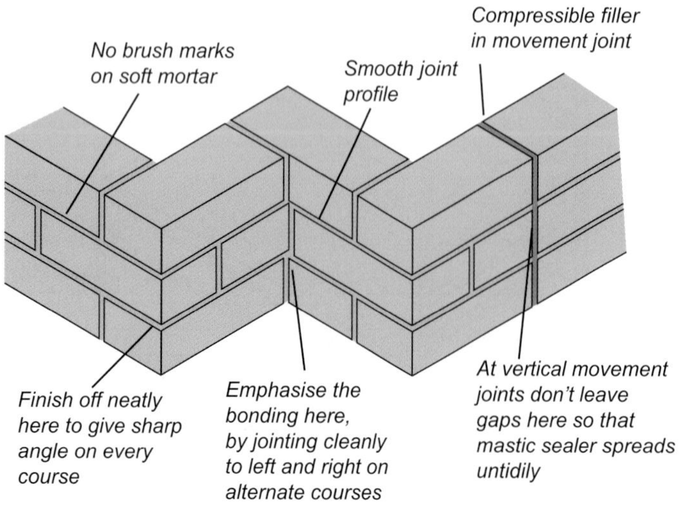
Compressible filler in movement joint
No brush marks on soft mortar
Smooth joint profile
Finish off neatly here to give sharp angle on every course
Emphasise the bonding here, by jointing cleanly to left and right on alternate courses
At vertical movement joints don't leave gaps here so that mastic sealer spreads untidily

FIGURE 13.9
Fine points of jointing and pointing

3. Brush the whole lift of brickwork using a stiff-bristle hand brush.
4. Wet the wall face generously if the bricks are very absorbent, less generously if the bricks have a lower suction rate.
5. Load the hand hawk with mortar flattened out to approximately 10 mm thickness.
6. Using the small pointing trowel or 'dotter', pick up joint-sized pieces of mortar from the hawk and press carefully and firmly into cross-joints, but see also item 14 below.
7. Completely fill each cross-joint with a second application if necessary, polish the mortar surface and indent on the left-hand side.
8. After completing approximately $\frac{1}{2}\,m^2$ of cross-joints, cut or trim the right-hand side of all these joints in the manner shown in Figure 13.10, so that they all look the same width on face.
9. Using a longer pointing trowel, pick up joint-sized pieces of mortar from the hawk and start pointing one bed joint, pulling the loaded trowel up to the last piece of mortar applied each time.
10. After filling a 500 mm length of bed joint, polish it with the pointing trowel and indent the top.
11. When half a dozen bed joints have been pointed in this way, cut or trim the bottom edge of each one (Figure 13.11) using a frenchman or the tip of a pointing trowel together with a feather edge pointing rule. The amount of mortar to be left projecting from the wall face after trimming or cutting joints should not exceed the thickness of the trowel blade.
12. Sensibly adjust areas of pointing to suit the drying conditions of bricks and weather, so that joints cut cleanly. Too early and the mortar will not fall away cleanly when trimmed. Too late and the joint edges will crumble and not leave a clean straight line when cut.

FIGURE 13.10
Weather struck and cut pointing: cutting the right-hand side of cross-joints

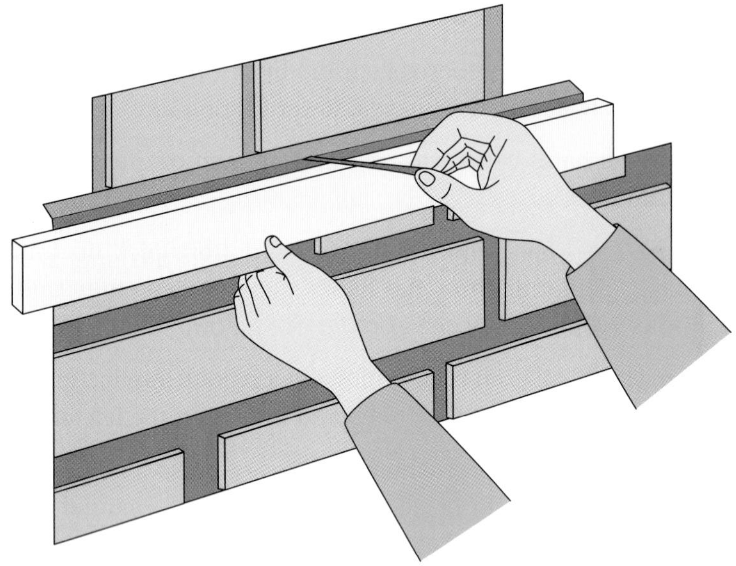

FIGURE 13.11
Weather struck and cut pointing: trimming the bottom of bed joints with straight edge and frenchman

13. Brush very lightly at the end of the day, or preferably on the following day if there is the slightest risk of marking the sharp cut edges of the pointing.
14. If a wall has alternate bands of class A and absorbent facing bricks, after wetting the whole wall, point absorbent facings first. When the wall has dried off, return and point the class A bands.

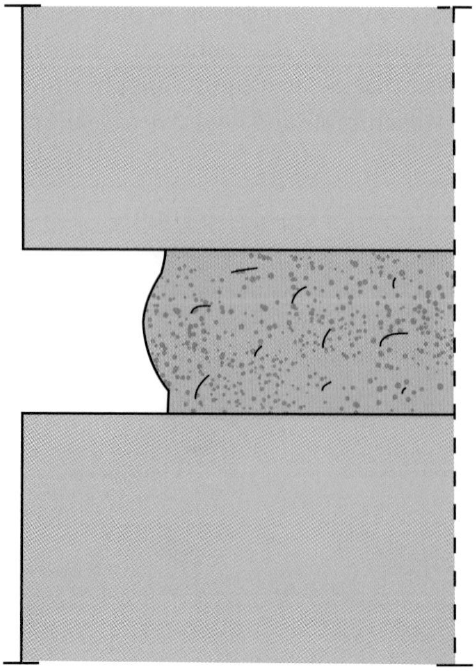

FIGURE 13.12
Correct way of raking out joints in preparation for repointing

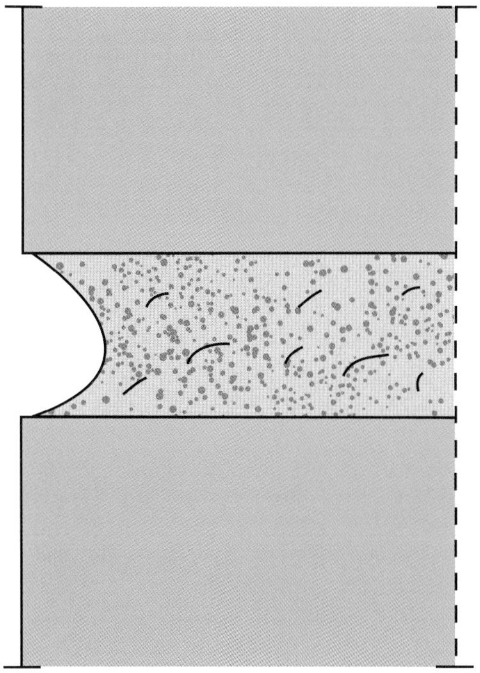

FIGURE 13.13
Incorrect way of raking out, which will lead to pointing failure due to poor adhesion

Summary

The great merit in surface finishing mortar as a jointing process is that the joint profile is an integral part of the mortar bed and there is no possibility of failure through insufficient adhesion between the main mortar bed and the surface finish.

Failure of pointing, i.e. its separation from the main mortar bed and consequent falling away, is caused by careless raking out and lack of suitable preparation. To overcome possible failure, joints in new brickwork should be raked out to a depth of at least 12 mm, as shown in Figure 13.12, and not as shown in Figure 13.13.

Multiple-choice questions

Self-assessment

This section of the book is designed to allow you to check your level of knowledge. The section consists of revision questions for this chapter. The questions are all multiple choice and have four possible answers. The answers are to be found at the end of the book.

The main type of multiple-choice question will be the four-option multiple-choice question. This will consist of a question or statement, known as the stem, followed by a choice of four different answers, called the responses. Only one of these responses is the correct answer; the others are incorrect and are known as distracters.

You should attempt to answer the questions by choosing either (a), (b), (c) or (d).

Example

The person employed by the local authority to ensure that the Building Regulations are observed is called the:

(a) clerk of works
(b) building control officer
(c) council inspector
(d) safety officer

The correct answer is the building control officer, and therefore (b) would be the correct response.

Jointing and Pointing

Question 1 When the joint finish is completed as the work proceeds it is known as:

(a) pointing
(b) jointing
(c) finishing
(d) flushing

Question 2 Identify the following joint finish:

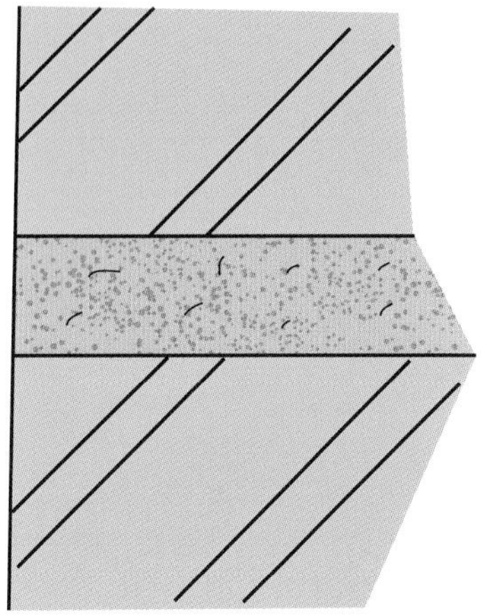

(a) flush joint

(b) half round joint

(c) recessed joint

(d) weather struck joint

Question 3 Where is the correct starting position when repointing a wall?

(a) around doors and windows

(b) highest point

(c) lowest point

(d) anywhere

Question 4 When the joint finish is applied after the whole area has been completed it is known as:

(a) pointing

(b) jointing

(c) finishing

(d) flushing

Question 5 What is the correct depth for raking out mortar joints before repointing a brick wall?

(a) 16 mm

(b) 14 mm

(c) 12 mm

(d) 10 mm

Question 6 Identify the tool used for finishing joints:

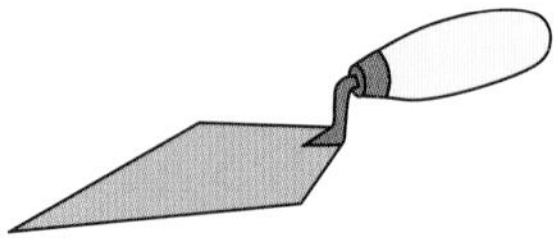

(a) jointing iron
(b) frenchman
(c) pointing iron
(d) pointing trowel

Question 7 Identify the following joint finish:

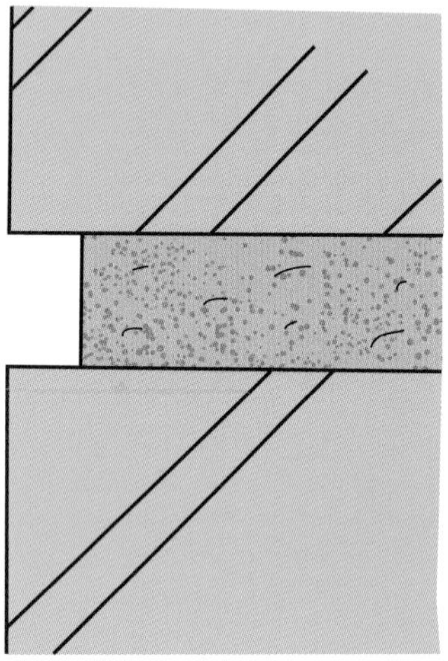

(a) flush joint
(b) half round joint
(c) recessed joint
(d) weather struck joint

Question 8 The main purpose of a joint finish is to:
(a) make the brickwork more decorative
(b) save on mortar
(c) compact the mortar
(d) make the brickwork stronger

CHAPTER 14

Answers to Multiple-Choice Questions

Chapter 1. The Construction Industry

1 (a); 2 (c); 3 (b); 4 (d); 5 (d); 6 (c); 7 (a); 8 (a)

Chapter 2. Health and Safety in the Construction Industry

1 (c); 2 (c); 3 (a); 4 (b); 5 (c); 6 (b); 7 (c); 8 (a)

Chapter 3. Communication

1 (a); 2 (b); 3 (a); 4 (c); 5 (a); 6 (b); 7 (c); 8 (b)

Chapter 4. Construction Technology

1 (d); 2 (a); 3 (b); 4 (b); 5 (a); 6 (b); 7 (d); 8 (c)

Chapter 5. Moving and Handling Resources

1 (b); 2 (c); 3 (d); 4 (a); 5 (a); 6 (c); 7 (d); 8 (d)

Chapter 6. Solid Walls

1 (b); 2 (d); 3 (a); 4 (c); 5 (a); 6 (d); 7 (a); 8 (c)

Chapter 7. Cavity Walls

1 (b); 2 (c); 3 (a); 4 (c); 5 (d); 6 (c); 7 (a); 8 (b)

Chapter 8. Cladding

1 (b); 2 (b); 3 (a); 4 (d); 5 (c); 6 (c); 7 (a); 8 (b)

Chapter 9. Thin Joint Masonry

1 (b); 2 (c); 3 (d); 4 (a); 5 (a); 6 (d); 7 (c); 8 (b)

Chapter 10. Bridging Openings

1 (a); 2 (b); 3 (c); 4 (d); 5 (c); 6 (a); 7 (b); 8 (c)

Chapter 11. Domestic Drainage

1 (a); 2 (c); 3 (a); 4 (b); 5 (c); 6 (d); 7 (d); 8 (b)

Chapter 12. Placing and Finishing Concrete

1 (b); 2 (d); 3 (a); 4 (b); 5 (a); 6 (b); 7 (c); 8 (b)

Chapter 13. Jointing and Pointing

1 (b); 2 (a); 3 (b); 4 (a); 5 (c); 6 (d); 7 (c); 8 (c)

Index